CRITICAL STATE IN SUPERCONDUCTORS

Proceedings of 1994 Topical International Cryogenic Materials Conference

CRITICAL STATE IN SUPERCONDUCTORS

Tokai University Pacific Center, Honolulu, Hawaii
24–26 Oct '94

Editors

Kyoji Tachikawa
Tokai Univ., Hiratsuka City, Kanagawa, Japan

Koichi Kitazawa
Univ. of Tokyo, Tokyo, Japan

Hiroshi Maeda
National Research Institute for Metals, Tsukuba City, Ibaraki, Japan

Teruo Matsushita
Kyushu Institute of Technology, Iizuka City, Fukuoka, Japan

World Scientific
Singapore • New Jersey • London • Hong Kong

Published by

World Scientific Publishing Co Pte Ltd
5 Toh Tuck Link, Singapore 596224
USA office: 27 Warren Street, Suite 401-402, Hackensack, NJ 07601
UK office: 57 Shelton Street, Covent Garden, London WC2H 9HE

British Library Cataloguing-in-Publication Data
A catalogue record for this book is available from the British Library.

CRITICAL STATE IN SUPERCONDUCTORS
Proceedings of 1994 Topical International Cryogenic Materials Conference

ISBN-13 978-981-02-2248-2
ISBN-10 981-02-2248-3

PREFACE

The International Cryogenic Materials Conference (ICMC) board together with the Cryogenic Engineering Conference (CEC) board organizes the well known biennial series of "CEC/ICMC" conferences. But between these large omnibus technical meetings (in the odd-number years) ICMC also organizes topical symposia. Over the years, in response to the wishes of particular board members, these have been held around the world. In the past they have taken place in Japan, China, Germany, Ukraine, and the United States. Individual topics have included superconductivity and structural materials (both metallic and nonmetallic). For 1994 it was decided again to hold the conference in the US; but on the island of Oahu in order to provide a joint US-Japan flavor. For the convenience of many of the prospective attendees this ICMC topical symposium was intended to follow directly after the Applied Superconductivity Conference (ASC). The primary topic to be discussed -- one that gave name to the symposium -- was **CRITICAL STATE IN SUPERCONDUCTORS**. Two compelling reasons guided this choice: (1) the desire to provide a forum for a topic which, although deeply rooted in classical low-Tc superconductivity, has been finding new voice within the framework of high-Tc superconductivity, combined with (2) the need to avoid conflict with the usual technical content of ASC.

The anisotropy and ambi-dimensionality of high-Tc superconductivity has stimulated many new approaches to the critical state. Accordingly it was considered timely to bring these together as a symposium and as a permanent record. The main theme of the conference was reasonably well adhered to, and was keynoted by a splendid invited talk by John R. Clem. However, to cater to the needs of the more general audience the symposium was flavored with some of the more traditional high-Tc topics such as: microstructure, critical current, fabrication, and so on. Nevertheless the symposium turned out to be dominated by discussions of the critical state in high-Tc superconductors; and we trust that these proceedings will provide a valuable record of the status of critical state theory as of October 1994.

E.W.Collings
Chairman, ICMC Board

PREFACE

The 1994 Topical International Cryogenic Materials Conference (ICMC) was held at Tokai University Pacific Center (TUPC) in Honolulu, Hawaii as requested from the ICMC Board. The topics of the conference was "Critical State in Superconductors" which was a very timely and important one especially for the critical current in high-Tc superconductors. Another topical conference of ICMC, non-metallic composite for cryogenic use, closely related to the application of superconductivity, was held simultaneously at the same site. TUPC was established in October 1990 by the Tokai University Educational System. The main activities of TUPC are international education and research, and conference service; TUPC is expected to contribute for the academic and cultural progress of the Pacific Rim.

The present topical conference was composed of five different sessions, i.e. vortex structure and dynamics, critical state phenomena, microstructure and critical current, fabrication of HTS bulk and wires with high-Jc, and thin films and new materials. We had 94 participants from 16 different countries and 84 scientific contributions. 21 key invited talks were given, and the fruitful exchange of informations was performed at the conference. Recently, researches on the critical state in high-Tc superconductors have made appreciable progress. The guiding principles to achieve high Jc in HTS are becoming clear. The fabrication of bulk, wire and thin film of HTS with improved performance is actively studied.

The present 1994 Topical ICMC was arranged by the local organizing committee composed of Prof. T. Matsushita (Kyushu Inst. Technology), Dr.K.Togano (National Res. Inst. Metals), Dr. H.Ogiwara (Toshiba Corp.), Dr. Y.Furuto (Furukawa Elec. Co.), Dr. R.Ogawa (Kobe Steel Ltd.), and Mr.T.Kurino (Soc. Non-Traditional Tech.). The chairmen of the conference are much grateful to the support of Tokai University Educational System, and that of the staffs of TUPC. Many arrangements of the conference secretary, the New Superconducting Materials Forum, are also greatly acknowledged. Finally, the chairmen cordially appreciate the participants of the conference for their contributions.

Kyoji Tachikawa
Hiroshi Maeda
Koichi Kitazawa

Chairmen of 1994 ICMC

Contents

Contributed Papers

1. Vortex Structure and Dynamics

2. Critical State Phenomena

4. Fabrication of HTS Bulk and Wires with High J_c

5. Thin Films, New Materials and Others

Invited Papers

VORTEX DYNAMICS IN THE HIGH-TEMPERATURE SUPERCONDUCTORS

JOHN R. CLEM
Ames Laboratory and Department of Physics and Astronomy
Iowa State University, Ames, IA 50011, USA

ABSTRACT

Some fundamental aspects of the structure and dynamics of vortices in the high-temperature superconductors are discussed. New methods for calculating the static and dynamic magnetic field profiles in thin samples in time-varying perpendicular applied magnetic fields are described.

1. Introduction

The subject of vortex dynamics is a broad area, and in this paper I discuss only a few topics. I first review in Sec. 2 some fundamentals about the vortex structure in high-temperature superconductors. It is best to visualize vortices in these materials as stacks of two-dimensional (2D) pancake vortices, connected by Josephson strings. Next, in Sec. 3, I summarize some recent theoretical advances that facilitate the calculation of critical-state magnetic field and current-density profiles in thin strips and disks of thickness of the order of the London penetration depth. Finally, in Sec. 4, I give a brief summary of the main ideas of this paper.

2. Fundamentals of Vortex Structure in the High-Temperature Superconductors

The starting point for most of our understanding of the vortex structure in type-II superconductors comes from Abrikosov's[1] famous 1957 paper, which was based upon the Ginzburg-Landau theory.[2] Contained within Abrikosov's paper was the idea that, for applied magnetic fields between a lower critical field H_{c1} and an upper critical field H_{c2}, magnetic fields, applied parallel to a slab of isotropic type-II superconducting material, penetrate not uniformly but inhomogeneously in the form of quantized vortices. The magnetic flux associated with each vortex is $\phi_0 = h/2e$. There are two characteristic lengths of an isotropic type-II superconductor, the penetration depth λ and the coherence length ξ, and their ratio is the Ginzburg-Landau parameter $\kappa = \lambda/\xi$. κ has a value greater than $1/\sqrt{2}$ in type-II superconductors and is of the order of 10^2 in the high-temperature cuprate superconductors. The bulk thermodynamic critical field H_c is related to λ and ξ via $H_c = \phi_0/2\pi\sqrt{2}\lambda\xi$, the upper critical field is $H_{c2} = \kappa\sqrt{2}H_c = \phi_0/2\pi\xi^2$, and, to good approximation when $\kappa >> 1$, the lower critical field is[3] $H_{c1} = (\phi_0/4\pi\lambda^2)(\ln\kappa + 0.50)$.

The high-temperature copper-oxide superconductors, however, are not magnetically isotropic. Anisotropy can be conveniently incorporated into the Ginzburg-Landau and

Abrikosov theories with the help of an effective mass tensor M_{ij}.[4-18] In the orthorhombic high-temperature cuprate superconductors, the mass tensor is diagonal when the coordinate axes are chosen to be along the crystallographic a, b, and c directions, which correspond to the principal axes of the mass tensor. In this case, the only elements of M_{ij} are the diagonal components M_a, M_b, and M_c; all the other components are zero. It is convenient to introduce a mass normalization factor and to work with dimensionless masses, which are normalized such that $m_a m_b m_c = 1$.

When a magnetic field is applied parallel to an anisotropic type-II superconducting slab described by the Ginzburg-Landau theory with an anisotropic effective mass tensor, magnetic flux still penetrates in the form of vortices. Even if the field is applied along a principal axis, however, the vortices are elliptical in shape. The exponential decay length for screening currents flowing along a principal axis i (i = 1, 2, 3 = a, b, c) is $\lambda_i = \lambda\sqrt{m_i}$, and the characteristic length over which the order parameter varies along this axis is $\xi_i = \xi/\sqrt{m_i}$. Because $m_a m_b m_c = 1$, we have $(\lambda_a\lambda_b\lambda_c)^{1/3} = \lambda$ and $(\xi_a\xi_b\xi_c)^{1/3} = \xi$. An isolated vortex oriented along the a axis (parallel to the layers of the high-temperature superconductors) has concentric elliptical contours of constant magnetic field, which coincide with the supercurrent streamlines. The ratio of the semimajor axis to the semiminor axis of any ellipse is $\lambda_c/\lambda_b = \sqrt{(m_c/m_b)}$. The Abrikosov vortex core (where the order parameter is strongly suppressed) is also elliptical, with the ratio of the semimajor axis to the semiminor axis being $\xi_b/\xi_c = \sqrt{(m_c/m_b)}$. The semimajor axes of the current-pattern ellipse and the vortex-core ellipse are both aligned parallel to the layers. The bulk thermodynamic critical field H_c is related to λ and ξ via $H_c = \phi_0/2\pi\sqrt{2}\lambda\xi$; the upper critical field along a is $H_{c2a} = \kappa_a\sqrt{2}H_c = \phi_0/2\pi\xi_b\xi_c$, where $\kappa_a = \lambda_c/\xi_b = \lambda_b/\xi_c = \kappa/\sqrt{m_a}$; and, to good approximation when $\kappa_a >> 1$, the lower critical field is $H_{c1a} = (\phi_0/4\pi\lambda_b\lambda_c)(\ln\kappa_a + 0.50)$. Similar results for vortices along the b and c axes can be obtained by cyclic permutation of the above results.

An anisotropic Ginzburg-Landau description of the vortex structure works well when all the relevant superconducting length scales λ_a, λ_b, λ_c, ξ_a, ξ_b, and ξ_c are much larger than the linear dimensions of the crystallographic unit cell. Although all these quantities are relatively large close to T_c (and actually diverge at T_c), the smallest of these, ξ_c, is usually much smaller than the others. In all the high-T_c cuprates, there in fact is a finite temperature T at which, for decreasing temperature, ξ_c becomes less than the crystallographic lattice parameter c, which characterizes the periodicity length perpendicular to the CuO_2 layers. (In $Bi_2Sr_2CaCu_2O_{8+\delta}$, this temperature is only a small fraction of 1 K below T_c.) When this happens, it is as if the vortex core slips between the CuO_2 layers of the cuprate superconductors. To describe vortices in this temperature regime, however, a continuum theory like the Ginzburg-Landau theory is no longer valid, and a theory that accounts for the discreteness of the atomic structure along the c direction is needed.

A theory that satisfies this requirement is the Lawrence-Doniach theory,[19] which describes superconductivity within the layers via the Ginzburg-Landau theory and current flow between layers via the Josephson effect. For the high-temperature superconductor $Bi_2Sr_2CaCu_2O_{8+\delta}$, for example, one regards the SrO and BiO layers as insulating tunnel barriers between superconducting counterelectrode CuO_2 bilayers (which sandwich a Ca layer). The relevant periodicity length s is the center-to-center distance between successive CuO_2 bilayers. For weak coupling between layers, when $\xi_c << s$, an isolated vortex aligned along the a axis has current patterns far from the core that are identical with those of the anisotropic Ginzburg-Landau theory. However, the vortex core is defined by the pattern of Josephson currents around the vortex axis, not by a region of suppressed order parameter.[20-22] The semimajor axis, where the interlayer Josephson current density has its maximum, has the value approximately $(\lambda_c/\lambda_b)s/2$, while the semiminor axis of the Josephson core is roughly s/2, so that the ratio of the semimajor to semiminor axis is $\lambda_c/\lambda_b = \sqrt{(m_c/m_b)}$, the same as in an Abrikosov core.[20]

When the magnetic induction has a component that is perpendicular to the layers of the high-temperature superconductors, it is convenient to think of the vortex lines threading through the material as consisting of two-dimensional (2D) pancake vortices,[23] whose associated supercurrents are confined to the layers, connected by interlayer Josephson strings. The latter are segments of Josephson vortices, characterized by patterns of interlayer Josephson currents circulating around a line of gauge-invariant phase difference $\Delta\gamma = \pi$. This line, which threads through the insulating layer between two superconducting layers, connects a 2D pancake vortex in one superconducting layer, to the nearest 2D pancake vortex in the other layer. The energy associated with the Josephson string is dominated by the Josephson coupling energy. Expressions for the Josephson interaction energies between the pancake vortices are given in Refs. 24 and 25.

3. Flux Penetration into Thin Films

While the penetration of magnetic flux into slabs and cylinders subjected to parallel fields has been reasonably well understood for some time,[26-28] the quantitative behavior of flux penetration into thin films or flat samples subjected to perpendicular magnetic fields or current-induced self fields has not been studied in detail until recently. An important paper published in 1970 by W. T. Norris[29] recently has been shown to be the key to analytic calculations of the critical-state electrical-current and magnetic-field distributions in long superconducting strips, and another advance by Mikheenko and Kuzovlev[30] opens the way to corresponding analytic solutions for thin superconducting disks. These theoretical results can be compared with experiments performed using magneto-optics,[31-35] miniature Hall sensors,[36-39] scanning Hall probes,[40-42] Hall-sensor arrays,[43,44] and scanning SQUIDs.[45,46]

Approaches following that of Norris[29] have been used by Brandt et al.,[47,48] Darwin et al.,[39] and Zeldov et al.[49] to obtain detailed solutions for the current and field distributions for many possible sequences of applied transport currents and perpendicular magnetic fields. Reference 49 emphasizes comparisons with the corresponding solutions for the more familiar slab geometry; although there are many similarities, there also are striking differences in behavior. For example, in slab geometry in a parallel field, changes in the applied field or transport current produces regions where the current density is either zero or has magnitude J_c, the critical current density. In the film geometry, however, because of demagnetizing effects, which permit the magnetic field to wrap around the sample, the current density associated with the discontinuity of the tangential magnetic field between the top and bottom of the film can flow with magnitudes less than J_c. An important aspect of the thin-film geometry is that the current density arises chiefly from the curvature of the magnetic field across the sample thickness, rather than from the field gradient, which is the dominant term in slab geometry in a parallel field.[50-53]

The above approach has been used to study the behavior of films with inhomogeneous pinning, including the case for which J_c is much higher at the edges of the film than in the middle. There also is a geometric barrier to the penetration of magnetic flux into thin films. These effects are discussed in recent papers by Schuster et al.[54] and Zeldov et al.[55] In addition to studies of the critical state, there recently have been further studies of the time dependence and frequency dependence of flux penetration into thin-film strips.[56]

It is surprising that the solutions of the critical-state profiles in thin circular disks are mathematically very similar to those in an infinitely long strip. An elegant method for the solution of these profiles was presented by Mikheenko and Kuzovlev.[30] Their application of this method to calculate the ac response, however, was shown by Zhu et al.[57] to be incorrect. Further corrections were pointed out by Clem and Sanchez,[58] who have shown how to calculate the real and imaginary parts of the complex ac susceptibility of thin disks in the critical state. Further studies of the time dependence and frequency dependence of flux penetration into thin disks have been carried out by Brandt.[59]

4. Summary

In this paper I have reviewed some of the fundamentals of vortex structure and dynamics in the high-temperature superconductors. Many of the basic properties of the vortex structure can be obtained from the Ginzburg-Landau or London theories, provided they are modified to incorporate anisotropic by including an anisotropic effective mass tensor. Such an approach, however, does not give a detailed account of the layered nature of the cuprate superconductors, in which superconductivity evidently is localized in the CuO_2 layers, bilayers, and trilayers. To account for the layering and how it affects flux pinning, pinning energies, and flux creep, it is best to think of vortex lines as stacks of 2D pancake vortices, connected by Josephson strings.

The dynamics of magnetic flux penetration into thin superconducting films recently has benefited from several theoretical advances, which permit the calculation of current-density and magnetic-field profiles for strips and disks. This theoretical work has been stimulated by the development of new experimental methods for quantitatively determining microscopic magnetic-field distributions at the superconductor's surface.

5. Acknowledgments

I thank E. H. Brandt, V. G. Kogan, J. Lockhart, A. Malozemoff, M. McElfresh, A. Sanchez, J. Sun, E. Zeldov, and J. Zhu for helpful discussions and correspondence. Ames Laboratory is operated for the U.S. Department of Energy by Iowa State University under Contract No. W-7405-Eng-82. This research was supported by the Director for Energy Research, Office of Basic Energy Sciences.

6. References

1. A. A. Abrikosov, *Soviet Phys. JETP* **5** (1957) 1174.
2. V. L. Ginzburg and L. D. Landau, *Zh. Eksperiim. i Teor. Fiz.* **20** (1950) 1064.

3. C.-R. Hu, *Phys. Rev. B* **6** (1972) 1756.
4. M. von Laue, *Theory of Superconductivity* (Academic Press, New York, 1952).
5. C. Caroli, P. G. de Gennes, and J. Matricon, *Phys. Kondens. Mater.* **1** (1963) 176.
6. L. P. Gor'kov and T. K. Melik-Barkhudarov, *Sov. Phys. JETP* **18** (1964) 1031.
7. D. R. Tilley, G. J. Van Gurp, and C. W. Berghout, *Phys. Lett.* **12** (1964) 305.
8. D. R. Tilley, *Proc. Phys. Soc. London* **85** (1965) 1977.
9. D. R. Tilley, *Proc. Phys. Soc. London* **86** (1965) 289.
10. E. I. Katz, *Sov. Phys. JETP* **29** (1969) 897.
11. E. I. Katz, *Sov. Phys. JETP* **31** (1970) 787.
12. K. Takanaka, *Phys. Status Solidi B* **68** (1975) 623.
13. R. A. Klemm and J. R. Clem, *Phys. Rev. B* **21** (1980) 1868.
14. V. G. Kogan, *Phys. Rev. B* **24** (1981) 1572.
15. V. G. Kogan and J. R. Clem, *Phys. Rev. B* **24** (1981) 2497.
16. V. G. Kogan and J. R. Clem, *Jpn. J. Appl. Phys.* **26 (Suppl. 26-3)** (1987) 1159.
17. A. V. Balatskii, L. I. Burlachkov, and L. P. Gor'kov, *Sov. Phys. JETP* **63** (1986) 866.
18. L. N. Bulaevskii, V. L. Ginzburg, and A. A. Sobyanin, *Sov. Phys. JETP* **68** (1988) 1499.
19. W. E. Lawrence and S. Doniach, in *Proceedings of the Twelfth International Conference on Low Temperature Physics,* ed. E. Kanda (Academic Press of Japan, Kyoto, 1971), p. 361.
20. J. R. Clem and M. W. Coffey, *Phys. Rev. B* **42** (1990) 6209.
21. J. R. Clem, in *Physics and Materials Science of High Temperature Superconductors,* ed. R. Kossowsky, S. Methfessel, and D. Wohlleben (Kluwer, Dordrecht, 1990), p. 79.
22. J. R. Clem, M. W. Coffey, and Z. Hao, *Phys. Rev. B* **44** (1991) 2732.
23. J. R. Clem, *Phys. Rev. B* **43** (1991) 7837.
24. J. R. Clem, *Physica A* **200** (1993) 118.
25. J. R. Clem, in *The Vortex State, Proceedings of the NATO Advanced Study Institute, Vortices in Superconductors, July 19-31, 1993, Cargese, Corsica,* ed. N. Bontemps (Kluwer, Dordrecht, 1994), p. 25.
26. C. P. Bean, *Phys. Rev. Lett.* **8** (1962) 250.
27. A. M. Campbell and J. E. Evetts, *Adv. Phys.* **21** (1972) 199.
28. J. R. Clem, *J. Appl. Phys.* **50** (1979) 3518.
29. W. T. Norris, *J. Phys. D* **3** (1970) 489.
30. P. N. Mikheenko and Y. E. Kuzovlev, *Physica C* **204** (1993) 229.
31. H. Theuss, A. Forkl, and H. Kronmüller, *Physica C* **190** (1992) 345.
32. L. A. Dorosinskii, M. V. Indenbom, V. I. Nikitenko, Y. A. Ossip'yan, A. A. Polyanskii, and V. K. Vlasko-Vlasov, *Physica C* **203** (1992) 149.
33. C. A. Duran, P. L. Gammel, R. Wolfe, V. J. Fratello, D. J. Bishop, J. P. Rice, and Ginsberg, *Nature* **357** (1992) 474.
34. V. K. Vlasko-Vlasov, M. V. Indenbom, V. I. Nikitenko, A. A. Polyanskii, R. L. Prozorov, I. V. Grekhov, L. A. Delimova, I. A. Liniichuk, A. V. Antonov, and M. Y. Gusev, *Superconductivity* **5** (1992) 1582.
35. M. V. Indenbom, A. Forkl, H. U. Habermeier, and H. Kronmuller, *J. Alloys Compounds* **195** (1993) 499.
36. M. Konczykowski, F. Holtzberg, and P. Lejay, *Supercond. Sci. Technol.* **4** (1991) 8331.

37. M. Konczykowski, L. I. Burlachkov, Y. Yeshurun, and F. Holtzberg, *Phys. Rev. B* **43** (1991) 13707.
38. D. A. Brawner, A. Schilling, H. R. Ott, R. J. Haug, K. Ploog, and K. von Klitzing, *Phys. Rev. Lett.* **71** (1993) 785.
39. M. Darwin, J. Deak, L. Hou, M. McElfresh, E. Zeldov, J. R. Clem, and M. Indenbom, *Phys. Rev. B* **48** (1993) 13192.
40. D. A. Brawner and N. P. Ong, *J. Appl. Phys.* **73** (1993) 3890.
41. T. Tamegai, Y. Iye, I. Oguro, and K. Kishio, *Physica C* **213** (1993) 33.
42. A. M. Chang, H. D. Hallen, L. Harriott, H. F. Hess, H. L. Kao, J. Kwo, R. E. Miller, R. Wolfe, J. van der Ziel, and T. Y. Chang, *Appl. Phys. Lett.* **61** (1992) 1974.
43. T. Tamegai, L. Krusin-Elbaum, P. Santhanam, M. J. Brady, w. T. Masselink, C. Feild, and F. Holtzberg, *Phys. Rev. B* **45** (1992) 2589.
44. S. T. Stoddart, H. I. Mutlu, A. K. Geim, and S. J. Bending, *Phys. Rev. B* **47** (1993) 5146.
45. A. Mathai, D. Song, Y. Gim, and F. C. Wellstood, *IEEE Transactions on Applied Superconductivity* **3** (1993) 2609.
46. C. C. Tsuei, J. R. Kirtley, C. C. Chi, L. S. Yujahnes, A. Gupta, T. Shaw, J. Z. Sun, and M. B. Ketchen, *Phys. Rev. Lett.* **73** (1994) 593.
47. E. H. Brandt, M. Indenbom, and A. Forkl, *Europhys. Lett.* **22** (1993) 735.
48. E. H. Brandt and M. Indenbom, *Phys. Rev. B* **48** (1993) 12893.
49. E. Zeldov, J. R. Clem, M. McElfresh, and M. Darwin, *Phys. Rev. B* **49** (1994) 9802.
50. D. J. Frankel, *J. Appl. Phys.* **50** (1979) 5402.
51. M. Däumling and D. C. Larbalestier, *Phys. Rev. B* **40** (1989) 9350.
52. L. W. Conner and A. P. Malozemoff, *Phys. Rev. B* **43** (1991) 402.
53. J. R. Clem, in *Proceedings of the 7th International Workshop on Critical Currents in Superconductors, Alpbach, Austria,* ed. H. W. Weber (World Scientific, Singapore, 1994), p. 117.
54. T. Schuster, M. V. Indenbom, H. Kuhn, E. H. Brandt, and M. Konczykowski, *Phys. Rev. Lett.* **73** (1994) 1424.
55. E. Zeldov, A. I. Larkin, V. B. Geshkenbein, M. Konczykowski, D. Majer, B. Khaykovich, V. M. Vinokur, and H. Shtrikman, *Phys. Rev. Lett.* **73** (1994) 1428.
56. E. H. Brandt, *Phys. Rev. B* **49** (1994) 9024.
57. J. Zhu, J. Mester, J. Lockhart, and J. Turneaure, *Physica C* **212** (1993) 216.
58. J. R. Clem and A. Sanchez, *Phys. Rev. B* **50** (1994) 9355.
59. E. H. Brandt, *Phys. Rev. B* **50** (1994) 4034.

REAL-TIME OBSERVATION OF VORTICES IN SUPERCONDUCTORS WITH COHERENT ELECTRON BEAM

AKIRA TONOMURA

Advanced Research Laboratory, Hitachi, Ltd.
Tonomura Electron Wavefront Project, ERATO, JRDC
Hatoyama, Saitama 350-03, Japan

ABSTRACT

Phase distribution in an electron wavefunction can be precisely measured to within 1/100th of the electron wavelength using a "coherent" field-emission electron beam. This technique allows for the ultra-fine measurement of material structures and electromagnetic field distributions on a microscopic scale. Especially, vortices in a superconducting thin film can be observed quantitatively by electron-holographic interference microscopy and dynamically by Lorentz microscopy.

1. Introduction

The intensity of an electron beam is used to observe specimens in electron microscopy. The phase distribution of an electron beam can be observed as well thanks to the development of a coherent field-emission electron beam[1]; Electron interference patterns can be directly observed on a fluorescent screen with the naked eye. Before this, interference patterns could only be recorded on film with conventional electron microscopes requiring a long exposure time. In addition, the electron phase can be measured within an accuracy of 1/100th of the wavelength using electron holography[2]. Because of these developments, the structure of materials at atomic scales and the microscopic distributions of electromagnetic fields can be observed by precisely measuring the phase distribution[3]. This paper introduces progressive work in electron interference experiments using a coherent electron beam and the applications to real-time observation of vortices in superconductors.

2. Electron holography

Holography is a two-step imaging process. In the first step of electron holography, an interference pattern is formed between an object wave and a reference wave in a field-emission transmission electron microscope, and recorded on film as a hologram. This hologram is subsequently illuminated by a collimated laser beam and an exact image is produced three-dimensionally in a diffracted beam. An additional image called a "conjugate image" is also produced in holography with the same amplitude in magnitude, but with the opposite sign.

An interference micrograph, or a contour map of the wavefront, can be obtained by simply overlapping an optical plane wave with this reconstructed wave. If a conjugate image instead of a plane wave overlaps this wavefront, the phase difference becomes twice as large, as if the phase distribution were amplified two times. By repeating this technique, phase shifts can be detected as small as 1/100 of a wavelength. This phase-amplified interference electron microscopy provides direct information about the microscopic distribution of electromagnetic fields.

Optical reconstruction using a laser light is simple, but is off-line as a result of the time consumed in developing film. Therefore, on-line reconstruction techniques are being developed using computers and optical devices. An image can be numerically reconstructed from a hologram recorded on a charge-coupled device (CCD) attached to an electron microscpe. By doing this, an amplitude image, and interference image can be displayed. These images can be obtained in a fairly short time, depending on the performance of the computer used, but not yet in real-time.

A real-time method using a liquid crystal panel as a phase hologram has been developed[4]; The image signal of the hologram detected with a TV camera attached to the electron microscope is transferred to the liquid-crystal panel, where the intensity distribution is transformed into a phase shifting function for an illuminating light beam. The time resolution of 1/30 s is determined by the TV system. Dynamic phenomena therefore can be observed in real-time with this method. Lorentz microscopy can be used as a simpler method for the real-time observation of special phase objects, though the phase distribution cannot be measured quantitatively but only the locations with sudden phase changes can be observed.

3. Applications

When a parallel electron beam is incident to electromagnetic fields, the electron beam is deflected, or phase-shifted. The phase shift $\Delta S / \mathrm{h}$ is calculated from the Schrödinger equation as,

$$\begin{aligned}\Delta S / \mathrm{h} &= (1/\mathrm{h})\oint(m\boldsymbol{v} - e\boldsymbol{A}) \cdot d\boldsymbol{s} \\ &= (1/\mathrm{h})\oint\left(\sqrt{2meV} - e\boldsymbol{t} \cdot \boldsymbol{A}\right)ds, \end{aligned} \tag{1}$$

where the integral is carried out along the route connecting two electron trajectories and t is the unit tangent vector of the electron trajectory. This equation shows that electromagnetic potentials (A, V) can be detected by measuring the phase shift in an electron beam, although what we detect is not the electromagnetic potentials themselves but their integrals along the electron beam trajectory.

3.1 Specimen Thickness Distribution in Atomic Dimensions

An interference micrograph for a thin film consists of uniform material displays the thickness contours of the film[2]. An example is shown in Fig. 1. The phase distribution here is displayed as deviations from regular fringes, i.e., an interferogram. Steps A, B, and C in the micrograph correspond to one, three, and five layers of atomic surface steps. The thickness change at step A is only 6.2 Å (one-half of the c-axis spacing), and produces a phase shift of $2\pi/50$.

In the transmission mode, surface topography can only be investigated by measuring thickness. In the reflection mode, thickness can be measured with a high degree of sensitivity because surface height differences are directly measured in units of extremely short electron wavelengths as geometrical path differences[5].

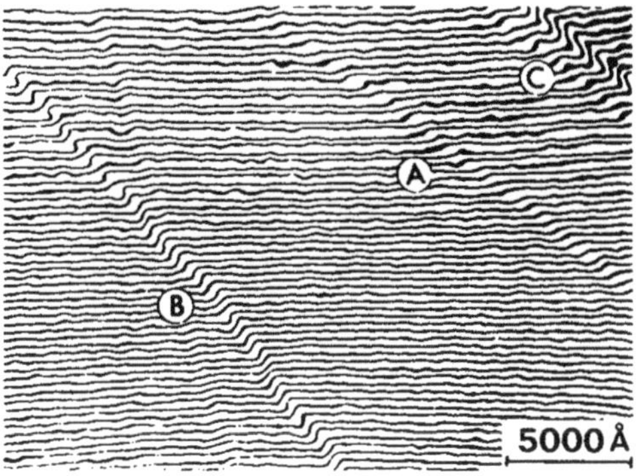

Fig. 1. Interference micrograph of molybdenite film (×24).

3.2 Magnetic Field Observation

An interference micrograph of a magnetic object displays projected magnetic lines of force. The phase difference between two electron beams passing through a magnetic object is given by

$$\Delta S / \hbar = -(e/\hbar)\oint \boldsymbol{A}\cdot d\boldsymbol{s} = -(e/\hbar)\int \boldsymbol{B}\cdot d\boldsymbol{S}, \tag{2}$$

where the first integral is carried out along a closed path along two electron trajectories, and the second integral is carried out over the surface determined by the two paths.

The following conclusions can be obtained from this equation[6].

(1) Contour fringes in the interference micrograph indicate magnetic lines of force, because the phase difference $\Delta S / \hbar$ vanishes between two beams passing through arbitrary points along a magnetic line.
(2) A magnetic flux for h/e flows between adjacent two contour fringes.

An example is shown in Fig. 2. The specimen used here is a fine cobalt particle. When observed with an electron microscope which displays the intensity of the transmitted electron beam, only a triangular outline can be seen. In this interference micrograph, however, two kinds of interference fringes appear. The fringes parallel to the three edges indicate that thickness increases linearly from the edges to 550 Å, and the fringes in the inner region where the thickness is uniform indicate the magnetic lines of force. Smoothly rotating magnetization can be observed at a glance even in such a fine particle.

3.3 Vortices in Superconductors

Magnetic vortices in superconductors can be observed quantitatively by interference microscopy[7,8] and Lorentz microscopy[9] with our 350 kV holography electron microscope[10]. In the experiments we conducted, a superconductive thin film was tilted with respect to both the electron beam and the magnetic field.

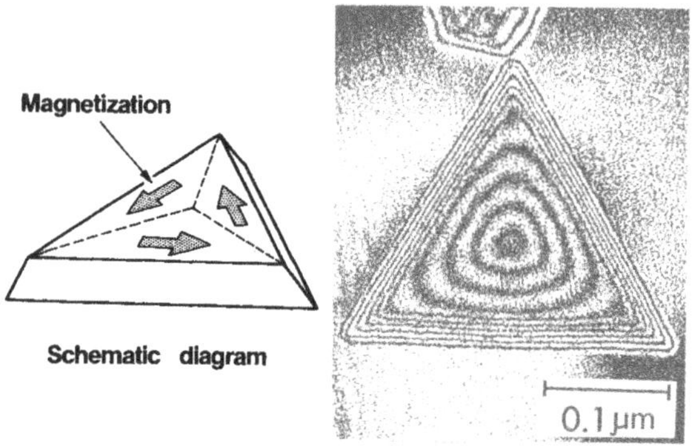

Fig. 2. Interference micrograph of Co particle.

The experimental arrangement is shown in Fig. 3. A Nb thin film, set on a low-temperature stage, was tilted 45° to an incident beam of 300-kV electrons so that the electrons could be influenced by the vortex magnetic fields. An external magnetic field of up to 150 gauss was applied horizontally. An example of a vortex array in a single-crystalline Nb thin film[7] is shown in Fig. 4. In this interference micrograph, projected magnetic lines of force can be observed. They become dense in the localized regions indicated by circles in the photograph, which correspond to individual vortices.

Although interference microscopy is highly resolved and quantitative, Lorentz microscopy is more convenient for observing the dynamic behavior of vortices. In this experiment, the sample was first cooled down to 4.5K and the magnetic field B was gradually increased. As it increased, vortices suddenly began to penetrate the film at B = 32 gauss, and the number increased as B was increased further. Their dynamic behavior was quite interesting. At first, only a few vortices appeared here and there in a 15×10-μm^2 field of view. They oscillated around their own pinning centers and occasionally hopped from one center to another. These movements continued until arriving at an equilibrium state.

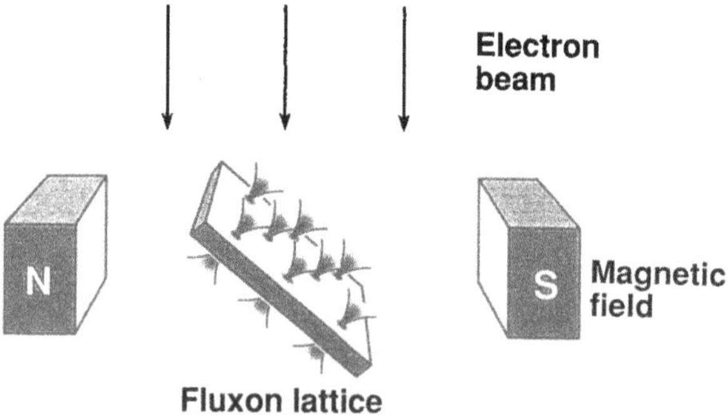

Fig. 3. Schematic diagram for vortex lattice observation.

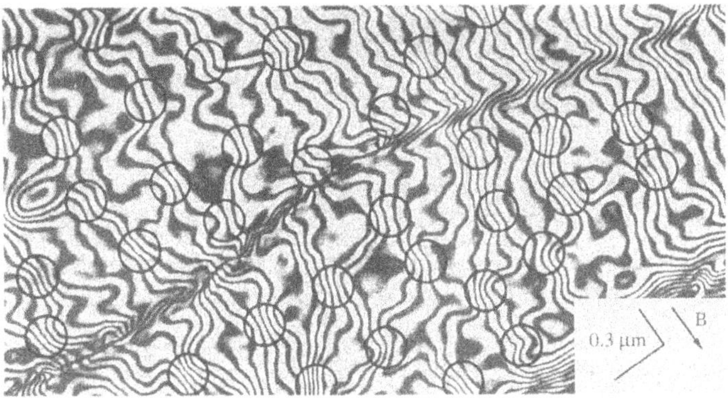

Fig. 4. Interference micrograph of a superconducting Nb film at $B = 100$ gauss (Phase amplification $\times 16$).

An equilibrium Lorentz micrograph at $B = 100$ gauss[9] is shown in Fig. 5. The film has a fairly uniform thickness in the region shown, but is bent along the black curves, called bend contours, caused by Bragg reflections at the atomic plane brought to a favorable angle by bending. Each spot showing a black and white contrast is an image of a single vortex. This contrast reversed, as expected, when the applied magnetic field was reversed. The tilt direction of the sample can be read from the line dividing the black and white parts of the spots. Because the black part is on the same side of all the spots, the polarities of all the vortices seen in the region are the same. At a low B, i.e., up to 30 - 50 gauss, the vortices are too sparse to form a lattice, even in equilibrium. At $B = 100$ gauss, the vortex density is so high that they cannot form anything but a hexagonal lattice.

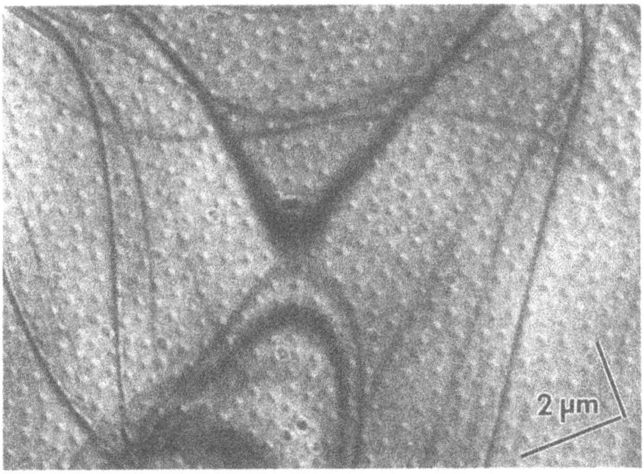

Fig. 5. Lorentz micrograph of a two-diamentional array of vortices in a superconducting Nb film.

A high-T_c superconductor has been investigated by means of Lorentz microscopy[11]. High-T_c superconductors are difficult to use practically, bacause the critical current vanishes at high temperatures and high magnetic-field regime even when the temperature is well below the critical temperature T_c. This phenomenon most probably arises from the behavior of vortices but has not yet been proven concretely. It is believed that these vortices melt like molecules in a liquid, and, as a result it is difficult to fix vortices at some pinning sites[12]. Evidence for vortex melting was provided by a Bitter BSCCO figure in which the vortex image was blurred even at 15 K and 20 gauss (T_c = 85 K)[13]. Accordingly, the practical use temperature would not be T_c but rather a much lower melting temperature T_m. Others, however, disagree with this, and attribute this phenomenon to weak pinning effects.

The vortices were dynamically observed to test whether vortices begin to move under such conditions. The observation was made under a fixed magnetic field B increasing the sample temperature from 4.5K to above T_c. A Lorentz micrograph at T = 4.5K and B = 20 gauss is shown in Fig. 6 (a). Vortices are distributed at random. When the temperature was raised stepwise by a few K, vortices moved. After a few minutes, vortices arrived at an equilibrium state and became still. They did not melt even at 20K. The vortex configuration changed between 40K to 50K. Vortices form a regular lattice (c) above this transition region. The vortex lattice persisted at higher temperatures though the image contrast gradually decreases and then disappears above 77K.

a

b

c

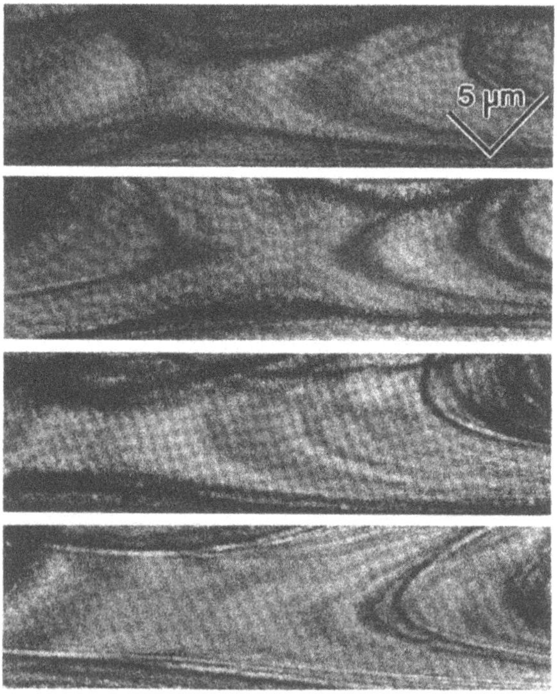

Fig. 6. Lorentz micrograph of a BSCCO (2212) film;
(a) T = 4.5 K, (b) T = 20 K, (c) T = 56 K. and (d) T = 68 K.

4. Conclusion

The performance of electron interferometry has been improved thanks to the development of a coherent electron beam and electron holography: Phase distribution in an electron beam can be measured to within $2\pi/100$, and has opened a way to measure microscopic objects and fields with ultra-high precision allowing for the direct observation of individual vortices in a superconductor. Lorentz microscopy has also been used for observing the dynamics of vortices in real time. Electron interferometry will be used to clarify the fundamental and practical applications of superconductivity, especially in the field of high-T_C superconductors.

5. References

1. A. Tonomura et al., *J. Electron Microsc.* **28** (1979) 1.
2. A. Tonomura et al., *Phys. Rev. Lett.* **54** (1985) 60.
3. A. Tonomura, *Electron Holography,* (Springer Verlag, Heidelberg, 1993).
4. J. Chen et al., *Optics Letters* **18** (1993) 1887.
5. N. Osakabe et al., *Phys. Rev. Lett.* **62** (1989) 2969.
6. A. Tonomura et al., *Jpn. J. Appl. Phys.* **19** (1980) 97.
7. J. Bonevich et al., *Phys. Rev. Lett.***70** (1993) 2952.
8. J. Bonevich et al., *Phys Rev.* **B49** (1994) 6800.
9. K. Harada et al., *Nature* **360** (1993) 51.
10. T. Kawasaki et al., *Jpn. J. Appl. Phys.* **29** (1990) 508.
11. K. Harada et al., *Phys. Rev. Lett.* **71** (1993) 3371.
12. For example, see D. J. Bishop et al., *Science* **255** (1992) 165.
13. R. N. Kleiman et al., *Phys. Rev. Lett.* **62** (1989) 2331.

MAGNETO-OPTICAL OBSERVATION OF AVALANCHES IN THE CRITICAL STATE OF SUPERCONDUCTING NIOBIUM FILMS

C. A. DURAN, P. L. GAMMEL, R. E. MILLER and D. J. BISHOP
AT&T Bell Laboratories, Murray Hill, New Jersey 0797

ABSTRACT

Using a high resolution magneto-optical technique, we have studied the penetration and expulsion of magnetic flux in superconducting Niobium thin films. Over a wide region of the phase diagram, the picture of uniform flux fronts described by a critical state model breaks down and the penetration of the field takes place through the growth of magnetic dendrites. Magnetic decoration studies with single vortex resolution show that the dendrites have a complex structure and consist of walls of both high and low magnetic field. Some of the dendrites are seen to grow by nucleating regions of antivortices in the film.

The penetration of magnetic flux into type-II superconductors is universally described in terms of a critical state model, in which the force driving the field into the interior of a sample is balanced by the pinning force. The analogy between this problem and that of a sandpile was first pointed out by de Gennes. In both cases, the equilibrium profiles are simple and well understood. What remains more controversial is the nature of the dynamical processes. In sandpile models, the role of avalanch dynamics is well established. Here, we show time and spatially resolved images which demonstrate that the superconducting profile also evolves through the formation of avalanches. The detalied nature of the avalanch structure and the interplay between interactions and disorder are easily varied just by changing the temperature in the superconducting state, making this an ideal system for systematic study. Indeed, the analysis of data from bulk measurement techniques such as magnetization and susceptibility experiments often makes assumptions about the structure of the magnetic response which can be unwarranted in the face of careful, spatially-resolved studies[1-4].

In this paper, we report on a detailed study of the flux penetration into thin, superconducting Niobium films. We find that the universally accepted description in terms of a critical state model breaks down over a wide region of the phase diagram and that rather than the penetration of flux by uniform flux fronts, as described by a Bean model[5], the field penetrates through the growth of magnetic dendrites whose structure and growth appear similar to that seen for dielectric breakdown in insulators. The dendrites are found to have a complex morphology which ranges from quasione-dimensional structures at T/T_c ~0.35 with little branching to highly-branched, "sea-weed" like structures when approaching T/T_c~0.65. Above T/T_c~0.65 the penetration is via relatively smooth flux fronts whose gross pattern can be described by a critical state model. A careful examination of the images shows evidence of coarsening of the fronts as they propagate. The dendrites themselves are wondrously complex structures with walls of alternating high and low magnetic field regions. Even less expected than the dendrites themselves, is the fact that the walls of the dendrites can have large regions containing several hundred antivortices. To our knowledge, the observation that a Type II superconductor can respond to an applied magnetic field by nucleating a large number of vortices opposite in sign to the applied field is unprecedented. Finally, the dendrites appear to grow very quickly in comparison to our temporal resolution of several milliseconds. This also strengthens the analogy of the phenomenon we observe to the dielectric breakdown of insulators.

The majority of the experimental results to be presented here were obtained using a high resolution magneto-optical technique which allows the observation of flux structures in real time. The technique works by placing a magneto-optical material in close proximity to the superconductor to be studied and using polarized light to measure the local magnetic field in the garnet which mirrors the magnetic response of the superconductor. The maximum spatial resolution afforded by the technique is clearly given by a convolution of the distance between the superconductor and the garnet, the garnet thickness and the wavelength of the light we use to probe the film. In the present experiments, to optimize this, we have followed the lead of Schuster[4], et al. and evaporated a thin film of EuSe directly onto the sample. Our cryogenic system is a conventional optical cryostat with both liquid helium and liquid nitrogen jackets. The sample is mounted on an OFHC copper cold finger with appropriate thermometry and temperature control mounted close to the sample. The magneto-optical film is imaged using a conventional optical microscope modified for polarization analysis of the film. The microscope is mounted horizontally with positioning stages for scanning the experiment. The cryostat was carefully designed to allow the sample to remain isothermal and aligned, independent of the level of the cryogens in the apparatus allowing for good long-term stability. The entire system was mounted on a high quality optical table to minimize the influence of vibrations at even the highest optical magnifications. The images were directly digitized using a computer system and a cooled CCD chip[6] with 12 bit dynamic range gray scale and excellent linearity and low dark current figures. Our final spatial resolution is $\sim 0.5\mu m$. We have also used Bitter decoration imaging using the same apparatus as described previously[7] to allow us to resolve individual vortices within the dendrite structure itself. This technique is complementary to the magneto-optical technique as it has a higher spatial resolution but obviously cannot look at the dynamics of the process of dendrite formation.

The samples studied in this experiment were pure Niobium thin films grown by DC magnetron sputtering in Argon from 8 inch diameter targets with the substrates positioned 230 mm above the target[8]. The system was cryopumped to a base pressure of $5x10^{-9}$ Torr and the evaporations were done with an Argon partial pressure of 3.5 mTorr with a Niobium deposition rate of 1.7 nm/s. These were high quality films used to produce excellent Nb-Al_2O_3-Nb junctions with $T_c(0)$=9.3 K and a critical current of ~$2x10^6$ A/cm^2 at 4.2K. We have carried out observations on films grown on three different kinds of substrates, glass microscope cover slips, single crystal quartz plates and oxidized Silicon wafers. The films grown on the Silicon substrates are known to be strain free. We find the same features for a number of films grown on all three substrates[9].

Shown in Fig. 1 is the main result of this paper. In Fig. 1a) is shown the remnant magnetic field distribution for a sample which was cooled from above T_c in the presence of a field of 135 Oe. At 7.42 K the temperature was stabilized and the field was subsequently removed. In the photo, the darker regions represent more intense local fields while the lighter regions correspond to lower fields. The observed distribution is qualitatively what would be expected from conventional critical state models. The vortices leave the sample from the edges giving a "roof-like" field distribution whose slope is given by the critical current density J_c. There are some inhomogeneities in the distribution due to the intrinsically random nature of the pinning but in general, the gross features agree with one's expectation. A more careful study of the flux front does show evidence of a coarsening of the front as it propagates (this is visible in the photo).

Fig. 1b) shows the response of the same film at 5.97K. The contrast with 1a) is evident. The pattern is much more complex with a critical state component near the edges of the sample but also with a large dendritic structures with significant branching. These structures are not at all predicted in a critical state model and suggest that there exists a magnetic instability as the field goes to zero. Shown in 1c) is the result of a measurement at 3.30K where the character of the dendrites has changed to more one-dimensional structures. The evolution of these patterns with temperature is continuous with it being rapid near 5.8K and still evolving, albeit slowly, at our lowest temperature of 3.3K.

Fig. 1 (opposite). Shown is the magnetic response of a thin Niobium film as measured using the magneto-optical technique. In a) is shown the response at 7.42K, in b) is shown the response at 5.97K and in c) is shown the response at 3.30K. As the temperature is decreased the penetration of the magnetic flux changes from a critical state like response to one in which the penetration proceeds through the growth of magnetic dendrites.

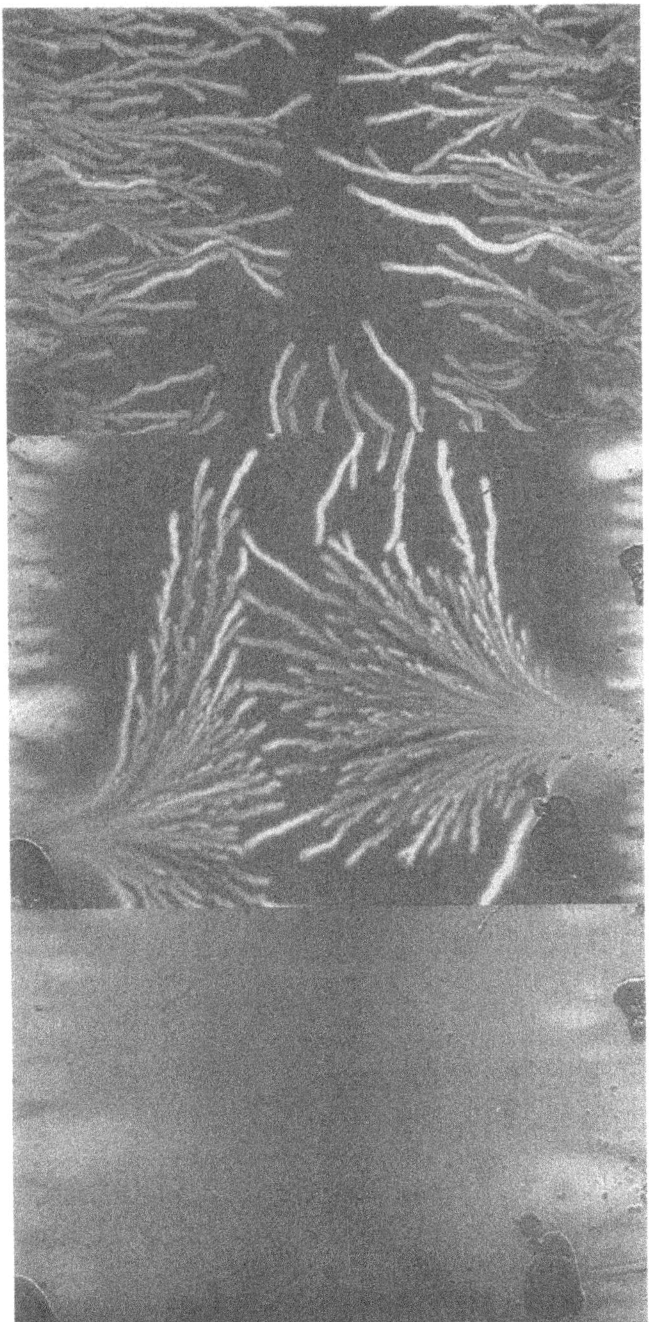

The observed dendritic structures have a number of remarkable properties. They always nucleate very rapidly regardless of the rate that the external field is changed. We were not able to measure the nucleation time but an upper limit is several milliseconds which is the frame refresh time of our CCD. We believe that the actual time is much faster[2]. Once the nucleation takes place, the whole structure is frozen in place and does not evolve upon further changes in the applied field. When the applied field is changed sufficiently, a new dendrite will appear similar in character to the others but different in detail. It is possible to cover the sample with a dense, superimposed tangle of dendrites. The new dendrites do not lie directly on the pattern of previous ones copying their shape but have complicated shapes of their own. This explicitly shows that the phenomena is not due to a macroscopic pattern of weak links in the sample. As further evidence for this, we find that identical experimental conditions always produce patterns which are similar in their overall features but different in detail showing that it is not an extrinsic effect. Patterns obtained on increasing or decreasing the field show some differences in the amount of branching but the dendrites occur for both types of field histories.

We have also done experiments in which we have slowly ramped the field at rates of a few hundredths of an Oe/sec which allows us to individually monitor each nucleation process. By subtracting subsequent frames just before and after a nucleation we have observed that the nucleation does not necessarily occur at the edge of the sample. However, each branch stops near the middle of the sample. This rules out simple surface barrier effects as the origin of the dendrites. The stopping of the dendrites near the middle of the sample shows that demagnetization is the driving force. A simple statistical analysis seems to indicate that the number of events $N\Phi$ with a given amount of flux Φ is roughly independent of Φ up to a certain cutoff.

The dendrites themselves have a very complex structure. The branches have walls of high and low field regions and are filled with a relatively uniform, high field region. This further suggests a dynamic origin for the dendrites. For some of the dendrites, there are even regions in which sizable numbers of antivortices are nucleated. The observation of an inverse local field in a superconductor in response to an applied field which was monotonically increased from zero at a constant temperature is completely outside of the framework of a critical state model.

Fig. 3 (opposite). Shown are the results of magnetic decoration studies of the dendrite growth. The upper figure shows a large area of the film and the lower figure is an expanded scale in which it is possible to see the individual vortices which make up the dendrite.

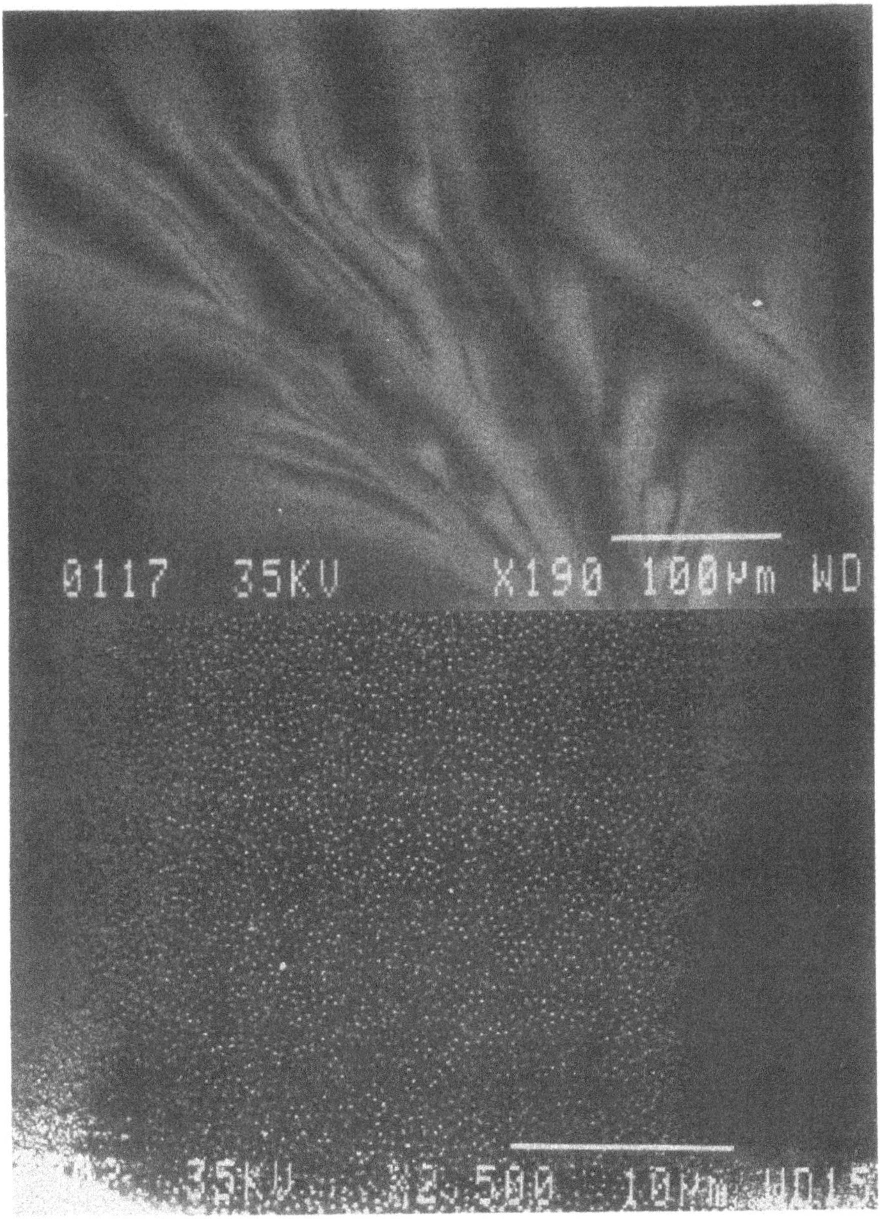
0117 35KV X190 100µm WD
35KV X2,500 10µm

Shown in Fig. 2 are two images of dendrites taken using Bitter decoration. The upper frame shows a large scale picture of the overall dendrite structure and the lower shows a dendrite at a larger magnification where one can see the individual vortices which comprise the structure. The decorations also show that the magnetic field in the dendrites is penetrating the sample as vortices and that differing field intensities correspond to differing vortex densities as is usually the case for Type II superconductors. We also do not see any features with a characteristic size comparable to the film thickness although we have the resolution to see them if they existed. The lower frame of Fig. 3 shows clearly that the vortices in the dendrites are disordered on a very short length scale. The Fourier transforms show a liquid-like structure factor with no six-fold modulation evident and rapidly decaying correlation functions. However, we cannot argue that this lack of order is evidence that the dendrites are due to the sudden quench of a penetrating vortex fluid as similarly disordered patterns are seen in these films for decorations done in FC, homogeneous magnetic fields.

Leiderer et al[2]. have seen magnetic structures in thin films of YBCO that are similar to those reported here. In those experiments, they nucleated the structures by using a laser pulse fired at a localized spot on the sample and presumably heating it up above T_c. The data in ref. 2 was explained by a model in which the normal state propagated at the expense of the superconducting one. Such a explanation does not work for our results in which the dendrites are nucleated by the application of an external magnetic field and the sample remains superconducting throughout the entire process. Our observation of singly quantized vortices in the dendrites make it clear that this is so. In addition, such a model cannot explain the appearance of antivortices. Either the origin of that effect and this one are different or the model they propose is incorrect.

In conclusion, we have used both high resolution magneto-optical imaging and Bitter decoration to study the magnetic penetration into thin films of Niobium. We have found that over a wide region of the phase diagram the critical state Bean model breaks down and that the field penetration proceeds via dendritic growth. We are able to rule out a number of conventional mechanisms to explain our observations. In addition to dendritic growth we have also observed roughening of the flux fronts in the region of the phase diagram approximately described by the Bean model. We also see that the dendrites can grow by nucleating large regions of antivortices.

We would like to thank B. Shraiman, D. Huse, D. Nelson and P. Littlewood for many helpful discussions and J. Garno for technical support.

References

1. C. A. Duran, et al., Nature **357**, 474 (1992).
2. P. Leiderer, et al., Phys Rev. Lett. **71**, 2646 (1993).
3. C. A. Duran, et al., Phys. Rev. **B49**, 3608 (1994).
4. Th. Schuster, et al., Cryogenics **31**, 811 (1991).
5. C. P. Bean, Phys. Rev. Lett. **8**, 250, (1962).
6. Photometrics, Ltd., Tucson, Arizona 85706.
7. D. J. Bishop, et al. Science **255**, 165 (1992).
8. R. E. Miller, et al., Appl. Phys. Lett. **63**, 1423 (1993).
9. Some variation of the temperature scales for the effects was observed with changing substrates. The details will be published elsewhere.
10. U. Essman, Physics Letters **41A**, 477 (1972).
11. P. P. Huebener, "Magnetic Flux Structures in Superconductors", (Springer Verlag, Berlin, 1979).

CRITICAL CURRENT DENSITIES ALONG THE C–AXIS OF SINGLE CRYSTALLINE $Bi_2Sr_2CaCu_2O_{8+\delta}$

KAZUO KADOWAKI, TAKASHI MOCHIKU, HIROYUKI TAKEYA AND KAZUTO HIRATA

National Research Institute for Metals, 1-2-1, Sengen, Tsukuba-shi, Ibaraki 305, Japan

and

TERUKAZU NISHIZAKI AND NORIO KOBAYASHI

Institute for Materials Research, Tohoku University, 2-1-1, Katahira, Aoba-ku, Sendai Miyagi 980-77, Japan

ABSTRACT

Experimental results of the resistivity and the critical current densities along the c–axis of single crystalline $Bi_2Sr_2CaCu_2O_{8+\delta}$ are presented in magnetic fields up to 20.1 T applied to the c–axis. The resistivity in high fields continuously increases as the temperature is decreased below T_c. It shows a broad peak, then comes back sharply to the zero level even at 20.1 T. This behavior indicates that the resistivity in this system is essentially insulating along the c–axis and the superconductivity is induced by a proximity effect. From the current–voltage ($I-V$) characteristics measuerments the c–axis critical current densities, J_c^c, are suppressed by the magnetic field and found to obey the exponential behavior similar to those in conventional SNS junctions. The dc–Josephson effect observed in the current voltage ($I-V$) characteristics in this system also supports this picture. These experimental results strongly suggest that the superconducting pair condensation occurs locally only near the conducting CuO_2 plane within a few atomic distances along the c–axis. This naturally leads to a conclusion that the superconductivity in this system is intrinsically inhomogeneous. This fact has to be bear in mind, when one interprets various kinds of experimental results in this system. Furthermore, anomalous features observed in the $I-V$ characteristics are discussed in terms of the new phase diagram of the vortex state in this system.

1. Introduction

Superconductivity of highly two dimensional systems such as in $Bi_2Sr_2CaCu_2O_{8+\delta}$ has attracted much attention due partly to fundamental interest in physics and partly to demand for practical applications of high temperature superconductors, which are, at the present stage, limited by a current carrying capacity in wires, tapes and bulk materials, etc. High temperature superconductors commonly have the two dimensional layered structure of the square coordinated CuO_2 planes, which are responsible for the superconductivity. Among others, the $Bi_2Sr_2CaCu_2O_{8+\delta}$ system is ideal for this type of study, since the two dimensional anisotropy, $\Gamma=\gamma^2=m_c/m_{ab}$, is huge, of the order of 10^5–10^6, estimated by various experimental results such as torque and transport measurements, etc[1-2]. Although this value is not fully established yet until now because of experimental difficulties, it is perhaps a record high value among previously known superconductors. The coupling strength between superconducting CuO_2 layers are extremely weak and has been thought to be due to semiconducting or even insulating nature of Bi_2O_2 layers in the structure. This high anisotropy as well as the short coherence length, especially along the c-direction, has been considered to be the cause of many unusual phenomena observed in this system.

In particular, study of resistivity in magnetic fields have demonstrated the importance of the superconducting fluctuation effect in such a system.[3-6] Here it is noteworthy to stress that the short coherence length in this compound, especially the c–axis coherence length, ξ_c, is responsible for the strong superconducting fluctuation effect. As a result, the phase diagram of the system is modified drastically in magnetic fields from the one for the conventional superconductors.[7-8] The schematic picture of the new phase diagram is presented in Fig. 1 together with the sketch of the in–plane resistivity behavior. Although this complicated phase diagram has not fully been understood and still issues of controversy, it has been established theoretically and experimentally some of the features that, for example, there is no sharp phase transition at $B_{c2}(T)$ defined as a mean field sense in conventional superconductors.[9-11] In stead, the physical parameters are continuous around this region until the irreversibility line, $B_{irr}(T)$ is met, where drastic changes in the physical parameters take place. The resistivity, for instance, seems to become zero, at least in a practical sense, indicating the occurrence of the true superconductivity. At lower temperatures below $B_{irr}(T)$, superconductivity manifested by the existence of supercurrents sets in with peculiar behaviors of the vortex state owing to the interplay between low dimensionality and various pinning centers.

The purpose of this paper is to show further modification of the phase diagram shown in Fig. 1 and to propose a physical picture associated with it on the basis of fundamental underlying concept of Josephson coupling existing between the superconducting CuO_2 layers. Experimental results shown here are all related to the transport phenomena along the c–axis of high quality single crystalline $Bi_2Sr_2CaCu_2O_{8+\delta}$: the recent high field resistivity measurements up to 20.1 T applied parallel to the c–axis and the $I-V$ characteristics in magnetic fields up to 6 T applied to the c–axis.

It is shown that all these results are consistent with the picture based on the Josephson coupled model given previously. Some results predict new phenomena, which need further experimental proof.

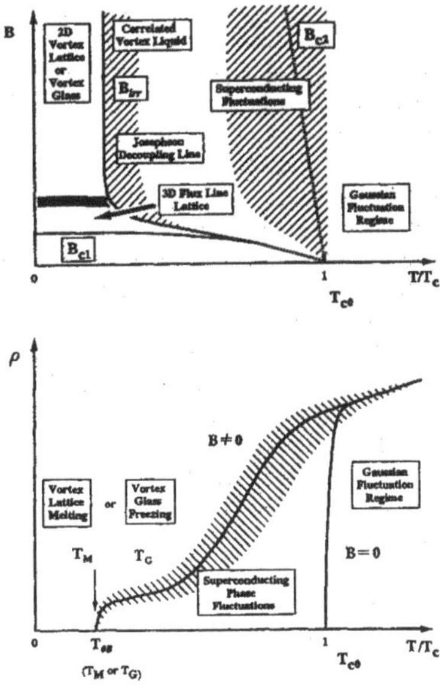

Fig. 1. The sketch of the phase diagram of the vortex state in highly two dimensional superconductors (above). This will be modified due to the existence of the Josephson coupling energy scale (see, the text below). The corresponding in–plane resistivity in magnetic fields is schematically shown.

2. Experimental

2.1. *Sample Preparation and Characterization of Single Crystallinity*

The sample used in the present study was grown by the traveling solvent floating zone technique with an infrared image furnace. The quality of the sample was checked carefully from both crystallographic and superconductivity points of view. The single crystallinity was measured by the double crystal diffractometer and the rocking curve of the (0,0,18) line showed a typical value between 70~150 sec., which directly represent the degree of the mosaic spread of the c–axis orientation. Multiple peaks with the width of the order of 0.1° in the rocking curve are often observed in poor crystals. This means that the crystal along the c–axis has low angle microtwins. This has also been seen in the microanalysis of the growth front of the crystal by SEM and is understood by the cellular growth mechanism due to highly anisotropic growth speed of the crystal.

From the results of the intensive study of HERM (High Resolution Electron Microscopy) of the single crystals, it has been recognized that the crystal of the $Bi_2Sr_2CaCu_2O_{8+\delta}$ phase contains inclusions of the isomorphous phases such as $Bi_2Sr_2CuO_{6+\delta}$ and $Bi_2Sr_2Ca_2Cu_3O_{10+\delta}$. Practically speaking, it is very difficult to avoid them. These phases intergrow as a stacking faults and intervene continuity of the layer stacking along the c-axis. The $Bi_2Sr_2CuO_{6+\delta}$ and $Bi_2Sr_2Ca_2Cu_3O_{10+\delta}$ phases are often seen together. Because the thermodynamical growth kinetics of these phases are all very akin to each other as expected from the chemical composition as well as the peritectic melting points close to each other, which lie within about 40 °C, no established method is known so far to avoid these inclusions. As far as our crystals used in these experiments is concerned, the inclusion rate is estimated to be less than 0.1 % in average, which is about one to two orders of magnitude smaller than that in the flux-grown single crystals. It is observed that in most cases the inclusion is driven by the spontaneous growth of the very low angle twin boundaries along the a-axis, which give rise to the mosaic spread of the rocking curve described above.

2.2. Characterization of Superconductivity

As for superconductivity the transition temperature, T_c, as well as the transition width, ΔT_c, was checked by a SQUID magnetometer in low fields (0.1~1 mT). The width is sharp and it is typically within 1 K or less at T_c of about 86 K, which varies a few degree from batch to batch.

It has been recognized that for further characterization of the superconductivity ac-susceptibility technique with a variable amplitude of oscillating ac-magnetic field is quite helpful and provides valuable information on the quality of the sample. This is, by no means, a complicated technique and it has resolutions beyond the commercial SQUID magnetometer in low frequencies. An example of this is presented in Fig. 2, in which the temperature variation of both real and imaginary components, $\chi'(\omega)$ and $\chi''(\omega)$, of the complex ac-susceptibility are shown. The amplitude of the ac-magnetic field with a frequency of 90 cycle/sec. is applied perpendicular to the thin plate-like single crystal (parallel to the c-axis) with a rectangular shape of approximately 2 mm × 2.5 mm and with a thickness of about 60 μm. As clearly seen in Fig. 2, the transition width becomes narrower as the amplitude of the oscillating field becomes smaller, and it no longer changes the width below the ac-field level of 30 μT. The width is about 400 mK at the mid-transition point of 88.850 K in this crystal. Furthermore, the shape of the transition in both $\chi'(\omega)$ and $\chi''(\omega)$ begins to have fine structures as the ac-field is decreased. This fact can be attributed to the multiple superconducting transitions in the sample, perhaps corresponding to the several platelets of single crystals with slightly different T_c stacked together. This fact indicates that if the sample is ideal, the intrinsic superconducting transition width may be much sharper.

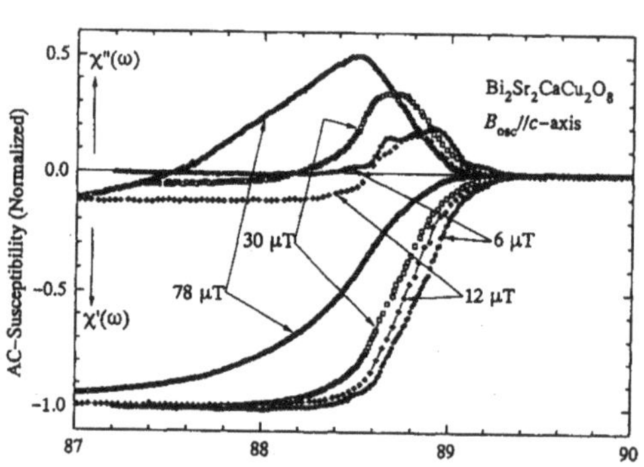

Fig. 2. Experimental signals of real and imaginary parts, $\chi'(\omega)$ and $\chi''(\omega)$, of the ac-susceptibility in various different ac-oscillating fields in single crystalline $Bi_2Sr_2CaCu_2O_{8+\delta}$. The sample can easily be characterized by looking at the width and the shape of the superconducting transition.

2.3 Experiments

The samples used in high field and $I-V$ characteristics measurements are cut into rectangular shape with 0.8 mm × 1.0 mm with a thickness of 13 μm and 0.25 mm × 0.3 mm with a thickness of 7–10 μm, respectively. Both sides of the sample surfaces are coated by silver and gold wires with 25 μm in diameter are attached by silver paint. For the high field measurements dc four probe technique is used with 1 mA, while for the $I-V$ characteristics measurements, two terminal method is applied with currents sweeping up and down linearly with time. The corresponding voltage response synchronized with the current is stored in the digital storage scope (Iwatsu DS–9121).

3. c–Axis Resistivity in High Magnetic Fields

3.1 Normal State

The normal state resistivity along the c–axis of $Bi_2Sr_2CaCu_2O_{8+\delta}$, ρ_c, has been shown to be increasing with decreasing temperature, whereas the in–plane resistivity, ρ_{ab}, is more or less linear in temperature. The absolute values of the resistivity at room temperature are approximately a few Ωcm and 50–100 μΩcm, respectively, which give rise to an extremely large anisotropy value of the resistivity, ρ_c/ρ_{ab}~10^{-5}.[3] This value even increases as temperature is decreased, suggesting that the electrical conduction along the c–axis in this material is no longer metallic but may be semiconducting or insulating.

3.2 Superconducting State in Magnetic fields

The resistivity along the c–axis even with such a high resistivity value undergoes to the zero resistivity state below T_c in zero magnetic field. This has been thought to be a result of manifestation that the superconducting pairs can even penetrate through the semiconducting or the insulating region.

In our previous study,[7–8,12] it was shown that the resistivity is induced even in a relatively weak magnetic field. This phenomenon is interpreted by the effect of superconducting fluctuations in this highly anisotropic system with an extremely short coherence length. The dissipation in this understanding is originated by the fluctuations of the superconducting order parameter. Furthermore, it should be mentioned that this understanding covers the conventional understanding of the resistive state in superconductors such as flux flow and flux creep.

The main interest here is how the resistivity disappears as temperature is decreased. In $YBa_2Cu_3O_{7-\delta}$, this has been proved to be accompanied by a phase transition from vortex liquid to vortex solid state in lower fields (B<7 T) and from vortex liquid to vortex glass state in higher fields (B>7 T).[13–14] In both cases below the phase transitions the pinning mechanism plays an important role for the occurrence of the vortex state.

In contrast to the case of $YBa_2Cu_3O_{7-\delta}$, there has been no clear evidence of such phase transitions at least, in the resistivity, in $Bi_2Sr_2CaCu_2O_{8+\delta}$, although more microscopic techniques such as μ–SR and small–angle neutron scattering experiments have shown disappearance of the flux–line lattice, suggesting the existence of the melting transition also in this system.[15]

It is certainly important to know the relevant energy scales governing the phenomena. The interactions involved and competing each other are the magnetic coupling of the vortex pancakes, the Josephson coupling and the pinning energy. Since the magnetic coupling is expected to be important only in a low field regime near B_{c1}, it may be neglected. It is very interesting to ask how the Josephson coupling is destroyed in a sufficiently high field region.

In Fig. 3, we demonstrate the results of the c–axis resistivity measurements of $Bi_2Sr_2CaCu_2O_{8+\delta}$ up to 20.1 T.[16] The resistivity continuously increases with decreasing temperature through the zero field transition region, following the results in low fields up to 6 T reported

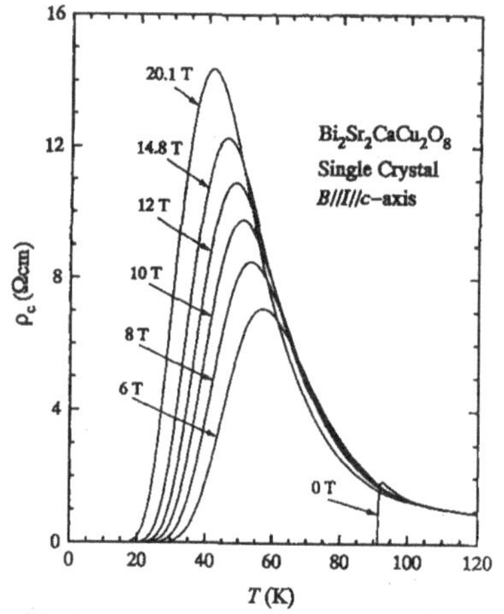

Fig. 3. High field c-axis resistivity behavior of single crystalline $Bi_2Sr_2CaCu_2O_{8+\delta}$.

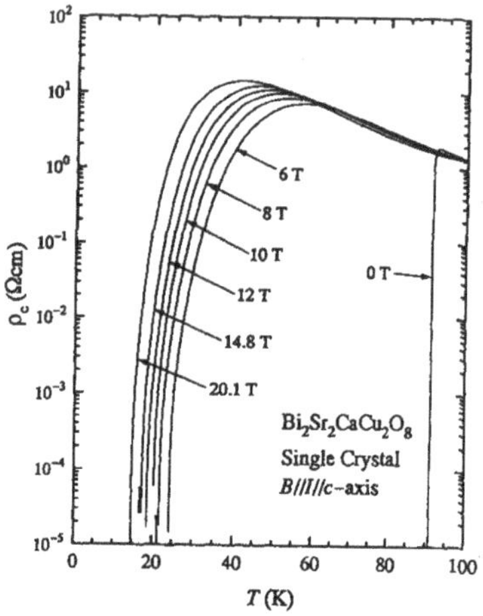

Fig. 4. The logarithmic plot of ordinate of Fig. 3 to closer look at the lower resistivity region.

previously. A steeper rise of the resistivity in higher fields also confirms the previous understanding that the resistivity along the c-axis is essentially semiconducting or even insulating. The resistivity sharply drops through a sharper maximum in higher fields towards zero resistivity. The behavior in lower resistivity region can more clearly be seen in Fig. 4, where the ordinate of Fig. 3 is re-plotted by a logarithmic scale. Although the resistivity curve shifts to the lower temperature side, there is no obvious difference in the temperature dependence of the resistivity even in 20.1 T, indicating that the resistivity undergoes zero resistance as seen in lower fields. If the shift of the resistivity in temperature is extrapolated linearly, it would be zero at a field between 40 and 50 T. This behavior is presented in Fig. 5 together with the behavior of the irreversibility line in the temperature-field diagram. As a criterion of the zero resistivity level, a value of resistivity of 5×10^{-4} Ωcm is chosen. Although the agreement with $B_{irr}(T)$ determined by static magnetization measurements is not excellent, the high field behavior of $B_{irr}(T)$ can be reasonably be extrapolated.

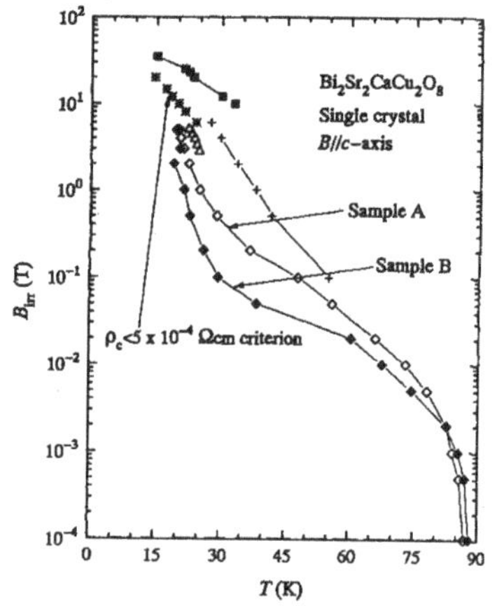

Fig. 5. The irreversibility lines in single crystalline $Bi_2Sr_2CaCu_2O_{8+\delta}$ determined by various experimental techniques. Two samples, A and B, which had a different oxygen doping level are measured by SQUID magnetometer. Symbols (■), (+), (▵), and (*) indicates data obtained in fields parallel to the c-axis by pulsed high field magnetization measurements, ac-susceptibility with low frequencies, vibration reeds and the present high field resistivity measurements with a criterion of 5×10^{-4} Ωcm.

In Fig. 3, it is also noteworthy to point out that the resistivity in a region where it is continuously rising above the broad peak shows a considerable magnetic field dependence. This negative magnetoresistance in this region can not be explained by the suppression effect of superconducting fluctuations, because it is expected to give rise to the opposite sign. Since the conduction mechanism in high T_c superconductors, especially, along the c–axis is highly controversial, more precise experiments as well as more quantitative analysis are needed and may shed light on the problem.

4. Critical Current densities and the *I–V* Characteristics

Since the CuO_2 plane is considered to be responsible for the superconductivity, the rest of the constituents of the structure must possesses semiconducting or insulating properties in a microscopic scale in order to reconsile the c–axis resistivity of the semiconducting or insulating nature. Considering the short coherence length along the c–axis in this system in addition, this fact simply implies that the superconducting state must be highly inhomogeneous even within a scale of the unit cell due to the local nature of the superconductivity.

As shown above the $Bi_2Sr_2CaCu_2O_{8+\delta}$ system at low temperatures still remains superconducting even in 20.1 T. If the system consists of superconducting layers stacking along the c–axis, which interact via the Josephson weak coupling, it is interesting to ask how the Josephson interaction is influenced by magnetic fields. To answer this question, we have performed the *I–V* characteristic measurements along the c–axis in magnetic fields.

In our previous work of the c–axis *I–V* characteristic measurements, we have experimentally proved and shown the clear evidence of the Josephson type of coupling existing along the c–axis.[18) This weak Josephson coupling turns out to be not single– but multiple–connected, at least of the order of the number of the CuO_2 layers, with an elemental coupling strength, which give rise to a minimum steps in the *I–V* characteristics. This experimental fact proves for the first time that the weak coupling exists ***in a scale of a unit cell dimension*** and the underlying assumption previously postulated was confirmed experimentally. It is worthwhile pointing out that the coupling strength obtained was about 0.1–0.2 meV, which is much smaller than the superconducting energy gap of 20–30 meV.

In Fig. 6, a typical example of the *I–V* characteristic curves in $Bi_2Sr_2CaCu_2O_{8+\delta}$ along the c–axis are shown in magnetic fields.[19) The main features are as follows. When magnetic field is lower than a characteristic field B^*~0.2 T at T=78.88 K, the *I–V* characteristic behavior is essentially the same as that of in zero field except that the critical current at which the sharp transition in voltage takes place shifts toward the lower current side. This results in the reduction of the size of the hysteresis. The sharpness of the transition, the hysteretic nature of the *I–V* curve as well as the sharp peak associated with the transition are also very similar. The effect of the magnetic field on the *I–V* curve in this field region seems to be the shift of the sharp jump at the transition. Above B^*, on the contrary, the transition becomes no longer sharp and is broadened considerably. Neither sharp peak nor large hysteresis is observed.

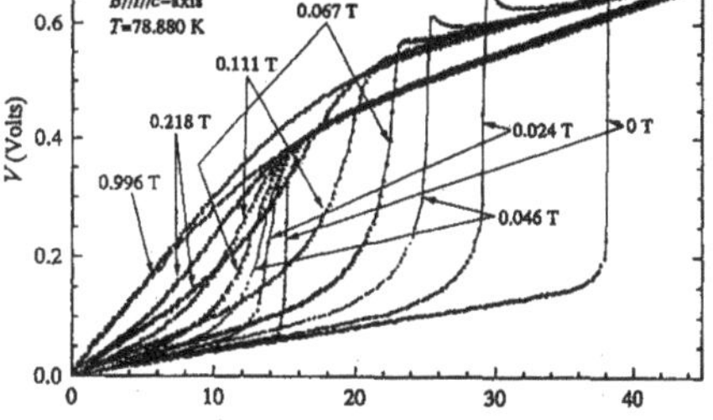

Fig. 6. *I–V* characteristic curves in single crystalline $Bi_2Sr_2CaCu_2O_{8+\delta}$ along the c–axis at 78.880 K in various magnetic fields.

The *I–V* characteristics at 4.2 K are shown in Fig. 7. As the temperature is lowered, a new feature becomes significant: a new broad peak begins to grow below about 50 K and shifts to the lower current regime with increasing magnetic fields. As a consequence, the dynamic resistivity becomes negative in this region. This peculiar behavior cannot be understood by a simple Josephson coupling model. A model which involves the anomalous conduction mechanism in the normal state is proposed as a speculation.

As is seen in Fig. 8 the critical current density, J_c^c, at 29.984 K is rapidly suppressed as the field is increased. It is relatively easier to define J_c^c below B^* but it becomes rather ambiguous because of lack of the sharp transition above B^*. The field dependence of $J_{c,up}{}^c(B,T)$ shown in Fig. 8 are defined by the most stringent criterion of 5 mΩcm and approximately follows the $\exp(-\alpha B)$ relation, where α is a constant. This behavior is in good agreement with that of conventional SNS junctions and is also consistent with the weak Josephson coupling picture of this compound along the c–axis.

Although $J_c^c(T,B)$ in magnetic fields grows rapidly with decreasing temperature below the irreversibility line, $B_{irr}(T)$, the characteristic field, B^*, does not have a steep temperature dependence. It is rather surprising that B^* even extends above $B_{irr}(T)$ at higher temperatures (T>30 K), because this region was thought to be the uncorrelated vortex liquid state. As seen above B^* characterizes the $I-V$ curves and plays an important role to measure $J_c^c(B,T)$ along the c–axis. Therefore, B^* is attributed to the characteristic energy scale of the Josephson coupling in this system.

In Fig. 9, the characteristic energy scale boundary, B^*, is added to the phase diagram. Here, two regions are newly discovered: one is the region below B^* and above $B_{irr}(T)$, and the other is the region below B^* and above B_{2d}, where the vortex lattice is destroyed by the 2D to 3D transition.[15] The former is interpreted as a region where the vortex pancakes are still coupled with short range correlations. Since the long range correlation is lost above $B_{irr}(T)$ the resistivity is finite in this region. On the other hand, the latter can be interpreted as a region where the static Josephson correlation begin to be broken due to strong magnetic fields. Accordingly, above this line the critical current density may drop sharply and diminishes asymptotically in higher fields as shown in Fig. 8 as an example at 29.984 K.

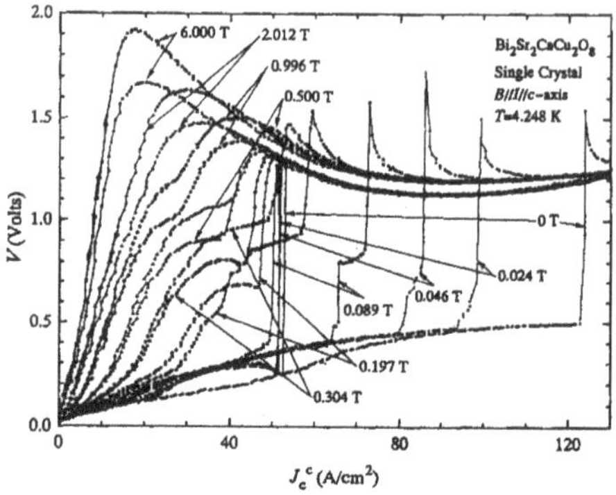

Fig. 7. $I-V$ characteristic curves in single crystalline $Bi_2Sr_2CaCu_2O_{8+\delta}$ along the c–axis at 4.248 K in various magnetic fields.

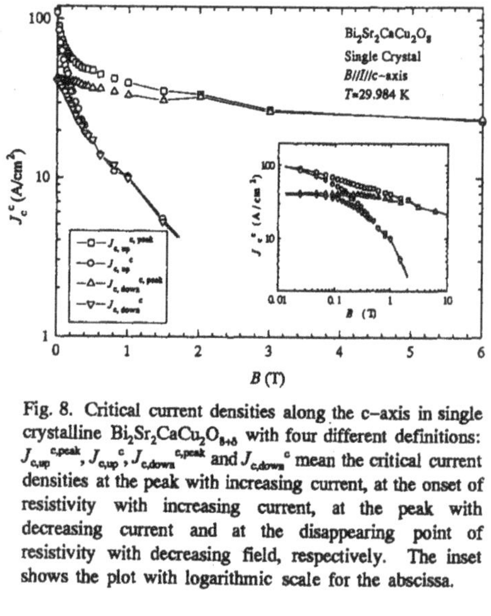

Fig. 8. Critical current densities along the c–axis in single crystalline $Bi_2Sr_2CaCu_2O_{8+\delta}$ with four different definitions: $J_{c,up}{}^{c,peak}$, $J_{c,up}{}^{c}$, $J_{c,down}{}^{c,peak}$ and $J_{c,down}{}^{c}$ mean the critical current densities at the peak with increasing current, at the onset of resistivity with increasing current, at the peak with decreasing current and at the disappearing point of resistivity with decreasing field, respectively. The inset shows the plot with logarithmic scale for the abscissa.

5. Conclusion

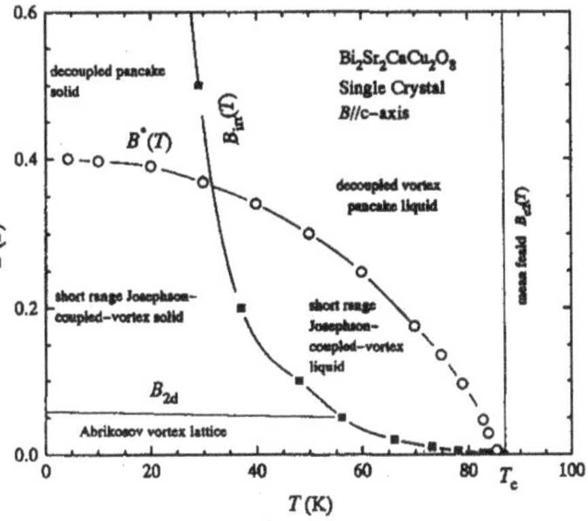

Fig. 9. The new phase diagram proposed in single crystalline $Bi_2Sr_2CaCu_2O_{8+\delta}$ in low magnetic fields along the c–axis.

We have shown the c–axis resistivity behavior of single crystalline $Bi_2Sr_2CaCu_2O_{8+\delta}$ in high magnetic fields up to 20.1 T for the first time and presented the results of the $I-V$ measurements in magnetic fields. All these phenomena observed can reasonably be explained by the weakly coupled superconducting layer model. The nature of the coupling in this system is experimentally proved to be of Josephson type and the field dependence was studied. A new boundary, B^*, is discovered, which characterizes the Josephson coupling strength. The significance of the boundary, B^*, is argued in terms of the vortex phase diagram. A physical picture that the new phase determined by B^* is a measure of the vortex pancake short range correlations at high temperatures above $B_{irr}(B,T)$ and the static short range correlation in low temperature region below $B_{irr}(B,T)$ is proposed.

References

[1]. K. Okuda *et al., J. Phys. Soc. Jpn.* **60** (1991) 3226.
[2]. Y. Iye *et al., Physica C* **199** (1992) 154.
[3]. K. Kadowaki *et al., Physica B* **165&166** (1991) 1159.
[4]. K. Kadowaki *et al., Physica C* **170** (1990) 298.
[5]. K. Kadowaki *et al., Supercond. Sci. Technol.* **7** (1994) 519.
[6]. K. Kadowaki *et al., Phys. Rev. B* **50** (1994) 7230.
[7]. K. Kadowaki *et al., Advances in Superconductivity IV* eds. by H. Hayakawa and N. Koshizuka, Springer Verlag, 1992, p395.
[8]. K. Kadowaki, *Electronic Properties and Mechanisms of High T_c Superconductors*, eds. by T. Oguchi, K. Kadowaki and T. Sasaki, Elsevier Science Publishers, 1992, p209.
[9]. R. Ikeda *et al., J. Phys. Soc. Jpn.* **59** (1990) 1740.
[10]. E. Brézin *et al., Phys. Rev. Lett.* **65** (1990) 1949.
[11]. R. Ileda *et al., J. Phys. Soc. Jpn.* **61** (1992) 254.
[12]. K. Kadowaki *et al., Advances in Superconductivity VI*, eds. by T. Fujita and Y. Shiohara, Springer Verlag, 1994, p545.
[13]. H. Safar *et al., Phys. Rev. Lett.* **69** (1992) 824.
[14]. W. K. Kwok *et al., Phys. Rev. Lett.* **69** (1992) 3370.
[15]. R. Cubitt *et al., Nature* **365** (1993) 407.
[16]. J. H. Cho *et al., Phys. Rev. B* **50** (1994) 6493.
[17]. K. Nakao *et al., Advances in Superconductivity VI*, eds. by T. Fujita and Y. Shiohara, Springer Verlag, 1994, p205.
[18]. K. Kadowaki and T. Mochiku, *Physica B* **194–196** (1994) 2239.
[19]. K. Kadowaki *et al.*, submitted to *Phys. Rev. Lett.*

CONTROL OF ANISOTROPY AND IRREVERSIBILITY LINES BY CARRIER DOPING IN HTSC SINGLE CRYSTALS

K. KISHIO, J. SHIMOYAMA, K. KITAZAWA
Department of Applied Chemistry, University of Tokyo, Tokyo 113, Japan
and
K. YAMAFUJI
Department of Electronics, Kyushu University, Fukuoka 812, Japan

ABSTRACT

Irreversibility lines (H//c) in single crystals of $(La,Sr)_2CuO_4$, $Bi_2Sr_2CaCu_2O_{8+\delta}$ and $YBa_2Cu_3O_{7-\delta}$ are summarized and discussed with paying particular attention on their carrier-doping states. As a function of Sr or oxygen contents, δ, systematic and dramatic widening of the irreversibility region in the $B-T$ phase diagrams was observed in all the systems. In particular, we have found a very good correlation of the irreversibility fields B_{irr} with the electrical anisotropy $\gamma^2(\equiv \rho_c/\rho_{ab})$. When compared at several reduced temperatures (T/T_c), B_{irr} scaled surprisingly well with γ^2 over four decades. This suggests a universal feature that all the HTSC cuprates display weak Josephson interlayer coupling and that the carrier doping is quite effective in improving the coupling and the flux pinning strength.

The layered crystal structure and associated anisotropy are the most critical factors determining the behavior of the reversible and irreversible magnetic properties of HTSC materials. It seems clear that essentially very weak flux-pinning properties in HTSC materials are caused by a weak interlayer coupling strength due to the large electromagnetic anisotropy. While the electrical anisotropy is determined principally by the interlayer distance of the CuO_2 planes and the inherent chemical nature of the "blocking layer", it is also expected to be affected by carrier doping. However, how the doping state of a given compound affects its magnetic properties has not been investigated in detail.

In this paper, irreversibility lines of high-quality single crystals of $(La_{1-x}Sr_x)_2CuO_{4-\delta}$ (LSCO), $Bi_2Sr_2CaCu_2O_{8+\delta}$ (Bi2212) and $YBa_2Cu_3O_{7-\delta}$ (YBCO) over wide ranges of carrier doping states, are summarized and the universal features, common in these layered HTSC cuprates, are addressed. LSCO crystals (x=0.05, 0.0625, 0.085 and 0.10) have been prepared by the TSFZ technique and their electrical anisotropy as well as reversible magnetic properties have been extensively investigated as described elsewhere[1,2]. Bi2212 crystals were prepared similarly by the FZ technique[3] and their oxygen contents were carefully controlled by low temperature annealing in sealed quartz tubes[4]. The electrical anisotropy of these crystals have been also described in Ref. 3. Irreversibility lines (H//c) were determined from the splitting points of the ZFC and FC magnetization curves measured by a SQUID magnetometer. The similar data for the YBCO crystals were taken from the literature[5].

Figure 1 summarizes the observed irreversibility points, B_{irr} (H//c), of LSCO, Bi2212 and YBCO single crystals with various chemical compositions as a function of their reduced temperature with respect T_c; $t \equiv T/T_c$. First of all, the four curves (h) through (k) with closed data symbols correspond to the crystals for x=0.05, 0.0625, 0.085 and 0.10, respectively. It should be noted that T_c's of these crystals are 26.2, 32.4, 35.0 and 29.8 K with an increasing order in x. It is immediately found that the irreversibility lines are located at higher fields in the crystals with larger Sr contents. Since the temperature slope of the mean field B_{c2} values, dB_{c2}/dT, has been found to be approximately same, $-1 \sim 1.5$ T/K, in these samples[2], Fig. 1 clearly suggests that the widening of the irreversible region in the $B-T$ plane is taking place as a function of x. The curves (l), (m) and (n) are for the YBCO crystals with estimated oxygen deficiency[5] of δ ~0.2, 0.1 and 0.05. It is clear that B_{irr} is located at much higher temperature in YBCO than in LSCO. Another feature is that B_{irr} increases as the oxygen deficiency is decreased. Therefore, in both the LSCO and YBCO systems, B_{irr} is understood to move to higher fields by increasing carrier-hole doping.

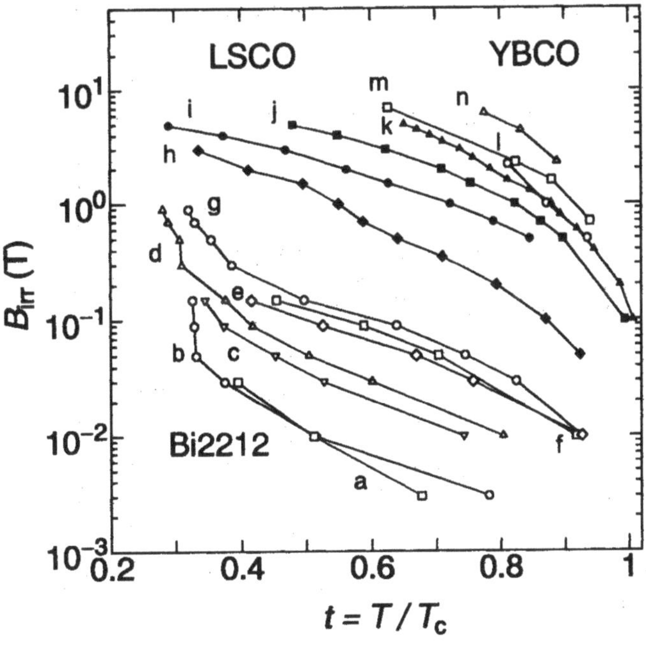

Fig. 1 : Irreversibility field, B_{irr} (H//c), of three cuprate single crystals with various doping states. Curves (h) through (k) are for LSCO with different Sr contents; (l) through (n) are for YBCO with various oxygen deficiency δ; (a) through (f) are for Bi2212 with various oxygen contents. See more details in the main text.

Let's further examine the Bi2212 system similarly. The curves (a) through (g) in Fig. 1 were obtained from seven pieces of single crystals of the identical batch but with different oxygen annealing treatment. The estimated difference in oxygen content[3] in $\Delta\delta$ is $\sim$ (a)-0.055, (b)-0.049, (c)-0.043, (d)-0.021, (e)-0.013, (f)-0.002, respectively, with reference to the crystal (g). Therefore, the sample (a) corresponds to the most carrier-underdoped while (g) is the most overdoped crystal. Now it is immediately noticed in Fig. 1 that B_{irr} in the present Bi2212 system is also a strong function of oxygen content; namely, underdoped samples lie at lower fields while the most overdoped crystal shows the highest irreversibility line. This behavior is perfectly identical to the LSCO and YBCO cases and it is concluded that the carrier doping significantly shift the irreversibility lines in all the systems examined here.

The universal behavior in Fig 1 can be understood in terms of the composition dependence of the carrier doping states and electrical anisotropy in these compounds. In both systems, it has been well-demonstrated that normal state resistivity both along the c-axis, ρ_c, and the ab-plane, ρ_{ab}, as well as their anisotropy ratio, ρ_c/ρ_{ab}, are monotonically decreased[1,3] with the Sr content in LSCO and with the oxygen content in Bi2212. Based on this consideration, we have plotted B_{irr} at $t \equiv T/T_c = 0.7 \sim 0.9$ of both systems as a function of $\gamma^2 \equiv \rho_c/\rho_{ab}$ evaluated at T_c onset temperatures.

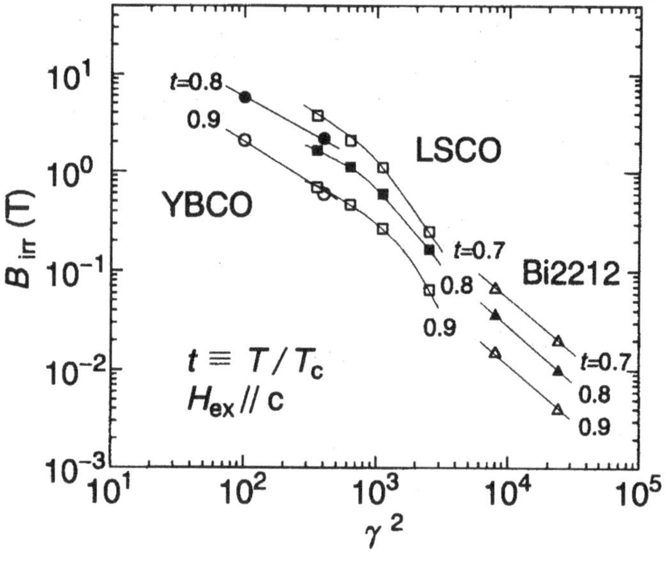

Fig. 2 : Irreversibility field, B_{irr} (H//c) at various reduced temperatures of LSCO, YBCO and Bi2212 single crystals plotted as a function of their electrical anisotropy, $\gamma^2 (\equiv \rho_c/\rho_{ab})$.

The results are shown in Fig. 2, including the B_{irr} of YBCO single crystals[5], where their anisotropy values were estimated from Ref. 6. It is particularly interesting to see here that B_{irr} changes quite rapidly with γ^2 and all the curves are correlated well in all the three systems extending over an extremely wide range of γ^2 values.

It has been believed that the weak pinning behavior of the Bi2212 compound originate from the short correlation length, ℓ_{44}, of the flux line lattice along the layer stacking direction because of the weak interlayer coupling (Josephson coupling) of the CuO_2 planes across the Bi_2O_2 layers. A recent microwave penetration measurement[7] has also shown that the interlayer coupling in the LSCO system is also Josephson-like. Figure 2 seems to suggest that such a picture should be correct in all of the layered HTSC cuprate systems and it is concluded that strengthening of the interlayer coupling by carrier doping is extremely effective, in order to improve the flux pinning properties of these materials.

REFERENCES

1. T. Kimura and K. Kishio *et al*, *Physica* **C192** (1992) 247.
2. Q. Li, M. Suenaga, T. Kimura and K. Kishio, *Phys. Rev.* **B47** (1993) 11384.
3. Y. Kotaka *et al*, Physica **C235-240** (1994) 1529.
4. J. Shimoyama and K. Kishio *et al*, in *Proc. 6th US-Japan Workshop on High-Tc Superconductors*, World Scientific, Singapore, pp.245-250 (1994).
5. M. A. Angadi *et al*, *Physica* **C185-189** (1991) 2159.
6. K. Takenaka, K. Mizuhashi, H. Takagi and S. Uchida, *Phys. Rev.* **B50** (1994) 6534.
7. T. Shibauchi *et al*, *Phys. Rev. Lett.* **72** (1994) 2263.

INTERGRANULAR AND INTRAGRANULAR CRITICAL STATES IN HIGH-TEMPERATURE SUPERCONDUCTORS

K.-H. MÜLLER and C. ANDRIKIDIS

CSIRO, Division of Applied Physics, Sydney 2070, Australia

ABSTRACT

Applying a magnetic field to a granular superconductor induces both intergranular and intragranular currents. Critical state models can be used to describe in great detail the magnetic properties of such a superconductor, to extract values for the intergranular and intragranular critical current densities and to study the correlation between the intergranular and intragranular critical states. Significant differences in the intergranular and intragranular behaviour are found between sintered polycrystalline samples and PBSCCO tapes.

1. Introduction

Soon after the discovery of the high-temperature superconductors it was noticed that the *inter*granular (transport) critical current density J_c^J in polycrystalline samples is limited by the weak-link behaviour of the grain boundaries. This deleterious influence of grain boundaries is believed to be due to the shortness of the coherence length in these cuprate oxides which is much smaller than the coherence length in conventional superconductors. Experiments on bicrystal grain boundaries demonstrated that a current across a grain boundary in the highly anisotropic cuprate oxides decreases very rapidly with increasing grain misorientation [1] and thus the random orientation of the grains is blamed for the low J_c^J in polycrystalline samples. In contrast, a very high *inter*granular critical current density J_c^J is achieved for PBSCCO/Ag tapes because of the strong alignment of the grains achieved by pressing and rolling.

The magnetic properties of these granular high-temperature superconductors can be modelled using critical state equations for both the *inter*granular and *intra*granular magnetic field distributions. While the *inter*granular critical state originates from Josephson currents circulating between grains where pinning is related to network inductances and defects in the network [2], the *intra*granular critical state, as in conventional superconductors, is related to the pinning of Abrikosov vortices inside grains. Using the critical state model most of the experimental magnetic data can be understood in great detail. One finds that in polycrystalline sintered materials the *intra*granular critical current density J_c^G is roughly three to four orders of magnitude larger than J_C^J while in PBSCCO/Ag tapes $J_c^G \simeq 10$ - $100\ J_c^J$.

2. Polycrystalline Sintered High-Temperature Superconductors

The magnetic moment is defined as

$$m = \frac{1}{2}\int \vec{\xi} \times \vec{J}(\vec{\xi})\, d^3\xi \tag{1}$$

where $\vec{\xi}$ is the space vector and $\vec{J}$ the local current density. One can split $\vec{J}$ into an *inter*granular current density $\vec{J}^J$ which flows between grains across the grain boundary Josephson junctions over the entire sample and an *intra*granular current density $\vec{J}^G$ which is flowing inside grains ($\vec{J} = \vec{J}^J + \vec{J}^G$) . Therefore

$$m = m^J + \sum_i m_i^G \tag{2}$$

where m^J is the *inter*granular magnetic moment and m_i^G the *intra*granular one originating from grain i.

Consider a sintered polycrystalline high-temperature superconducting sample in the form of a cylinder of radius R and length L with a field H_a applied parallel to the axis of the cylinder. The induced *inter*granular current density $\vec{J}^J$ is flowing in azimuthal direction. Using Faraday's law one can express m_J in terms of the *inter*granular magnetic field $H_J(r)$ where r is the radial coordinate

$$m^J = 2\pi L\, [\int_0^R H_J(r)\, r\, dr - \frac{1}{2}H_a R^2\,] \quad . \tag{3}$$

The *inter*granular magnetic field $H_J(r)$ is the solution of the *inter*granular critical state equation [3,4]

$$\frac{dH_J(r)}{dr} = \pm J_c^J(H_J) \tag{4}$$

with the boundary condition $H_J(r = R) = H_a$ where J_c^J is the *inter*granular critical current density. The + sign refers to the region were flux has moved into the sample and the − sign where flux has moved out of the sample. Because the *inter*granular critical field H_{c1J} is small compared to fields commonly used in measurements, no *inter*granular equilibrium magnetization has to be taken into account in the above *inter*granular critical state equation.

The *intra*granular magnetic moment is more difficult to calculate because the microscopic *inter*granular field meanders around grains in a complicated way where the microstructure causes demagnetizing effects. In addition, the lower critical field of grains, H_{c1G}, is anisotropic. Therefore the boundary condition for the *intra*granular critical state is not well defined. In spite of these complications a reasonable approach seems to treat the grains as "cylindrical grains" of length L (sample length) and radius

R_G aligned parallel to the applied field. Because $| dH^J/dr | << | H_J(r) | /R_G$ one can replace $\sum_i$ in Eq.(2) by an integral and obtains

$$\sum_i m_i^G = \frac{4\pi Lgf_S}{\mu_o R_G^2}\int_0^R [\int_0^{R_G} B_G(r,r')r'\, dr']\, rdr - 2\pi Lgf_S \int_0^R H_J(r)\, r\, dr \quad . \qquad (5)$$

Here r' is the radial coordinate of a grain, f_S the volume fraction of grains and g the fraction of the complete Meissner state[4]. The *intra*granular magnetic induction $B_G(r,r')$ is given by the *intra*granular critical state equation [5,6]

$$\frac{dB_G(r,r')}{dr'} = \pm J_c^G(B_G)\frac{dB_G^{eq}}{dH_G} \qquad (6)$$

with the boundary condition $B_G(r,r'=R_G) = \mu_o H_J(r) + \mu_o M_G^{eq}(H_J)$. Here B_G^{eq} and M_G^{eq} are the equilibrium induction and magnetization of a grain.

Figure 1 shows the calculated magnetization $M = m/\pi R^2$ of a polycrystalline YBCO sample of cylindrical shape ($R = 2$ mm) for a maximum field $\mu_o H_m = 5$ mT. The calculation uses

$$J_c^J(H_J) = \frac{J_c^J(0)}{1+(H_J/H_{oJ})^2} \qquad (7)$$

with $J_c^J(0) = 10^6$ A/m^2 and $\mu_o H_{oJ} = 1$ mT. The loop is due to the *inter*granular critical state behaviour defined by Eq.(4). The tilting of the loop originates from the diamagnetic moment of the grains represented by the last term in Eq.(5).

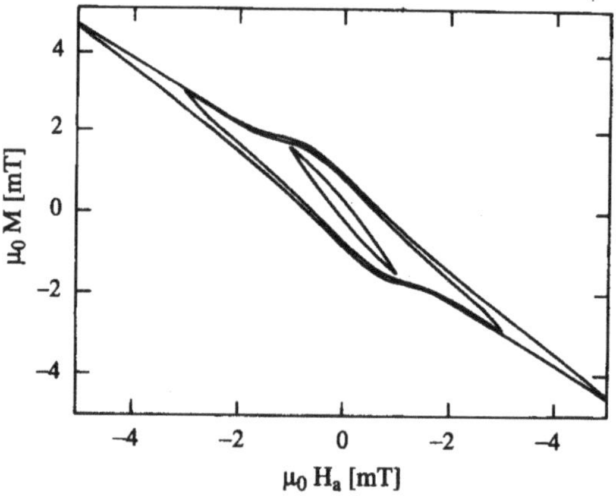

Fig. 1. Calculated *inter*granular magnetization of a sintered polycrystalline sample as a function of the applied field H_a.

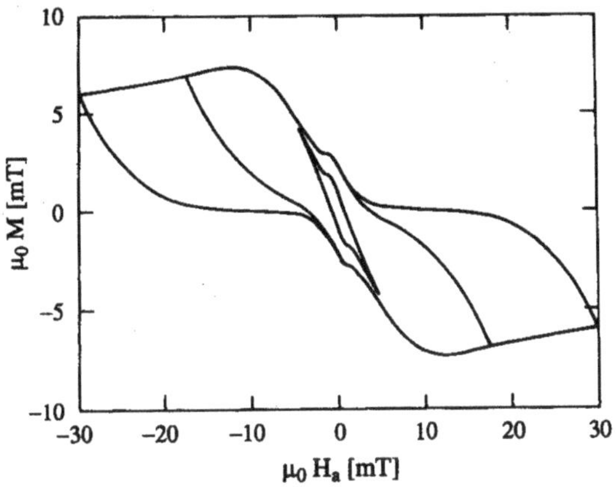

Fig. 2. Calculated magnetization of a sintered polycrystalline sample as a function of the applied field H_a. The *inter*granular contribution is signficant between - 5 mT and + 5 mT. For $| \mu_o H_a | >$ 5 mT the *intra*granular magnetization dominates.

Figure 2 shows the calculated magnetization M for a maximum field $\mu_o H_m = 30$ mT. The central loop is the *inter*granular loop. In higher fields the magnetic moment is dominated by the *intra*granular critical state contribution given by Eq.(5). The calculation uses

$$J_c^G(B_G) = \frac{J_c^G(0)}{1+ | B_G/B_{oG} |} \tag{8}$$

with $J_c^G(0) = 10^9$ A/m^2, $B_{oG} = 50$ mT and $R_G = 10$ μm. The z-shape of the loop is caused by the equilibrium magnetization of the grains [5,6].

The hysteretic ac loss W which is the energy dissipated per volume per cycle when an ac magnetic field of amplitude H_m is applied, is given by

$$W = \frac{\mu_o}{\pi R^2 L} \oint m(H_a)\, dH_a \tag{9}$$

and thus W measures the magnetic moment loop area. Figure 3 shows W as a function of $\mu_o H_m$ over 5 orders of magnitude [7]. The low field region is dominated by *inter*granular losses while in strong fields the *intra*granular losses dominate.
The imaginary part χ'' of the ac susceptibility is given by

$$\chi'' = \frac{W}{\pi \mu_o H_m^2} \ . \tag{10}$$

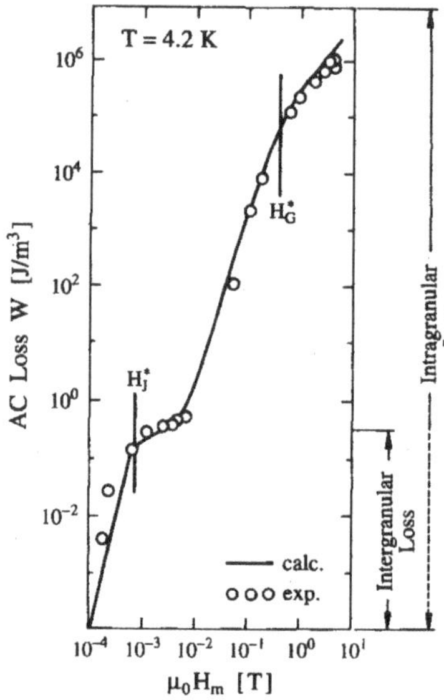

Fig. 3. *Inter*granular and *intra*granular ac losses W of a polycrystalline sintered superconductor as a function of the ac field amplitude H_m at 4.2 K. The experimental data is taken from Xu et al.[8] H_J^* and H_G^* are the fields of full penetration.

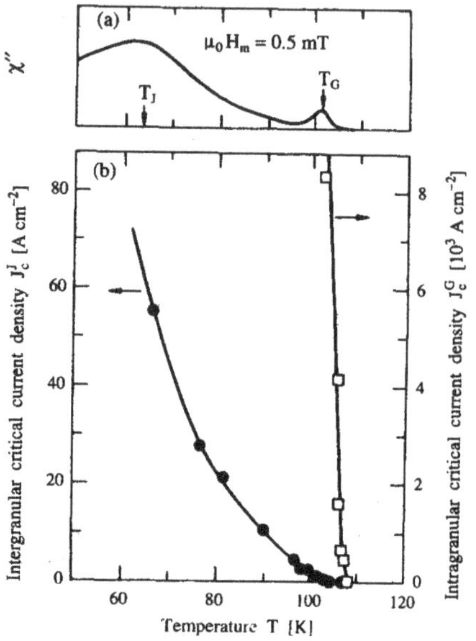

Fig. 4. (a) *Inter*granular and *intra*granular susceptibility χ'' with peaks at temperatures T_J and T_G. (b) *Inter*granular and *intra*granular critical current densities as a function of temperature extracted from χ'' data.

χ'' is commonly measured with a susceptometer as a function of temperature T and ac field amplitude H_m [9]. For constant H_m, χ'' peaks at two temperatures T_J and T_G as indicated in Fig. 4 (a) [10]. At T_J, the field H_m just penetrates to the centre of the sample while at T_G the field just penetrates to the centre of the grains. Because $H_m(T_J) \simeq J_c^J(T_J)\, R$ and $H_m(T_G) \simeq J_c^G(T_G)\, R_G$, the *inter*granular and *intra*granular critical current densities as a function of temperature (for zero applied dc field) can be estimated. Figure 4 (b) shows J_c^J and J_c^G for a polycrystalline PBSCCO sample. As can be seen, J_c^G is many orders of magnitude larger than J_c^J.

3. PBSCCO/Ag Tapes

Based on the work of Norris [11], Brandt et al. [12] have recently derived analytical expressions for the magnetic moment of a thin superconducting strip (type II) in a perpendicular applied field, H_a. The critical state equation in this case is

$$\frac{\partial H_{Jx}}{\partial z}-\frac{\partial H_{Jz}}{\partial x}=J^J \quad \text{and} \quad |J^J|\leq J_c^J \tag{11}$$

where z is perpendicular to the tape and J^J flows in the y-direction along the tape. The term $\partial H_{Jx}/\partial z$ does not vanish as in the case of a cylindrical sample in a parallel field but instead becomes the dominant term. One finds that the absolute value of the current density in the central region of the strip is less than J_c^J unlike in a cylinder (or slab) where J^J is either $\pm J_c^J$ or zero. In the following, the length of the monofilamentary tape is L, the half-width of its superconducting core is a and its thickness d with $d << a$. Assuming that J_c^J is independent of the local magnetic induction, one obtains for the *inter*granular virgin magnetic moment m_V^J [12,13]

$$m_V^J=-\pi a^2 L H_c \tanh(H_a/H_c) \tag{12}$$

with $H_c=J_c\, d/\pi$, and for the upper (↑) and lower (↓) branch of the *inter*granular magnetic moment [12,13]

$$m_{\uparrow\downarrow}^J=\pm\pi a^2 L H_c\,[\tanh(H_m/H_c)+2\tanh((\pm H_a-H_m)/2H_c)]\ . \tag{13}$$

In the case of a weak applied field where the field inside of the superconductor is everywhere less than the lower critical field of grains, H_{c1G}, only Meissner shielding currents are being induced in the grains and

$$(\sum_i m_i^G)_{Meissner}\simeq-\frac{f_s}{1-n}2\,a\,L\,d\,H_a \tag{14}$$

where n is the effective demagnetizing factor of the grains.

Figure 5 shows the magnetic moment m of a PBSCCO/Ag tape (#1) at 5 K for $\mu_o H_m$ = 25 mT. The tape's electrically measured J_c^J is about 5000 A/cm^2 at 77 K. The loop represents the *inter*granular magnetic moment m_J where the loop tilting is due to the Meissner signal of the grains. To fit the data, Eqs. (12)-(14) were used with $J_c^J = 2.8\times10^8$ A/m^2, L = 5.9 mm, a = 1.35 mm and d = 35 μm.
Fig. 6 shows the remanent magnetic moment m_R of the tape. After bending the tape (#2) to the diameter D = 10 mm the tape was straightened and the magnetic moment was remeasured. The bending produced microcracks in the material which diminishes the *inter*granular current. As can be seen from Fig. 6, bending the tape to the small diameter of 1.25 mm causes the *inter*granular current to disappear almost entirely such that m_R now represents only the *intra*granular magnetic moment. For strong fields one obtains

$$(\sum_i m_i^G)_R(H_m \to \infty) = \frac{2}{3}\, a\, L\, d\, R_G\, J_c^G \quad . \tag{15}$$

Assuming a grain radius of $R_G = 10\ \mu$m for tape #1 one obtains $J_c^G(5\text{K}) = 5.4 \times 10^{10}$ A/m^2 and therefore for tape #1 the *intra*granular critical current density is about 200 times larger than the *inter*granular one. For a different tape (#3) which had $J_c^J(77\text{K}) \simeq 20000$ A/cm^2, it was found that $J_c^G(5\text{K}) \simeq 50 J_c^J(5\text{K})$. The dashed curve in Fig. 6 shows the experimental m_R obtained by subtracting from the "no-bend" data the "D $\simeq$ 1.25 mm" data. The drop of m_R^J in high fields is due to the hysteretic behaviour of J_c^J because trapped flux in grains weakens the Josephson current across grain boundaries [14,15].

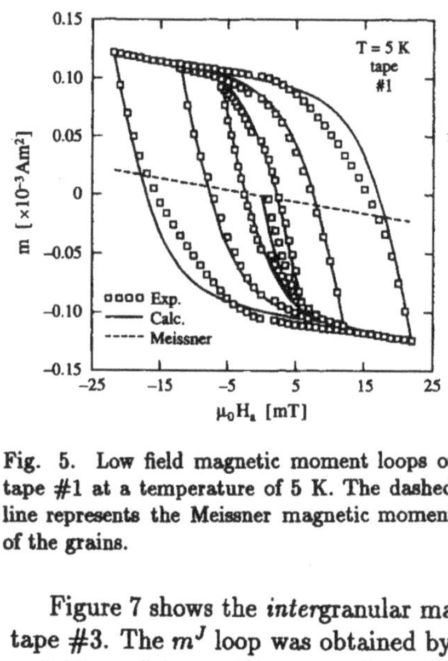

Fig. 5. Low field magnetic moment loops of tape #1 at a temperature of 5 K. The dashed line represents the Meissner magnetic moment of the grains.

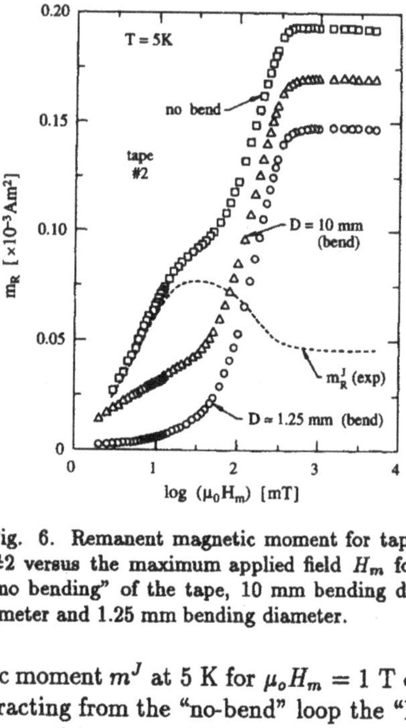

Fig. 6. Remanent magnetic moment for tape #2 versus the maximum applied field H_m for "no bending" of the tape, 10 mm bending diameter and 1.25 mm bending diameter.

Figure 7 shows the *inter*granular magnetic moment m^J at 5 K for $\mu_o H_m = 1$ T of tape #3. The m^J loop was obtained by subtracting from the "no-bend" loop the "D $\simeq$ 1.25 mm" loop. The unusual maximum at positive applied field originates from the hysteretic behaviour of J_c^J. The hysteretic *inter*granular critical current density $J_c^J(H_a, \mu_o H_m = 1\ \text{T})$ is shown in Fig. 8 where J_c^J was obtained from data in Fig. 7 using Eq. (13). Because $H_c << H_m$ one derives

$$J_c^J \simeq \frac{|\, m^J \,|}{da^2 L} \quad . \tag{16}$$

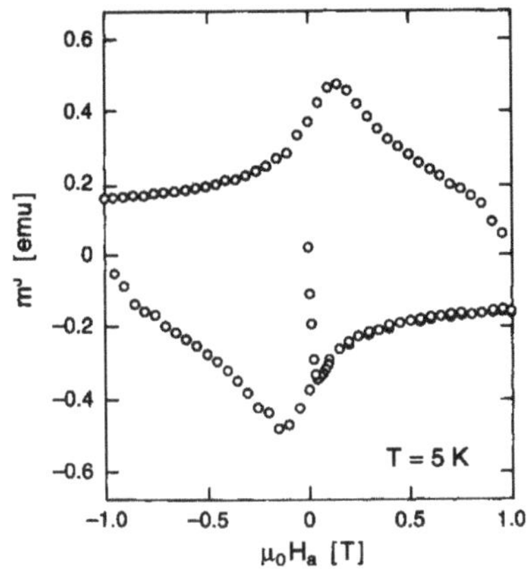

Fig. 7. *Intergranular* magnetic moment loop at 5 K for $\mu_o H_m = 1$ T for tape #3.

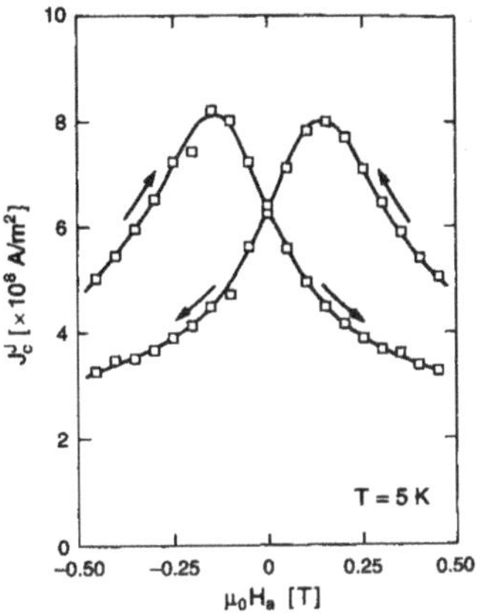

Fig. 8. Hysteretic *intergranular* critical current density J_c^J at 5 K for $\mu_o H_m = 1$ T extracted from Fig. 7.

4. References

1. D. Dimos, P. Chaudhari and J. Mannhart, *Phys. Rev. B* **41** (1990) 4038.
2. T. Wolf and A. Majhofer, *Phys. Rev. B* **47** (1993) 5383.
3. C.P. Bean, *Phys. Rev. Lett.* **8** (1962) 250.
4. K.-H. Müller, in *Magnetic susceptibility of superconductors and other spin systems*, ed. R.A. Hein et al. (Plenum Press, New York, 1991) p. 229.
5. D.D. Stancil, T.E. Schlesinger, A.K. Stamper and D. Wong, *J. Appl. Phys.* **64** (1988) 5899.
6. C. Andrikidis, K.-H. Müller, N. Savvides, S.X. Dou and H.K. Liu, *Cryogenics* **32** (1992) 411.
7. K.-H. Müller, *IEEE Trans. on Magnetics* **27** (1991) 2174.
8. Y. Xu, W. Guan and K. Zeibig, *Appl. Phys. Lett.* **54** (1989) 1699.
9. K.-H. Müller, *Physica C* **159** (1989) 717
10. N. Savvides, A. Katsaros and S.X. Dou, *Physica C* **179** (1991) 361.
11. W.T. Norris, *J. Phys. D* **3** (1970) 489.
12. E.H. Brandt and M. Indenbom, *Phys. Rev. B* **48** (1993) 12893.
13. K.-H. Müller, C. Andrikidis, H.K. Liu and S.X. Dou, Phys. Rev. B (Oct. 1994) (in press).
14. A. I. D'yachenko, *Physica C* **213** (1993) 167.
15. K.-H. Müller and D.N. Matthews, *Physica C* **206** (1993) 275.

IMPROVEMENT OF CRITICAL CURRENTS BY RADIATION-INDUCED DEFECTS

H. W. WEBER
Atomic Institute of the Austrian Universities
Schüttelstrasse 115, A-1020 Wien, Austria

ABSTRACT

New results on neutron-irradiated high temperature superconductors are presented. Following a summary of TEM data on the nature of neutron-induced defects in YBCO single crystals as well as a presentation of recently evaluated mixed state parameters of some high-T_c superconductors, special emphasis will be placed on J_c-enhancements and shifts of the irreversibility lines in more "two-dimensional" systems, such as Tl-2223 single crystals. Furthermore, transport measurements on Bi-2223/Ag tapes are presented which show clear improvements in the field dependence of J_c at 77 K. Since the 2D-character is retained as demonstrated by the angular dependence of J_c, these results as well as those on the single crystals indicate significant flux pinning of pancake vortices by the radiation-induced defects.

1. Introduction

Radiation-induced defects have proved to be highly efficient flux pinning centers, if the irradiating particles are chosen to transfer sufficient energy (> 1 keV) to the primary knock-on atoms either by nuclear (e.g. fast neutrons) or by electronic mechanisms (e.g. high-energy heavy ions), thus creating extended defects[1,2]. Clearly, irradiation experiments are useful for analyzing flux pinning effects in a more fundamental way, since the pin density can be varied systematically[3] and the flux pinning properties assessed as a function of particle fluence. However, in order to take full advantage of these possibilities, detailed information on many parameters of the system is required, which is not available in general. With regard to radiation effects, we need to know the size of the defects, their physical properties, and their volume concentration as a function of neutron fluence (cf. section 2). With regard to the superconductor itself, we need to know its reversible mixed state parameters such as the GL-coherence length ξ, the penetration depth λ, and the thermodynamic critical field $H_c(T)$, which are usually difficult to assess because of flux pinning due to "as-grown" crystalline defects. An example for a consistent evaluation of these parameters in the widely reversible Tl-2223 system will be presented in section 3. The rest of this contribution will be devoted to a presentation of results on critical current densities and irreversibility lines in neutron-irradiated Tl-2223 single crystals and Bi-2223/Ag tapes.

2. Neutron-induced Defects

The nature of neutron-induced defects has been assessed by transmission electron microscopy only for $YBa_2Cu_3O_{7-\delta}$ single crystals so far, either from in-situ simulation experiments using ions of appropriate mass and energy[4] or from exposure of TEM-samples to a fission neutron spectrum[5]. The results show that a significant number of collision cascades are formed through a direct transfer process. The defects are roughly spherical in shape with a diameter of ~2.5 nm and surrounded by a strain field of approximately the same size. Hence, the size of these defects is ideally suited for core-pinning in YBCO, especially at higher temperatures, where the GL-coherence length is of the same order of magnitude. The defect concentration varies linearly with neutron fluence and amounts to 5×10^{21} m^{-3} per 10^{21} neutrons m^{-2} ($E > 0.1$ MeV). Of course, similar studies of other high temperature superconductors would be highly desirable, especially in view of recent theoretical work, which predicts a significant dependence of the cascade size on the crystal structure[6].

In addition to these TEM-visible cascades, smaller defects are produced by the less energetic neutrons of the reactor spectrum, which accordingly transfer less energy (< 1 keV) to the primary knock-on atoms. Damage calculations for YBCO show[5] that significant amounts of atoms are displaced in this way (with oxygen being the easiest to displace), which form point defects as well as smaller and larger defect clusters. The first category (mainly O-displacements) manifests itself in a certain decrease of the transition temperature, which can be recovered by annealing[7,8], whereas the second will contribute to flux pinning, but in a less pronounced way than the cascades due to their smaller pinning potential.

3. Reversible Mixed State Parameters

Many attempts have been made to determine the fundamental mixed state parameters of high temperature superconductors, but most of them failed either because of the onset of irreversibility or because of inadequate theoretical treatments. Obviously, the low-lying irreversibility lines of some systems, such as Bi-2212 or Tl-2223, offer some advantages in this respect, which can be combined with improved theories[9-11], in order to tackle this problem at least in some special cases, e.g. for H||c in the above mentioned materials. As an example, the analysis of magnetization measurements on Tl-2223 single crystals in terms of the vortex fluctuation theory[11] will be presented in the following.

The magnetic moments, which were measured in a SQUID magnetometer up to 8 T (H||c), can be analyzed directly in terms of this theory for reduced temperatures, $t = T/T_c$, in the range from ~0.5 to 1.0, in order to obtain the temperature dependence of λ_{ab} and of the upper critical field H_{c2} (cf. Fig. 1). The fit of $H_{c2}(T)$ to the isotropic BCS-theory in the clean limit is excellent, the result for $T = 0$ is $\mu_0 H_{c2}(0) = 120$ T. For $\lambda(T)$, the agreement with BCS theory[12] is not as good, the best fit is obtained for $\lambda(T) = \lambda(0)(1-t^2)^{-a}$ with $a = 0.36$ instead of 0.5, which represents the best fit to the BSC results ($\lambda_{ab}(0) =$ 192 nm). Next, the coherence length is obtained from $\mu_0 H_{c2}(T) = \phi_0/2\pi\mu_0\, \xi^2(T)$. The

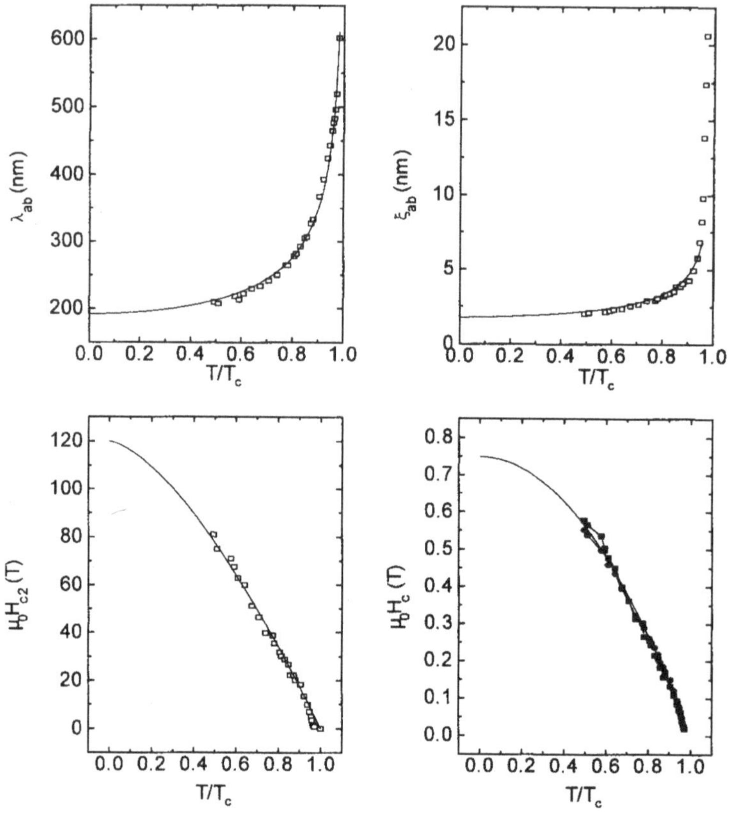

Figure 1: Reversible mixed state parameters of Tl-2223 for H||c.

solid line represents again the best fit to BCS theory $[H_{c2}(T) = H_{c2}(0)(1-t^{3/2})]$ without further assumptions and leads to $\xi_{ab}(0) = 1.76$ nm. Taking $\kappa = \lambda/\xi$ near T_c results in a nearly temperature independent value of 86 for the GL-parameter. Finally, the thermodynamic critical field $H_c(T)$ can be calculated in two ways, firstly by integrating the magnetization curve from 0 to H_{c2} (squares in Fig. 1) and secondly by employing the GL-equation $\mu_0H_c(T) = \phi_0/[2\sqrt{2}\,\pi\lambda(T)\,\xi(T)]$ with the experimental data for ξ and λ. Both procedures lead to nearly identical results. The fit to the usual parabolic law (solid line) leads to $\mu_0H_c(0) = 748$ mT. The fit to clean-limit BCS is equally good, but results in a somewhat higher value for $H_c(0)$. Hence, a consistent set of mixed state parameters is available for Tl-2223[13,14], which can be used for flux pinning considerations. A similar situation prevails for Bi-2212[14,15].

4. Radiation Effects on Tl-2223 Single Crystals

Several Tl-2223 single crystals were subjected sequentially to reactor neutron irradiation up to a fluence of 2.4×10^{22} m^{-2} (E > 0.1 MeV). The measurements were done in SQUID magnetometers either at fixed temperature (hysteresis loops) or at fixed field in the field-cooled or zero-field-cooled mode to determine the transition temperature T_c and the irreversibility lines (all for H||c).

The results for T_c show a small plateau at the lowest fluence, but generally a linear decrease of T_c from 117.5 K with a slope of -4.5 K per 10^{22} neutrons m^{-2}, i.e. the degradation of T_c is clearly more pronounced than in YBCO[3]. After the last irradiation step (2.4×10^{22} m^{-2}) T_c was only 106.8 K.

The critical current densities flowing in the a,b-plane were evaluated in terms of an extended Bean model[16], which takes demagnetization effects into account. Typical results for J_c as a function of the induction B within the sample, are shown in Fig. 2 for various neutron fluences. As a general trend, we observe a continuous enhancement of J_c with neutron fluence up to 1.6×10^{22} m^{-2} followed by a decrease of J_c at all temperatures after the next step. Obviously, the reversible material parameters have degraded too much at this point to allow for a further improvement of flux pinning. (Note that T_c has dropped by ~11 K at this fluence.) At low temperatures (e.g. 5 K), the enhancements of J_c are rather moderate (~ factor of 2) and the J_c-values themselves about one order of magnitude

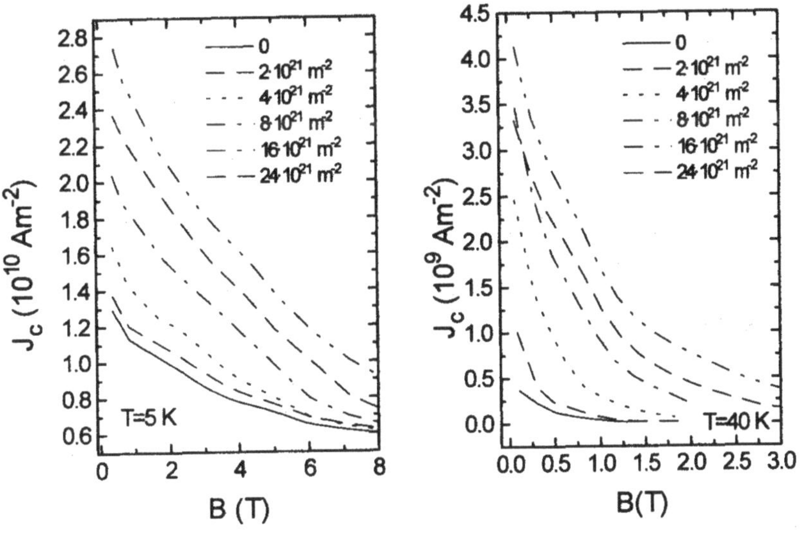

Figure 2: Critical current densities at 5 and 40 K as a function of magnetic field at various neutron fluences (Tl-2223, H||c).

smaller than in YBCO[3]. However, the fluence dependence of J_c is strongly different from YBCO and shows saturation at higher fluence levels. The enhancements of J_c at 40 K are much larger (~ factor of 60 at 1 T) and comparable to the situation in YBCO, but the absolute values of J_c are again smaller by about one order of magnitude. It will be noted, however, that irreversible behavior extends far into the previously reversible field range. This is emphasized in Fig. 3, where the irreversibility lines are plotted for all fluences. The shifts are dramatic and amount, e.g. at 1 T, to 30 K. Even at 77 K (t ~ 0.7), irreversibility is observed up to ~0.5 T. At the highest fluence, we note a cross-over of the

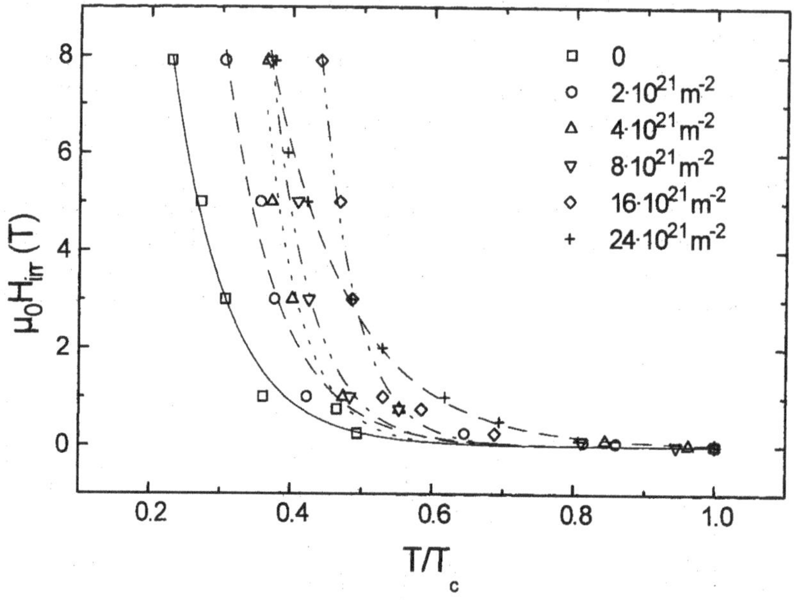

Figure 3: Irreversibility lines (H||c) of Tl-2223 at various neutron fluences.

irreversibility lines at ~3 T, i.e. flux pinning is still improved at high temperature and low field inspite of the degradation of the material parameters.

In summary, the results clearly demonstrate that efficient flux pinning of strongly and/or weakly coupled pancakes by defect cascades is feasible.

5. Transport Measurements on Neutron Irradiated BiSCCO-2223/Ag Tapes

Similar conclusions can be drawn from transport measurements on neutron irradiated BiSCCO-2223 tapes[17]. In this case, J_c and the angular dependence of J_c were measured at 77 K prior to and following neutron irradiation to the relatively low fluence of 4×10^{21} m^{-2}. The results, which are presented in Fig. 4 in terms of the bulk pinning force F_p, versus field for H||a,b and for H||c, show some deterioration of pinning at low fields, where the critical current is dominated by the weak link structure, but clear enhancements in the flux pinning dominated high field regime for both field orientations. The complete angular dependence of J_c was measured as well, with the current always flowing perpendicular to the field direction. If this data is evaluated as a function of $\mu_0 H\sin\theta$, where θ is the angle between the external field and the plane of the broad surface of the tape ("a,b-plane"), we obtain the results shown by the open symbols in Figure 5. For comparison, the field dependence of J_c measured for B||c on the same tape is shown by the full symbols. We notice that both curves collapse onto a single one in the high field region, but separate at low fields. According to our previous analysis[18], this collapse provides evidence for two-dimensional superconducting behavior of Bi-2223, i.e. the field dependence of the critical current density is entirely caused by the field component along the c-direction of the superconductor. The low field separation of the curves results from the misalignment of the grains from the a,b-plane of the tape. An "effective" misalignment angle ϑ_{eff} can be defined, which was found to be consistent with the metallographic misalignment angle observed by SEM. The results of Fig. 5 confirm that fast neutron irradiation (at least at this fluence level) does not affect the anisotropy of the superconductor, i.e. the original effective misalignment angle ($\vartheta_{eff} \sim 6.8°$) is retained ($\vartheta_{eff} \sim 7°$) in the irradiated state. Hence, pinning of pancakes and even of Josephson vortices by suitable defects is proved directly by this transport measurement.

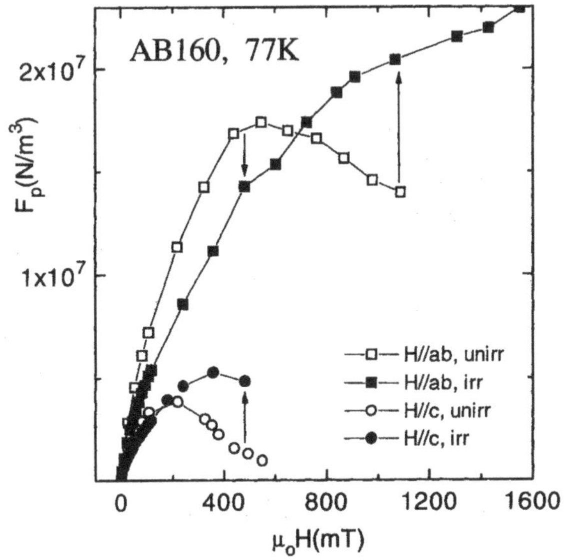

Figure 4: Bulk pinning forces F_p of a Bi-2223/Ag tape versus magnetic field (H||c, H||a,b) prior to and following neutron irradiation to $4x10^{21}$ m^{-2} (E > 0.1 MeV).

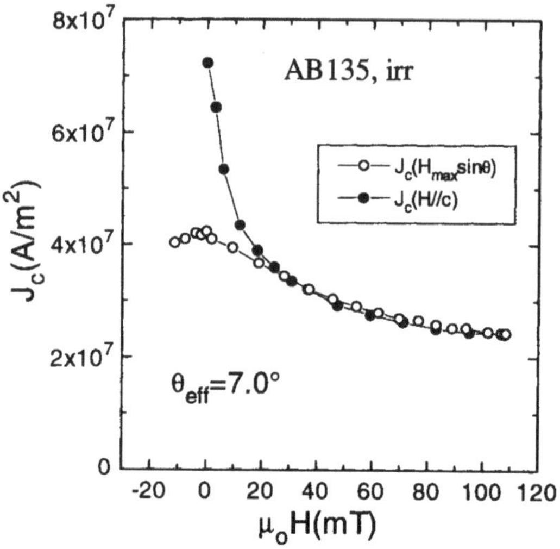

Figure 5: Critical current densities versus applied field for H||c (full symbols) and for the two-dimensional case (open symbols). Bi-2223 tape after fast neutron irradiation.

6. Summary

In conclusion, I wish to point out that some progress towards an improved analysis of flux pinning in high temperature superconductors has been achieved. Basic requirements for this purpose are, firstly, the predominance of a certain defect type with strong pinning characteristics, secondly, the assessment of its size, shape and volume concentration, and thirdly, basic information on the mixed state properties of the superconductor, which can be derived from improved theories. From an applied point of view, irradiation experiments on two-dimensional systems either with fast neutrons or with heavy ions are certainly encouraging. However, a lot of additional work will have to be done to consistently describe the data on various classes of high-T_c superconductors.

7. Acknowledgements

I wish to acknowledge significant contributions to this work by Drs. F.M. Sauerzopf, M.C. Frischherz, G. Brandstätter, R.M. Schalk, and Q.Y. Hu as well as technical support by Mr. H. Niedermaier. This work is supported in part by Fonds zur Förderung der Wissenschaftlichen Forschung, Wien, under contract numbers 8849 and 9194.

8. References

1. H. W. Weber, and G. W. Crabtree, in *Studies of High Temperature Superconductors*, ed. A. V. Narlikar (Nova Science Publishers, New York, 1992), Vol. 9, p. 37.
2. M. A. Kirk and H. W. Weber, *ibid.*, Vol. 10, p. 253.
3. F. M. Sauerzopf, H. P. Wiesinger, H. W. Weber, and G. W. Crabtree, submitted to *Phys. Rev. B* (1994).
4. M. C. Frischherz, M. A. Kirk, J. P. Zhang, and H. W. Weber, *Phil. Mag.* **A67** (1993) 1347.
5. M. C. Frischherz, M A. Kirk, J. Farmer, L. R. Greenwood, and H. W. Weber, *Physica C* (in press).
6. V. V. Kirsanov, N. N. Musin, and A. Wisniewski, *Phys. Lett.* **182** (1993) 169.
7. H. P. Wiesinger, F. M. Sauerzopf, H. W. Weber, H. Gerstenberg, and G. W. Crabtree, *Europhys. Lett.* **20** (1992) 541.
8. B. M. Vlcek, M. C. Frischherz, S. Fleshler, U. Welp, J. Z. Liu, J. Downey, K. G. Vandervoort, G. W. Crabtree, M. A. Kirk, J. Giapintzakis, and J. Farmer, *Phys. Rev.* **B46** (1992) 6441.
9. Z. Hao and J. R. Clem, *Phys. Rev. Lett.* **67** (1991) 2371.
10. L. N. Bulaevskii, M. Ledvij, and V. G. Kogan, *Phys. Rev. Lett.* **25** (1992) 3773.
11. V. G. Kogan, M. Ledvij, A. Y. Simonov, J. H. Cho, and D. C. Johnston, *Phys. Rev. Lett.* **70** (1993) 1870.
12. B. Mühlschlegel, *Z. Physik* **155** (1959) 313.
13. G. Brandstätter, F. M. Sauerzopf, H. W. Weber, F. Ladenberger, E. Schwarzmann, *Physica C* (in press).
14. G. Brandstätter, F. M. Sauerzopf, and H. W. Weber, submitted for publication.
15. J. C. Martinez, P. J. E. M. van der Linden, L. N. Bulaevskii, S. Brongersma, A. Koshelev, J. A. A. J. Perenboom, A. A. Menovsky, and P. H. Kes, *Phys. Rev. Lett.* **72** (1994) 3614.
16. H. P. Wiesinger, F. M. Sauerzopf, and H. W. Weber, *Physica* **C203** (1992) 121.
17. Q. Y. Hu, H. W. Weber, F. M. Sauerzopf, G. W. Schulz, R. M. Schalk, H. W. Weber, H. W. Neumüller, and S. X. Dou, *Appl. Phys. Lett.* (in press).
18. Q. Y. Hu, H. W. Weber, H. K. Liu, S. X. Dou, and H. W. Neumüller, *J. Alloys and Comp.* **195** (1993) 515.

FLUX PINNING IN MELT-GROWN $REBa_2Cu_3O_y$ (RE: Nd, Sm, Eu, Gd) SUPERCONDUCTORS

M. MURAKAMI, S. I. YOO, T. HIGUCHI, T. OYAMA, N. SAKAI
International Superconductivity Technology Center,
Superconductivity Research Laboratory,
1-16-25 Shibaura, Minato-ku, Tokyo 105 Japan

ABSTRACT

$REBa_2Cu_3O_y$ superconductors with rare earth (RE) ions with large radii (RE: La, Nd, Sm, Eu) exhibit relatively low T_c due to the presence of RE-Ba solid solution. We have found that this solid solution can be suppressed if these superconductors are melt processed in a reduced oxygen atmosphere. We have also found that critical current densities of these superconductors are higher than those of melt processed $YBa_2Cu_3O_y$ with fine Y_2BaCuO_5 inclusions in a high field region. The irreversibility line was also shifted toward the higher H-T region. We believe that flux pinning in these superconductors is ascribable to a finely distributed $RE(Ba_{1-x},RE_x)_2Cu_3O_y$ phase in a good superconductive matrix.

1. Introduction

It is known that melt processed $YBa_2Cu_3O_y$ exhibits a large critical current density even at liquid nitrogen temperature[1-3]. This is attributed to a combination of a less anisotropic crystal structure and the presence of effective pinning centers. Many candidates of pinning centers have been proposed in this material. Those are twin planes, stacking faults, oxygen defects, dislocations and second phases[4]. Among these, the effect of Y_2BaCuO_5 (Y211) inclusions is still controversial, although it is generally accepted that fine dispersion of Y211 inclusions is effective in raising the flux pinning force. It is also known that columnar defects produced by heavy ion irradiation are good pinning centers even in a high temperature region[5]. However, the columnar defects will not be widely used because of both technological and economic problems. Therefore, it has long been hoped that new pinning centers which can act at liquid nitrogen temperature and can be produced by economically feasible processes will be discovered.

Recently, we found that a new type of pinning center can be introduced in $NdBa_2Cu_3O_y$ (Nd123) superconductors, which show larger flux pinning forces than Y123 containing finely distributed Y211 inclusions in high field regions[6-8]. In this system, magnetization hystereses exhibit a peak effect in intermediate fields and is open to 8T at 77K for H//c axis, which is the highest field ever reported. In order to explain flux pinning in this system, we proposed the following mechanism.

There is a wide range of composition in the Nd-Ba solid solution. Thus the chemical composition of the Nd123 system can be described as $Nd(Ba_{1-x},Nd_x)_2Cu_3O_y$. If we synthesize this system in air, x lies in the range of 0 to 0.3, and therefore T_c also varies over a wide range, and zero resistivity temperature decreases to lower than 90K. When this compound is synthesized in a reduced oxygen atmosphere, x shifts to lower values, which means that the chemical composition becomes close to 123, and thereby T_c is improved. However, even in such a case, a slight chemical variation will be present. In low fields, regions with slight chemical variation will exhibit good superconducting transition, and therefore the flux pinning force will be small. As the strength of magnetic field is increased, some of these regions will become normal, and can then contribute to flux pinning. The anomalous peak observed in the OCMG processed Nd123 can be explained by this mechanism.

If such assumption is correct, the same phenomena will be observed in other RE123 systems where RE ions substitute Ba site. In the present study, we show that

RE 123 systems (RE: Nd, Sm, Eu) show high T_c and J_c when they are processed in a reduced oxygen atmosphere.

2. Experimental

RE123 (RE: Nd, Sm, Eu, Gd) samples were prepared by the oxygen-controlled-melt-growth (OCMG) process, the details of which are described elsewhere[7]. In this method, the final melt-growth process is performed in oxygen-controlled atmospheres by changing the ratio of O_2 and Ar gas. After the final melt growth, the samples were annealed in flowing oxygen.

T_c values were measured by both resistive and magnetization measurements. We also measured magnetization hysteresis using both a vibrating sample magnetometer and a SQUID magnetometer.

The irreversibility field was determined by the peak temperatures in the loss part of AC susceptibility in the presence of DC field. The amplitude and frequency of the AC field were 0.001T and 1000Hz, respectively.

3. Results and discussions

Figure 1 shows temperature dependence of magnetization for Sm, Eu, and Gd123 samples melt processed in 100%O_2, air, and reduced oxygen atmospheres (0.1% and 1%O_2). T_c values of Sm and Eu 123 samples were enhanced by employing low oxygen partial pressures at the melt-growth stage as in the case of Nd123. This improvement in T_c is ascribable to the suppression of RE-Ba solid solution. It is also notable that the T_c enhancement is less drastic as the RE ion radius becomes smaller, and therefore, in the case of Gd123, T_c enhancement is not clearly visible compared to the other systems. This is because the range of Gd-Ba solid solution is very small.

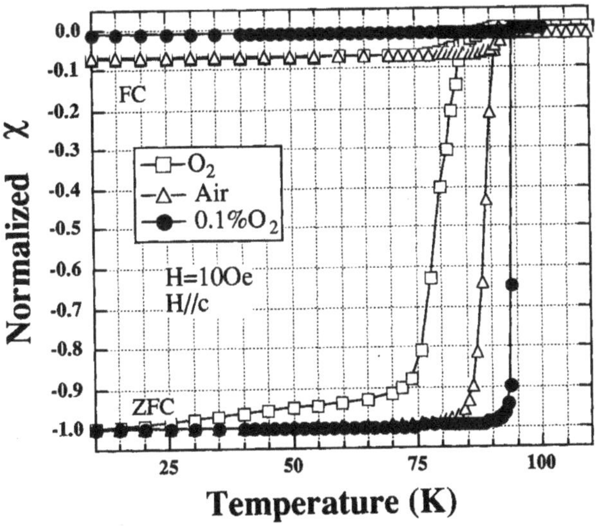

Fig. 1(a). Temperature dependence of magnetization for Sm123.

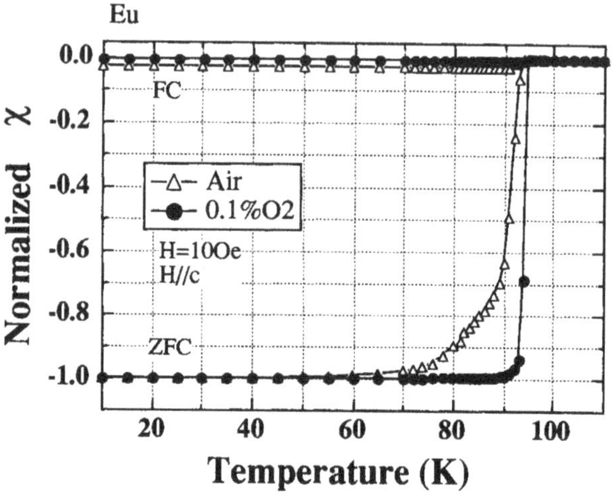

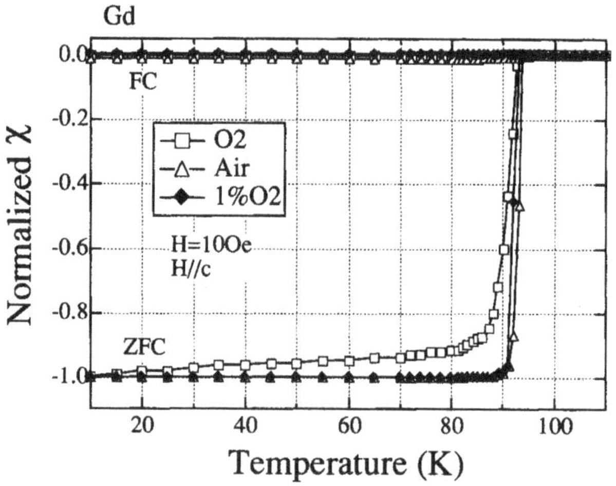

Fig. 1 (b). Temperature dependence of magnetization for Eu and Gd123 melt-processed in various oxygen partial pressures.

Figure 2 shows the magnetization hysteresis (*M-H* loop) of Eu123 and Gd123 samples at 77K for fields parallel to the *c* axis. It is clear that the hysteresis is also greatly improved by employing the OCMG process in Eu123. A dramatic enhancement was also observed in Sm123. In the case of Gd123, the samples melt-processed in air and 1%O_2 showed larger hysteresis than that processed in pure O_2, however, the benefit with the OCMG process was very small. This behavior is also explainable by the fact that the range of RE-Ba solid solution is small in Gd123.

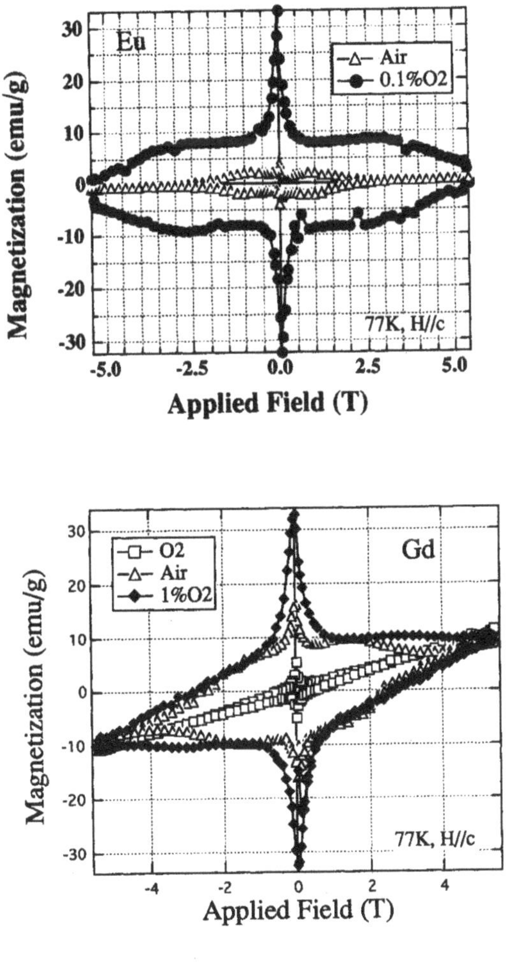

Fig. 2. *M-H* loops (77K, *H*//*c* axis) for Eu and Gd123 melt-processed in various oxygen partial pressures.

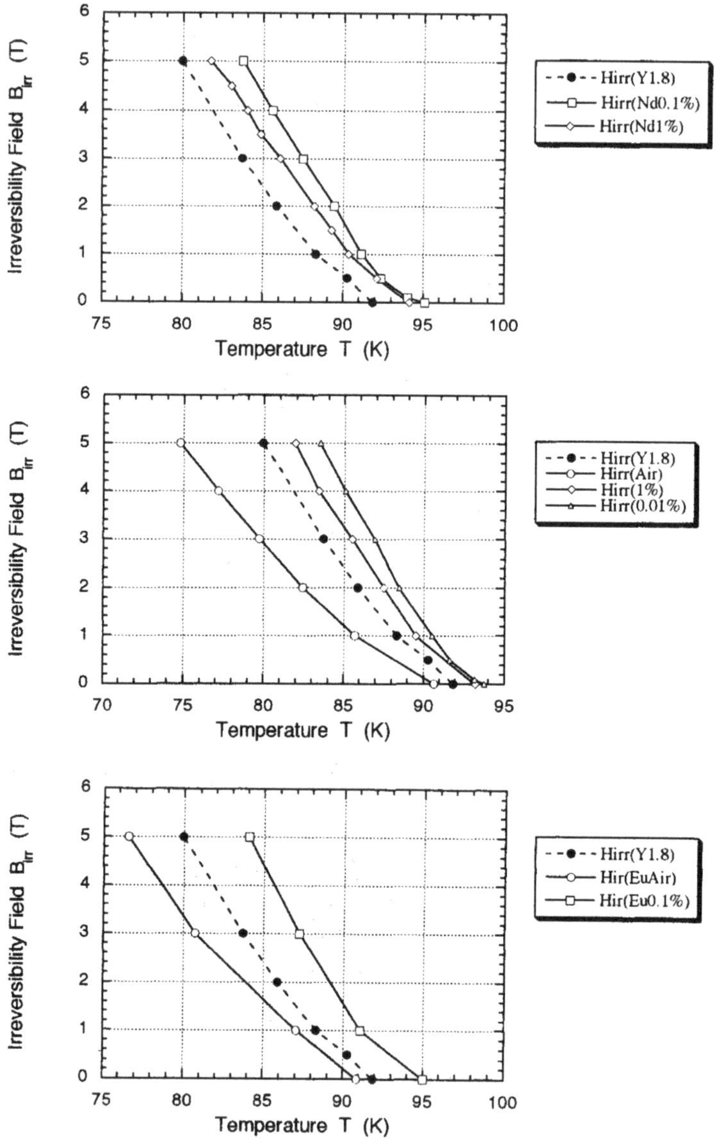

Fig. 3. Temperature dependence of the irreversibility field for OCMG-processed Nd, Sm, and Eu123. For comparison, the data of MPMG-processed Y123 with Y211 are shown in the figure.

The increase in magnetization hysteresis or the improvement in flux pinning is in part due to the increase in T_c. The samples in which T_c was improved through the OCMG process always exhibited better flux pinning at 77K. However, all the changes are not simply explained by the change in T_c. As in the case of Nd123, partially substituted regions seem to be responsible for flux pinning in all these systems. This may be supported by the fact that the magnetization curves show the so-called peak effect. However, the distribution, the amount and the size of these substituted regions are dependent on both the kind of RE and the processing conditions, and these variables will strongly affect flux pinning. The optimization of these parameters will be our future work. The general tendency is that T_c and the peak field are higher as the oxygen partial pressure is lower.

Figure 3 shows temperature dependence of the irreversibility field with *H*//*c* axis for Nd, Sm and Eu123 samples. For comparison, the data of Y123 with 30%211 is also shown in the figure. It should be noted that the irreversibility field was greatly improved in these systems by employing the OCMG process. This result clearly indicates that effective pinning centers are surely introduced in RE123 systems when fabricated by the OCMG process.

4. Conclusions

Although RE123 superconductors with large RE ionic radius (RE: Nd, Sm, Eu, Gd) were believed to be unattractive systems because of the presence of RE-Ba solid solution, the present study has shown that they are highly attractive systems, since they exhibit better flux pinning than the Y123 system when properly treated. It should also be noted that processing conditions are not yet optimized for these systems and further improvement is possible.

It is also interesting to note that RE123 has a different melting point. This fact may provide a number of engineering benefits from viewpoints of processing conditions.

This work was partially supported by NEDO.

References

1) S. Jin, T. H. Tiefel, R. C. Sherwood, M. E. Davis, R. B. van Dover, G. W. Kammlott, R. A. Fastnacht and H. D. Keith: *Appl. Phys. Lett.* **52** (1988) 2047.

2) K. Salama, V. Selvamanickam, L. Gao, and K. Sun: *Appl. Phys. Lett.* **54** (1989) 2352.

3) M. Murakami, M. Morita, K. Doi and M. Miyamoto: *Jpn. J. Appl. Phys.* **28** (1989) 1189.

4) M. Murakami: "*Melt Processed High Temperature Superconductors*" (ed. Murakami, World Scientific, Singapore, 1993) p. 21.

5) L. Civale, A. D. Marwick, T. K. Worthington, M. A. Kirk, J. R. Thompson, L. Krusin-Elbaum, Y. Sun, J. R. Clem and F. Holtzberg: *Phys. Rev. Lett.* **67** (1991) 648.

6) M. Murakami, S. I. Yoo, T. Higuchi, N. Sakai, J. Weltz, N. Koshizuka and S. Tanaka: *Jpn. J. Appl. Phys.* **33** (1994) L715.

7) S. I. Yoo, N. Sakai, H. Takaichi, T. Higuchi and M. Murakami: *Appl. Phys. Lett.* **65** (1994) 633.

8) S. I. Yoo, M. Murakami, N. Sakai, T. Higuchi, and S. Tanaka: *Jpn. J. Appl. Phys.* **33** (1994) L1000.

PROCESSING AND PROPERTIES OF HIGH J_c GRAIN BOUNDARIES IN MELT TEXTURED $YBa_2Cu_3O_x$

K. SALAMA AND A. S. PARIKH.
Texas Center for Superconductivity and Mechanical Engineering Department,
University of Houston, Houston, Texas 77204-4792 USA

and

M. N. KUNCHUR AND D. K. CHRISTEN
Solid State Division, Oak Ridge National Laboratory, Oak Ridge, TN 37831-6061 USA

ABSTRACT

Conventional polycrystalline HTS possess extremely low transport critical current density, which has been attributed to grain boundary related deficiencies. We have developed a processing method to eliminate secondary phases and to reduce the amount of microcracks at the grain boundaries in melt-textured YBCO superconductors. In this process, a fraction of the liquid phase is removed during melt-texturing which leads to cleaner grain boundaries. These boundaries have been found to carry high currents (3×10^4 A/cm^2 at self field and 1.5×10^4 A/cm^2 at 1.5 T) at 77 K in spite of the relatively high misorientation angles. Also, the weak dependence of J_c on magnetic field indicates that these grain boundaries are strongly coupled. In addition, the linear I-V characteristics of these samples suggest that flux flow along these boundaries may be the J_c limiting mechanism. These results give promise to manufacturing of polycrystalline YBCO superconductor with high current carrying capability.

1. Introduction

Bulk polycrystalline superconductors contain large number of grain boundaries, inhomogeneities and other defects which lead to a poor current carrying capability. This is a major hurdle to many practical applications of these materials.[1] The fact that grain boundaries act as weak links stresses the need for developing new processing methods to manufacture superconductors capable of transporting high currents. Grain boundaries have been found to be weakly linked because of the presence of secondary phases, micro-cracks and atomic disorder which gets manifested in the form of misorientation at the grain boundary interface. While the first two problems can be overcome by processing methods to a large extent, the last one can be alleviated by optimizing the processing conditions such that the severity of the problem is considerably reduced by decreasing misorientation angles between grain boundaries.[2]

In a preliminary work carried out by Willén and Salama,[2] a fraction of the liquid phase formed during melt-texturing was removed and the grain boundary microstructure was improved. Following these encouraging results, further efforts were directed to test and characterize the grain boundaries in samples processed using this method.[3,4] It was found that grain boundaries with misorientation angles up to 27° are capable of transporting large current densities ranging from 10 kA/cm^2 to 30 kA/cm^2 in self field and also in the presence of external magnetic field of 1.5 T. This indicates that the grain boundaries, in samples produced using the above method, are relatively clean and therefore are strongly coupled.

2. Processing

During melt texturing sintered YBCO undergoes a peritectic reaction and decomposes into 211 and liquid phase and is then slowly cooled to allow the formation of textured 123. The 211 is rich in Yttrium while the liquid phase is rich in Barium and Copper. Upon slow cooling the textured 123 consists of peritectic 211 which has not recombined and the excess liquid phase which is trapped at the grain boundaries. This excess liquid phase is usually composed of $BaCuO_2$ and CuO.[5]

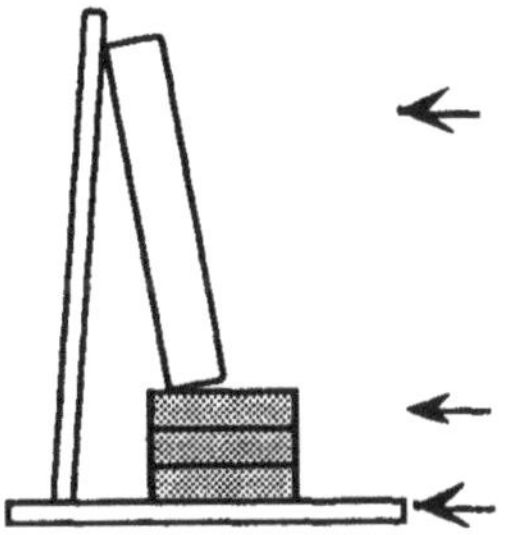

Figure 1. Assembly used to fabricate grain boundaries free of secondary phases in melt textured YBCO.

If a fraction of this liquid phase is removed, there will not be any excess liquid phase trapped at the grain boundaries and the 211 content in the melt textured 123 will increase. This increase in 211 content serves as a mechanical reinforcement between the 123 grains and platelets, and reduces the amount of micro-cracks in the bulk specimen. This improvement can be achieved by absorbing some liquid phase, using 211 as absorbant, during melt texturing. The arrangement used to manufacture 123 samples with high current carrying capability is shown in Figure 1. In this arrangement the 211 is compressed in the form of blocks and placed in contact with the 123 beam during melt texturing. Because of the porosity in the 211 blocks the liquid phase is absorbed easily with the high capillary forces. A removal of 40 % liquid phase is found to produce samples with grain boundaries free of secondary phases.[2] The samples fabricated using this approach consists of large domains with length extending over 15 to 20 mm, width from 8 to 10 mm and thickness of about 5 mm. Samples in the form of bars of lengths ranging from 50 mm to 100 mm and cross sectional area from 10×5 mm^2 to 10×10 mm^2 have been successfully processed.

Specimens for grain boundary characterization are cut from these samples after selecting the regions of interest by observing the microstructure. The final dimensions of the grain boundary specimens after cutting and polishing usually are 10-15 mm $\times$ 0.5 mm $\times$ 0.4 mm.

3. Characterization

Macroscopic studies of the samples processed revealed that these bars possessed several large domains. These domains are misoriented to each other and therefore the bars are polycrystalline in nature. The transport current carrying capability of these bars is tested at 77 K in self field and in magnetic field by the four-point probe method. Current and voltage leads are applied using silver wires and silver colloidal liquid. The bars and the electrical contacts are annealed in flowing oxygen atmosphere at 500 °C for 2-6 days.

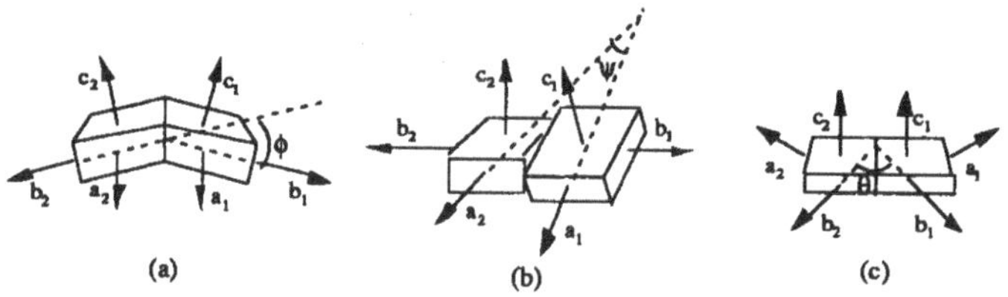

Figure 2. Description of misorientation angles at a grain boundary; (a) a-axis tilt, (b) b-axis twist and (c) c-axis tilt.

The misorientation between two adjacent grains where stacks of parallel plates meet edge to edge is determined by optical microscopy using a combination of three different misorientation angles shown in Figure 2.[3] The misorientation between two grains is also characterized by X-ray pole figure analysis using two angles namely the kink angle γ between the c-axes of both the grains, and the angle β between the b-axis of the first grain and the projection of the b-axis of the second grain on the a-b plane of the first grain.[3,6,7] The transport critical current density across the grain boundaries is measured in self field and also in the presence of magnetic field. The DC current is transported through the specimen and the voltage drop across the voltage leads, that are bonded on either side of the boundary, is measured.

4. Results and Discussion

4.1 Self field measurements of grain boundaries

Table 1 contains the critical current density in self field at 77 K for 13 specimens studied in the previous work. The misorientation angles ϕ, ψ and γ characterizing the boundary through which the current was passing varies from 0° to 47° for the a-axis tilt ϕ and 0° to 83° for the b-axis twist ψ and the kink angle γ. The grain boundary specimens with misorientation angles lower than 27.8° are found to transport large currents and have current densities up to 26,300 A/cm^2 while for those with higher misorientation angles the critical current density drops by approximately two orders of magnitude.

4.2 Self field measurements of bars

The current-voltage characteristics of the melt-textured bars of dimensions 50 mm × 10 mm × 5 mm were measured in self field at 77 K. The bars are found to carry currents up to 30-40 Amps before losing superconductivity. As the current is increased further, the voltage across the voltage leads increases. The reason for studying this dissipative regime was to observe how the resistivity changes with higher currents. These results are displayed in Figure 3. From Figure 3 it can be seen that the I-V behavior is ohmic. The voltage increases linearly as a function of the current. Using the Pulse DC current measurement technique, the bars were tested up to 1000 Amps.

Table 1 Misorientation angles of grain boundary specimens and the critical current density in self field at 77 K.

Specimen	No. of grain boundaries	Misorientation Angles			J_c (A/cm^2)
		a-tilt $\phi°$	b-twist $\psi°$	kink ang $\gamma°$	
1	0	0.0	0.0	0.0	47,600
2	1	4.5	4.5	6.4	11,100
3	2	3.4	9.1	9.7	26,300
		11.0	3.2	11.5	
4	2	7.2	7.1	10.1	22,600
		9.3	4.4	10.3	
5	1	3.1	10.2	10.7	10,300
6	2	4.2	9.5	10.4	7,600
		15.8	4.2	16.3	
7	1	16.2	2.5	16.4	11,300
8	1	23.0	16.0	27.8	19,300
9	1	30.4	32.1	43.0	115
10	1	40.0	20.0	44.0	216
11	1	9.7	45.0	45.8	304
12	1	47.0	18.4	49.7	605
13	1	42.3	81.5	83.7	213

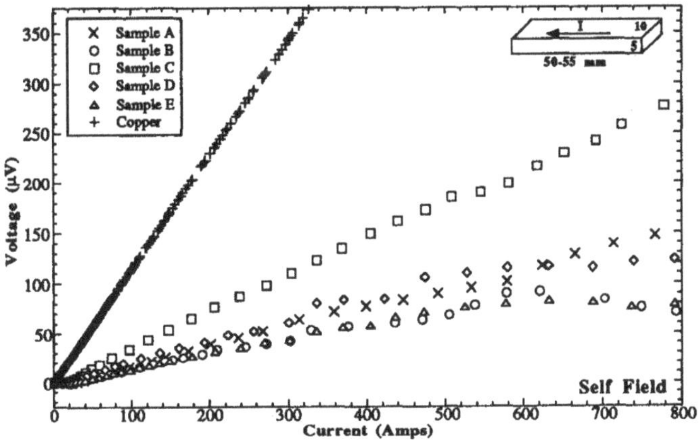

Figure 3 Self field measurements of bulk polycrystalline melt textured bars at 77 K.

The decrease in voltage at higher currents is due to the inductive voltage which is negative in nature. In Figure 3, the voltage across a copper bar of dimension 50 mm × 10 mm × 5 mm, of grade B187, at 77 K is also plotted. The resistance of this bar at 77 K for these dimensions is 1.14 μΩ.[8] This shows that these superconducting bars have a lower resistance than that of an industrial grade electrical copper conductor. Therefore it is possible to reach currents much larger than the critical value and still maintain a resistivity that is very low compared with that of copper at the same temperature. Some of these bars have resistance as low as 0.15 μΩ. Hence the heat losses can be reduced by as much as 1/7th of that of copper. Also since the thermal conductivity of YBCO is 1/50th that of copper, these bars are very appealing as current leads and bus bars.

The low critical current density of these bars can be attributed to the high angle grain boundaries between the grains of these polycrystalline samples. The low resistance, however, demonstrates the low resistivity of the boundaries. Many applications like current leads and interconnects can benefit from these bars due to their low electrical resistance in addition to their low thermal conductivity.

4.3 Magnetic field measurements of grain boundaries

In order to study the effect of magnetic field on the current passing through a grain boundary, measurements were made with the current flowing along the length of the specimen and the magnetic field applied in the plane perpendicular to the current direction. The critical current density is then measured when the field is applied at various orientations. The critical current density is found to behave anisotropically and has maximum and minimum values corresponding to certain orientations of the magnetic field.[3]

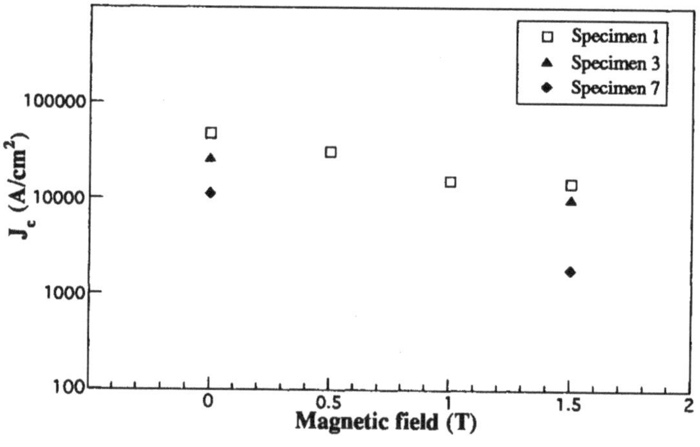

Figure 4 Maximum critical current density observed during magnetic field measurements of grain boundaries at 77 K.

The dependence of maximum J_C on magnetic field applied perpendicular to specimen axis for specimens Nos. 1, 3 and 7 is shown in Figure 4. The drop in the value of J_C is about 66 % which is similar to that usually observed in melt-textured single grain samples.[7] Specimen 3 ($\gamma = 9.7°$) shows a similar drop of only 62 % at 1.5 T from its zero field value even with the presence of two grain boundaries. Specimen 7 ($\gamma = 16.4°$) shows a drop of almost 80 % but still has a J_c of 2,000 A/cm^2 at 1.5 T which is one order of magnitude higher than J_c in polycrystalline sintered samples. These results indicate that the grain boundaries in these specimens are strongly coupled and are not easily penetrated by magnetic flux even in high magnetic fields. The dependence of minimum J_c on magnetic field applied perpendicular to the specimen axis in these three specimens is shown in Figure 5. The drops in the J_c values are comparable or even higher than that observed in single grain specimens and again indicates that these grain boundaries are strongly coupled and are capable of carrying relatively high currents in the presence of large external magnetic fields.

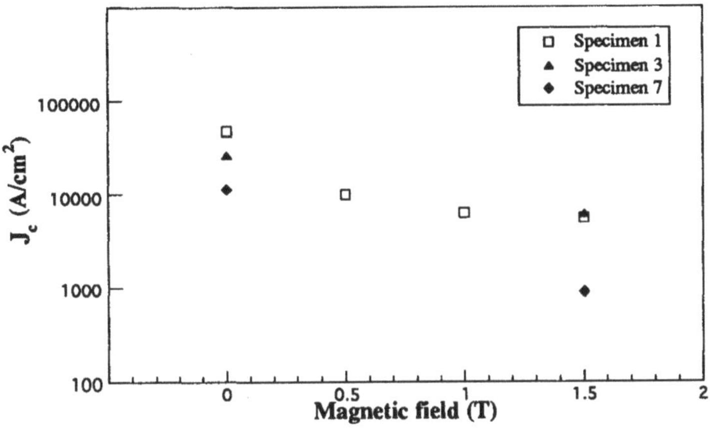

Figure 5 Minimum critical current density observed during magnetic field measurements of grain boundaries at 77 K.

4.4 *Magnetic field measurements of bars*

In the presence of magnetic field, the bars exhibit an ohmic behavior very similar to the self field. Due to the magnetic field, the critical current density across the bars decreases. However as the magnetic field increases, no substantial change is observed in the I-V behavior (see Figure 6). This indicates that the I-V behavior is insensitive to the increasing magnetic field and the bars behave like a normal electrical conductor. The increase in resistance of these bars in the presence of magnetic field is apparent because of the poor quality of current contacts. As the current is transported continuously, heat is generated at the contacts between the current leads and the superconductor bars. This heat generation gives rise to additional resistance. This is absent in the case of Pulse DC measurements. Improving the quality of current contacts will decrease this apparent resistance and the I-V behavior in magnetic field should follow the I-V behavior in self field as it does for sample B. For this sample, the I-V behavior at 0.5 T, 1.0 T and 1.5 T is almost similar and the resistance is equal to that measured in self field using the Pulse DC current measurement technique. These observations give promise to the use of these bars for carrying high currents with minimal heat loss in the presence of large magnetic fields.

Since the single grain specimens possess power-law I-V characteristics, it can be assumed that the linear I-V characteristics observed in these samples represent the dissipation mechanism in the boundary rather than in the grains themselves. This suggests that the mobility of the flux lines is higher along these grain boundaries due to lower pinning forces in these boundaries.[9]

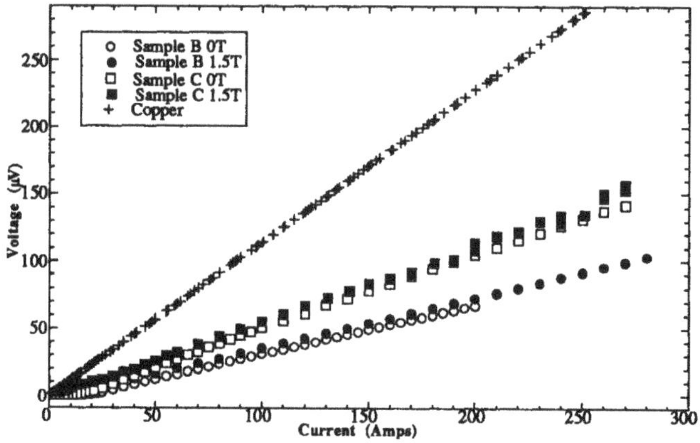

Figure 6 Magnetic field measurements of bulk polycrystalline melt textured bars at 77 K.

The reason for the decreased pinning energy along the boundary can be understood by considering that the pinning energy arises from fluxoids which typically penetrate the superconductor where the superconductivity is weakened. The creation of a normal vortex inside a superconductor requires additional energy which is equal to the difference between the energy of the normal state and the energy of the superconducting state. This difference in energy is smaller if the superconducting region is weakly superconducting. When a fluxoid is to be moved, the activation energy will then correspond to the energy that must be added to the system in order to move the normal vortex into the strongly superconducting region. If the boundary consists of a two-dimensional region with weaker superconductivity than the adjacent grains, there will be no strongly superconducting barriers between the pinning centers along the boundary. Consequently, the activation energy necessary for flux flow to occur is reduced along the boundary.

5. Summary

Melt-textured samples are processed by a liquid phase removal method to minimize the amount of secondary phases in the grain boundaries and also reduce the micro-cracks present at the interface. The critical current density is found to decrease very sharply, almost by two orders of magnitude where the misorientation angles exceed above 27.8°. Specimens capable of carrying high currents in self field at 77 K, are also found to carry high currents in 1.5 T at 77 K. Bulk polycrystalline samples 50-100 mm long and 50-100 mm^2 cross-section are found to transport currents up to 1000 Amps in self field and magnetic field of 1.5 T with ohmic resistance much less than that of copper at 77 K. These results give promise to using the method of liquid phase removal for processing bulk YBCO superconductors capable of carrying high currents.

6. Acknowledgements

This work was supported by the Texas Center for Superconductivity at the University of Houston under Prime Grant MDA 972-88-G-0002 from the Defense Advanced Research Projects Agency and the State of Texas.

7. References

[1] S. Jin, T. H. Tiefel, R. C. Sherwood, R. B. van Dover, M. E. Davies, G. W. Kammlot and R. A. Fastnacht, *Phys. Rev. B* **37** (1988) 7850.

[2] Dag W. A. Willén and K. Salama, *Physica C* **201** (1992) 311.

[3] A. S. Parikh, B. Meyer and K. Salama, *Superconductor Science and Technology* **7** (1994) 455.

[4] A. S. Parikh, M. S. Thesis, University of Houston (1993).

[5] K. Salama, V. Selvamanickam, and D. F. Lee, *Processing and properties of High T_C Superconductors*, ed. S. Jin (World Scientific, Singapore, 1993) 155.

[6] B. Meyer, M. S. Thesis, University of Houston (1993).

[7] M. B. Field, X. Y. Cai, S. E. Babcock, and D. C. Larbalestier, *IEEE Trans. on Appl. Supercon.* **3(1)** (1993) 1479.

[8] D. G. Fink and H. W. Beaty, *Standard Handbook for Electrical Engineers* (McGraw-Hill, London, 1987), p. 4-5.

[9] M. Nilsson-Mellbin, A. Parikh and K. Salama, *Physica C* **223** (1994) 19.

SOME ASPECTS OF CRITICAL CURRENTS IN Ag SHEATHED $Bi_2Sr_2Ca_2Cu_3O_{10}$ COMPOSITE TAPES: A SUMMARY

M. SUENAGA, QIANG LI, Y. FUKUMOTO*, K. SHIBUTANI**, Y. L. WANG, AND H. J. WIESMANN
Department of Applied Science, Brookhaven National Laboratory
Upton, NY 11973, USA

and

P. HALDAR
Intermagnetics General Corp.
450 Old Niskayuna Road, Latham, NY 12110 USA

and

L. MOTOWIDLO
IGC/Advanced Superconductors, Inc.
1875 Thomaston Avenue, Waterbury, CT 06704 USA

ABSTRACT

Detailed and high sensitivity ($E \geq 10^{-9}$ V/cm) measurements of J-E (I-V) characteristics were made for $Bi_2Sr_2Ca_2Cu_3O_{10}$/Ag composite tapes at 4.2, 27, 55-90, and 64-77 K in liquid cryogens and at H = 0 - 17 T. The results have helped clarify the critical current limiting factors as well as revealing a new low voltage dissipation phenomena in these composites. A brief summary of the results are presented.

1. Introduction

Since the discovery of high T_c Bi-cuprates,[1] extensive efforts to develop practical conductors utilizing these materials have been initiated. As the result, a number of encouraging values of critical current densities[2,3] J_c as well as of the lengths of the conductors have been reported. In addition, the usefulness of these conductors for winding magnets are also being demonstrated. However, these composites with $Bi_2Sr_2CaCu_2O_8$ and $Bi_2Sr_2Ca_2Cu_3O_{10}$ [Hereafter Bi(2:2:1:2) and Bi(2:2:2:3), respectively] require substantial improvement in J_c before they become useful for wide areas of practical applications at elevated temperatures, e.g. ≥ 50 K. In order to access

*On leave from Kobe Steel Ltd., Kobe, Japan
**Present address, Kobe Steel Ltd., Kobe, Japan

the J_c limiting factors in these composites, we have carried out a detailed examination of I-V curves of these tapes over a wide temperature range and in high magnetic fields.[4-6] An account of this investigation is given below, and we focus our discussion only on Bi(2:2:2:3)/Ag tapes in the interests of space.

2. Experimental Procedure

Bi(2:2:2:3)/Ag composite tapes were fabricated using the conventional powder-in-tube method both at Intermagnetics General Corp. (IGC) and Brookhaven National Laboratory (BNL). An approximately 4.0 cm long tape was mounted on a rotatable Cu platform by soldering to a pair of current leads. The voltage leads were approximately 1.0 cm apart and ~1.0 cm away from the inside edge of the current leads. All of the I-V curves were taken by having the specimen immersed in a liquid cryogen. For the measurements at 4.2 and 27 K, liquid He and Ne[5] were used at or near the atmospheric pressure while the temperature ranges, 55-90 and 64-77 K, were achieved using liquid O_2 and N_2, respectively, at reduced pressures, and the temperature of the bath was monitored by a Si-diode temperature sensor which was mounted on the specimen platform. The voltage from a specimen as a function of current was measured using a Keithley 2001 multimeter in conjunction with a 1801 preamplifier. This combination allowed us to achieve a 1 nV sensitivity while our earlier measurements were limited to a 10-20 nV sensitivity level using a Keithley 182 nanovoltmeter. The I-V characteristics were determined up to 8 T at BNL and some of the measurements were carried out up to 17 T at 4.2 and 27 K at the National High Magnetic Field Laboratory at Florida State University.

3. Results and Discussion

3.1 logE vs. logJ and $J_c(H)$

One of the key aspects of making a detailed measurement of the I-V characteristics is the fact that the shapes of the I-V curves from a specimen can provide very important information about the nature of the dynamics of the vortex motion under the Lorentz force. An example of this is illustrated in Fig. 1 as a logE vs. logJ plot for one of the earlier Bi(2:2:2:3)/Ag tapes measured at 64 K as a function of applied magnetic field H perpendicular to the tape face.[4] The very notable characteristic of the E-J curves in the figure is that all of the curves below a certain value of magnetic field H_g, in this case 0.7 T, are concave down while those above this field are concave up. [It will be shown below that this particular field H_g represents the vortex solid-to-liquid phase transition boundary at 64.5 K for Bi(2:2:2:3).] Furthermore, those curves below H_g = 0.7 T can be described by an expression, $E = J\rho_o \exp(-J_o/J)^{\mu}$, where ρ_o and J_o are the characteristic resistivity and current density, respectively, as predicted by both the vortex glass-liquid transition[7] and the collective creep theory.[8] The values of the exponent μ are expected to be ≤ 1 for the former and 1/7, 3/2, and 7/9 for the latter depending on the magnetic field regions. It was found that the values of μ at $T \geq 55$ K were mostly less than 1 and the description of the vortex state of a Bi(2:2:2:3) as a glassy solid appears to be appropriate while at low temperatures, e.g. 27 K, $\mu \cong 3/2$ for a large range of H and thus, the collective creep model appears to describe the vortex motion.[9]

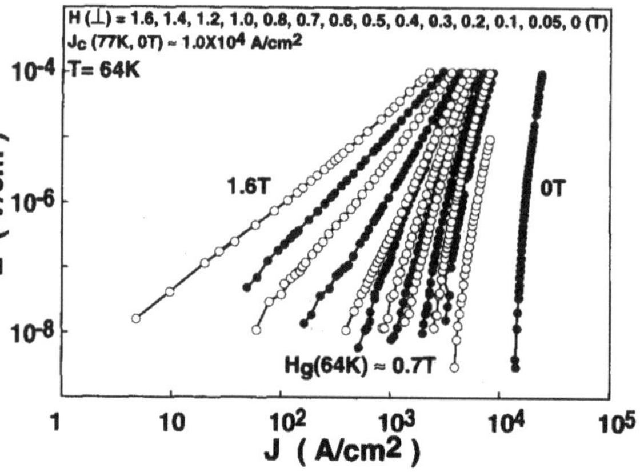

Fig. 1. An example of the resistive transition curves for H = 0-1.6 T at 64 K for a Bi(2:2:2:3)/Ag tape.

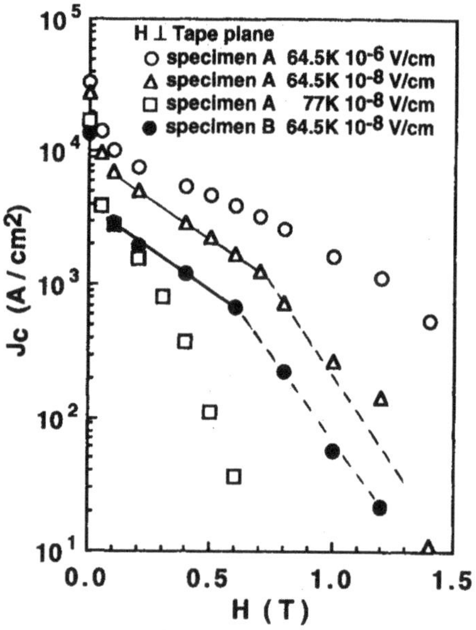

Fig. 2. Critical current densities as a function of H at 64.5 K are shown for the two tapes with $E_c = 10^{-6}$ and 10^{-8} V/cm.

From a figure such as Fig. 1, one can determine the values of J_c as a function of applied magnetic field by selecting an appropriate criteria E_c. Shown in Fig. 2 are J_c for Bi(2:2:2:3)/Ag tapes as a function of H at 64.5 K using $E_c = 1$ µV/cm and 10 nV/cm. As noted by many investigators, J_c drastically (by a factor of 3-4 or more) decreases upon the application of a small magnetic field. This is attributed to the fact that the weakly coupled grain boundaries in the composite can no longer transport high currents. On further increase in H, J_c follows an exponential decrease[10] with H up to H_g (= 0.7 T) and beyond this point J_c decreases at an increasing rate. As shown in the figure, this change in the rate of decrease is more clearly observed at $E_c = 10^{-8}$ V/cm than at $E_c = 10^{-6}$ V/cm. Such a change in the slope in J_c vs. H is expected since the sign of the curvature across H_g changes in the logE vs. logJ curves.

From a number of measurements such as shown in Figs. 1 and 2,[4-6] one can construct a simplified picture of the factors controlling J_c for a Bi(2:2:2:3)/Ag tape at elevated temperatures as follows: (1) at a very low field, the fraction of weakly coupled grain boundaries in the cross section of a composite tape primarily controls the initial drop in J_c; (2) then, for an increasing magnetic field, the current is transported through the grains by the fraction of the strongly coupled grain boundaries. However, the limiting cause for J_c in this region is the Lorentz force driven vortex motion in the interior of the grains, and thus J_c is limited by the vortex pinning strength of the grain; and (3) upon further increase of H beyond H_g, the vortex state in the grains is driven into a liquid state and the motion of the vortices becomes increasingly easier with increasing H and J_c is reduced.

3.2 *Phase Diagram of Bi(2:2:2:3) in the Mixed State*

In the above discussion, it was assumed that the value of H, at which a change in the sign of the curvature takes place in the logE vs. logJ is the vortex solid-liquid transition point in the H-T plane of the mixed state of a type II superconductor. In this section, a brief summary of a study[6] is given to show that there exists a second order phase transition in the H-T plane of a Bi(2:2:2:3) by an application of the scaling argument proposed by Fisher,[7] and it is shown that $H_g(T)$, is indeed the phase boundary $T_g(H)$.

Since the application of the scaling is based on the E-J curves taken at various temperature at a fixed field, pumped liquid oxygen was employed to measure the I-V curves at a number of applied field values. An example of such measurements is shown in Fig. 3(a). These E-J curves can also be collapsed into two branches by the use of the scaling variables, $(E/J)|1-T/T_g|^{\nu(1-z)}$ and $(J/T)|1-T/T_g|^{-2\nu}$ with an appropriate choice of the exponents, ν and z, and the solid-to-liquid transition temperature $T_g(H)$ as shown in Fig. 3(b). By determining $T_g(H)$ for several values of H by this procedure, the phase diagram for Bi(2:2:2:3) was constructed as shown in Fig. 4. Moreover, it was confirmed that $T_g(H)$ is identical to $H_g(T)$ as assumed in Sec. 3.1. Thus, this also confirms that the upper boundary of J_c in the grains is determined by the phase boundary in the T-H plane. Only improved strength of the vortex pinning can shift this boundary upward as illustrated by a high energy heavy ion irradiation on a Tl(2:2:1:2) film.[11]

At lower temperatures, it is thought that the improved connectivity of the a-b planes along the c-direction causes a stiffening of the vortex lattice tilt modulus. This results in a drastic upward change in the phase boundary as are shown in Fig. 4 and associated with it are large increases in J_c as the temperature is lowered.[6]

3.3 Effects of the Grain Boundaries and Low Level Residual Resistivity

In the above discussion of J_c at elevated temperatures, it was assumed that the weakly coupled grain boundaries do not contribute to the transport current beyond the lowest

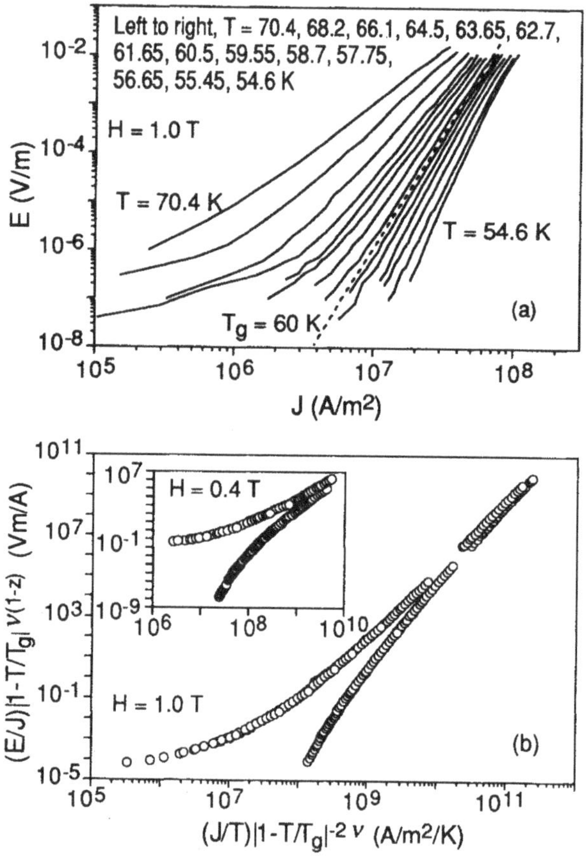

Fig. 3. (a) An example of electric field vs. current density in a *log*E vs. *log*J plot indicating the $T_g(H)$ by a dotted line for H = 1.0 T and (b) A scaling plot of all E-J isotherms for a Bi(2:2:2:3)/Ag tape for H = 1.0 and 0.4 T (the inset) with $\nu = 1.25 \pm 0.15$.

magnetic fields. It has been shown, however, that at low temperatures, e.g. 4.2 K, the boundaries can transport a substantial current under a certain condition of the boundary geometry and it is commonly believed that the observed hysteresis in J_c for the increasing and decreasing field cycle is a manifestation of the presence of such current carrying boundaries.[12] Furthermore, it was shown that, in general, the higher the value of J_c the smaller the size of the hysteresis. Also, theories were developed to predict the shape of the hysteresis in $J_c(H)$.[13,14] These theories are based on Yanson's nonuniform junction current density model[15] and the effect of the circulating intragranular currents on the junction surface currents. Although it is possible to match the theoretical hysteresis curves in critical currents with the experimental curves, it is difficult to conclusively show that these junctions are the source of the critical current limiting dissipation since these models do not predict explicit expressions for the E-J relationship for comparison with the experimental results and the nearly Lorentz force independent $J_c(H)$ has been observed in epitaxial Bi(2:2:1:2) films[16] where an application of Yanson's model is not necessary.

More recently, we have observed a new phenomena in the E-J curves which appear to originate from grain boundaries or microscopic cracks.[17] In a number of specimens, it was found that there exist low level resistivity of an order 0.1 nΩ. An example of such low level dissipation is shown in Fig. 5. In this case, the "tail" in the E-J curves was not originally detected down to E = 1 nV/cm. However, after repeated thermal cycling procedures, the low level resistive tail was observed before the initiation of the massive vortex motion causing a rapid rise in E. Upon further thermal cycling, degradation in J_c was clearly observed, as shown in Fig. 5, indicating initiation of macroscopic cracks in the core of the composite. Interestingly, the linear plot of the "tail" shown in the inset of Fig. 5 reveals not only is the dynamic resistivity nearly constant with J, but also the "tail" is associated with a very small critical current. Thus, it is presumed that the "tail" is induced by microscopic cracks which exhibit the E-J curves as if it consisted of a series of resistively shunted junctions.

Occasionally, we have observed substantially larger resistance tails than those shown. From the fact that this specimen was not knowingly treated to induce mechanical damage in the Bi(2:2:2:3) core, we assume that the source of this "tail" is likely to be the grown-in weak grain boundaries. These boundaries appear to be quite different from those assumed to contribute to the hysteretic behavior in J_c as no hysteresis in this part of the E-J curves as observed as H is cycled. Although the resistivity of the tail is quite small and not likely to be a concern for most applications of the tapes, it will be a very significant problem for a persistent mode of operation of a magnet employing these conductors.

4. Summary

It has been shown that at elevated temperatures, the critical current in the Bi(2:2:2:3)/Ag tapes were determined by: (1) first, the fraction of the strongly coupled boundaries, and (2) second the intragranular vortex pinning strength as it determines the vortex glass-liquid phase boundary. At low temperatures, some of the grain boundaries

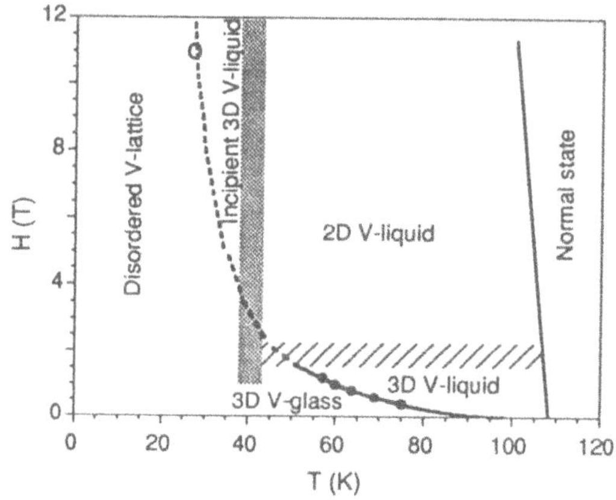

Fig. 4. The vortex phase diagram in the mixed state of a Bi(2:2:2:3) superconductor and the solid line is a theoretical curve of $H_g(T_g) = H(0)(1-T_g/T_{co})^{2\nu}$ and $\nu = 1.25$

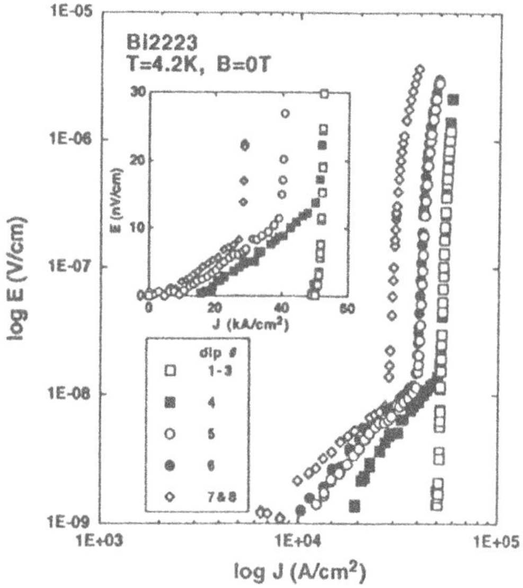

Fig. 5. An example of the E-J characteristics showing the thermal cycle induced low level residual resistivity for a Bi(2:2:2:3)/Ag tape.

could transport substantial supercurrents although detailed mechanisms for it is not clearly understood yet. Another type of boundary and/or cracks which cause appearance of a low voltage "tail" in the E-J curve also appear to exist. In most cases, this voltage can be described by an expression $V \sim (I'_c - I)^n$ where I'_c is the field dependent critical current of the boundary and n is 1-1.2. Further studies are required to clarify the roles of the grain boundaries in determining critical currents in these composites.

5. Acknowledgments

The authors greatly appreciate constant and helpful discussions with D. O. Welch. This research was performed under the auspices of the U.S. Department of Energy, Division of Materials Sciences, Office of Basic Energy Sciences under Contract No. DE-AC02-76CH00016.

6. References

1. H. Maeda, Y. Tanaka, M. Fukutomi, and T. Asano, *Jpn. J. Appl. Phys.* **27** (1988) L209.
2. Y. Yamada, M. Satou, S. Murase, T. Kitamura, and Y. Kamisada, in *Advances in Superconductivity V*, (Springer, Tokyo, 1993), p. 172.
3. Q. Li, K. Brodersen, H. A. Hjuler, and T. Freltoft, *Physica* **C217** (1993) 360.
4. K. Shibutani, Q. Li, R. L. Sabatini, M. Suenaga, L. Motowidlo and P. Haldar, *Appl. Phys. Lett.* **63** (1993) 3515.
5. K. Shibutani, H. J. Wiesmann, R. L. Sabatini, M. Suenaga, S. Hayashi, R. Ogawa, and Y. Kawate, L. Motowidlo, and P. Haldar, *Appl. Phys. Lett.* **64** (1994) 924.
6. Qiang Li, H. J. Wiesmann, M. Suenaga, L. Motowidlo and P. Haldar, *Phys. Rev.* **B50** (1994) 4256.
7. M. P. A. Fisher, *Phys. Rev. Lett.* **62** (1989) 1415; D. s. Fisher, M. P. A. Fisher, D. A. Huse, *Phys. Rev.* **B43** (1991) 130.
8. M. V. Feigel'man, v. B. Geshkenbein, A. I. Larkin, and V. M. Vinokur, *Phys. Rev. Lett.* **63** (1989) 2303.
9. Qiang Li, H. J. Wiesmann, M. Suenaga, L. Motowidlo, and P. Haldar, in print.
10. K. Yamafuji, T. Fujiyoshi, K. Toko, T. Matsuno, T. Kobayashi, J. Shimayama, and K. Kishio, to be published.
11. R. Budhani, W. L. Holstein, and M. Suenaga, *Phys. Rev. Lett.* **72** (1994) 566.
12. K. Sato, T. Hikata, H. Mukai, M. Ueyama, N. Shibuta, T. Kato, T. Masuda, M. Nagata, K. Iwata, and T. Mitsui, *IEEE Trans. Magn.* **MAG-27** (1991) 1281.
13. A. I. Dyachenko, *Physica* **C213** (1993) 167.
14. G. Ries, H. W. Neumüller, W. Schmidt, and C. Struller, to be published.
15. I. K. Yanson, *Sov. Phys. JETP* **31** (1970) 800.
16. P. Schmidt, P. Kummeth, L. Schultz, and G. Saemann-Ischenko, *Phys. Rev. Lett.* **67** (1991) 267.
17. Y. Fukumoto, Qiang Li, and M. Suenaga, to be published.

PERCOLATIVE CURRENT FLOW IN AG-SHEATHED BSCCO-2223 TAPES: MANIFESTATIONS, VISUALIZATION AND ORIGINS

D. C. Larbalestier, X.Y. Cai, H. Edelman, Y. Feng, M. Karuna, J. Parrell, A. Pashitski, and A. Polyanskii*.

Applied Superconductivity Center, University of Wisconsin, Madison, WI 53706 USA

ABSTRACT

Recent work at the University of Wisconsin on the understanding of the factors controlling the critical current density of Ag-sheathed BSCCO-2223 tapes is summarized. Two new techniques (microslicing of the tape and magneto optical imaging) have been used to measure the local critical current density and to visualize the effective current path in the superconducting state. We conclude that local variations of the grain-to-grain electromagnetic connectivity cause percolative current flow and greatly limit the overall critical current. Therefore, increasing the active cross-section of tapes by understanding and controlling the defects that inhibit good electromagnetic coupling is the most promising way to raise the overall critical current density of BSCCO tapes.

1. Introduction

Ag-sheathed Bi-Pb-Sr-Ca-Cu-O (BSCCO) superconducting tapes of the nominal $Bi_2Sr_2Ca_2Cu_3O_x$ composition made by the powder-in-tube process demonstrate great promise as conductors for magnetic field use[1-7]. However, extensive study of such conductor-form tapes has not yet yielded a proper understanding of how to reproduce the best properties in a consistent manner. High phase purity, a well textured grain structure and good grain-to-grain connections appear to be crucial to the development of high transport critical current density (J_{ct}) in BSCCO tapes, yet relatively little is understood about the mechanism(s) by which the 2223 phase forms, the texture develops and the complex role that the silver sheath plays. Defects, such as cracks and non-superconducting second phases are common in the tapes, producing permanent barriers to current flow and some high angle grain boundaries exhibit a weak electromagnetic coupling which is field- and temperature-dependent. This spatially variable electromagnetic coupling produces percolative current flow, meaning that not all the cross section is useful for carrying transport current. We here summarize recent work from our programs[7-18] which concern the phase formation, properties, connectivity and the effective current path of BSCCO-2223 tapes.

2. Phase Conversion Reactions

A study of the reactions that occur when powder having a nominal overall 2223 composition is reacted within a Ag tube illustrates many of the complexities inherent to this system[9]. Many phases are simultaneously present and multiple parallel reactions

*On leave from Institute of Solid State Physics, 142432 Chernogolovka, Moscow District, Russia

form undesired non-superconducting phases, interfering with the desired 2212 to 2223 phase conversion reaction, which has a significantly larger reaction energy[15,16] (~ 1.5 MJ/mol) than that measured for the undesired reactions (0.27-0.51 MJ/mol). The 14:24 AEC (alkaline earth cuprates) showed little growth with time and the growth was principally in 2:1 AEC and SrO content. The 2:1 AEC has an activation energy of formation of ~370kJ/mol, a value not too different from that determined for the combined second phase reaction rate. Reaction to 2:1 AEC clearly inhibits completion of the 2212 to 2223 reaction by denying the growing 2223 phase essential Ca, Sr, and Cu.

3. Structure of the Ag-BSCCO Interface

Preferential reaction to the 2223 phase occurs near the Ag interface, at temperatures about 20°C below that at which 2223 forms in the interior of the BSCCO core[7,9]. This 2223 reaction layer aligns itself such that its basal planes are parallel to the Ag, thus appearing to use the Ag as a template (inset to Fig. 1). The preferential formation of the 2223 phase at the interface is most reasonably explained by the depression of the melting point observed when Ag is present[15,19]. This is consistent with the model of liquid-assisted-transformation proposed by Morgan et al.[20]. However, there cannot be much liquid formed, because the reaction zone is confined to the vicinity of the Ag-oxide interface, particularly for reactions in air. Because the growing front of the 2223 phase tends to align itself to the Ag, any sausaging or irregularity of the Ag tends to propagate into the superconductor core, misaligning the grain-to-grain connections and interfering with potentially good "railway switches"[5,7].

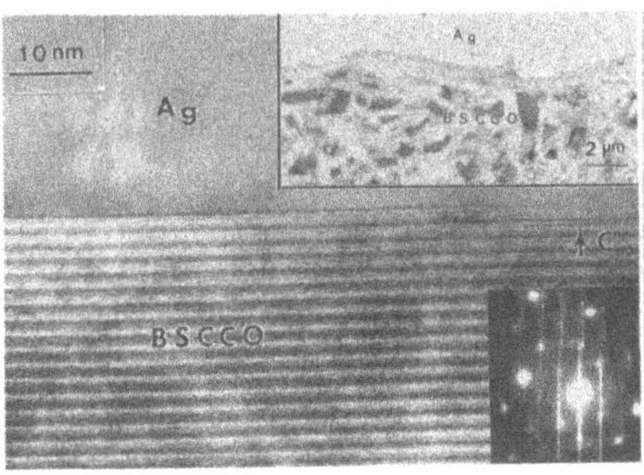

Fig. 1. HRTEM image showing the preferential formation of 2223 phase at the interface.

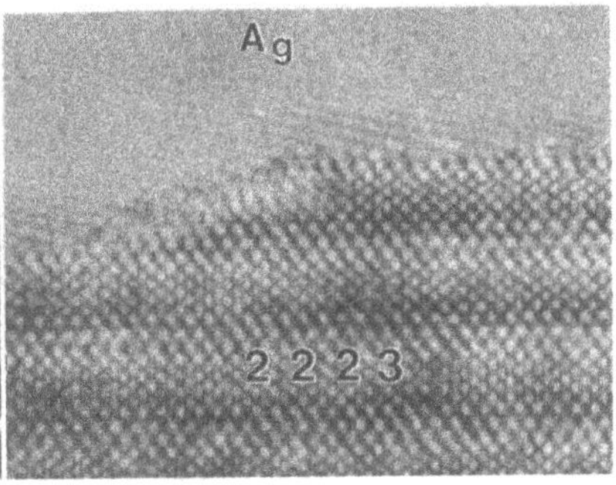

Fig. 2. HRTEM image of a typical Ag/BSCCO-2223 interface which shows good bonding.

For both 2212 and 2223 tapes, the interface between the Ag and the BSCCO is well bonded, free of non-superconducting phases and appears abrupt in all HRTEM images[7,11,21]. In most cases, a Bi-O layer abuts the Ag. A typical example is shown in Fig. 2. The (001) BSCCO planes lie parallel to the Ag, regardless of the Ag crystal orientation. Small misalignments are accommodated by facetting to half unit cell steps of the BSCCO. BSCCO-2212 tapes have their own peculiarity: whether processed by complete melting or by sintering below the melting point, there is always a residual half-layer of 2201 at the Ag interface[21]. By contrast, no 2201 or 2212 layers of any thickness were observed at the Ag/BSCCO interface in 2223 tape[11].

4. Residual 2212 Layers at the (001) Twist Boundaries and Their Influence on J_c

Field-dependent AC or DC susceptibility measurements were correlated with transport J_C and microstructure[10,24] for some 20 Ag-clad 2223 tapes having J_{ct} (0T, 77K) values ranging from 0-20,000 A/cm^2. Fields of 0.1-1mT induced a lower transition, $T'(H)$, in the real component of the susceptibility. $T'(H)$ exhibits a marked kink at 80-81K, being insensitive to H below 80K and very sensitive above 80K, indicating electromagnetically granular behavior (Fig. 3) at higher temperature.

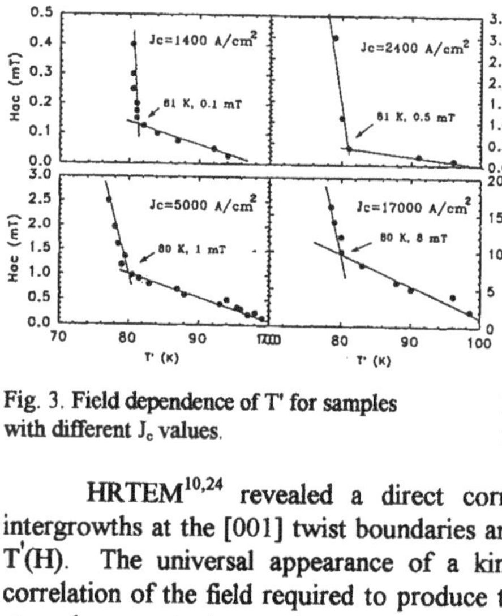

Fig. 3. Field dependence of T' for samples with different J_c values.

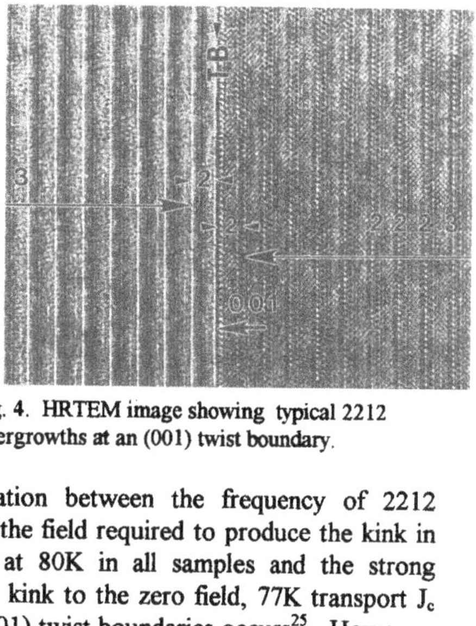

Fig. 4. HRTEM image showing typical 2212 intergrowths at an (001) twist boundary.

HRTEM[10,24] revealed a direct correlation between the frequency of 2212 intergrowths at the [001] twist boundaries and the field required to produce the kink in $T'(H)$. The universal appearance of a kink at 80K in all samples and the strong correlation of the field required to produce the kink to the zero field, 77K transport J_c strongly suggests that c-axis transport across (001) twist boundaries occurs[25]. However, this current should be suppressed by weak fields and, indeed, it is commonly observed that there is an anomalous drop of ~50% in J_c (77K) in only mT fields[5]. Presumably only the strong linked grain-to-grain connections[5] without intergrowths remain active in higher fields.

5. Local J_c Variation in Tapes as Revealed by Microslicing

Direct proof that the J_c is highly variable from place to place within even one filament was first provided by an experiment in which we used a laser to cut microslices from different sections of one filament of prototype Ag-sheathed Bi-2223 tape[8]. The results shown in Fig. 5 contrast the properties of two edge and three central slices of a filament taken from a tape having an overall J_c (0T, 77K) of 14,000 A/cm^2. The central slice (3) and the intermediate position slices (2 and 4) had a J_{ct} (0T, 77K) value of 13,000-14,000 A/cm^2. These three slices had well-rounded V-I characteristics, typical of a stable, reversible transition to the flux flow state. By contrast, the two outside slices (1 and 5) had abrupt transitions at much higher J_{ct} values of 32,000 and 38,000 A/cm^2. These transitions were unstable (and therefore underestimate the true J_{ct}), no doubt due to contact and/or lead heating. There was also much less field sensitivity of J_c for the higher J_c slices: the edge slices exhibited no field sensitivity in a field of 8 mT, while the same small field reduced the central slice J_c to less than half its zero field value.

A major cause of the variation is shown in the scanning electron micrographs of Fig. 6. As Fig. 6a shows, there is a marked tendency for the grains of the 2223 phase to be both larger and better aligned, and for there to be less obstructive second phase in the extreme edge slices. The phase purity of the 2223 grains is also greater there, a result which is consistent with the beneficial effect of the Ag sheath on the phase conversion reaction[10,11], the Ag being in closer proximity to the tape core for the edge than for the center slices. In the center slice, ~10μm diameter "rocks" of non superconducting alkaline earth cuprate are obvious, as is the disruptive effect that they have on the local grain shape and on the grain-to-grain orientation.

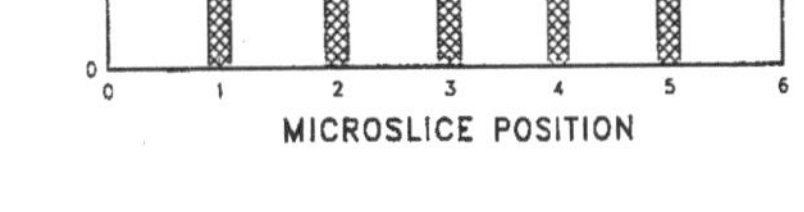

Fig. 5. J_{ct}(77K,0T) vs. position for different laser-cut microslices of a 2223 tape.

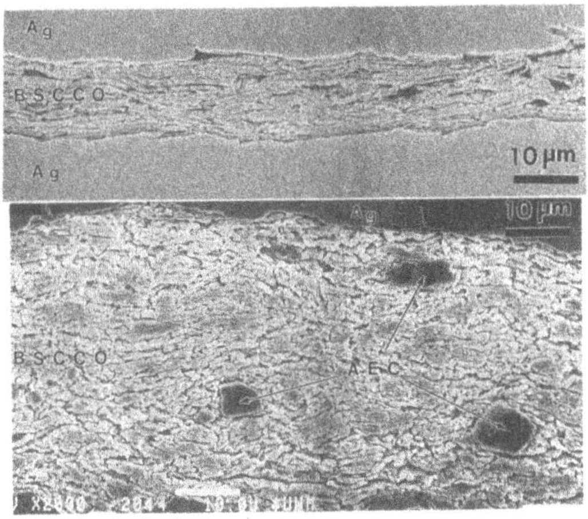

Fig. 6. SEM images of (a) slice 1 and (b) slice 3 of the 2223 tape of Fig. 5.

Microslice experiments have been made on several types of Bi-2223 sample. The highest J_c (0T, 77K) values found exceed 76,000 A/cm^2, a value higher than any whole filament J_c yet reported[7]. However, many microslices have J_c less than the J_c measured on the whole filament. Such a result can be explained both by the difficulty of performing the experiment without inducing damage into the microslice and/or by the presence of many actual or incipient cracks which open when the Ag is removed from the BSCCO core.

6. Magneto Optical Imaging of BSCCO tapes

The Magneto Optical (MO) imaging technique has great potential for visualizing the percolative current flow paths directly[26,27]. The technique images the field immediately above the superconductor as it carries a current, thus allowing direct comparison of the microstructure with the shielded field. With appropriate geometry and calibration the current path can be directly observed with a resolution of order 5μm or better. An example of an MO image on a BSCCO 2223 tape is shown in Fig. 7, where polarized light microstructural and MO images are placed side by side. The microstructural image (top) shows contrast both from misaligned grain blocks and from non superconducting second phase particles. Although the BSCCO cross section is rather uniform, the MO image (bottom) shows that the shielded field is highly variable from place to place, from which we can infer that the current flows very non-uniformly. In the imaged condition (40 mT and 12K), the regions at either end of the image are darker because they resist the penetration of the field better than regions in the middle.

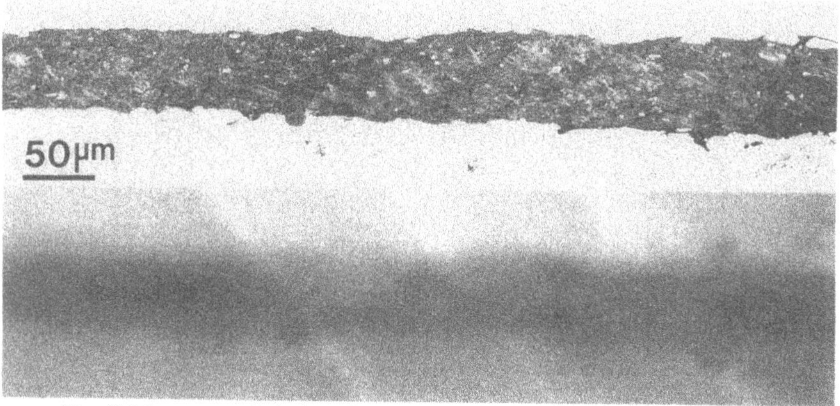

Fig. 7 Microstructural (top) and magneto optical image (bottom) of a half slice of a monofilament BSCCO-2223 Ag-sheathed tape. The tape is viewed in longitudinal section at 12K in a 40 mT field applied perpendicular to the page. The three diagonal stripes in the MO image are artifacts due to scratches in the film.

It is noticeable that there is a marked penetration of the field in pockets along the center line of the tape in the central region of the image. Such penetration appears to result from weakly coupled grains and/or cracks, rather than from weak intragrain flux pinning. By carefully calibrating the intensity images in terms of the temperature, field and settings of the polarizer and analyzer of the microscope, it is possible to produce detailed 2D images of the current flow in the tape. Detailed studies of the influence of microstructural parameters on the percolation properties are now in progress[27].

7. Summary

In summary, we believe that there is great potential in combined study of the transport, the microstructural and the magneto optical properties of BSCCO-2223 tapes as a function of their process conditions for addressing the very important issue of why the current flow is percolative in these tapes. Providing a reliable answer to this question is vital for the further development of applications of these important materials.

Acknowledgments

We are very grateful to W. Starch and A. Squitieri for much experimental assistance, to W.L. Carter and G.N. Riley Jr. of American Superconductor Corporation and S.E. Dorris of Argonne National Laboratory for certain samples studied in portions of the work summarized here and to E.E. Hellstrom and S.E. Babcock for many discussions. Different aspects of the work were supported by ARPA (N00014-90-J-4115), EPRI(RP8065-6) and NSF(DMR-9214707).

References

1. M. Ueyama, T. Hikata, T. Kato and K. Sato, *Jap. J. Appl. Phys.*, **30**, L1384 (1991)
2. Y. Yamada, M. Satou, S. Murase, T. Kitamura, and Y. Kamisada, *Proceedings of the ISS'92*, 717, Nov. 1992, Kobe
3. Q. Li, H. Brodersen, H.A. Hjuler and T. Frelhoft, *Physica C*, **217**, 360, (1993)
4. A. Otto, L.J. Masur, J. Gannon, E. Podtburg, D. Daly, G.J. Yurek and A. P. Malozemoff, *IEEE Trans. on Appl. Supercond.*, **3** (1993) 915
5. B. Hensel, J.-C. Grivel, A. Jeremie, A. Perin, A. Pollini and R. Flukiger, *Physica C*, **205**, 329 (1993)
6. P. Haldar, J. G. Hoehn, Jr., M. S. Walker, L. R. Motowidlo, U. Balachandran and C. A. Youngdahl, *IEEE Trans. on Appl. Supercond.*, **3** (1993) 1127
7. Y. Feng, Y. E. High, D. C. Larbalestier, Y. S. Sung, E. E. Hellstrom, *Appl. Phys. Lett.* **62**, 1553 (1993)
8. D.C. Larbalestier, X.Y. Cai, Y. Feng, H.S. Edelman, A. Umezawa, G.N. Riley and W.L. Carter, *Physica C*, **221**, 229 (1994)
9. Y.E. High, Y. Feng, Y.S. Sung, E.E. Hellstrom and D.C. Larbalestier, *Physica C*, **220**, 81 (1994)
10. A. Umezawa, Y. Feng, H.S. Edelman, T.C. Willis, J.A. Parrell, D.C. Larbalestier, G.N. Riley and W.L. Carter, *Physica C*, **219**, 378 (1994)
11. Y. Feng and D.C. Larbalestier, *Interface Science*, **1**, 397 (1994)
12. H.S. Edelman and D.C. Larbalestier, *Adv. in Cryo. Eng.*, **40**, (1994)
13. S.E. Dorris, B.C. Prorok, M.T. Lanagan, N.B. Browning, M.Hagen, J.A. Parrell, Y. Feng, A. Umezawa and D.C. Larbalestier, *Physica C*, **223**, 163 (1994)

14. J.A. Parrell, S.E. Dorris and D.C. Larbalestier, *Adv. in Cryo. Eng.* **40**, 193 (1994)
15. J.A. Parrell, S.E. Dorris and D.C. Larbalestier, Submitted to *IEEE Trans. on Appl. Superconductivity*, to appear 1995
16. M. Karuna, J.A. Parrell and D.C. Larbalestier, Submitted to *IEEE Trans. on Appl. Superconductivity*, to appear 1995
17. D.C. Larbalestier, Y. Feng, X.Y. Cai, H. Edelman, E.E. Hellstrom, Y.H. High, J.A. Parrell, Y.S. Sung and A. Umezawa, *Proc. of the 7th Int. Workshop on Critical Currents,* Alpbach, Austria, Jan. 1994, to appear World Scientific 1994
18. J.S. Luo, N. Merchant, E. Escorcia-Aparicio, V.A. Maroni, D.M. Gruen, B.S. Tani, G.N. Riley Jr., W.L. Carter, *IEEE Trans. on Appl. Supercond.* **3**, 972 (1993)
19. P.E.D. Morgan, R.M. Housley, J.R. Porter and J.J. Ratto, *Physica C*, **176**, 279 (1991)
20. P.E.D. Morgan, J.D. Piche and R.M. Housley, *Physica C*, **191**, 179 (1992)
21. Y. Feng, D.C. Larbalestier, S.E. Babcock and J.B. Vander Sande, *Appl. Phys. Lett.* **61**, 1234 (1992)
22. T. Ito, H. Takagi, S. Ishibashi, T. Ido and S. Uchida, *Nature*, **350**, 596 (1991)
23. J. Halbritter , *J. Applied Physics*, **68**, 6315 (1990)
24. A. Umezawa, Y. Feng, H.S. Edelman, Y.E. High, D.C. Larbalestier, Y.S. Sung, E.E. Hellstrom and S. Fleshler, *Physica C*, **198** (1992) 261
25. L. Bulaevski, L.L. Daemen, M.P. Maley, J.Y. Coulter, *Phys. Rev.* **B48,** (1993)
26. L.A. Dorosinskii, M.V. Indenbom, V.I. Nikitenko, Yu. A Ossipyan, A. Polyanskii, V.K. Vlasko-Vlasov, *Physica C*, **203**, 149, (1992)
27. A. Polyanskii, A. Pashitski, A. Gurevich, J.A. Parrell and D.C. Larbalestier, submitted

FLUX PINNING CHARACTERISTICS IN SUPERCONDUCTING Bi-2223 TAPE WIRES

T. MATSUSHITA, M. KIUCHI and N. IHARA
Department of Computer Science and Electronics, Kyushu Institute of Technology, 680-4 Kawazu, Iizuka 820, Japan

ABSTRACT

The flux pinning characteristics in superconducting Bi-2223 tape wires are numerically analyzed by using the flux creep model. The irreversibility field is shown to be much smaller than the upper critical field due to the significant thermally activated flux creep, although it depends on the flux pinning strength. The attainable irreversibility field by introduction of strong pinning centers such as finely dispersed normal particles is also analyzed. It is empirically known that the pinning force density shows a scaling behavior when it is plotted versus the magnetic field normalized by the irreversibility field. Such a scaling behavior is found to be derived from the analysis based on the flux creep model. These theoretical results are compared with the experimental results.

1. Introduction

The critical current density is one of the most important parameters of high temperature superconductors in their application. However, the critical current density in these materials is significantly degraded at high temperatures and/or magnetic fields. This is caused by the thermally activated flux creep. Such a degradation is more prominent in two-dimensional Bi-2223 superconductors than in more three-dimensional Y-123. This effect can be clearly seen from the drastic reduction in the irreversibility field from the upper critical field. It is desired, therefore, to improve the critical current characteristics by introduction of strong pinning centers also in Bi-2223 superconductors. For this purpose, it is necessary to understand the critical current characteristics in present superconductors such as silver sheathed tape wires.

In this paper the irreversibility field and the pinning property in the vicinity of this characteristic field are theoretically analyzed in terms of the flux creep model. The scaling behavior of the pinning force density is also investigated. It will be shown that this model exactly describes the pinning property under the significant thermal activation. In terms of this theoretical model the critical current density in Bi-2223 superconductors in which finely dispersed normal particles are artificially introduced as pinning centers is estimated.

2. Flux Creep Model

According to the flux creep model[1] the pinned fluxoids are thermally activated and sometimes hop over the pinning potential barrier in the form of a bundle, resulting in appearance of the electric field. The probability of the hopping is given in the form of the Arrehnius expression, $\exp(-U/k_BT)$, where U is the activation energy and k_B is the Boltzmann constant. The induced electric field due to the activated fluxoid motion is given by[2]

$$E = Ba_f\nu\exp\left[-\frac{U(j)}{k_BT}\right]\left[1-\exp\left(-\frac{\pi U_0 j}{k_BT}\right)\right], \tag{1}$$

where a_f is the fluxoid spacing that approximately gives the hopping distance, ν is the attempt frequency of the flux bundle, U_0 is the pinning potential depth and j is the current density normalized by the virtual critical current density J_{c0} in the creep-free case. In Eq. (1) the sinusoidally varying washboard potential is assumed. In this case the activation energy is expressed as

$$U(j) = U_0[(1-j^2)^{1/2} - j\cos^{-1}j]. \tag{2}$$

The attempt frequency is given by the damping oscillation frequency in the pinning potential and is[3]

$$\nu = \frac{\zeta\rho_f J_{c0}}{2\pi a_f B}, \tag{3}$$

where ζ is a constant of the order of 2π dependent on the kind of pinning centers and ρ_f is the flow resistivity. In this paper the Bardeen-Stephen model[4], $\rho_f = (B/B_{c2})\rho_n$ with $\rho_n = (T/T_c)\rho_n(T_c)$ denoting the normal resistivity, is assumed for the flow resistivity.

According to the flux creep theory of the collective pinning type[5] the pinning potential in bulk superconductors is given by

$$U_0 = \frac{0.835g^2k_BJ_{c0}^{1/2}}{\zeta^{3/2}B^{1/4}}, \tag{4}$$

where g^2 is the number of fluxoids in the flux bundle. If it is assumed that the transverse flux bundle size is given by the transverse elastic correlation length of the fluxoid lattice, we have[5]

$$g^2 = \frac{C_{66}a_f}{\zeta J_{c0}B} \equiv g_e^2, \tag{5}$$

where C_{66} is the shear modulus of the fluxoid lattice. It has been clarified[5] that the practical g^2 values estimated from the observed irreversibility field are smaller than g_e^2 given above except the case where g_e is close to unity. The difference becomes larger for the larger g_e^2 value. Such a deviation means that the practical cluster size is generally smaller than the correlation length. This can be understood as the result of the irreversible thermodynamics so as to minimize the energy dissipation. According to this interpretation[6] we have

$$g^2 = g_e^2\left[\frac{5k_BT}{2U_e}\ln\left(\frac{Ba_f\nu}{E_c}\right)\right]^{4/3}, \tag{6}$$

where U_e is the virtual pinning potential depth for the case of $g = g_e$ and E_c is the electric field criterion for the definition of the critical current density.

If the temperature and magnetic field dependences of J_{c0} are given, the pinning potential depth U_0 is obtained from Eq. (4) as a function of temperature T and magnetic field B. As for such dependences of J_{c0} on T and B, we assume the relation:

$$J_{c0} = A\left[1 - \left(\frac{T}{T_c}\right)^2\right]^m B^{\gamma-1}\left(1 - \frac{B}{B_{c2}}\right)^{\delta}, \tag{7}$$

that is empirically known to hold for metallic superconductors. In the above A, m, γ and δ are pinning parameters.

In this paper the pinning property in Bi-2223 superconductor is analyzed. We treat the case where the magnetic field is applied parallel to a wide surface of a silver-sheathed tape wire, i.e., normal to the c-axis. Since the dominant pinning centers are not known in Bi-2223 tape wires, the theoretical expression of J_{c0} cannot be given. Hence, the experimental results of the critical current density measured at sufficiently low temperatures where the effect of flux creep is not significant are approximately used for J_{c0}. In practical wires, however, normal particles of Bi-free phase of about 5 μm in diameter are found to be distributed by macroscopic observation. We assume that these normal particles also contribute to the flux pinning. As for the contribution from the normal particles, the direct summation of the individual elementary pinning forces of the condensation energy interaction gives[6]

$$J_{c0} = \frac{\pi B_c^2 f \xi}{4\mu_0 B a_f D}\left(1 - \frac{B}{B_{c2}}\right), \tag{8}$$

where D and f are the mean size and the volume fraction of the normal particles and B_c is the thermodynamic critical field. The coherence length ξ in the above equation is that in the direction of the current, and hence, that in the a–b plane in the present geometry. If the contributions from the dominant pinning centers and normal particles to U_0 are represented by U_{01} and U_{02}, respectively, the resultant U_0 is expected to be given as

$$U_0 = (U_{01}^2 + U_{02}^2)^{1/2}. \tag{9}$$

From the above equations the current–voltage curves are numerically calculated at all the temperature and the magnetic field. The critical current density and the irreversibility field can be obtained in a similar manner as in experiments. That is, the critical current density is estimated by the offset method from the E–J curve at the point of $E = E_c = 1.0 \times 10^{-5}$ V/m. The irreversibility field is defined by the magnetic field at which the critical current density is reduced to 1.0×10^5 A/m^2.

3. Results and Discussion

The parameters of the Bi-2223 tape wire used for the numerical calculation are listed in Table 1. The number of fluxoids in the flux bundle, g^2, is estimated as[7] 4.0 around the observed irreversibility field of B_i(85 K)=4.0 T, while g_e^2 is estimated as 6.7. Although g^2 changes slightly with temperature and magnetic field, we assume for simplicity that g^2 is approximately given by the above value in the vicinity of the irreversibility line. The numerically calculated irreversibility line is shown in Fig. 1, in which the corresponding experimental result[7] is also shown for comparison. The agreement seems to be satisfactory. This means that the flux creep model describes the phenomenon correctly.

Table 1. Superconducting and pinning parameters in Bi-2223 tape wire used for calculation. The numbers in parentheses are for the pinning by normal particles.

T_c (K)	$B_{c2\parallel}(0)$ (T)	$\rho_n(T_c)$ ($\mu\Omega$m)	A	m	γ	δ	ζ	g_e^2	g^2
107.7	1000	100	2.54×10^9	3.0	0.63	2.0	2π	6.7	4.0
			(6.57×10^8)	(1.5)	(0.5)	(2.0)	(4)	(1.55)	(1.55)

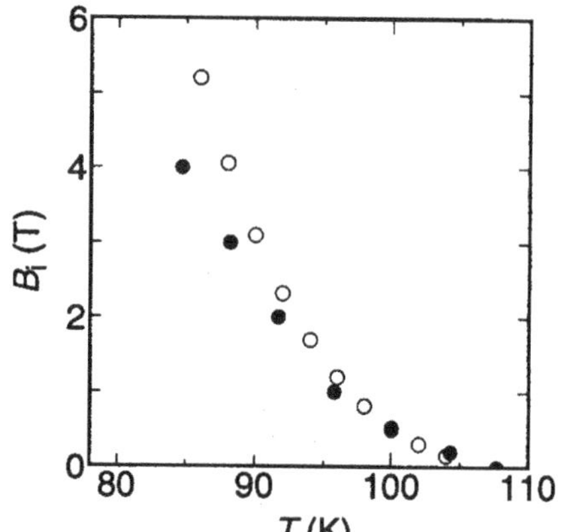

Fig. 1. Irreversibility line of Bi-2223 tape wire under the magnetic field normal to the c-axis : theoretical result (○) and experimental result(●).

In Fig. 2, the obtained pinning force density $F_p = J_c B$ at 77.3 K is compared with the virtual pinning force density $F_{p0} = J_{c0} B$. It is found that the practical pinning property is severely degraded from the virtual one even for the silver-sheathed tape wire with a relatively high critical current density. For example, the irreversibility field is estimated as only 2.7 % of the upper critical field at 77.3 K. Such a significant effect of the flux creep is mostly caused by the two-dimensional property of the Bi-2223 superconductor, that can be seen in the relatively small g^2 value.

It is interesting, therefore, to estimate the attainable irreversibility field of the Bi-2223 superconductor by introducing strong pinning centers such as finely dispersed normal particles. In this case Eq. (9) can be used. We assume that the normal particles of 0.5 μm in diameter are successfully introduced up to 20 % in volume fraction. In this case A amounts to 1.32×10^{10}. Since g_e^2 is estimated as 1.1 at 77.3 K, we use this value as g^2. The irreversibility line in this virtual specimen is shown in Fig. 3. The result given in Fig. 1 is also shown for comparison. The irreversibility field is found to be drastically improved by the strong pinning centers. For example, the irreversibility field at 77.3 K is estimated as 27.5 T. The virtual critical current density J_{c0} at 77.3 K and 1.0 T is estimated as 4.8×10^9 A/m^2, and hence, this is found to be the case where the pinning strength is improved by a factor of 15 or so from the present champion data. Nevertheless, the obtained irreversibility field at 77.3 K is as large as only 5 % of the upper critical field. This value is also smaller than the value of melt-processed bulk Y-123. Such an inferior property in the irreversibility field in Bi-2223 comes from its two-dimensional characteristic.

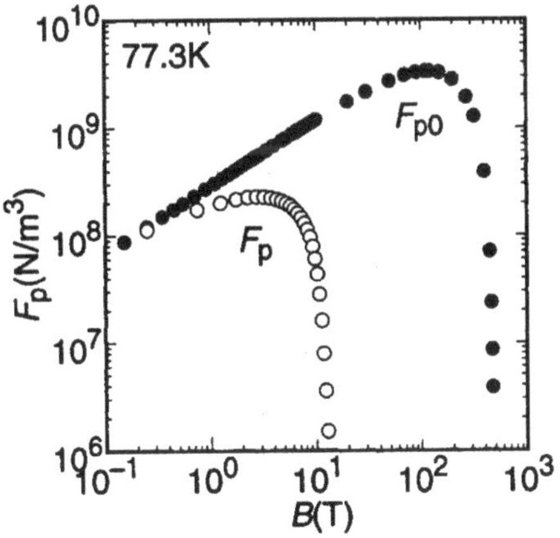

Fig. 2. Pinning force density (○) and virtual pinning force density in the creep-free case (●) for Bi-2223 tape wire at 77.3 K under the magnetic field normal to the c-axis.

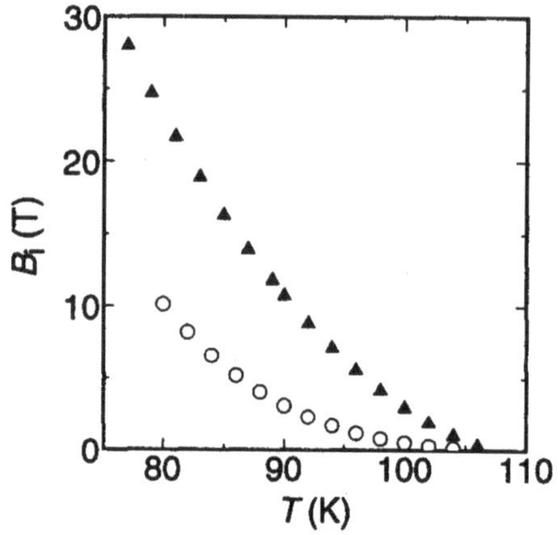

Fig. 3. Theoretically estimated irreversibility line of strongly pinned Bi-2223 tape wire under the magnetic field normal to the c-axis(▲). the circles show the result for a usual tape wire given in Fig. 1.

It is empirically known that the pinning force vs magnetic field curves meet on a single curve when the pinning force and the magnetic field are normalized by the maximum pinning force and the irreversibility field, respectively. The temperature dependence of the maximum pinning force density can be expressed in terms of the temperature dependent irreversibility field as $F_{p(max)} \propto B_i^{m'}$. Thus, the above scaling law is given by

$$F_p = A' B_i^{m'} b_i^{\gamma'} (1 - b_i)^{\delta'} . \tag{10}$$

Figure 4(a) and 4(b) show experimental results[8] on a Bi-2223 tape wire under the magnetic field normal to the c-axis. Such a scaling is confirmed also in this specimen. The scaling parameters of this specimen obtained from the scaling relation are m'=1,4, γ'=0.7 and δ'=2.5, while the parameters obtained at sufficiently low temperatures are $A = 2.10 \times 10^9$, m=3.96 and γ=0.73 and δ is expected to be 2.0. The numerical calculation of the pinning force density can also be done with the use of the above parameters. The theoretically obtained results of the scaling relation are shown in Fig. 5(a) and 5(b), where g^2 is estimated as 2.54. It is found that the agreement is qualitatively good, although the values of the calculated maximum pinning force density and irreversibility field are larger than the experiments. For example, the peak field is approximately $b_i = 0.2$ for the both cases. The parameters obtained from the numerical calculation are $m' = 1.2$, $\gamma' = 0.45$ and $\delta' = 1.80$. The slightly longer tail of the observed scaling curve in Fig. 4 (a) than in Fig 5 (a) is considered to be caused by a distribution of B_i due to the distributed direction of the c-axis. The theoretically obtained B_i is larger

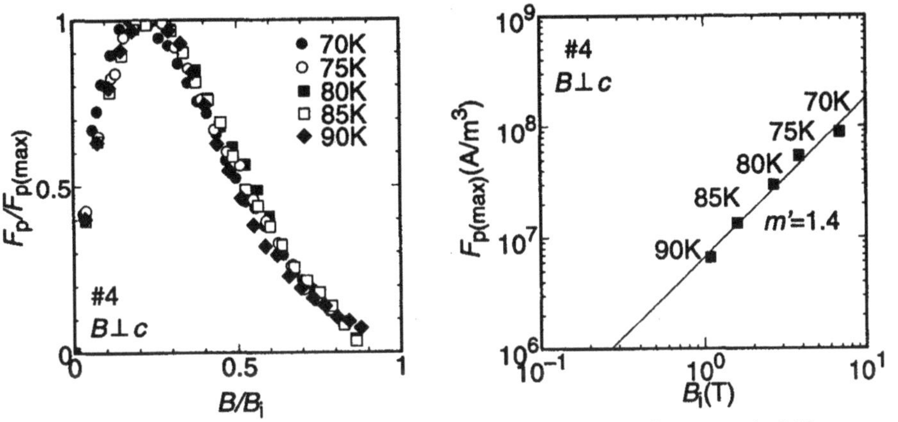

Fig. 4. Observed scaling relation in a Bi-2223 tape wire under the magnetic field normal to the c-axis: (a) normalized pinning force density vs normalized magnetic field at various temperatures and (b) relation between the maximum pinning force density and the irreversibility field.

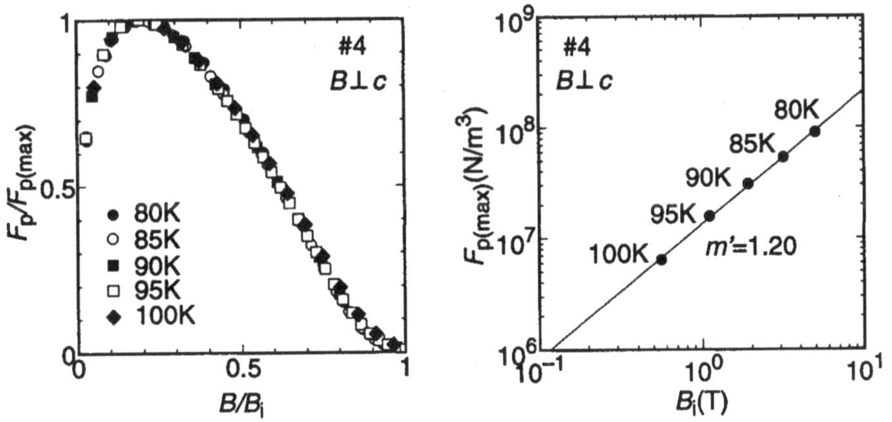

Fig. 5. Theoretically obtained scaling relation in a Bi-2223 tape wire under the magnetic field normal to the c-axis corresponding to the experimental results in Fig. 4(a) and 4(b).

by a factor of about 1.9 than the observed one at 80 K. This seems to be attributed to the ambiguity in the estimation of g^2. That is, a slight overestimation of g^2 may lead to the deviation. It can be expected, therefore, that the more precise estimation of the pinning characteristics will be accomplished by the more exact evaluation of g^2.

4. Summary

The irreversibility field and the pinning force density in Bi-2223 tape wires under the magnetic field normal to the *c*-axis are theoretically calculated with the aid of the flux creep model and experimentally obtained parameters. The obtained results agree well with the experimental results and this suggests that the flux creep model describes correctly the flux pinning phenomena under the thermal agitation on the fluxoids. The attainable irreversibility field in Bi-2223 superconductor by introduction of strong pinning centers is calculated. This value is found to be much larger than the present level. However, it is still smaller than that in common melt-processed Y-123 due to the two-dimensionality of this material.

5. References

1. P. W. Anderson and Y. B. Kim, *Rev. Mod. Phys.* **36** (1964) 39.
2. T. Matsushita, A. Matsuda and K. Yanagi, *Physica C* **213** (1993) 477.
3. K. Yamafuji, T. Fujiyoshi, K. Toko and T. Matsushita, *Physica C* **159** (1989) 743.
4. J. Bardeen and M. J. Stephen, *Phys. Rev.* **140** (1965) A1197.
5. T. Matsushita, T. Fujiyoshi, K. Toko and K. Yamafuji, *Appl. Phys. Lett.* **56** (1990) 2039.
6. T. Matsushita, *Physica C* **217** (1993) 461.
7. T. Matsushita and N. Ihara, *Proc. 1st Europ. Conf. Appl. Supercond.*, Göttingen, 1993, p. 779.
8. M. Kiuchi and T. Matsushita, unpublished data.

CRITICAL CURRENT DENSITIES AND MAGNETIZATION OF $Bi_2Sr_2Ca_1Cu_2O_8$ BASED WIRES

H. KRAUTH, J. GERSTER*), J. TENBRINK
Vacuumschmelze GmbH, Grüner Weg 37, D-63450 Hanau, Germany
*) also Universität Tübingen, Phys. Inst. Experimentalphysik II, D-72076 Tübingen

ABSTRACT

Monocore and multifilamentary Ag sheathed $Bi_2Sr_2Ca_1Cu_2O_8$ round wires were produced by the powder in tube technique. Optimization of the critical current density of these wires requires special temperature and atmosphere schedules for final heat treatment. Resistively and inductively measured critical current densities do agree ordisagree dependent on temperature and filament diameter, respectively. Creep measurements with test coils in persistent mode reveal a logarithmic time dependence of the transport current.

1. Introduction

Soon after the discovery of high-T_c superconductivity in oxides in 1986 worldwide efforts led to the development of oxides exhibiting superconductivity at temperatures above the boiling point of liquid nitrogen (77 K). The development of technical conductors in the form of round wires or tapes presently mainly concentrates on Bi-Sr-Ca-Cu-O (BSCCO) as with this system it was for the first time possible to achieve high critical currents in polycrystalline HTSC wire material[1,2] applying the so called "powder in tube" (PIT) technique. Since then conductor performance has greatly improved, especially with respect to current carrying capacity along longer lengths of conductor. Although the present worldwide efforts concentrate on the development of technical $(Bi,Pb)_2Sr_2Ca_2Cu_3O_{10+x}$ (Bi-2223) thin textured tapes, $Bi_2Sr_2Ca_1Cu_2O_{8+x}$ (Bi-2212) wires may be interesting at temperatures of 20 K or lower because the influence of T_c is negligible, production cost of round wires are much smaller and Bi-2212 shows no sign of weak link behaviour, whereas Bi-2223 exhibits increasing weak link limitations with decreasing temperature. Therefore Bi-2212 should be suited for use in very high field superconducting magnets at about 4 K and in laboratory size magnets operated at e.g. 20 K. This article describes the manufacturing process of Bi-2212 wires, their transport critical current density and magnetization properties as well as first measurements of field decay in test windings operated in persistent mode.

2. Wire Manufacturing Technology

2.1 Powder Preparation

The most commonly used technique for the preparation of powder is the so called "mixed oxide"-technique. Thereby appropriate amounts of oxides or carbonates of the metallic constituents are mixed and calcined several times in order to produce the desired BSCCO powder with low residual carbon content. After the final calcination treatment and an additional milling the average particle size of the Bi-2212 powder typically is about 1 to 5 μm depending on the raw materials used and on the intensity and duration of milling during the intermittent grinding steps.

2.2 Wire Drawing

Wires are produced applying the conventional "powder in tube" technique. As the most simple technique the powder is filled into a pure Ag or Ag-alloy (e.g. AgMg, AgNiMg) tube. Due to insufficient densification this technique yields rather poor filling densities of the order of 50 %. Higher filling densities of 70 to 75 % are obtained by filling the tube with cold isostatically pressed powder rods. Besides a much higher filling density another advantage of this technique is the fact that the filling is more uniform compared to the case of filling with a loose powder. The filled tubes are swaged and drawn to their final diameter of about 1 mm. Multifilamentary wires are produced by drawing the wire to a diameter of 2 to 4 mm, cutting the wire, and then bundling these wire pieces into a 2nd tube. This composite again is swaged and drawn to a final diameter of about 1 mm. Wire lengths of more than 400 m (at a diameter of ϕ 1.2 mm) have been fabricated, the length being limited only by the size of the tube used for manufacture. The process is easily scaleable to larger quantities.

2.3 Annealing Procedure

After wire drawing the HTSC wire core consists of a densely packed and compressed powder. In order to restore the superconducting properties and especially to gain connectivity between neighboured grains a final annealing has to be performed leading at least to a sintering of the material. Only very poor transport currents could be observed with sinter processing. A substantial breakthrough was achieved by introducing partial melting of the HTSC wire core as a means of processing[3,4] . With this technique it was possible for the first time to have large superconducting transport currents with HTSC wire material. The final annealing therefore consists of a short term partial melting at a temperature slightly beyond 900 °C in air followed by a long term annealing of the order of 100 h at e.g. 840 °C in order to form the superconducting Bi-2212 phase. Later it was shown that additionally a quick cooling has to be performed at temperatures below about 550 °C (in air) in order to prevent a decomposition of the Bi-2212 as Bi-2212 becomes unstable in this temperature range[5]. Rapid cooling at the end of the annealing procedure is accomplished by enforced cooling with streaming air. With air as the annealing atmosphere it is necessary to have temperatures of about 900 °C in order to reach the partial melting region. As the Bi-2212 melt is able to dissolve substantial amounts of the Ag matrix this sometimes leads to a melting through the Ag sheath at this high temperature. This effect can be overcome by performing the partial melting step under a reduced oxygen partial pressure. The temperature of partial melting thereby can be reduced by up to 50 K[6].

3. Measurement Techniques

The critical current I_c of Bi-2212/Ag short specimens usually is measured on short straight samples of 3.5 cm length, the distance between the voltage taps being about 1 cm. The critical current I_c is determined at a criterion of 1 μV/cm except where otherwise stated. By fitting a power law $U = I^n$ to the transition curve between $U = 0.1$ μV/cm and $U = 1$ μV/cm n-values are determined describing the transition behaviour to the normal state, the n-values being an indicator of flux creep effects and, at a given temmperature, of wire homogeneity. The transport critical current density j_{ct} is evaluated always with respect to the cross sectional area of the HTSC wire cores determined on micrographs of the as drawn wires.

Magnetization measurements are performed at temperatures between 4.2 K and 77 K using a commercial Oxford 12 T vibrating sample magnetometer (VSM). Specimens usually are cut into pieces of about 6 mm length and glued to the sample holder. The critical current density j_{cm} is evaluated from the magnetization data applying the Bean-model yielding $j_{cm} = 3\pi/4 \cdot \Delta M/Vd$ in case of magnetic field B aligned perpendicular to the wire axis and $j_{cm} = 3 \cdot \Delta M/Vd$ in case of B parallel to the wire axis, ΔM being the width of the hysteresis loop, V the HTSC volume, and d the filament diameter. Rate dependent magnetization is measured by changing the external field sweep rate between 2 mT/s and 20 mT/s.

Creep measurements j(t) at 4.2 K are performed on test windings of multifilamentary wires short circuited by a monolithic Bi-2212 contact with about the same critical current as the multifilamentary wire at the conductor ends. After charging of such a small coil the magnetic field decay is observed with a Hall probe. In order to avoid external disturbances the first measurements reported here were performed in zero external field[7].

4. Results

4.1 Optimization of the critical current density

The critical current density j_c of Bi-2212/Ag wire specimens at 77 K very much depends on the critical temperature T_c of the material. If specimens for example are simply furnace cooled in air at a rate of the order of 100 K/h j_c of monocore wires usually is of the order of 0.5 to 1 kA/cm^2, T_c (R=0) being about 85 K, whereas multifilamentary wires exhibit very poor superconductive behaviour with a critical current density below 0.1 kA/cm^2 due to excessive pick-up of oxygen. A higher critical temperature of just below 90 K is obtained by rapidly cooling the samples e.g. with streaming air. This procedure leads to an enhanced critical current density of the order of 2 kA/cm^2 (maximum 2.5 kA/cm^2). Even higher critical current densities can be obtained if the specimens are cooled under a reduced oxygen partial pressure. The maximum critical current density thus obtained was 5.1 kA/cm^2, the average value for multifilamentary conductors being about 3 kA/cm^2.

Quick cooling and/or the cooling under a reduced oxygen partial pressure also prevents a decomposition occuring due to a metastability of the $Bi_2Sr_2Ca_1Cu_2O_{8+x}$ phase below about 550 °C (in air)[5]. A low cooling rate of e.g. 100 K/h leads to a huge scatter of critical currents with poorer values and few extremely high values (especially at 4.2 K, see below) in case of measuring several short samples of a monocore wire of one and the same batch. Enforced cooling with streaming air or cooling at lower oxygen partial pressure obviously prevents this decomposition resulting in very much improved critical current homogeneity from sample to sample at 77 K as well as at 4.2 K.

At 4.2 K all samples exhibit the now well known flat dependence on magnetic field. It has to be noted that with the above mentioned and discussed quick cooling no improvement of the maximum values of j_c at 4.2 K can be observed but applying this enforced cooling with streaming air leads to a distinct improvement of critical current homogeneity at 4.2 K. This can be seen from Fig. 1 where the critical currents I_c at 4.2 K, 9 T of different monocore and multifilamentary wire specimens are shown as function of the cooling rate. It can be seen that lower cooling rates still allow for high critical currents at least in monocore wires but the scatter of I_c does not allow for high critical currents along longer wire lengths. With multifilamentary wires with their smaller filament diameter only cooling rates exceeding 500 K/h allow for sufficiently high and uniform critical currents. From this result it is obvious that high critical currents along longer wire lengths can only be achieved by avoiding the decomposition of Bi-2212 by e.g. a sufficiently high cooling rate. It has to be clarified in further investigations why some slowly cooled monocore short specimens yield distinctly higher critical current densities as compared with quickly cooled multifilamentary wire specimens.

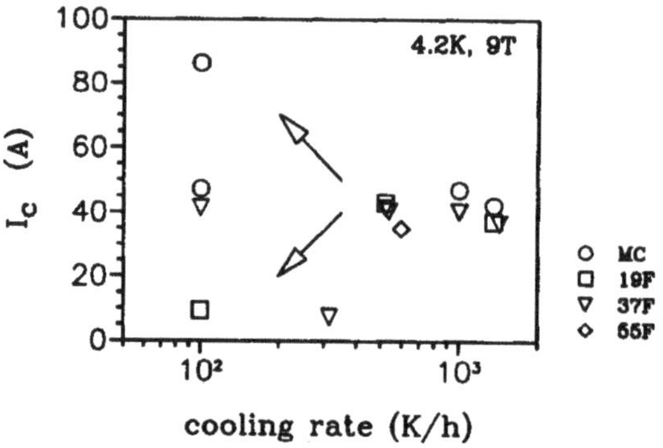

Fig. 1: Critical current I_c of monocore (MC), 19-filamentary (19F), 37-filamentary (37F), and 55-filamentary (55F) Bi-2212 Ag sheathed wire specimens (diameter 1.0 mm) as function of the cooling rate (from 650 to 400 °C).

4.2 Comparison of j_{ct} and j_{cm}

Comparisons between j_{ct} and j_{cm} are helpful in understanding the limitations of current transport in a polycrystalline material for several reasons. First of all they are effective at very different sensitivities i.e. electric field levels. And secondly they are depending differently on connectivity and texturing of grains and on local defects like voids.

The results of j_{cm} measurements between 4.2 K and 30 K are shown in Fig. 2 as a function of the ramp rate of the external field dB/dt. A linear dependence of j_c from lndB/dt can be seen[8] in agreement with the following formula derived from a flux creep model[9]:

$$j_{cm} = j_{co}[1 + \frac{kT}{U} \ln (\frac{2a}{vB} \cdot \frac{dB}{dt})] \quad \text{for} \quad \frac{2a}{vB} \cdot \frac{dB}{dt} < 1$$

with U being the pinning potential, v the flux jump velocity and 2a the thickness of the superconductor. The ratio of μ_0dM(2 mT/s) to μ_0dM(20 mT/s) is in general agreement with the resistively determined n values between 0.1 μV/cm and 1 μV/cm in the sense that similar values of "n" and its field and temperature dependence result from these ratios of magnetization.

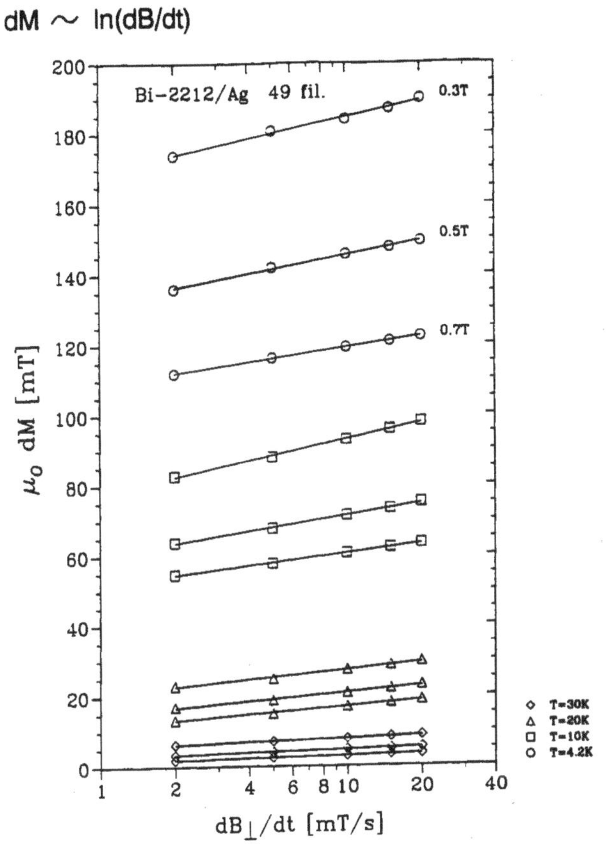

Fig. 2: Rate dependence of magnetization of a 49 filament Bi-2212 wire at different fields and temperatures.

Fig. 3 shows a comparison of j_{ct} and j_{cm} of monocore wires for different temperatures. It can be seen that at low temperatures the results from transport and magnetization are very close to each other, reflecting a difference of j_{cm} and j_{ct} in accordance with the n-value of 20 - 25 for these monocore wires at 4.2 K. With increasing temperature j_{ct} becomes considerably larger compared to j_{cm} because of strong flux creep effects in Bi-2212. At 77 K the n values measured by transport are about 5 and the magnetization curves are nearly reversible.

Monocore wires have an essentially untextured microstructure as indicated by metallographic investigations and by neutron diffraction. With multifilamentary wires an additional effect arises. j_{ct} and j_{cm} measurements performed at 4.2 K on wire specimens having different filament numbers are shown in Fig. 4 as a function of the total fila-

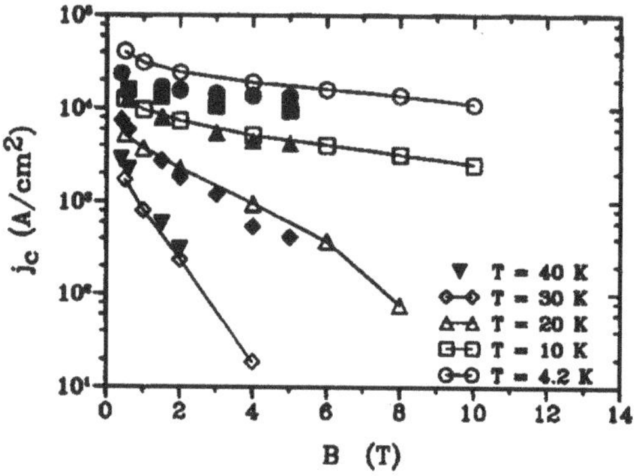

Fig. 3: Temperature dependence of the critical current density j_c of Bi-2212 Ag sheathed wires as determined from transport (j_{ct}, closed symbols) and magnetization (j_{cm}, open symbols) measurements of monocore Bi-2212 Ag sheathed wire specimens.

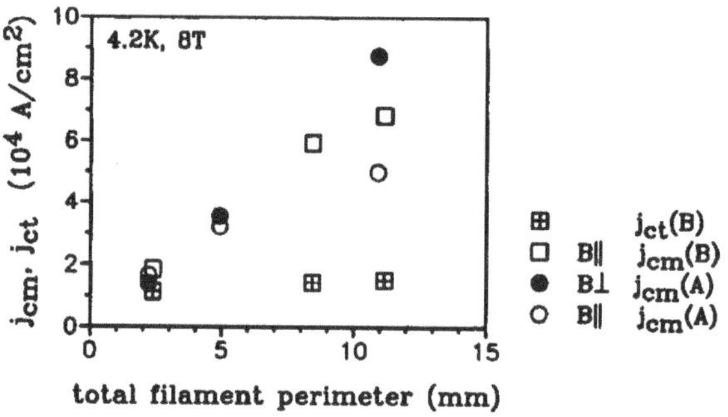

Fig. 4: Critical current density j_{cm} compared with j_{ct} for Bi-2212 wire specimens with different filament numbers as function of the total filament perimeter. B means an orientation of B parallel, and B an orientation of B perpendicular to the wire axis during magnetization measurements. A and B are two different Bi-2212 powder batches.

ment perimeter. For monocore wires the values of j_{ct} and j_{cm} are in agreement taking into account the experimental error. With increasing filament number and therefore with increasing total filament perimeter the transport critical current density increases only moderately whereas the magnetization critical current density increases significantly by e.g. a factor of about 7 in case of a 37-filamentary wire. The linear increase of j_{cm} in case of B aligned perpendicular to the wire axis led to the assumption that an increased amount of textured material associated with the increased size of the Ag/HTSC interface area is responsible for the observed increase of j_{cm} with filament number or total filament perimeter[10]. For magnetic field B aligned parallel to the wire axis the increase of j_{cm} is less than linear. As for this orientation of B the screening currents have to flow in a circle along the filament perimeter this situation becomes more and more unfavorable with decreasing filament diameter as the tilt angle between neighboured Bi-2212 grains will become larger.

The fact that j_{ct} is much lower than j_{cm} with multifilamentary wires may be related to filament inhomogeneities like e.g. pores and large alcaline earth cuprates which severly degrade j_{ct} whereas j_{cm} is nearly unaffected by such local inhomogeneities. If filament inhomogeneities can be avoided more effectively an increase of transport critical current density j_{ct} of up to a factor 7 seems possible to a value of about 100 kA/cm^2 starting from a present level of j_{ct} of 15 kA/cm^2 at 4.2 K, 9 T for a 37-filamentary Bi-2212 Ag sheathed wire.

4.3 Creep measurements $j_t(t)$ in persistent mode

A typical result of creep measurements in the very low j_t regime as determined from field decay measurements of small short circuited test coils at 4.2 K[7] is shown in Fig. 5. The decay can reasonably well be fitted to a logarithmic law

$$S = \Delta j_t / j_{to} \Delta \ln(t)$$

with S ≈ 0.13 for times smaller than about 5 min. At times larger than about 50 min the decay becomes zero within the accuracy of the measurement apparatus.

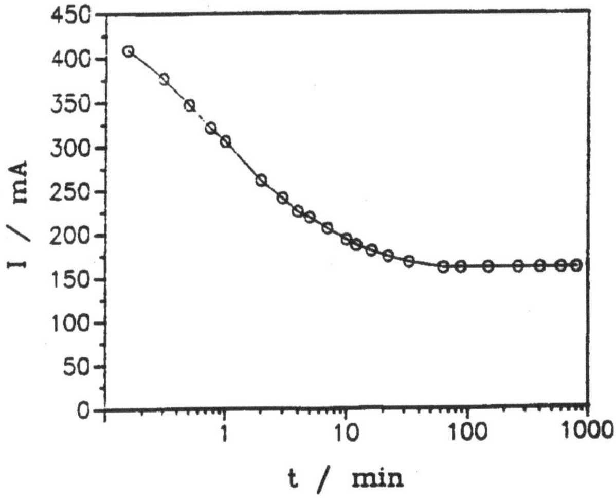

Fig. 5: Decay of transport current of a test coil wound from multifilamentary Bi-2212 wire in persistent mode.

The measured value of S ≈ 0.13 for Bi-2212 is much larger than the typical values of S ≈ 0.02 to 0.04 measured for $Y_1Ba_2Cu_3O_7$ and $Bi_2Sr_2Ca_2Cu_3O_{10}$[11]. More measurements, also in the high current density regime, will be performed in order to verify this high creep rates even at 4.2 K.

5. Conclusion

Optimized heat treatment of as drawn Bi-2212/Ag wires result in high critical current densities at low temperatures. The optimum heat treatment depends on the atmosphere and on the geometry of the wire. Especially multifilamentary wires require fast cooling below 650 °C in order to avoid too much oxygen pick-up and decomposition of Bi-2212. Due to the increase of textured regions close to the Bi-2212/Ag interface[12] multifilamentary wires have the potential of very high critical current densities. This could be demonstrated by j_{cm} measurements. Exploration of this potential with respect to j_{ct} requires improved microstructure e.g. by avoiding pores and second phase particles. Flux creep effects on j_t seem to be pronounced in Bi-2212 even at low temperature and at very low current densities. This fact needs further investigations including comparison with Bi-2223.

References

1. J. Tenbrink, K. Heine, H. Krauth, A. Szulczyk, M. Thöner, *VDI-Berichte* **733** (1989) 399
2. K. Heine, J. Tenbrink, M. Thöner, *Appl. Phys. Lett.* **55** (1989) 2441
3. J. Tenbrink, H. Krauth, *Adv. Cry. Eng.* **40A** (1994) 305
4. J. Tenbrink, K. Heine, M. Wilhelm, H. Krauth, *IEEE Trans. Magn.* **27** (1991) 1239
5. J. Shimoyama, J. Kase, T. Morimoto, H. Kitaguchi, H. Kumakura, K. Togano, H. Maeda, Jpn. *J. Appl. Phys.* **31** (1992) L1167
6. J. Tenbrink, H. Krauth, BMFT Status-Seminar Supraleitung, Weimar, Germany, June 13-14, 1994
7. J. Gerster, Diploma Thesis, Univ. Tübingen, 1995, to be publ.
8. K. Heine, J. Tenbink, H. Krauth, *Cryogenics* **32 S** (1992) 504
9. R. Griessen, *Physica C* **172** (1991) 441
10. J. Tenbrink and H. Krauth, to appear in Proc. 7th Int. Workshop on Crit. Curr. IWCC'7, Alpbach, Austria, January 24-27, 1994
11. M. Mittag, R. Job, M. Rosenberg, *Physica C* **174** (1991) 101
12. R. Wesche, A.M. Fuchs, B. Jakob, G. Pasztor, *Cryogenics* **34** (1994) 805

CRITICAL CURRENTS AND CRITICAL STATE IN BSCCO WIRES

A. P. MALOZEMOFF
American Superconductor Corporation
2 Technology Dr., Westborough MA 01581 USA

ABSTRACT

Rapid progress has been made in the fabrication of long-length wires of flexible composite multifilamentary high-temperature bismuth-based superconductors, and a variety of prototypes have already been demonstrated. Wires over 1 km long carrying 12,700 A/cm^2 at 77 K have been fabricated, and conduction-cooled coils have now reached a performance level of 2.1 T at 27 K at a 1 μV/cm criterion. An area of special opportunity for high temperature superconductors involves ac applications. Significant progress in understanding ac losses has been made recently, and twisted wires have been fabricated which will be important for reducing eddy coupling losses in multifilamentary wires.

1. Introduction

High temperature superconducting (HTS) wires made of bismuth-based materials continue to show rapid progress towards commercialization. This progress is based on progress in improving the critical current density. While progress at other organizations around the world is significant, I review here specifically the progress at American Superconductor Corporation (ASC). As critical current density improves, attention can begin to shift to some subsidiary but ultimately very important additional properties, such as ac loss. Progress in this area is also reviewed here.

The most promising method for fabricating HTS superconducting wires[1-6] involves composite structures of silver or a silver-alloy with a bismuth-strontium-calcium-copper-oxide (BSCCO) high-temperature superconductor. For applications in the temperature range above about 20 K, BSCCO-2223 is usually preferred and is the main vehicle for wire manufacturing at ASC with a powder-in-tube technique. Two main variants use either oxide or metallic precursor powders. Both are practiced at ASC, and both are at the level of enabling construction of significant magnets and multistrand conductors. The oxide technique presently shows the highest current densities while the metallic precursor technique has the best mechanical properties. Usually the wire is deformed by drawing and then rolling into a tape shape, which is favorable for maximizing the HTS platelet texture and hence the current density, but round BSCCO-2223 wire is also possible, as will be described below.

It is important to recognize at the outset that multifilamentary composite structures have so far proven to be essential for achieving the mechanical properties required for practical wires.[1,6] Monofilamentary wires, usually with HTS thickness of 50 μm or more, are susceptible to cracking under the mechanical stresses of handling, thermal stresses of cooling and warming, and Lorentz forces of magnet operation. Multifilamentary structures appear to suppress crack formation. Therefore work at American Superconductor has concentrated on such structures. Even with 19 filaments, critical strains for the onset of irreversible loss of current density can be as high as 0.5%, and with filament counts above 1000, attained for example with a metallic precursor technology,[2] they can reach 1%. As reported recently by Minot et al.,[6] these multifilamentary wires also show less than a few percent decrease in critical current over 10 cycles from 300 to 77 K, and after 7 days in a 95% humidity and 70 C environment.

2. Critical Current Density Progress

A basic point about the critical current density is that the transition to normal behavior is not infinitely sharp. In the low-voltage range, the voltage usually goes as current to a power n, which is usually in a range of 10-25 for the good-quality BSCCO superconducting composite wires in the operating ranges of interest. As pointed out earlier,[3] there are two mechanisms for this behavior. One stems from local inhomogeneity in critical current density, and the other from flux creep; both appear to play a role in BSCCO wires. This has made it necessary to define a voltage criterion whenever one discusses a current density. Typically 1 μV/cm or 10^{-11} Ωcm criteria have been used, with values at 1 μV/cm higher by about 20% compared to values at 10^{-11} Ωcm.

Another basic point about the critical current density is to distinguish two types: One is the filament current density J_c, for which the total wire current is normalized by the filament cross-sectional area; this is of value to the materials scientist for understanding the performance of the superconducting material itself. The other is the engineering current density J_e, for which the total wire current is normalized by the total composite wire cross-section, including the silver or silver-alloy sheath. The fill-factor, or ratio of superconducting filament cross-section to the total wire cross-section, is typically in the range of 20-30% for the multifilamentary wires.

The current status ASC championship 77 K critical current densities[4] at the various voltage criteria is shown in Table I. The wire is either 19 or 85-filament (the 10 and 70 m wires are 85). While there is an obvious fall-off in performance with longer length in Table I, it is expected that the long-length performance will continue to improve since the long lengths reflect a less optimized processing path than the shorter lengths at this point in time. There is also the expectation that these multifilamentary wire critical current numbers will continue to improve rapidly toward the levels achieved in short monofilament pressed samples. The highest value at 77 K for a full wire cross-section, achieved by Li et al.,[5] is 69,000 A/cm2 at a 1 μV/cm criterion. Levels of 76,000 and most recently over 110,000 have been reported for portions of the tape, either sliced from the center of a monofilament[7] or measured from a surface layer adjacent to the silver.[8] These results are very encouraging for further progress in this technology. Furthermore, the engineering critical current density above 9000 A/cm^2 in Table I, although still in a short length sample, is significant because it approaches the level required for the onset of commercial applications.

Table I. Filament and engineering critical current densities of tape-shaped 19-filament silver/BSCCO-2223 composite wire at 77 K and self field.

Length, m	J_C, A/cm^2		J_e, A/cm^2	
	1μV/cm	10^{-11} Ω-cm	1μV/cm	10^{-11} Ω-cm
0.01	32,600	---	9,100	---
10	22,600	19,100	6,300	5,300
70	22,000	17,800	6,100	5,000
400	17,000	13,500	4,100	3,200
1,100	12,700	8,800	3,000	2,100

Recently, a magneto-optic technique[9,10] has been developed which permits probing the current paths in multifilamentary BSCCO wire. A superimposed yttrium-iron-garnet film gives a magneto-optic signal determined by the local field, which is in turn generated by the local current paths in the underlying sample. The pictures, with close to 1 μm resolution, show evidence for nonuniformity and some tendency for current to concentrate near the silver interfaces. Microstructures also show an increased secondary (non-2223) phase content in the center of the filaments. One path to improved current density will therefore be to eliminate these secondary phases. Another will be to decrease filament size so that the interfacial regions will dominate the filament. The challenge in this case is to do this without degrading filament uniformity,[1] which has so far limited progress.

Important also is the ability to manufacture significant quantities of this wire with consistent properties. This has been demonstrated at ASC with standard deviations of order 15% in the critical current.[1] This, coupled with the good performance characteristics, has enabled significant new prototypes.

3. Applications prototypes

A significant prototype[11] has been announced by Reliance Electric Company and ASC, a 5 hp synchronous motor with a rotor cooled to 77 K containing four salient poles operating at 1800 rpm. Four racetrack coils, each containing about 170 m of conductor and wound with a react-and-wind method, drive the iron poles.

A 1 meter-long multistrand conductor has been developed by ASC and Pirelli Cavi SpA of Milan, Italy as a prototype for one phase of a three-phase power transmission cable.[12] It consists of 140 individual helically-wound, react-and-wind tapes, in six layers around a cylindrical phenolic (G-10) former and carries 4200 A of direct current at a 1 μV/cm criterion. Its ac performance will described below.

A solenoidal magnet has been developed by ASC and Oak Ridge National Laboratory for quenching studies.[13] The magnet consists of 575 m of metallic precursor HTS BSCCO-2223 wire wound with a react-and-wind method. It has an inner winding diameter of 2.5 cm and an outer diameter of 8.7 cm, and it generates 1.1 T at 4.2 K at 1 μV/cm across the entire magnet including all joints. Measurements show that the stability margins of this coil are more than two orders of magnitude larger than would be expected for a corresponding coil made out of low-temperature superconductors. This stability is expected to come in part from the increased critical temperature and correspondingly increased heat capacity of the composite wire and in part from the moderate n-value referred to earlier.

Another solenoidal conduction-cooled magnet system has been developed by ASC with technology development supported by the Advanced Technology Program of the Department of Commerce.[14] The magnet consists of approximately 1000 m of conductor in 7 double pancakes wound with a wind-and-react method. It has an inner winding diameter of 2.85 cm and generates 2.16 T with 183,000 amp-turns at 27 K at 1 μV/cm across the entire magnet including all joints. The magnet is conduction-cooled in a vacuum chamber with a two-stage Gifford-McMahon mechanical cryocooler, and the system has a warm bore of 1.75 cm diameter. In liquid nitrogen, the magnet generates 0.43 T with the same voltage criterion. This significant new magnet demonstrates the ability of HTS wire technology to enable magnets generating field levels above the saturation level of iron, and without liquid cryogen.

A variety of other magnets delivered recently by ASC address future applications in Maglev, an aerospace generator and a homopolar motor for electric boat propulsion. These will be described elsewhere.

4. Ac applications

In an ac environment, superconductors generate losses. This has been a severe limitation on low temperature superconductors, because the 4.2 K environment is intolerant of power losses. HTS materials have an advantage in this regard because at higher temperatures, their heat capacity and the ability of cryogenic systems to handle the heat load is significantly increased. This creates opportunities for ac superconducting applications, such as inductors in a power electronics system,[15] transformers and power transmission.

Nevertheless, to take optimal advantage of these opportunities, ac losses in HTS materials will have to be minimized. Significant progress has been made recently in understanding these losses in wires and making the necessary modifications to reduce the losses.

The classical theory of ac-losses in multifilamentary composite superconductors treats three main terms: the hysteretic loss coming from vortex motion inside the Type II superconductor, eddy current loss in the metal sheath and eddy current coupling loss generated by currents circulating between the superconducting filaments across the normal metal matrix. In the simplest case, for a slab of thickness 2a, the hysteretic power loss P per unit volume at a frequency f in mks is[16]

$$P = \eta(fB_m{}^2/2\mu_0)\beta/3\,, \qquad \beta < 1\,, \tag{1}$$

$$P = \eta(fB_m{}^2/2\mu_0)\{(1/\beta)-(2/\beta^2)\}\,, \quad \beta > 1\,, \qquad \beta = B_m/2\mu_0 J_c a\,, \tag{2}$$

where η is the superconducting fill factor, J_c is the filament critical current density (assumed in the ideal model to be independent of field) and B_m is the peak to peak applied magnetic field. Crudely, the parameter a corresponds to the filament width. A series of other models for other shapes, developed by Wilson and others, show the same trends, with only modest numerical factors separating the different results. Therefore we use the simple slab model to get an initial estimate of the relevant parameters. If the current density varies with field, one can use $J_c = J_{c,ave}$ averaged over the field cycle for a first approximation.

The loss consists essentially of a prefactor $fB_m{}^2/2\mu_0$ times a "loss-factor" containing β; the loss-factor rises to a maximum of 1/3 and then falls. In the low field regime, $\beta < 1$, there is only partial flux penetration; this is the regime of relevance to power transmission applications. The regime $\beta > 1$ is relevant to most magnet applications, where full penetration is required to reach full current carrying capacity. The maximum loss, corresponding to the prefactor discussed above, occurs at the point where the flux front reaches the middle of the filament during each cycle.

In the high field regime, the loss per volume can be written

$$P_{hyst} = \eta f B_m J_{c,ave} a\,. \tag{3}$$

Thus, to cut the losses to an acceptable level, it is necessary to have a small filament size 2a. To roughly estimate the hysteretic loss level, we first estimate the prefactor $fB_m{}^2/2\mu_0$ which, even for the modest field of 0.1 T rms (0.28 peak-to-peak = B_m) gives 2000 kW/m^3, a very substantial loss level, roughly 10 times what one might expect for copper in typical power applications. However, with a fill factor of 1/3, and $J_{c,ave}$ = 10000 A/cm^2, one can estimate that a filament size of 89 μm would be required to reduce the loss level to 25 kW/m^3 at this field level. Such filament sizes are well within reach for HTS multifilament wire technology.

Transport measurements of the losses in such tapes by Gannon et al.[12] have shown losses increasing roughly as the cube of the ac current and linearly with frequency, in agreement with the hysteretic loss model. More recent measurements by Kwasnitza and Clerc, using ac hysteresis loops, have also confirmed the presence of hysteretic losses in ASC 19-filament tapes, both in the low and high field regimes.[17] By contrast, measurements[12] on the ASC 4200 A multistrand conductor show a very different loss behavior, with a quadratic increase with current and frequency, suggesting an eddy current loss mechanism. The loss level, however, is consistent with targets for a first generation power transmission line.

As long as the sheath thickness is small compared to the superconductor/metal composite dimensions, one can neglect the conventional eddy current loss in comparison to the eddy coupling loss which comes from currents traveling in part through the superconducting filaments. The coupling loss per cycle per volume in a sinusoidally varying field of peak-to-peak amplitude B_m has been calculated[16] for a multifilamentary round wire with twisted filaments

$$Q_e = (B_m^2/2\mu_0)\ \pi\omega\tau/(\omega^2\tau^2+1)\,, \tag{4}$$

where $\omega=2\pi f$ is the angular frequency and τ is a characteristic time which can be written in terms of a characteristic twist pitch L_c for a given frequency f:

$$\tau_c = \omega^{-1}(L/L_c)^2\,, \tag{5}$$

$$L_c = (4\pi\rho_{et}/\mu_0 f)^{1/2}\,. \tag{6}$$

Eq. 6 has just the same form as the conventional eddy current penetration depth formula, but with ρ_{et} the effective transverse resistivity.[16] To carry full current, the conductor must have a twist pitch L less than L_c. Then Eq. (4) can be simplified to:

$$P_{coup} = \pi(fB_m^2/2\mu_0)(L/L_c)^2 = (fB_mL)^2/8\rho_{et}\,. \tag{7}$$

Once again the characteristic prefactor $fB_m^2/2\mu_0$ appears. To reduce the power loss to an acceptable level, it is necessary to reduce L and increase ρ_{et}. For pure silver, taking into account a 1/3 fill factor to calculate ρ_{et},[16] one can estimate L_c for pure silver at 60 Hz to be 1.6 cm at 77 K and 0.7 cm at 20 K. Thus, to obtain a significant reduction of the coupling loss, it will be necessary to twist to pitches of order only a few millimeters, although the requirement can be somewhat relaxed if the transverse resistivity of the sheath can be increased.

Kwasnitza and Clerc[17] have verified the presence of coupling losses following the classical formulae by varying frequency and comparing different length samples of multifilamentary ASC tape. They show losses increasing as the square of frequency and of sample length, which corresponds approximately to half a twist pitch. At $\omega=\tau_c^{-1}$ which is about 0.8 Hz for their 2 cm sample and 1.2 Hz for their 1.4 cm sample at 4.2 K, they show the peak in loss per cycle predicted by Eq. 4.

The existence of these very substantial losses creates a challenge for the BSCCO-2223 multifilamentary technology: Is it possible to twist the internal filaments of such tape shaped conductors to the required level? Recently Christopherson and Riley[18] at ASC have indeed reported 85-filament tape conductors with internal twist pitches down to 3.7 mm. An analogous set of conductors, untwisted and twisted, showed a filament critical current density at 77 K which only dropped from 18,500 to 13,500 A/cm^2, and a field dependence which was identical in spite of the fact that a component of the current in the twisted conductor must travel along the perpendicular to the tape plane. This is very encouraging for ac applications. They also fabricated an initial round conductor of 85-filament BSCCO-2223 wire with a 4.6 mm twist pitch and J_c of 6,900 A/cm^2.

5. Conclusion

BSCCO-2223 composite wire technology has proven to have remarkable flexibility in meeting the requirements for practical superconducting wire. In multifilament form it is robust and demonstrates the required mechanical properties. The critical current density is improving rapidly to levels which permit significant prototypes. The powder-in-tube process has been developed to a level of manufacturing which makes it possible to sustain a wide range of prototype development. And most recently, additional properties such as ac-losses are being successfully addressed. The pace of progress is expected to continue.

The author thanks his many colleagues at American Superconductor referenced in the papers below. He also thanks Prof. Kwasnitza for permission to refer to his recent work on ASC samples.

6. References

1. G. N. Riley Jr., J. J. Gannon Jr., P. K. Miles and D. R. Parker, *Applied Superconductivity* **2** (1994) 155.
2. A. Otto, L. J. Masur, J. J. Gannon Jr., E. Podtburg, D. Daly, G. J. Yurek and A. P. Malozemoff, *IEEE Transactions on Applied Superconductivity* **3** (1993) 915; A. Otto, L. J. Masur, C. Craven, D. Daly, E. R. Podtburg and J. Schreiber, submitted to Applied Superconductivity Conference, Boston, Oct. 17-21, 1994.
3. A. P. Malozemoff, W. L. Carter, J. J. Gannon Jr., C. H. Joshi, P. K. Miles, M. J. Minot, D. R. Parker, G. N. Riley Jr., E. Thompson and G. J. Yurek, *Cryogenics, ICMC Supplement* **32** (1992) 478.
4. W. L. Carter, G. N. Riley Jr., A. Otto, D. R. Parker, C. J. Christopherson, L. J. Masur and D. Buczek, submitted to Applied Superconductivity Conference, Boston, Oct. 17-21 (1994).
5. Q. Li, K. Brodersen, H. A. Hjuler and T. Freltoft, *Physica C* **217** (1993) 360.
6. M. J. Minot, D. Buczek, J. J. Gannon Jr., P. K. Miles, D. R. Parker and P. Metra, submitted to Applied Superconductivity Conference, Boston, Oct. 17-21. 1994.
7. D. C. Larbalestier, X. Y. Cai, Y. Feng, H. Edelman, A. Umezawa, G. N. Riley Jr. and W. L. Carter, *Physica C* **221** (1993) 299.
8. M. Lelovic, P. Krishnaraj, N. G. Eror and U. Balachandran, preprint.
9. G. N. Riley, D. R. Parker, C. J. Christopherson, P. K. Miles, J. J. Pickett, S. E. Hughson, J. D. Schreiber, A. Polyanskii, A. Pashitskii and D. C. Larbalestier, *Physica C* (1994), to be published.
10. U. Welp, D. O. Gunter, G. W. Crabtree, J. S. Luo, V. A. Maroni, W. L. Carter, V. K. Vlasko-Vlasov and V. I. Nikitenko, submitted to *Appl. Phys. Lett.*

11. C. H. Joshi, C. B. Prum, R. F. Schiferl and D. I. Driscoll, submitted to Applied Superconductivity Conference, Boston, Oct. 17-21 (1994).
12. J. J. Gannon Jr., M. J. Minot, D. Buczek, G. Vellego and P. Metra, submitted to Applied Superconductivity Conference, Boston, Oct. 17-21, 1994; J. J. Gannon, Jr., A. P. Malozemoff, M. J. Minot, F. Barenghi, P. Metra, G. Vellego, Orehotsky and Suenaga, *Advances in Cryogenic Engineering* **40** (1994) 45..
13. J. W. Lue, L. Dresner, S. W. Schwenterly, D. Aized, J. M. Campbell and R. E. Schwall, submitted to Applied Superconductivity Conference, Boston, Oct. 17-21, 1994.
14. M. Manlief, B. Bent, M. Navarro and R. E. Schwall, private communication; ASC press release Oct. 13, 1994.
15. E. Schempp and C. Russo, *IEEE Transactions on Applied Superconductivity* **3** (1993) 563.
16. M. N. Wilson, *Superconducting Magnets* (Clarendon Press, Oxford, England, 1983).
17. K. Kwasnitza and S. Clerc, *Physica C*, to be published.
18. C. J. Christopherson and G. N. Riley Jr., submitted to *Appl. Phys. Lett.*

DEVELOPMENT OF Ag-SHEATHED Bi-BASED SUPERCONDUCTING WIRES AND APPLICATION

T.Hikata, N.Saga, S.Kobayashi, K.Muranaka, J.Fujikami, M.Ueyama, T.Kato, T.Kaneko, H.Mukai, K.Ohkura, N.Shibuta, K.Hayashi and K.Sato
Osaka Research Laboratories, Sumitomo Electric Industries, Ltd.
1-1-3 Shimaya, Konohana-ku, Osaka, 554 Japan

ABSTRACT

It is required for actual application to fabricate homogeneous long length wires having high Jc. The Jc of 61-filamentary wire reached 40,000 A/cm^2 at 77.3 K by the rolling process. We have developed 1 km class flexible superconducting wires having a length of 623 m (Jc=15,900 A/cm^2) and 1,080 m (Jc=4,020 A/cm^2). Various prototypes, such as 10 kA busbar (with JAERI), 5 m/ 3 kA flexible cable conductor (with TEPCO), 2.5 kA current leads (with KEPCO) and 2.1 T refrigerator-cooled magnets were demonstrated. The refrigerator cooled magnet operated stably generating 1.9 T at 18 K for over 50 hours.

INTRODUCTION

Since the discovery of Bi-2223 oxide superconductor[1)], much effort have been exerted to increase critical current density (Jc) and fabricate flexible long length wires[2,3)]. Jc over 50,000 A/cm^2 at 77.3 K and 10^5 A/cm^2 at 4.2 K have already been achieved through the homogeneous 2223 phase, alignment of c-axis and strong contact at grain boundaries[4)]. The flexibility has been improved by multifilamentary techniques. In this paper we present the development of long length wires having high-Jc and various prototypes for application.

EXPERIMENTS

Bi-2223 silver sheathed wires were fabricated through the powder-in-tube process. Bi_2O_3, PbO, $SrCO_3$, $CaCO_3$ and CuO powders with 4 to 5N purity were mixed into the composition ratio of the 2223 phase, sintered and ground[4)]. The ground powder was put into silver tubes, and drawn into round wires. The monofilamentary round wires were stacked and put into silver tubes, then drawn again into a multifilamentary round wire. The multifilamentary wires were rolled to tape wires with typical dimensions of 0.2–0.3 mm^t and 2.5–3.0 mm^w, and thereafter sintered. The rolling and sintering process was repeated again. Ic was measured by 4-probe dc method with a criterion of 1 μV/cm for short samples and 10^{-13} Ω·m for long wires[5)] and prototypes.

RESULTS AND DISCUSSIONS

DEVELOPMENT OF LONG LENGTH WIRES

For fabricating long length wires, the rolling process is indispensable to homogeneous deformation of processing tape shapes. We have developed the rolled samples of 61-multifilamentary wire having Jc=40,000 A/cm^2 at 77.3 K and 0 T. Figure 1 shows the Jc-B properties at 77.3 K for the 61-multifilamentary and monofilamentary wires with Jc(77.3 K,0 T) = 3.5x10^4 A/cm^2 by rolling process. There was no difference in the Jc-B properties between the mono- and 61-multifilamentary wires having the same Jc.

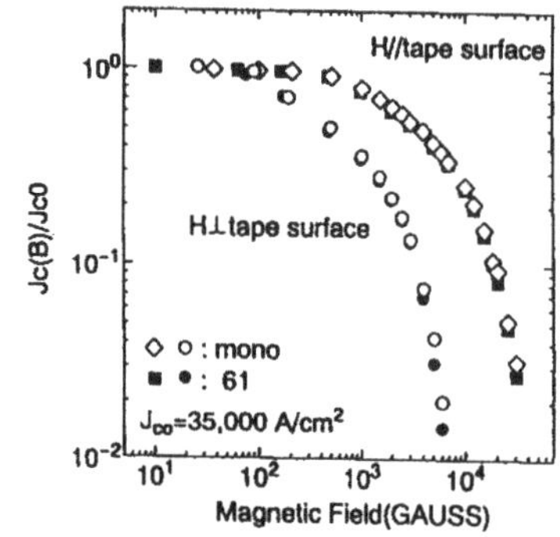

Figure 1 Jc-B properties of mono- (open marks) and 61-multi (solid marks) filamentary wires.

Figures 2 shows the morphologies of cross sections in the 61-multifilamentary wire with Jc = 3.5x10^4 A/cm^2 at 77 K and 0 T. Figure 2(1) and 2(2) show a photograph through a scanning electron microscope (SEM) and the contrasted non-superconducting phases with the high-Tc superconducting phases, respectively. It was found that the Bi-2223 oxide superconductor assembled densely as shown in Fig.2(1) and shape of non-superconducting phases such as Ca_2CuO_4 and $(Ca,Sr)_2PbO_4$ became very fine and thin with sub-μm as shown in Fig.2(2). The 61-multifilamentary wire has the filament size below about 10μm thickness. Therefore it is important for enhancement of Jc to decrease the thickness of non-superconducting phases into sub-μm level because disorder of aligned Bi-2223 superconducting grains is decreased.

Figure 3 shows the bending strain characteristics and the appearance of a cross section of a 61-multifilamentary wire. The measurement of Jc at 77.3 K was carried out by bending the wires on various-sized mandrels at room temperature. Multifilamentary wires have the advantage with higher flexibility with the increase of filament number. 90% of Jc was kept against 1% bending strain for a 61-multi filamentary wire. Figure 4 shows the cyclic bending characteristics

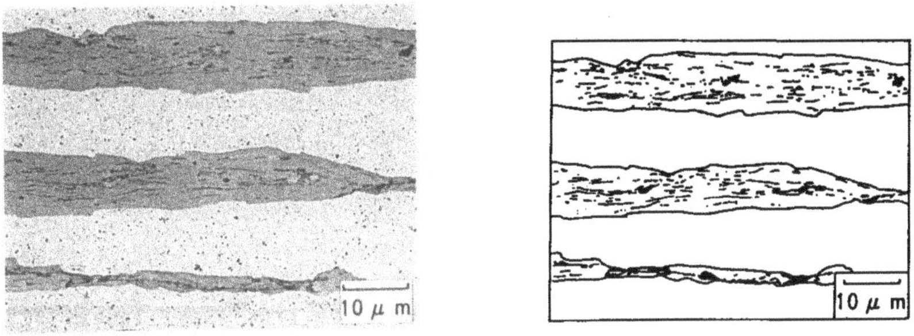

(1) (2)

Figure 2 Morphologies of a cross section of a Ag-sheathed Bi-2223 61-multifilamentary superconducting wire.
(1) SEM photograph.
(2) The contrasted non-superconducting phases with the high-Tc superconducting phases.

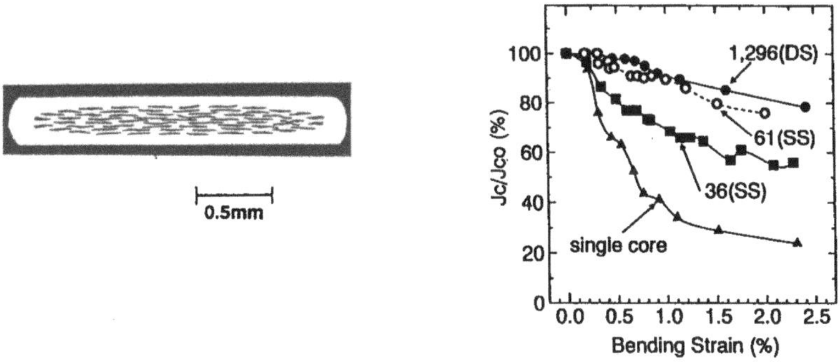

Figure 3 Bending strain characteristics of Bi-2223 Ag-sheathed tapes and an appearance of the cross section of a 61-multifilamentary wire.

undergoing four times bending strains including heat cycles of every bending procedure for the 61-multi filamentary wires. After bending cycles of 150 times for 0.18 % (120 mmϕ), 90 % of initial Jc was retained. These data indicate that the wires with filaments more than 61 have sufficient flexibility. Therefore the wires can be treated through react & wind techniques.

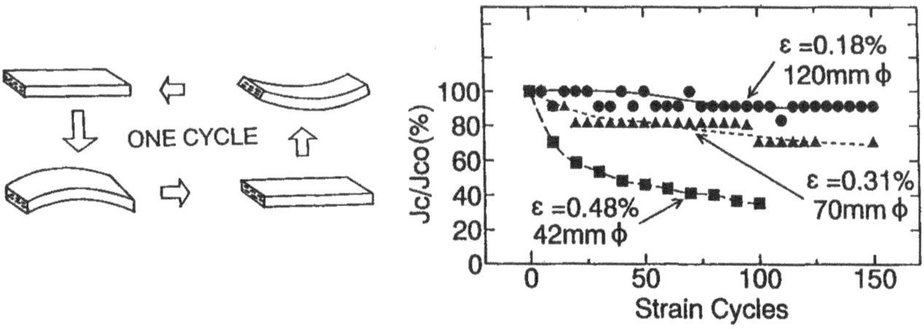

Figure 4 Cyclic bending characteristics of 61-multifilamentary wires.

For fabricating prototypes, it is important to develop the long length wires with high Jc. Table 1 shows typical properties of the long length wires from 114 m to 1,080 m. Jc=27,800 A/cm^2 was obtained at 77.3 K and 0 T in 114 m length wire. (Ic x L) is a important factor for development of actual application to indicate the total capability. The maximum (Ic x L) was obtained as 1.64x10^4 A•m in the 623 m length wire having 15,900 A/cm^2. Figure 5 shows the over view of a 1,000m-class wire.

Table 1 properties of long length wires.

	Jc [77.3 K]	Ic
	A/cm²	A
114 m	27,800	28.4
623 m	15,900	26.4
1,080 m	4,020	2.4

Figure 5 Over view of a 1,000 m-class wire.

Figure 6 shows the Ic distribution of a 623 m length wire at 77.3 K. Jc was measured for full length and for 50 m length distribution. The distribution was homogeneous along the wire. Jc exceeded 20,000 A/cm^2 at part along the entire 623 m.

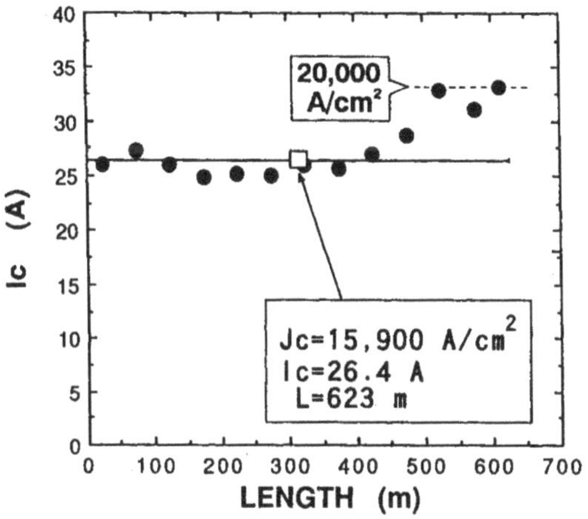

Figure 6
Jc distribution
of 623 m length wire.

APPLICATIONS

Various prototypes of power applications have been developed by using long length wires. Table 2 shows examples of the prototypes such as 2.5 kA current leads (with KEPCO)[6], 10 kA busbar (with JAERI)[7], 5 m/3 kA flexible cable conductor (with TEPCO)[8] and a refrigerator-cooled 2.1 T magnet[9,10]. It is important for the development of actual applications to obtain the homogeneous properties, high Jc and high Ic in long length wire.

Table 2 Various developed prototypes and features.

APPLICATION	CURRENT STATUS	FEATURES
CURRENT LEADS	500A (0.28w/kA) 2,500A (0.41w/kA)	REDUCE He CONSUMPTION (Cu: 1.0~1.2w/kA)
BUSBAR	10,000A/1m (GO&RETURN) RIGID	VERY LARGE CURRENT REDUCE SELF FIELD
CABLE CONDUCTOR	3,000A/5m FLEXIBLE MULTILAYER/SPIRAL WIND	INCREASE CAPACITY COMPACT (13,000A/cm²)
LN_2 MAGNET (64~77K)	0.56T/PANCAKE	EASY OPERATION MULTIFILAMENTARY WIRE
15~20K MAGNET	2.1T/40mm-BORE PANCAKE 0.35T/PANCAKE (B_{EX}:20T)	NO CRYOGEN/SWITCH ON HIGH Jc-B
LHe MAGNET (4.2K)	2.2T/PANCAKE 0.37T/PANCAKE (B_{EX}:23T)	SUPER-HIGH FIELD ANTI-HIGH STRESS

REFRIGERATOR-COOLED MAGNET

Bi-2223 high-Tc superconducting magnets can be applied in a wide temperature range from 4.2 K to 77.3 K. At 4.2 K, a super-high magnetic field can be generated through the excellent properties of Bi-2223 tapes as shown in Fig.7. At 77.3 K, it is operated in a relative low field below 1 T. While, the refrigerator cooled Ag-sheathed high-Tc superconducting magnet has excellent features compared to other superconducting magnets as follows;

(1) The magnetic field dependence of Jc at 20 K is almost the same or more excellent compared with metallic superconducting wires such as NbTi and Nb_3Sn as shown in Fig.7.

(2) The magnet can be operated without coolant such as liquid helium.

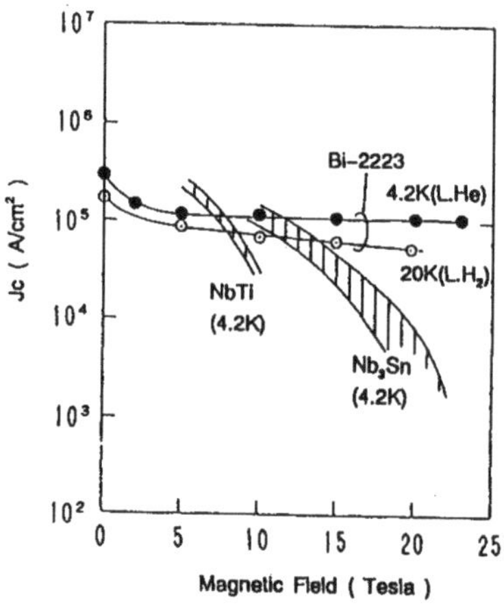

Figure 7 Jc vs. B properties for NbTi, Nb_3Sn and Bi-2223 superconducting wires.

(3) The cooling cost at 20 K is low because the cooling efficiency at 20 K is 5 times more than that at 4.2 K by a refrigerator.

(4) Structures of the cryostat is simpler than that of 4.2 K.

(5) The magnet becomes more stable against thermal disturbance because a thermal conductivity of silver at 20 K is large and the specific heat becomes about 100 times larger than that of NbTi at 4.2 K.

Figure 8 shows an appearance of a high-Tc superconducting magnet. This magnet was set into a Gifford-MacMahon (GM) refrigerator cooling system which has two stages with the first and the second stages.

Figure 8 Appearance of a refrigerator cooled high-Tc superconducting magnet.

Table 3 Parameters of the high-Tc superconducting magnet.

WIRE	
Thickness	0.2 mm×3
Width	3.0 mm
Number of filament	61
Total length	1,200 m
MAGNET	
Inner diameter	40 mm
Outer diameter	108 mm
Height	113 mm
Number of double pancake coils	17
Number of turn	1,565
Inductance	39 mH
Weight	10 kg

Table 3 shows parameters of high-Tc superconducting magnet. The magnet composed of 17 double pancake coils was attached to a copper flange at the second stage for the refrigerator-cooled magnet. A double pancake coil was wound through a react and wind technique by using a bundled wire with Bi-2223 high-Tc superconducting Ag-sheathed 61-multifilamentary tapes and an insulator using a thin polymide tape. Fiber reinforced plastic (FRP) sheets were inserted as an insulator between the pancake coils. This magnet has an inner bore with 40 mmϕ as the practical size. In order to decrease the heat load through the current leads in the refrigerator-cooled magnet, Ag-sheathed high-Tc superconducting tapes were used. Ag-sheathed tapes were bundled and wound on a 20 mmϕ FRP pipe and fixed by a Teflon sheet. The high-Tc superconducting magnet was contacted to the 2nd stage of a GM refrigerator and held in a 10^{-5} Torr vacuum state. A super-insulation was wound on it as a thermal insulation shield. Coil temperature and current lead temperature were monitored using AuFe-Ag thermocouples. Generating magnetic field was monitored using a Hall probe (Bell BHA 921).

Before measurement of coil Ic using the refrigerator, the Ic respectively at 4.2

K, 64 K and 77.3 K, were measured. Figure 9 shows the load line of the Bi-2223 high-Tc superconducting coil. The magnet generated the maximum magnetic field (B_{max}) of 2.2 Tesla with 40 mmϕ bore at the criterion of $10^{-13}\Omega \cdot m$ at 4.2 K in liquid He. The coil Ic was 178 A and Jc was 40,000 A/cm^2. At the criterion of 1μV/cm, coil·Jc generating 2.54 Tesla was fitted to Jc of the short sample. Generating voltage of the coil was 3.5 mV at B_{max}= 2.2 Tesla. Joint voltage was estimated as 1.4 mV. There was no fluctuation for the voltage in generating magnetic field.

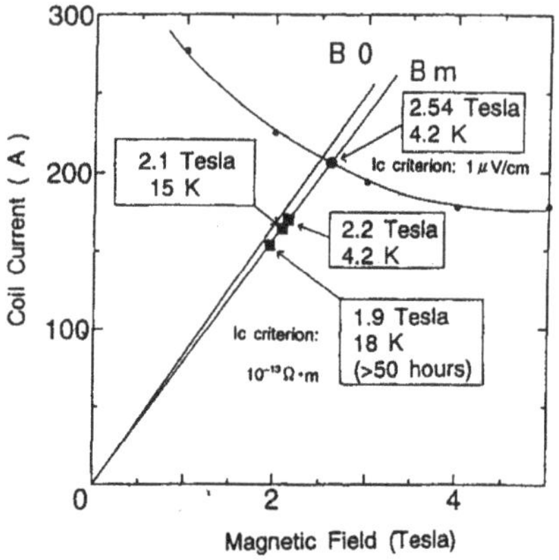

Figure 9 Central field load line B_0 and the maximum field load line B_{max} of the high-Tc superconducting magnet.

Table 4 Properties of high-Tc superconducting magnet at each temperature.

	L. He	refrigerator		L. N_2
Temperature (K)	4.2	15	18**	64
B_{max} (Tesla)	2.2	2.1	1.9	0.56
B_0 (Tesla)	2.1	2.0	1.8	0.54
Coil Ic* (A)	178	166	153	45
Jc (×10^4 A/cm^2)	40	37	34	11

* Criterion: 10^{-13} Ω·m
** Continuous operation (>50 hr.)

The magnet was cooled at 64 K in reduced liquid nitrogen at 100 Torr for about 3 hours. The magnet generated the magnetic field of 0.56 T (Ic=45 A). While, the generating magnetic field at 77.3 K was 0.27 Tesla with Ic=22 A. Therefore, it was found that the generating magnetic field and coil Ic at 64 K increased twice compared to 77.3 K.

The magnet was cooled from room temperature to 15 K at the lower part of the magnet by the refrigerator for about 14 hours. When the coil Ic was 166 A (Jc=37,000 A/cm^2) at the criterion of $10^{-13}\Omega \cdot m$, B_{max} was 2.1 T and B_0 was 2.0 T. For continuous operation, we realized to operate it at 1.9 T for over 50 hours at the coil temperature of 18 K. Table 4 shows a summary of the characteristics of the magnet operating at each temperature. This indicates that the high-Tc superconducting magnet can be available as high field magnets and easy operating magnets using a refrigerator.

CONCLUSION

Multifilamentary wires are available as the long and flexible wires with high Jc. The 61-multifilamentary wire with Jc=40,000 A/cm^2 at 77.3 K was developed through rolling process by improvement of the microstructure. 1,000 m–class length wires were developed. 623 m length wire with Jc=15,900 A/cm^2 was obtained at 77.3 K and 0 T. These results accelerate the development of various prototypes for actual application, such as 2.5 kA current leads, a 10 kA busbar, a 5m/3 kA flexible cable conductor and refrigerator–cooled magnets. The refrigerator–cooled magnets generated B_{max}=2.1 T (B_0=2.0 T) at 15 K. The magnet was operated stably with generating 1.9 T at 18 K for over 50 hours. At 64 K, the magnet generated B_{max}=0.56 T with twice magnetic field generating at 77.3K. The development of actual application will be accelerated through the confirmation of usefullness of prototypes.

ACKNOWLEDGEMENTS

The authors offer their appreciations to Drs. H. Tsuji, T. Ando and T. Isono of Japan Atomic Energy Research Institute for the joint developments of a busbar, and Drs. T. Hara and H. Ishii of Tokyo Electric Power Company for the joint developments of a cable conductor, and Drs. S. Terai T. Kishida and E. Haraguchi of Kansai Electric Power Company for the joint developments of current leads.

Part of this study was performed as "Cooperative Technology Development", being consigned by Research Development Corporation of Japan.

REFERENCES

1) H.Maeda, Y.Tanaka, M.Fukutomi and T.Asano; Jpn. J. Appl. Phys. 27(1988)L209.
2) T.Hikata, T.Nishikawa, H.Mukai, K.Sato and H.Hitotsuyanagi; Jpn. J. Appl. Phys., 28(1989)L82.
3) H.Mukai, K.Ohkura, N.Shibuta, T.Hikata, M.Ueyama, T.Kato, J.Fujikami, K.Muranaka and K.Sato; Adv. Superconductivity V (ISS'92)(Springer Verlag, Tokyo, 1993)679.
4) M.Ueyama, T.Hikata, T.Kato and K.Sato; Jpn. J. Appl. Phys., 30(1991)L1384.
5) K.Sato, N.Shibuta, K.Ohkura, H.Mukai, T.Hikata, T.Kaneko, T.Kato, M.Ueyama, J.Fujikami, K.Muranaka and S.Kobayashi; Proc. of 6th Int. Japan–U.S.A. High–Tc Superconductivity WorkShop (6,7–12–'93).
6) T.Kato, K.Sato, T.Masuda, N.Shibuta, Y.Hosoda, S.Isojima, S.Terai, T.Kishida and E.Haraguchi; Proc. of 6th Int. Symp. on Superconductivity (ISS'93)1273.
7) T.Kato, T.Hikata, K.Sato, T.Isono, T.Ando, and H.Tsuji; Proc. of 6th Int. Symp. on Superconductivity (ISS'93)1301.
8) J.Fujikami, N.Shibuta, K.Sato, H.Ishii and T.Hara; Proc. of 6th Int. Symp. on Superconductivity (ISS'93)1257.
9) K.Ohkura, H.Mukai, T.Hikata, M.Ueyama, T.Kato, J.Fujikami, K.Muranaka and K.Sato; Proc. of 6th Int. Symp. on Superconductivity (ISS'93)735.
10) K.Sato, K.Ohkura, H.Mukai, M.Ueyama, J.Fujikami; The Institute of Electrical Engineers of Japan, ASC–94–9 RM–94–9(1994)71.

HIGH CRITICAL CURRENT DENSITY OF Ag SHEATHED $(Bi,Pb)_2Sr_2Ca_2Cu_3O_x$ TAPE AND $YBa_2Cu_3O_x$ BULK CURRENT LEAD

Y. YAMADA, T. MASEGI, M. SATOU, K. YAMAMOTO, S. NOMURA, O. HORIGAMI, H. OGIWARA, S. KIMURA[1], T. FUJIOKA[2]
Toshiba R&D Center, Ukishima-cho 4-1, Kawasaki-ku, Kawasaki 210, Japan
[1]Toshiba Heavy Apparatus Engineering Laboratory, Suehiro-cho 2-4, Tsurumi-ku, Yokohama 230, Japan
[2]Toshiba Nuclear Fusion Development Department, Uchisaiwai-cho 1-1-6, Chiyoda-ku,Tokyo 100, Japan

T. HASEGAWA and Y. KAMISADA
Showa Electric Wire and Cable Co. Ltd., 2-1-1, Odasakae, Kawasaki-ku, Kawasaki 210, Japan

ABSTRACT

Two typical high temperature superconductors, Ag sheathed $(Bi,Pb)_2Sr_2Ca_2Cu_3O_x$ tape and unidirectionally solidified $YBa_2Cu_3O_x$ bulk sample, have been investigated in terms of critical current density. Especially, the latter is for the application to the current lead. Ag sheathed $(Bi,Pb)_2$ $Sr_2Ca_2Cu_3O_x$ tape exhibited non uniform and bulk-density-dependent current flow within the cross section. Higher bulk density resulted in higher critical current density. In rolling process, high-tensile Ag-Mg sheath caused high rolling pressure and is a promising method to obtain the high critical current density via the densification of oxide core. Unidirectionally solidified $YBa_2Cu_3O_x$ bulk sample resulted in highly oriented microstructure and thus high critical current density of the order of 10^4 A/cm^2 at 77K. The critical current was 3300 A at 77K. Gas cooled current lead was fabricated and successfully carried a large current of 7550 A.

1. Introduction

High critical current density is a key factor for the applications such as magnet conductor and current lead. Critical current density, J_c, of the Ag sheathed $(Bi,Pb)_2$ $Sr_2Ca_2Cu_3O_x$ (Bi-2223) is now in the range of 60000 A/cm^2 at 77K[1-4]. However, it is insufficient to expand the application. Values at least one order of magnitude higher are required. It is well known that many microstructural factors affect the critical current density: composition of packing powders, the degree of crystal orientation, the smoothness of the interface between oxide core and Ag sheath or so called sausaging and bulk density of the oxide core. Especially, high bulk density improves also the alignment of crystals and degree of interface smoothness. Then, high bulk density resulting from hard cold working produces high J_c[1,2]. Moreover, the bulk density depends on the position within the oxide core; the density is high in the center and near the interface but low at the edge[1]. Cutting the sample in the width direction, J_c was high in the center but low at the edge[4]. This implies that current flows dominantly in the high density region. Mechanical and/or thermal treatments causing high bulk density are important to increase over all J_c. In this context, we consider the effect of parameters in cold rolling on the oxide bulk density. In particular, high-tensile sheath materials such as Ag-Mg alloy seem to be effective to densify the oxide core and thus obtain a higher J_c. Ag-Mg sheathed tape had already been developed to tolerate large stress during magnet excitation. The obtained 0.2% yield strength was about 200 MP[4], which is double and one order of magnitude higher than the conventional Nb_3Sn conductor and Ag sheathed tape, respectively. In this paper, we stress that Ag-Mg sheathed tape is also the leading candidate material for increasing bulk density and thus J_c.

Another application where J_c is important is a current lead. The required properties are large current carrying capacity and small heat leak Q to evaporate liquid helium, which are limited by critical current and thermal conductivity of the sample, respectively. Heat leak Q is also affected by J_c as shown in the following equation;

$$Q=(A/L)\int\lambda dT= (I/LJ_c)\int\lambda dT \quad \text{..............(1)}$$

where A is the cross sectional area of current lead sample, L the length, λ the thermal conductivity, and I the operating current. λ is integrated from high temperature end to low temperature end, normally 4.2K, of the current lead. Until now, various types of high-temperature superconductors have been used for the current lead: sintered Bi-2223[5], melt casting Bi-2212[6], laser melting Bi-2212[7] and MTG YBCO[6]. Sintered Bi-2223 and laser melting Bi-2212 were used for the refrigerator cooled magnet system. Melt casting Bi-2212 carried 2000 A for current lead system. The YBCO has also been considered but has not been adopted for the lead system because of the relatively small transport critical current and large thermal conductivity compared to the Bi system[6]. This means a large heat leakage as supposed from the above equation. In the previous works, the Bi-system was mostly used. However, the current capacity of the Bi-system current lead was limited to 2000 A[5] at most due to its low J_c.

Very recently YBCO bulk samples with high J_c were developed by the unidirectional solidification method[8,9]. The sintered sample was slowly moved in the high temperature zone in the furnace, which produced a highly-aligned and well-connected microstructure. In this paper, we report large transport critical currents of this YBCO sample exceeding 3000 A at 77K and current lead test results which successfully carried 7550 A. This extends the application range of current lead system.

2. Experimental

Ag sheathed Bi-2223 tape was made by the powder-in-tube method[1-3]. The

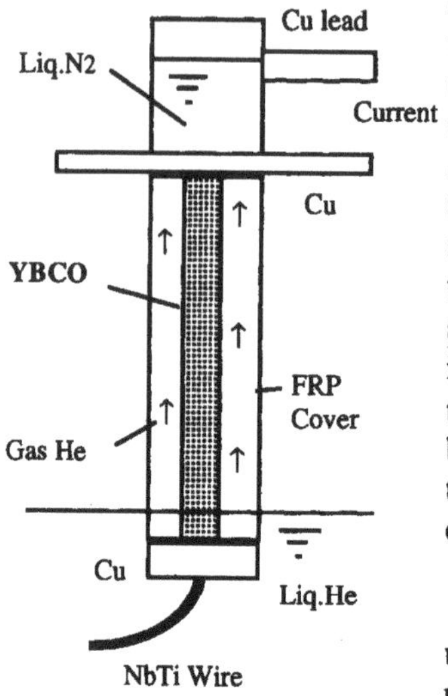

Figure 1 Schematic drawing of YBCO lead component.

calcined powder was packed into an Ag tube and then the composite was drawn and rolled to 0.1mm in thickness. The tape was heat-treated and pressed a few times. High-tensile conductor was also made using Ag-Mg tube. After the heat-treatment, Mg was oxidized and MgO was finely dispersed. This caused a high yield strength of 200 MPa[3], an order of magnitude higher than the normal pure Ag sheathed tape. J_c was measured at 77K and 4.2K in a field up to 20T. To study inhomogeneous distribution of J_c, magnetic field around the sample with carrying a given current was measured by a small Hall probe. The Hall probe was moved along the tape width direction. Then flowing current was calculated from the obtained magnetic field.

$YBa_2Cu_3O_x$ bulk sample was made by an unidirectionally solidified method[8,9]. The furnace with a narrow and steep high temperature zone with the maximum temperature above 1000 °C. The sintered YBCO sample was moved at 1 mm/hr. This formed long and aligned crystals along the moving direction during the solidification. Using this method, we obtained a rod sample with 5.5 mm diameter and a cylindrical sample with 9.5mm outer diameter, 6.5mm inner diameter and 150 mm length. Ag paint was pasted at both ends and heat-treated for the low-resistivity connection needed for the current lead. The contact resistance was in the order of 10^{-8} ohm at 4.2K. The transport critical current was measured at 77K. Thermal conductivity was measured by the constant heat flux method[7] for the 5 cm long sample cut from the same batch sample as for the current lead system. The details of the fabrication method are given in reference 9 by Kase et al.

Using the above cylindrical bulk sample, we made HTSC current lead component[11] with liquid nitrogen reservoir and FRP cylindrical cover as schematically shown in Fig. 1. The high-temperature end of the YBCO sample was cooled by the liquid nitrogen. The whole YBCO sample was cooled by evaporating gas helium. Cu parts connected to YBCO were designed large enough to reduce the joule heating up to 10 kA. Two current lead components were connected to each other through NbTi wires by soldering and the whole was inserted in the cryostat for current transport test. Liquid helium was filled a few cm above the lower end of YBCO. Outlet of the gas helium was connected to gas flow meters, which gave evaporation rate of liquid helium and thus the heat leak to liquid helium.

3. Results and Discussion

3.1. Ag sheathed $(Bi,Pb)_2Sr_2Ca_2Cu_3O_x$ tape

Critical current flows inhomogeneously in the Ag sheathed tape. Figure 2 shows the flowing current distribution within the cross section measured by the Hall probe method. The transport current, J, during the measurement. was 95% of J_{c0}, 30000 A/cm^2, obtained by the transport four probe method. As shown in Fig. 2, the current mainly flow in the center while about one third of the current is carried at the edge. This inhomogeneous flow was also confirmed by the method of cutting the sample in the width direction[4], which also showed the current flowing mainly in the center. As reported earlier[1,2], the density is varied within the oxide core. Especially, the density was high in the center and low at the edge. Therefore, J_c depends on the bulk density even in a small cross section. As shown in Fig. 2, the J_c in the center was about two times as high as that of the transport J_{c0}. To increase J_c, the bulk density needs to be high.

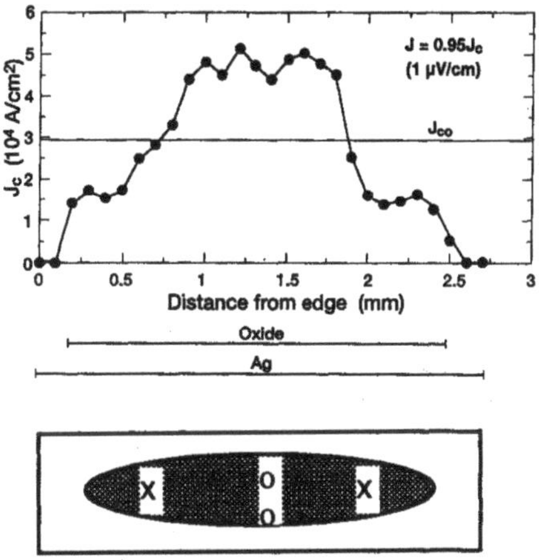

Figure 2 Current distribution in the cross section of Ag sheathed tape measured by Hall probe method. The magnetic field around the tape was detected by moving the small Hall probe in the width direction while 95% of transport critical current was carried. The transport J_c was $3x10^4$ A/cm^2 at 77K. The bulk density was high (o) in the center and low (x) at the edge[1].

The rolling parameters remain to be clarified because J_c of the rolled tape is still much lower than that of the pressed tape. This is due to the low bulk density resulting from the rolling method because of the lower pressure applied. Thus, it is important to pay careful attention to the parameters in the rolling process.

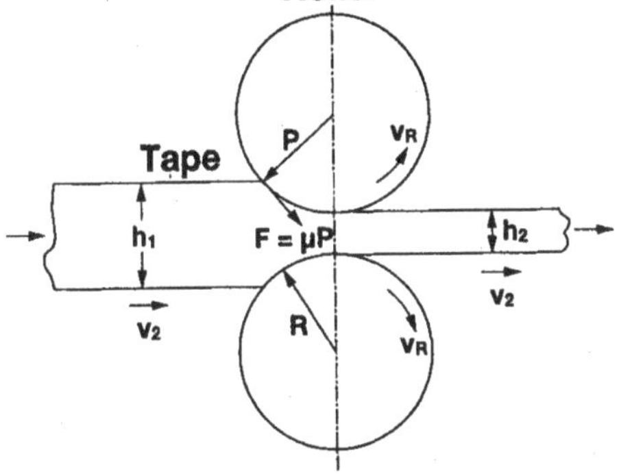

Fugure 3 Schematic drawing of rolling showing the rolling pressure, P. R, roller diameter; F, frictional force; h1, h2, thickness before and after rolling; v1, v2, tape speed; vR, roller speed.

Figure 3 shows the schematic drawing of the rolling with the pressure, P, during rolling. According to the conventional plastic deformation theory in the rolling[10], the pressure is expressed by the

the following equation;

$$P \propto k_f \cdot b_m \cdot (R \cdot \Delta h)^{1/2} \qquad (2)$$

where k_f is the deformation resistance of the rolled material, b_m the average width of the tape, R the radius of the roller and $\Delta h=h_1-h_2$ the thickness difference before and after the rolling. Therefore, the rolling pressure, P, is proportional to the deformation resistance, k_f, if the other parameters are the constant. This implies that high-tensile sheath material meaning high k_f results in high rolling pressure and thus densifies the oxide core. Figure 4 shows the stress-strain curves at room temperature of Ag-Mg sheathed tape and Ag sheathed tape before intermediate treatment combined of heat-treatment and rolling. Obviously, Ag-Mg sheathed tape has higher Young's modulus or deformation resistance than Ag sheathed tape. This means that the Ag-Mg sheath results in a higher bulk density than Ag sheathed tape. Actually the Vickers hardness of the oxide core, which was found to be proportional to the bulk density[2], was about 200 for Ag-Mg sheathed tape and about 100 for Ag sheathed tape. Due to this densification effect, higher values of J_c were obtained in Ag-Mg sheathed tape than that in Ag sheathed tape as shown in Fig. 5 at 0. 15 to 0.35 mm in thickness. Of course, J_c is affected by other factors such as the smoothness of the interface between the oxide core and the sheath. As result, for 0.07 mm thick tape in Fig. 5, the value, 45000 A/cm^2, obtained in Ag-Mg sheathed tape is not higher than 66000 A/cm^2 in Ag sheathed tape. The optimization besides the density will give

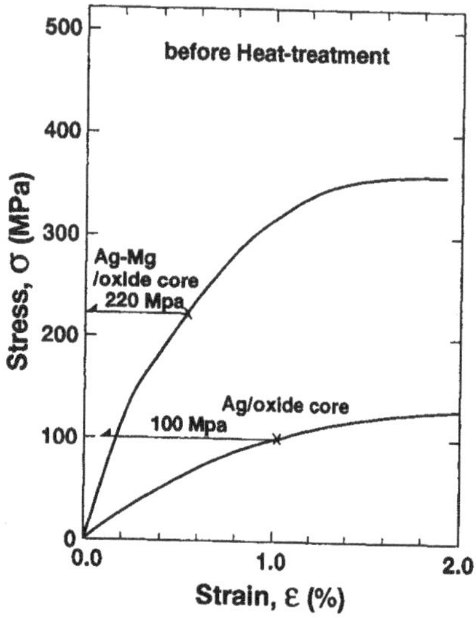

Figure 4 Stress-strain curves at room temperature for Ag sheathed and Ag-Mg sheathed Bi-2223 tapes, which are subsequently rolled and heat-treated.

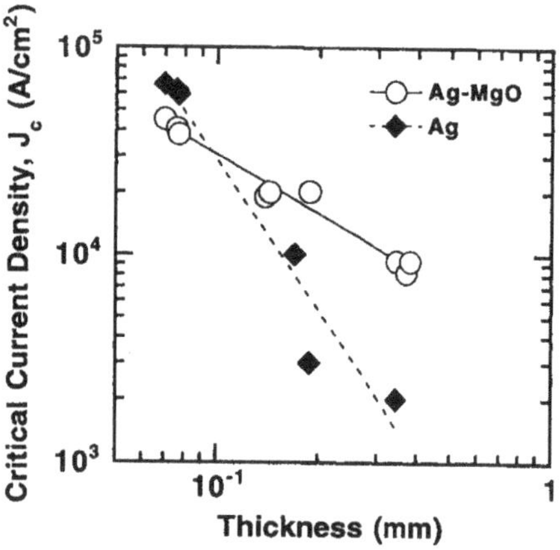

Figure 5 Comparison of J_c at 77K of Ag sheathed tape and Ag-Mg sheathed tapes 0.07 mm to 0.35 mm in thickness.

a higher value of J_c.

Figure 6 shows the field dependence of Ag sheathed tape together with that of Ag sheathed tape. Ag-Mg sheathed tape also exhibited high J_c above 10^5 A/cm^2 at 4.2K and high fields. This gives the possibility of the practical high field magnet conductor to tolerate large stress.

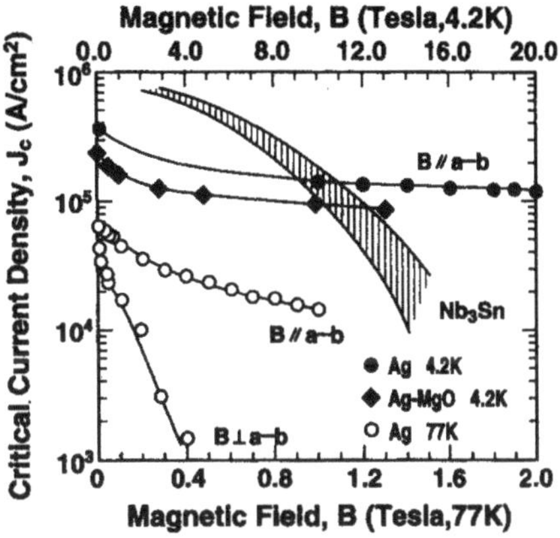

Figure 6 Field dependence of J_c of Ag-Mg sheathed and Ag sheathed tape.

3.2. unidirectionally solidified $YBa_2Cu_3O_x$

Unidirectionally solidified sample[8,9] showed very high critical current and high J_c as shown in Fig. 7. Two typical samples, 150 mm long rod and cylinder, carried large currents of 3100 A and 3350 A at 77K, corresponding to the critical current density around 10^4 A/cm^2. The temperature dependence of J_c was measured magnetically by vibrating magnetometer as shown in Fig. 8. The value at 77K is almost the same as that measured by the transport method in Fig. 7. J_c was increased with decreasing the temperature. The values reached 30000 A/cm^2 and 55000 A/cm^2 at 60 K and 40K, respectively. It is inferred that transport critical currents are more than 10000 A at 40 to 60 K. These high critical current densities are due to the less weakly linked microstructure. According to the observation by optical micrograph and SEM[9], the a-b planes of Y-123 crystals, a few cm long and a few mm wide, lay parallel to the cylinder axis and the crystals were parallel to each other.

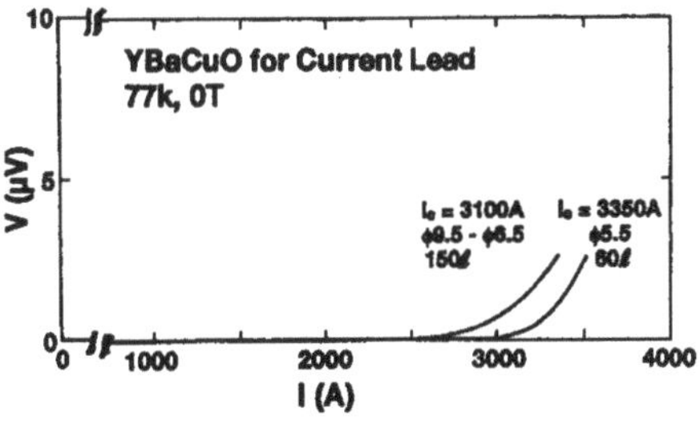

Figure 7 Transport critical currents of 150 mm long YBCO samples.

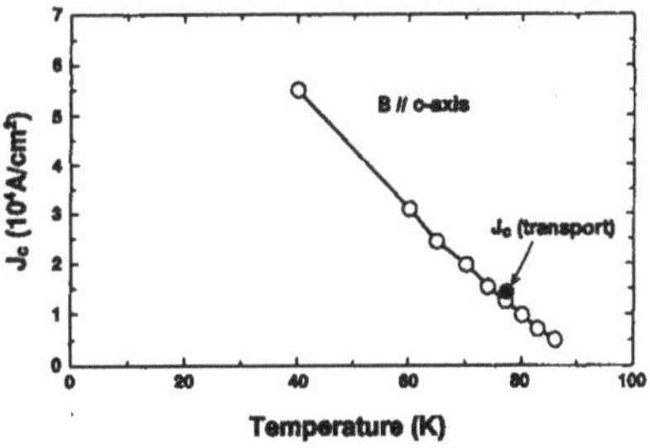

Figure 8 Temperature dependence of J_c by vibrating magnetometer. Solid circle is measured by the transport method for 150mm long sample.

Table 1 YBCO cylindrical sample for current lead. λ, thermal conductivity; ∫λ, temperature integral of λ.

Size			Electrical properties		Thermal properties	
length	outer diameter	inner diameter	I_c (77K)	J_C (77K)	∫λ (77-4.2K)	λ (77K)
150mm	9.5mm	6.5mm	3100A	8200A/cm2	3.5W/cm	76mW/cm/K

For the current lead application, thermal conductivity was also measured for the 5 cm long sample cut from the 150mm sample. The value was 76 mW/cm/K from the constant thermal flux method[7]. This is not so low as the values of Bi-2212[6,7] and Bi-2223[5] but much lower than that of metals. Meanwhile, the J_c is high enough to compensate the large thermal conductivity. Therefore, the unidirectionally solidified YBCO is a promising lead material as supposed from the equation (1). The integral of thermal conductivity from 77K to 4.2K was 3.5W/cm[11]. Thus, the conduction heat leak is calculated to be 110 mW for one YBCO cylindrical sample. Table 1 above summarizes the dimensions and properties of the sample for current lead test.

3.3. $YBa_2Cu_3O_x$ Current Lead Test

The above YBCO sample was constructed in the current lead component for the current transport test. Tow components were set in the cryostat. Both lower ends were connected with NbTi wires by soldering. After filling liquid helium a few cm above the lower end of YBCO, operating current was supplied. When the operating current reached 5560 A, liquid helium evaporation was measured by the flowmeter. The temperature of the YBCO high-temperature end was lowered by evaporating gas to 60 K. The heat leaks from the evaporation for both leads were 2.6 W and 3.6 W[11] at 5560 A, respectively, which are lower than the value of the conventional Cu current lead, 1 W/ 1kA. The measured heat leak consisted mostly of joule heating due to the contact resistivity including Cu-YBCO, Cu-NbTi and Cu-Cu. The conduction heat leak was calculated only 440 mW including FRP cover and measurement Cu wire leads. When increasing the operating current, heat leak or helium evaporation were increased by joule heating at the contacts. When the operating current reached 7550 A, the maximum in this experiment, the temperature at the high-temperature end was reduced to 50K. The details of transport current test will be reported elsewhere[11].

4. Conclusions

Ag sheathed Bi-2223 tape was found to have density-sensitive J_c. Higher density region showed higher J_c within the cross section. Ag-Mg sheathed tape, originally developed for high-tensile conductor, is adequate to increase J_c due to the high rolling pressure which densifies the oxide core. Unidirectionally solidified YBCO sample exhibited large critical current above 3000 A, J_c around 10^4 A/cm^2 and thermal conductivity of 76 mW/cm/K at 77K. The current lead using this sample successfully carried 5560 A with heat leak lower than that of the conventional Cu lead. The maximum transport current reached 7550 A.

5. Acknowledgements

We would like to thank Dr.Hamajima of Toshiba Heavy Apparatus Engineering Laboratory and Dr. Murase of Toshiba R&D Center for their discussion.

6. References

1. Y.Yamada, M.Satou, S.Murase, T.Kitamura and Y.Kamisada, in *Advances in Superconductivity V*, ed. Y.Bando and H.Yamauchi (Springer-Verlag, Tokyo, 1993), p.717.
2. M.Satou, Y.Yamada, S.Murase, T.Kitamura and Y.Kamisada, Appl.Phys.Lett. **10** (1994) 640.
3. Y.Yamada, M.Satou, T.Masegi, S.Nomura, S.Murase, T.Koizumi and Y.Kamisada in *Advances in Superconductivity VI*, ed. T.Fujita and Y.Shiohara (Springer-Verlarg, Tokyo, 1994) p.609.
4. Y.Yamada, M.Satou, T.Masegi, S.Nomura, S.Murase, T.Koizumi and Y.Kamisada in *Critical Currents in Superconductors*, ed.H.W.Weber (World Scientific, Singapore, 1994), p76.
5. Y.Yamada, T.Yanagiya, T.Hasebe, K.Jikihara, M.Ishizuka, S.Yasuhara and M.Ishihara, IEEE Trans. on Applied Superconductivity **3** (1993) 923.
6 P.F.Herrmann, E.Beghin, G.Bottini, C.Cottevieille, A.Leriche and T.Verhaege and J.Bock, Cryogenics **33** (1993) 555.
7. K.Yamamoto, T.Masegi, S.Nomura and Y.Yamada, presented at 6th US-Japan Workshop on High-Tc Superconductors,Dec. 1993, Houston, USA.
8. E.Yanagisawa, J.Kase, J.Shimoyama, T.Matsubara and T.Morimoto, in *Advances in Superconductivity V*, ed. Y.Bando and H.Yamauchi (Springer-Verlag, Tokyo, 1993), p.593.
9. J.Kase, J.Shimoyama, E.Yanagisawa and T.Morimoto, Jpn.J.Appl.Phys. **33** (1994) 3424.
10. For example, R.B.Smith, J.Iron & Steel Inst. **178-4** (1954) 19. R.Hill, *The Mathematical Theory of Plasticity*, (Oxford, 1960), p199
11. T.Masegi et al. preprint of *International Symposium on Superconductivity 1994*, Okayama, Japan November 6-8, 1994.

BRIDGING IN SUPERCONDUCTORS

M.D. SUMPTION AND E.W. COLLINGS

Battelle, 505 King Ave.
Columbus, OH 43201

ABSTRACT

A general treatment of the influence of mechanical filament bridging on superconductor magnetization has been developed. Bridging is treated from the perspective of an anisotropic continuum, and a comparison is given to the traditional d_{eff} treatment. Bridging induced loss is seen to be determined by an intrinsic parameter γ_2, describing the amount of connectivity perpendicular to the filamentary axis. Expressions for magnetization for samples as a function of twist pitch and sample length are given. The effects of weak connectivity along the filament axis are discussed. The expressions developed are applied to Nb_3Sn and HTSC strands.

1. Background

Mechanical filament bridging is seen frequently in Nb_3Sn multifilamentary strands where it can greatly enhance the magnetization and loss over that of unbridged strands[1-3]. Typically, bridging has been seen as a local agglomerating of initially separated filaments, creating one larger filament of diameter d_{eff}, but the recent work of Goldfarb et al.[4], has shown that loss due to ***mechanical bridging*** in Nb_3Sn strands has the same hallmarks of global coupling seen in PE coupled Nb_3Sn and NbTi, viz., L_p and L dependence. The above developments, together with our recent experience in HTSC conductors, prompted us to develop a generic bridging theory, which we have adapted from various models used to describe PE coupled low T_c conductors[5-9] as well as HTSC ones[10-12]. We assert that the mechanical bridging so frequently seen in Nb_3Sn (and now seen in HTSC, as shown in Fig. 1) conductors is one degenerate type of bridging, and PE coupling is another. We will also show that the less uniform HTSC filament morphology adds its own complications.

2. Anisotropic Continuum Model Approach to Mechanical Bridging

2.1 L_p Dependence

Bridging loss in Nb_3Sn conductors relies on the existence of small superconducting links between neighboring filaments. Locally, this produces a superconducting link which surrounds a region of fully normal stabilizer, similar in many ways to the eddy current problem, as noted in[4]. Another approach, valid if the bridging occurs on a sufficiently small scale, is to treat these intermittent superconducting links on an average basis. We can then see the bridged superconductor, ***on average***, as a superconductor with continuous, but anisotropic properties. Presumably these intermittent links have the same J_c as the filament J_c, but flowing through a smaller area. However, it is not necessary to

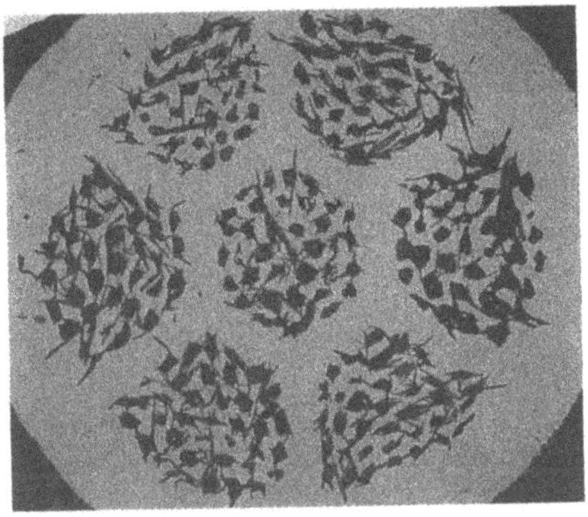

Fig. 1. Optical micrograph of Bi-2212 multifilamentary 7-bundle conductor with bridging evident.

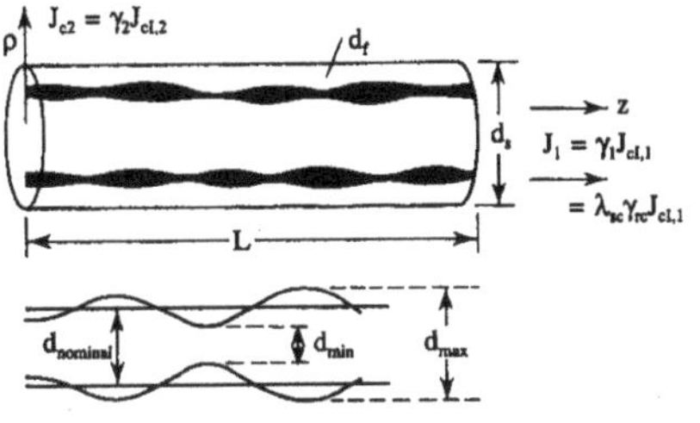

Fig. 2. Strand schematic with axis and parameter definitions.

make this assumption. In a sense, this represents a compliment to the PE coupling case, where the area for supercurrent flow perpendicular to the filament axis was assumed to be the entire matrix area, but the J_c across these regions was intrinsically smaller. The case for HTSC multifilamentary conductors follows from the Nb_3Sn treatment, since mechanical bridging induced loss should not be influenced by the kind of superconductor present, but only the bridging morphology. We will now proceed to adapt the theory used for PE coupling to the treatment of the mechanical bridging problem.

In Carr's original treatment of Nb_3Sn **proximity effect** induced coupling, several assumptions were made, the primary of which were; (1) In a transverse magnetic field, shielding and trapping currents flow along the z-axis of the conductor, and (2) currents which flow along the filament axis, and those which flow perpendicular to these, are independent of one-another. Proceeding with the Carr assumptions, we first define the filamentary J_c along the z-axis to be J_{c1}, and the filamentary J_c along the r and θ directions to be J_{c2}, as shown in Fig. 2. The relative connectivity along the filamentary axis, γ_{rc}, is defined as the transport critical current, $J_{cT,1}$, over the intrinsic critical current along this axis, $J_{cI,1}$. Multiplying this by the superconductor to total area ratio, λ_{sc}, we get the connectivity along this axis, γ_1. The connectivity along the radial direction is similarly defined, viz. $\gamma_2 = J_{cT,2}/J_{cI,2}$. This parameter clearly is some measure of the fraction of the area between the filaments which is bridged (i.e., in a longitudinal cross section). Using these assumptions, the critical current along the z-axis (as distinct from the filamentary axis) is[5,7,9] :

$$J_{cz} = \gamma_1 J_{cI,1}\left[1 + \left(\frac{2\pi\rho}{L_p}\right)^2\right]^{1/2} \qquad \alpha < \alpha_c \tag{1a}$$

$$J_{cz} = \frac{\gamma_2 J_{cI,2} L_p}{2\pi\rho}\left[1 + \left(\frac{2\pi\rho}{L_p}\right)^2\right]^{1/2} \qquad \alpha > \alpha_c \tag{1b}$$

where ρ is the radial distance in the strand,

$$\cos\alpha = \frac{L_p/2\pi\rho}{(1 + (L_p/2\pi\rho)^2)^{1/2}}, \quad and \quad \cos\alpha_c = \frac{\gamma_1 J_{cI,1}/\gamma_2 J_{cI,2}}{(1 + (\gamma_1 J_{cI,1}/\gamma_2 J_{cI,2})^2)^{1/2}}. \tag{2}$$

Here α is the angle between the filament and the z-axis, and α_c (the critical angle) is the parameter which determines whether Eq. 1 (a) or Eq. 1 (b) is valid[5,7,9].

We further define the dimensionless parameters β_p ($\equiv L_p/2\pi\rho$), β_R ($\equiv L_p/2\pi R$), and β_C ($\equiv J_{c1}/J_{c2} = \gamma_1 J_{cI,1}/\gamma_2 J_{cI,2} = \lambda_{sc}\gamma_{rc} J_{cI,1}/\gamma_2 J_{cI,2}$). Also, we introduce a critical L_p given by $L_{pcm} = (2\pi R\gamma_1 J_{cI,1})/(\gamma_2 J_{cI,2})$. Following the analysis in[9], this can be seen to give

$$\Delta\mu = 3\left\{\frac{(1+\beta_R^2)^{3/2}}{3\beta_C} - \frac{\beta_R^3}{8\beta_C^2}(1+\beta_C^2)^{1/2} - \frac{\beta_R^3}{12\beta_C^4}(1+\beta_C^2)^{3/2} - \frac{\beta_R^3}{8}\ln\left[1 + (1+\beta_C^2)^{1/2}\right] + \frac{\beta_R^3}{8}\ln\beta_C\right\} \tag{3}$$

for $L_p < L_{pcm}$ and

$$\Delta\mu = 3\left\{-\frac{\beta_R}{8}(1+\beta_R^2)^{1/2} + \frac{1}{4\beta_R}(1+\beta_R^2)^{3/2} - \frac{\beta_R^3}{8}\ln\left[1 + (1+\beta_R^2)^{1/2}\right] + \frac{\beta_R^3}{8}\ln\beta_R\right\} \tag{4}$$

for $L_p > L_{pcm}$, where we have normalized the magnetization by its saturation value ($\Delta M_{sat} = 4\gamma_1 J_{cI,1} R/15\pi$). For details of this calculation for the analogous PE case, see[9]. For large values of β_R, this expressions saturates to 1, but for $\beta_C \gg \beta_R \gg 1$ (i.e. for low to moderate levels of PE coupling), this expression has a linear approximation given by $\Delta\mu \approx (3\beta_R/2\beta_c)$. Written in cgs-practical units this is

$$\Delta M_s \approx \frac{J_{c2} L_p}{5\pi^2} \tag{5}$$

where the magnetization is normalized to the composite volume. If we assume that the intrinsic filamentary J_c is isotropic, $J_{cI,2}$ is replaced by J_{cI}, leading to a straightforward treatment of Nb_3Sn bridging (see below).

2.2 L Dependence

The equations for short samples can be adapted from[13] for the present need, however, the coordinate system used there is different, viz. J_1', and J_2' in the notation of[13] become J_2, and J_3' becomes J_1. Therefore, we set $J_{c3}' = \gamma_1 J_{cI,1}$, and $J_{c2}' = \gamma_2 J_{cI,2}$. We find that

$$\Delta M = \frac{4\gamma_1 J_{cI,1} d_s}{30\pi}\left(1 - \frac{3\pi d_s \beta_c}{32L}\right) \tag{6}$$

for $1/\beta_c > d_s/L$ (strong coupling), and

$$\Delta M = \frac{\gamma_2 J_{cI,2} L}{20}\left(1 - \frac{4L}{3\pi\beta_c d_s}\right). \tag{7}$$

for $1/\beta_c < d_s/L$ (intermediate coupling). Here $\beta_c = \gamma_1/\gamma_2$ (if J_{cI} is isotropic), as above. We additionally note that for coupling that is weak (but not in the very weak coupling regime), ΔM is given by the first (linear) term in Eq. 7.

3. Application of Theory

3.1 Application to Nb_3Sn

We can use the above expressions to fit Goldfarb et al.'s recent data[4]. We will make the approximation that $Q = 2\ \Delta MH$, and since H_m is 3 T, this leads to $\Delta M = Q/(6\ T)$. Data is presented in Fig. 3 (as taken from[4]) for loss (Q) both as a function of sample length and sample twist pitch, where we note the strikingly similar functional dependence. Starting first with the Q vs L data for $L_p = \infty$, and fitting to Eqs. 6 and 7, we find the curve shown in Fig. 3. Parameter values found from this least squares fits are; $\beta_c = 520$, with an offset loss (the filamentary loss itself) of 89.5 kJ/m^3. The β_c value extracted from fitting the initial slope of the data to Eq. 7 in the linear limit is 430. We note that it was necessary to re-normalize the loss to the total volume of the bundles, as opposed to the total strand volume (as in[4]), giving a conversion factor of 2.42. Additionally, it was necessary to replace d_s by d_b (the bundle diameter, 0.074 mm) wherever it occurred.

Eqs. 3 and 4 cannot be directly applied to these strands. The reason for this can be seen in Fig. 4, where a single bundle strand and a multiple bundle strand are shown. For the single bundle strand [4 (a)], flux is excluded (trapped) from (in) the entire strand, and the associated currents flow along the z-axis of the ***strand coordinates***. However, for the multiply bundled strand, flux is shielded (trapped) from the bundles, and therefor the associated currents flow along the ***z-axis of the bundle*** over large portions of the bundle,

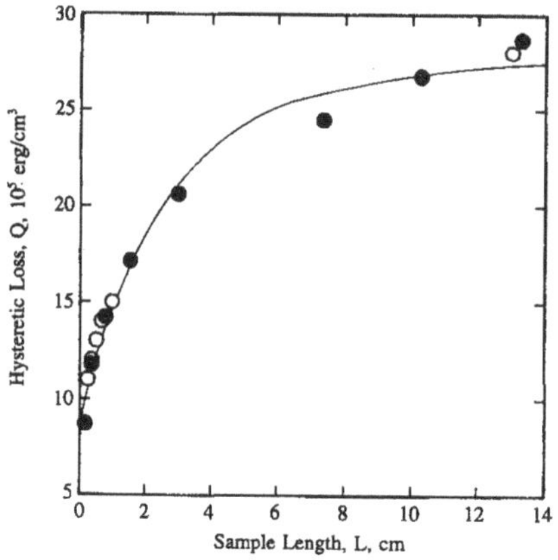

Fig. 3. Loss vs L and L_p for Nb_3Sn from[4], along with model fit.

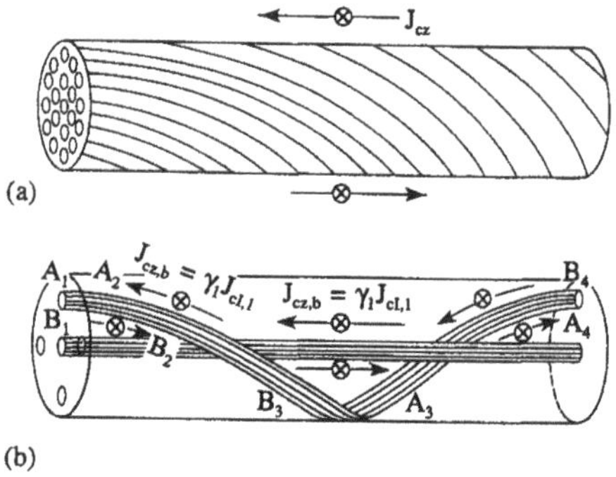

Fig. 4. Current paths in (a) single- and (b) multiple- bundle multifilamentary conductors.

as shown in Fig. 4(b). For strands with lengths significantly larger than their twist pitch, it is the angle that the shielding and trapping currents form with respect to the filamentary axis that determines the magnitude of the current. For strands of the configuration treated in Section 2 (single bundle strands) this is determined by a function of twist pitch and strand radius. For the strands with multiple bundles, however, the situation is more complicated. For a bundle centered at the strand center, the field axis has a relation to the bundle axis (and therefor the filament axis) which is invariant along the strand z-axis. For this reason the central bundle can be directly treated by Eqs. 3 and 4 in Section 2 (replacing d_s with d_b). For bundles near the strand perimeter, however, the relation of the field axis to the bundle axis varies periodically along the strand z-axis with the period of the twist pitch. Twice during each twist pitch, the shielding and trapping currents vary from perpendicular to the filamentary axis (A_1, B_1), to parallel to the filament axis (A_2, B_2) and again to perpendicular to the filament axis (A_3, B_3). Regions like A_2 and B_2 will saturate quickly, while regions like A_1, B_1 and A_3, B_3 will remain weakly coupled. In short, the bundle should behave like a series of short samples, of length similar to the twist pitch, rather than being described by an L_p dependence like that seen in single bundle conductors. Indeed, we have seen this in the data shown in Fig 3, where samples of a certain L_p behave almost exactly like those of a certain L. However, it is not clear to us why L_p, and not $L_p/2$, is equivalent to L.

In any case it is readily apparent that bridged Nb_3Sn strands are well described by the anisotropic continuum model, however, a more quantitative treatment of the multiple bundle case is necessary.

3.2 Application to HTSC

Bridging has also been seen to occur, as would be expected, in HTSC multifilamentary

Table 1. HTSC Multifilament Data.

Strand/Sample	BI6	BI7	BI8	BI9
J_{cT}, A/cm^2	20,765	90,163	168,958	57,043
L, cm	1.136	0.877	0.909	0.866
d_s, cm	0.0991	0.099	0.368	0.367
d_b, cm	0.0219	0.219	0.815	0.00815
d_b/L	0.0193	0.0250	0.896	0.00941
d_s/L	0.0872	0.113	0.0405	0.0405
λ_{sc}	0.25	0.25	0.25	0.25
$\lambda_{s,local}$	0.73	0.73	0.73	0.73
d_f, μm	30	30	11.66	11.66
ΔM^a (4.2K,H=0), erg/cm^3	199	321	67.14	176
d_{eff}, cm	0.309	0.115	0.0128	0.0998
d_{eff}/d_b	14.1	5.2	1.57	12.2

^{a} Normalized to total bundle volume.

strands. Measurements have been made on Bi-2212 based, round multifilamentary strands of the basic configuration shown in Fig. 1. Samples similar to that of Fig.1 were measured, and strand specifics are listed in Table 1.

These strands were produced by IGC-Advanced Superconductors, who also measured the transport J_c. ΔM was taken at H = 0 from M-H loops measured with a vibrating sample magnetometer. As discussed previously, although d_{eff} values have little quantitative meaning, they are qualitative indicators of bridging. As a first step in the analysis, then, we used the traditional expression for bridging, $d_{eff} = (30\pi\Delta M)/(4\lambda_{sc}J_{cT})$ to obtain the values of d_{eff} listed in Table 1. We notice that strong bridging is present, since $d_{eff} >> d_f$ in all cases. In fact, in all cases $d_{eff} > d_b$. One might interpret this as indicating bridging from bundle to bundle, such that the whole strand was coupled. However, such inter-bundle coupling is not present in a micrograph of a strand with 16 μm filaments shown in Fig. 1, and in view of the wide bundle-to-bundle spacing is not likely for the strands discussed here.

We interpret these results as indicating that not only is significant bridging present, but that γ_{rc} is significantly less than 1. This indicates that $J_{cT} < J_{cI}$. In order, then, for the magnetization to be as large as it is, it is necessary that "minor" current loops flow around lengths of the sample shorter than the total sample length, as in Fig. 5. If γ_2 is sufficiently large, then the bundle acts (magnetically) as several shorter bundles with larger "local" filamentary current, $J_{c,loc}$. This explains the otherwise unphysical d_{eff} values. And of course the fact that the connectivity along the z-axis is less than perfect is just what is expected.

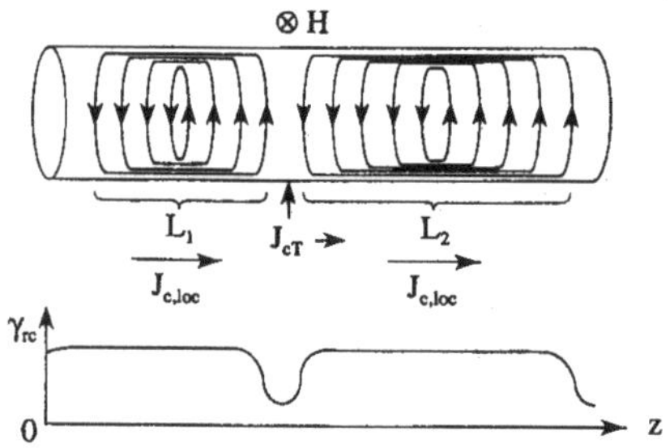

Fig. 5. Diagram of minor current loops and associated γ_{rc}.

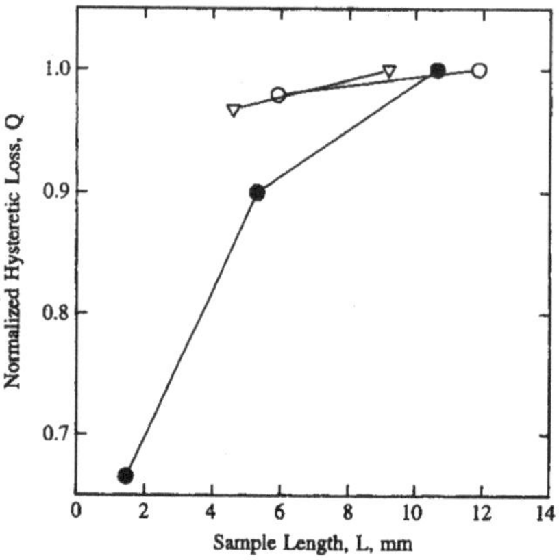

Fig. 6. Normalized loss vs L for Bi-2212 multifilamentary samples BI6 (○), BI7 (●), and BI8 (▽).

We have also measured the length dependence of the loss of these samples, which is displayed in Fig. 6, where each set of values are normalized to the highest value of the set. Consistent with the model of Section 2, the samples show a length dependence of the loss. For samples BI6 and BI8, the length dependence is very minor, but for BI7, it is quite pronounced. We argue that there is no intrinsic difference between the bridging in Nb_3Sn and that in these HTSC multifilamentary samples, or any other multifilamentary superconductors. The lack of significant length dependence in samples BI6 and BI8 can then be seen to reinforce the earlier indications of minor current loops. Unfortunately, this means that while the models derived in Sec 2 would apply to these strands, if the existence of minor loops is taken into account, practically speaking there are too many unknowns for its application to be useful with the limited data set available.

However, useful information can be inferred in the following way. Since each bundle is acting magnetically as a number of smaller (in length) bundles with a higher $J_{c,loc}$, which can then be seen (on average) to be at least d_{eff}/d_b greater than J_{cT}. This represents a minimum level of variation because any end effects associated with these shorter bundles would tend to lessen the d_{eff}/d_b ratio. The variation of $J_{c,loc}$ (which is an average value) is really a measure of γ_{rc}. The extent to which γ_{rc} is less than one would seem to indicate the amount of "headroom" available in J_c which can be obtained by improvement of extrinsic factors.

4. Conclusions

We have formed a generalized treatment of mechanical-bridging-induced magnetization and loss in superconductors. The model was then successfully applied to bridging in Nb_3Sn measurements[4], as well as HTSC multifilament data. Useful parameters γ_1 and γ_2 are derived which quantitatively determine the level of bridging (γ_2) and z-axis connectivity (γ_1) in particular samples. Additionally, the existence of minor currents loops is demonstrated for HTSC Bi multifilamentary conductors.

5. Acknowledgements

The analysis of bridging and its applications to multifilamentary Nb_3Sn was supported by the U.S. Department of Energy, Division of High Energy Physics, under Grant No. DE-FGO2-93ER40763. Its extension to Bi:2212/Ag was supported by IGC-Advanced Superconductors under SBIR subcontract. The Bi:2212/Ag samples and their J_c values were provided by L.R. Motowidlo, IGC-ASI.

6. References

1. S. Sakai, K. Miyashita, K. Kamata, et al., *Adv. Cryo. Eng. (Materials)* **38** (1992) 595.
2. A.I. Braginski and G.R. Wagner *IEEE Trans. Magn.* **17** (1981) 243.
3. R.B. Goldfarb and J.W. Ekin, *Cryogenics* **26** (1986) 478.
4. R.B. Goldfarb and K. Itoh, *J. Appl. Phys.* **75** (1994) 2115.
5. W.J. Carr Jr., *J. Appl. Phys.* **54** (1983) 6549.
6. V.M. Pan, S.I. Mukhin, V.S. Flis, M.G. Vasilenko, V.I. Latysheva, L.M. Fisher, and V.M. Dzugutov, *IEEE Trans. Magn.* **21** (1985) 408.
7. N. Harada, Y. Mawatari, O. Miura, Y. Tanaka, and K. Yamafuji, *Cryogenics* **31** (1991) 183.
8. M. D. Sumption and E. W. Collings, *Adv. Cryo. Eng. (Materials)* **38** (1992) 783.
9. M.D. Sumption and E.W. Collings, *Cryogenics* **34** (1994).
10. E.M. Gyorgy, R.B. van Dover, K.A. Jackson, L.F. Scheneemeyer, and J.V. Waszczak, *Appl. Phys. Lett.*, **55** (1989) 283.
11. F.M. Sauerzopf, H.P. Wiesinger, and H.W. Weber, *Cryogenics* **30** (1990) 650.
12. S. Hu, H. Hojaji, A. Barkatt, M. Boroomand, A.N. Thorpe, and S. Altersecu, *Phys. Rev. B* **43** (1991) 2878.
13. M.D. Sumption, *Applied Superconductivity* **2** (1994) 41.

Electronic Structure Calculations of Low-T_c Quaternary Intermetallic Compound, $LuNi_2B_2C$ and Related Compound, LuNiBC

Hanchul Kim, Chi-Duck Hwang, and Jisoon Ihm
Department of Physics and Center for Theoretical Physics,
Seoul National University, Seoul, 151-742, Korea

ABSTRACT
We carry out *ab initio* pseudopotential calculations for the electronic sturcture of low-T_c intermetallics $LuNi_2B_2C$ and related non-superconducting LuNiBC. The results imply that the Ni-based quaternary intermetallics are three-dimentional metals despite the apparent two-dimensionality in their geometric structure. The superconducting and the non-superconducting compounds are compared, especially in terms of the Fermi surfaces and the symmetry-decomposed density of states near the Fermi level.

1. Introduction

Recently Cava *et al.*[1] have investigated superconductivity in the Ni-based quaternary intermetallics $Ln$$Ni_2B_2C$ with Ln = Y, La, Ce, Sm, Tb, Dy, Ho, Er, Tm, and Lu. Superconductivity was found not only for the non-magnetic Y and Lu, but also for the magnetic rare-earth elements Tm, Er, and Ho. For other rare earths, $Ln$$Ni_2B_2C$ compounds were not superconducting above 4.2K. The highest T_c of $\sim$ 17K is recorded in the single-phase $LuNi_2B_2C$. Siegrist *et al.*[2] observed that the 17K superconducting compound is part of a homologous series of compounds of general formula $(LuC)_m(NiB)_n$, where the $m = 1$, $n = 2$ member is the 17K superconductor and $m = n = 2$ member is found to be non-superconducting. This series exhibits the layered structure, reminiscent of the high-T_c cuprates.

The superconducting compound $LuNi_2B_2C$ crystallizes in the body-centered tetragonal (*bct*) structure and its space group is $I4/mmm$. The LuC (NaCl-type) layers alternate with $(NiB)_2$ layers in a stoichiometry of 1:1. The $(NiB)_2$ layers contain a square-planar Ni_2 array sandwiched between the boron planes with the nickel atom being tetrahedrally coordinated to four boron atoms. One thing to note is that the Ni–Ni in-plane distance of 2.45 Å is slightly shorter than that found in the metallic nickel (2.50 Å). This suggests strong Ni–Ni metallic bonding. The geometric structure of LuNiBC is formed by substituting the rocksalt-type $(LuC)_2$ bilayer into the LuC monolayer in $LuNi_2B_2C$, and the Bravais lattice is changed from *bct* to simple-tetragonal (*st*) and the resulting nonsymorphic space group is $P4/nmm$.

In this report we make a comparative study between the superconducting $LuNi_2B_2C$ and the non-superconducting LuNiBC and the electronic structures are calculated via *ab initio* pseudopotential plane-wave method[3] within the local-density-functional theory (LDF).[4] The energy band structure, the total, the local (*or* site-projected) and the partial (*or* site-projected and symmetry-deomposed) densities of states (DOS), and the Fermi surfaces (FS) are presented for both $LuNi_2B_2C$ and LuNiBC †.

2. Computational Method

The soft norm-conserving pseudopotentials are generated following the scheme of Troullier and Martins.[6] The Ceperly-Alder correlation as parametrized by Perdew and Zunger[7] is used. The pseudopotentials are cast into the fully-nonlocal form of Kleinman-Bylander.[8] We then perform the *ab initio* pseudopotential calculations within the LDF using the plane-wave basis set. The diagonalization of the Hamiltonian matrices is achieved through the iterative method of Liu-Davidson[9] with the modified Jacobi relaxation scheme.

3. $LuNi_2B_2C$

We carry out the total energy calculation using the experimental structural parameters [$a = 3.4639 Å$, $c = 10.6313 Å$, and the atomic positions are Lu(0, 0, 0), Ni($a/2$, 0, $c/4$) and (0, $a/2$, $c/4$), B(0, 0, $\pm 0.3621c$), and C($a/2$, $a/2$, 0)].[2] The self-consistency is achieved with use of as many as 415 **k**-points in the irreducible Brillouin zone (IBZ, 1/16 of BZ) due to the expected complicated energy band structure near the Fermi level.

The calculated energy bands are presented in Fig. 1(a). There are three bands

---

†In preparing this paper, two works based on the LAPW method have been published.[5] The present work, however, provides more information on the electronic structure, *i.e.*, both the mapping of FS and the detailed analysis of partial DOS near E_F.

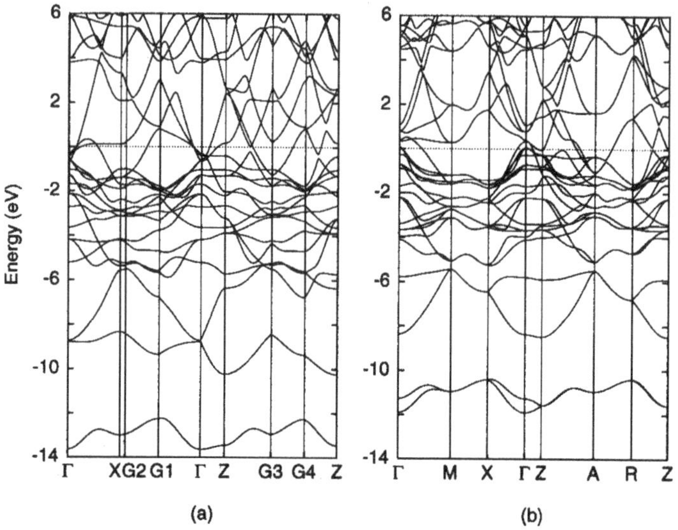

FIG. 1. Energy band structure of (a) $LuNi_2B_2C$ and (b) LuNiBC.

which cross the Fermi level. The energy bands near E_F exhibit an appreciable dispersion along the c-axis direction, which indicates that this compound is a three-dimensional metal in spite of the apparent two-dimensionality in the geometric structure. The C($2s$) band lies in the energy range between -14 and -12 eV. B($2s$)–C($2p$)

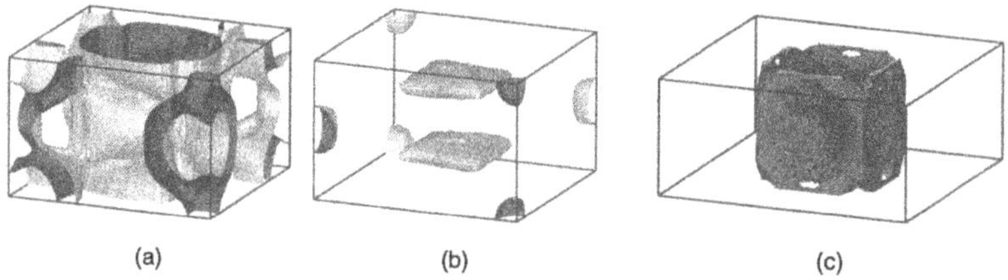

FIG. 2. Fermi surfaces of (a) most complex constricted cylindrical FS of $LuNi_2B_2C$, (b) square-planar pancake-like FS of $LuNi_2B_2C$, and (c) main hole-like FS of LuNiBC. The bounding box has the same volume as the Brillouin zone. The box centers are the X point in (a) and (b) and the Γ point in (c) respectively.

hybridized bands extend from -10 to -8.5 eV. The main valence band complex is composed of contributions from all atoms in the compound. One of Lu(5d) bands showing $d_{x^2-y^2}$ character dips below the Fermi level at Γ. The three bands crossing E_F produce four electronic FS's (shown in Figs. 2(a) and (b)): two prolate spheroids centered at Γ with their long axes parallel to the c axis, one square-planar pancake centered at P perpendicular to the c axis with sides parallel to (100) and (010) directions, and the most complex one whose main portion is a cylindrical surface (parallel to c axis and centered at X, a reflection of the two-dimensional nature of the geometric structure) constricted along (100) and (010) directions near the $k_z = \pi/c$ plane. One spheroid at Γ originates from the Lu(5d) band and exhibits the $dd\sigma$ character with some admixture of the C(p_x, p_y) character. The other spheroid at Γ has the character of Ni($d_{x^2-y^2}$, d_{yz}, d_{zx})–B(p_z)–C(p_z) hybridization. The square-planar pancake has the character of Ni $dd\sigma^*$ with some Ni($d_{3z^2-r^2}$), B(p_x, p_y), and Lu(d_{yz}, d_{zx}) character. The main character of the constricted cylinder is Ni($d_{x^2-y^2}$, d_{yz}, d_{zx})–B(p_z)–C(p_z) hybridization as for the second spheroid. The remaining portion attached to the cylinder has the same character as the square-planar pancake.

The total density of states (TDOS) and the local (*or* site-projected) density of states (LDOS) are shown in Fig. 3(a). The Fermi level lies just below a DOS peak where all atoms contribute, though the major contribution comes from Ni(3d) states. The TDOS at E_F, $N(E_F)$ is 3.88 states/(eV unit-cell) and the resulting DOS per Ni atom 1.94 is larger than that of La_2CuO_4, 1.37 states/(eV Cu) or that of $YBa_2Cu_3O_7$, 1.41 states/(eV Cu).[10] Under the assumption of the rigid band, additional 0.43 electron/unit-cell would shift the Fermi level to the DOS peak with $N(E_F) \simeq 4.83$ states/(eV unit-cell) ($\delta E_F \simeq 0.094$ eV). The Ni(3d) DOS peak around E_F is separated from the main body of the Ni(3d) DOS by ~ 1 eV. The partial (*or* site-projected and symmetry-decomposed) DOS (PDOS) are shown in Fig. 3(b) in expanded scale, and clearly exhibit the Ni(d_{xy}, $d_{3z^2-r^2}$)–Lu(d_{yz}, d_{zx})–B(p_x, p_y) hybridization and the Ni($d_{x^2-y^2}$, d_{yz}, d_{zx})–Lu($d_{3z^2-r^2}$)–B(p_z)–C(p_z) hybridization near E_F.

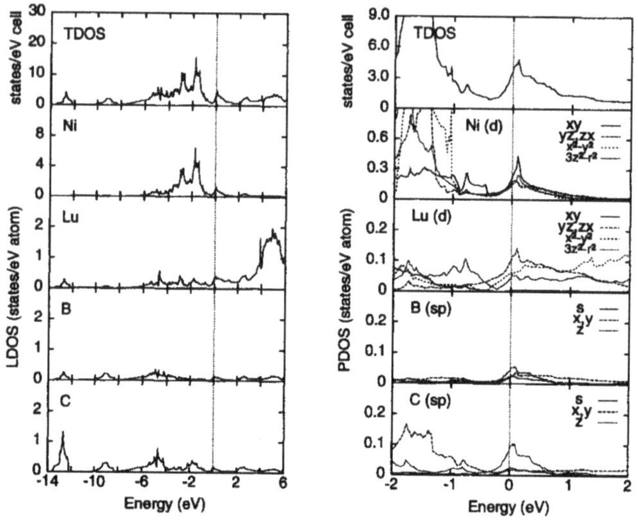

(a) (b)

FIG. 3. (a) TDOS and LDOS, (b) TDOS and PDOS in expanded scale of $LuNi_2B_2C$.

4. LuNiBC

The total energy calculation is performed with use of the experimental structural data (a = 3.4985Å and c = 7.7556Å) with 275 k-points in the *st* IBZ (1/16 of BZ). The calculated energy bands are shown in Fig. 1(b) and there are five bands crossing E_F. One outstanding feature of the band structure is that all the states are doubly degenerate on the zone-boundaries of *st* BZ (whose normal vectors are perpendicular to the c-axis direction) due to a nonsymorphic operation. Near the Fermi level, the energy bands are dispersive along the c-axis direction suggesting the three-dimensionality in the electronic structure similar to the case of $LuNi_2B_2C$. A pair of C(2s) bands lie from -12 to -10.5 eV and B(2s)–C(2p) hybridized bands range from -8.5 to -6.5 eV. The main valence band complex is formed by the contributions from all atoms as in $LuNi_2B_2C$. The existence of flat bands at ~ -0.3 eV along Γ-Z-A and at $\sim$ 0.3 eV along Γ-M is manifested in PDOS of Ni as peaks located at corresponding energies. The five E_F-crossing bands produce six FS's: a pair of hole pockets at Γ, one hole-like FS centered at Γ shown in Fig. 2(c), two electron pockets at A, and one tiny needle-like electron pocket at Z. The two hole pockets have the character of Lu($d_{x^2-y^2}$)–Ni($d_{x^2-y^2}$, d_{yz}, d_{zx})–B(p_x, p_y)–C(p_x, p_y) and Lu(d_{xy})–Ni(d_{yz}, d_{zx})–B(p_x, p_y)–C(p_x, p_y) respectively. The main character of the large hole FS (which is the main FS of this non-superconducting compound) shown in Fig. 2(c) is the Ni($d_{x^2-y^2}$)–C(p_x, p_y)–B(p_x, p_y) hybridization. Judging from the topology of the large hole FS, LuNiBC is more three-dimensional in its electronic properties than $LuNi_2B_2C$.

The TDOS and the LDOS are shown in Fig. 4(a). The Fermi level lies at a valley of the TDOS (not around a peak as in $LuNi_2B_2C$) resulting in a reduced DOS at E_F, $N(E_F) = 1.95$ states/(eV unit-cell) compared with 3.88 of the superconducting

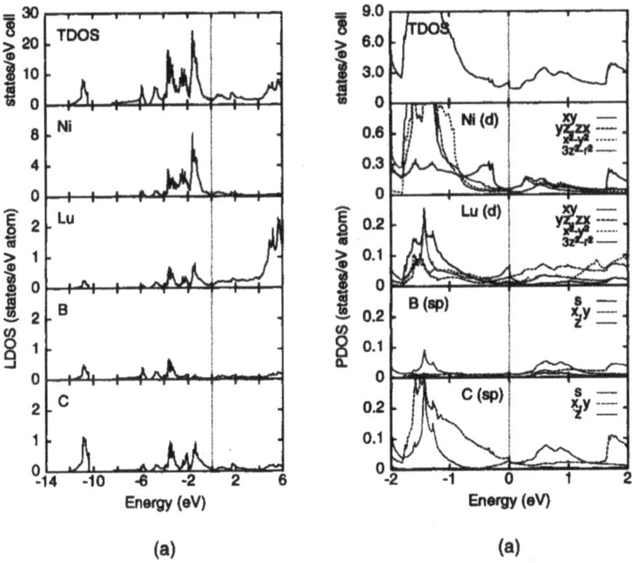

FIG. 4. (a) TDOS and LDOS, (b) TDOS and PDOS in expanded scale of LuNiBC.

$LuNi_2B_2C$. The DOS per Ni atom 0.98 is smaller than the DOS per Cu atom of high-T_c cuprates contrary to the superconducting $LuNi_2B_2C$. One notable feature is the overall structure of the LDOS of Ni. For Ni, the separated peak of LDOS above E_F is much broader than the corresponding peak in $LuNi_2B_2C$. More details can be obtained from the PDOS in Fig. 4(b) drawn in expanded scale near E_F. The main peaks of five Ni($3d$) PDOS above E_F do not coincide. (For $LuNi_2B_2C$, they are nearly coincident.) Ni(d_{xy}) subbands and Ni($d_{3z^2-r^2}$) subbands produce coincident PDOS peak at 0.3 eV above E_F with some additional hybridization of Lu(d_{yz}, d_{zx}) subbands, and this peak is related to the flat band along Γ-M slightly above E_F. The other three Ni(3d) subbands hybridize with B(p_x, p_y, p_z), C(p_z), and Lu(5d) subbands to form a pair of peaks at 0.6 and 0.9 eV. The high-energy peak reflects Ni($d_{x^2-y^2}$)-B(p_z)-C(p_z) hybridization with some additional contribution from Ni(d_{yz}, d_{zx}) and Lu($d_{3z^2-r^2}$) states. The low energy peak has the character of Lu(5d)-Ni($d_{x^2-y^2}$, d_{xy}, d_{yz}, d_{zx})-B(p)-C(p_z) hybridization. The flat band along Γ-Z-A manifests itself as a peak at −0.3 eV of Ni(d_{xy})-Ni(d_{xy}) character. Another tiny peak (of the Lu($d_{3z^2-r^2}$)-C(p_x, p_y) character) is found at E_F.

5. Conclusion

We report the *ab initio* pseudopotential density functional calculation within the local density approximation.[4] The band structure, the TDOS, the LDOS, the PDOS, and the FS are presented for both $LuNi_2B_2C$ and LuNiBC. The Fermi level states are dominated by the Ni(3d) character, but with considerable contributions from other atoms. Ni(3d), B(2p), and C(2p) states hybridize by a noticeable amount around E_F resulting in the three-dimensional electronic structure, and the strength of such hybridization is not so large as the Cu–O hybridization in CuO_2 plane of high-T_c cuprates. Hence, the energy band structure is very complicated near the Fermi level and quite dispersive along the c-axis direction. Existence of the much constricted cylindrical FS is a reflection of the two-dimensional geometric structure of these compounds. The reduced $N(E_F)$ of LuNiBC compared with $LuNi_2B_2C$ would be a possible reason for the failure of observing superconductivity down to 4K.

---

[1] R. J. Cava, H. Takagi, H. W. Zandbergen, J. J. Krajewski, W. F. Peck Jr, T. Siegrist, B. Batlogg, R. B. von Dover, R. J. Felder, K. Mizuhashi, J. O. Lee, H. Eisaki, and S. Uchida, Nature **367**, 252 (1994).

[2] T. Siegrist, H. W. Zandbergen, R. J. Cava, J. J. Krajewski, and W. F. Peck Jr, Nature **367**, 254 (1994).

[3] J. Ihm, A. Zunger, and M. L. Cohen, J. Phys. C **12**, 4409 (1979); **13**, 3095(E) (1980).

[4] P. Hohenberg and W. Kohn, Phys. Rev. **136**, B864 (1964); W. Kohn and L. J. Sham, Phys. Rev. **140**, A1133 (1965).

[5] L. F. Mattheiss, Phys. Rev. B **49**, 13279 (1994); W. E. Pickett and D. J. Singh, Phys. Rev. Lett. **72**, 3702 (1994)

[6] N. Troullier and J. L. Martins, Phys. Rev. B **43**, 1993 (1991).

[7] J. P. Perdew and A. Zunger, Phys. Rev. B **23**, 5048 (1981).

[8] L. Kleinman and D. M. Bylander, Phys. Rev. Lett. **48**, 1425 (1982).

[9] E. R. Davidson, J. Comput. Phys. **17**, 87 (1975); B. Liu, in *Report on the Workshop 'Numerical Algorithms in Chemistry: Algebraic Methods' of the National Resource for Computation in Chemistry*, edited by C. Moler and I. Shavitt (Lawrence Berkeley Lab., Univ. of California, 9–11 August 1978), p. 49.

[10] H. Kim and J. Ihm, Phys. Rev. B (to be published).

NEW HIGH-FIELD MAGNETS DEVELOPED AT NRIM, JAPAN

H. MAEDA, K. INOUE, T. KIYOSHI, T.ASANO, Y.SAKAI,
T.TAKEUCHI, K.ITOH, and G. KIDO
Tsukuba Magnet Laboratories - National Research Institute for Metals
Sengen, Tsukuba 305, Japan

ABSTRACT

We have been developing and installing several high field magnets, i.e, long-pulsed magnets, a 40T class hybrid magnet, a 21T superconducting magnet, etc. The 40T class hybrid magnet is composed of a 15T superconducting magnet with a room temperature clear bore of 400mm and a 25T polyhelix-type water-cooled magnet with a 30mm bore. For the 15T superconducting magnet a new structural Cu-housing $(Nb,Ti)_3Sn$ conductor is adopted to reduce the influence of transverse compressive stress. In the 21.1T superconducting magnet with a clear bore of 50mm, operated at 1.8K, a newly-developed bronze-processed $(Nb,Ti,Ta)_3Sn$ multifilamentary wire is used. Inside the magnet small Bi-2212 test coils have been installed and excited to generate a total field up to 21.8T in a clear bore of 13mm. Using newly developed Cu-Ag alloy composite wires with high strength/high conductivity, we have fabricated two types of pulsed magnet coils to generate 73.4 T in a pulse duration of 5ms and 64.5 T in a long 100ms duration.

1. Introduction

A new high magnetic field research center, the Tsukuba Magnet Laboratories(TML) of the National Research Institute for Metals, Japan will be opened next year. Several kinds of high field magnets, i.e. long pulsed magnets, a 40T class hybrid magnet(40THM)[1], a 21T superconducting magnet(21TSM)[2], etc., are installed in a newly constructed building of TML and are now progressing in test operations. By the end of 1994 fiscal year, next March, all the magnets will be ready for operation.

Last October we succeeded in generating 21.16T at 1.8K in a clear bore of 50mm of the 21TSM[3]. The 21TSM is now being used, for instance, for a test of high-Tc oxide coils. Under a back-up field of 20.9T of the magnet an inserted small Bi-2212 oxide double pancake coil have generated totally 21.8T in a 13mm clear bore[4], which is the highest field in a full superconducting magnet system. A test operation of the 15T superconducting magnet, a part of 40THM, was carried out[5]. We will finish the full operation test of the 40THM by the end of next March. Newly developed Cu-Ag alloy wires and sheets are very useful not only for pulsed magnets but also for water-cooled resistive magnets, because of its excellent combination of high electrical conductivity and high strength. Using the Cu-16%Ag alloy wire, we have fabricated two types of pulsed magnet coils; one is a generating field of 73.4T in a 10 mm bore with a short pulse duration of 5ms[6], and the other is a 64.5T in a 16mm bore with a 100ms long duration.

2. 21T Superconducting Magnet(21TSM)

The 21TSM developed can supply several kinds of experimental spaces by changing the combination of three component coils(outer, middle and inner coils), as shown in Fig. 1.

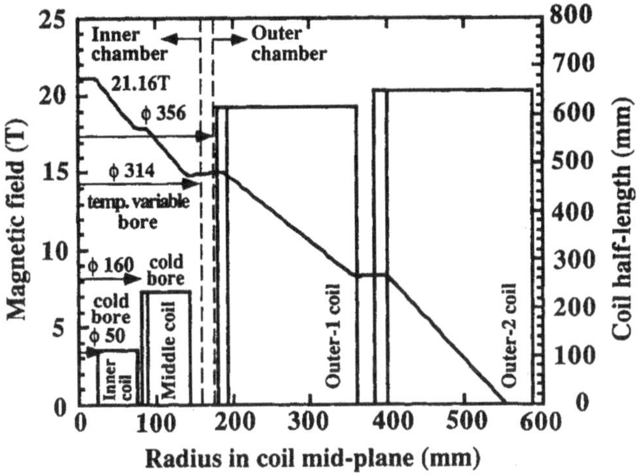

Fig. 1 Magnetic field distribution and bore sizes in the middle plane of 21T superconducting magnet

Table 1 Parameters for 21T superconducting magnet

Coil Parameter	Inner Coil(No. 3)	Middle Coil	Outer-1 Coil	Outer-2 Coil
Inner Winding Diameter(mm)	50	180	380	801
Outer Winding Diameter(mm)	151.2	285.2	717.4	1165.8
Winding Height(mm)	220.3	463.5	1230	1293.6
No. of Double Pancakes		31	40	41
No. of Turns	1760	1364	1520	2296
Superconducting Material	$(Nb,Ti,Ta)_3Sn$	$(Nb,Ti)_3Sn$	$(Nb,Ti)_3Sn$	Nb-Ti
Conductor Type	Monolith	Monolith	Cu-Housing+Al	Cu-Housing+Al
Whole Conductor Size	2 x 2.42	2 x 6	8.15 x 13.4	6 x 13.8
Copper (+Al) Ratio	0.48	1.5	5.15	7.31
Operation Current(A)	340.4	883	4717	4717
Maximum Field(T)	21.16	18.1	15.3	8.4
Stored Energy(MJ)		0.052	4.0	27.0

The specifications of the coils are listed in Table 1. The 21TSM is operated at 1.8K in saturated superfluid helium, and by three power supplies. The middle and inner coils are isolated thermally from the outer coils by another coaxial cryostat or chamber. When the middle and inner coils are removed out of the inner chamber, the magnet can supply a large temperature-variable bore of 314 mm with a field up to 15T. The middle coil generates 18T in a 160 mm bore. We have already fabricated three kinds of inner coils; two of them(Nos. 1 and 2)are made of a commercially available $(Nb,Ti)_3Sn$ conductor and the other(No. 3) listed in Table 1 is fabricated by a newly developed $(Nb,Ti,Ta)_3Sn$

conductor[7]. Using the No. 3 coil with a 50mm bore we succeeded in generating 21.16T, which is the highest field in metallic-superconducting magnets. To develop the new conductor with high critical current, I_c, improvements were made on following two points: One is a reduction of Cu ratio from 0.8 to 0.48. The other is the additions of both Ti and Ta to Nb_3Sn ; 0.6%Ti is added to bronze matrix and 0.5%Ta to Nb core. The both additions increase I_c in high field range not only due to the improvement of H_{c2} but also due to the uniform deformation of Nb cores. We are now fabricating a No. 4 inner coil with a 61mm bore, wound by $(Nb,Ti)_3Sn$ multifilamentary monolith conductor with an extremely reduced Cu ratio of 0.25. The coil is also designed to generate above 21T. The 21TSM will become more efficient and convenient.

Using the 21TSM we are now testing small coils, for instance, high-T_c oxide coils. Recently, a double pancake coil made of Bi-2212 tape, fabricated by the dip-coating process, was excited at 1.8K to generate 21.8T in a 13mm bore under a back-up field of 20.9T[4].

3. 40T Class Hybrid Magnet(40THM)

The 40THM is composed of a 15T superconducting magnet(15TSM-HM) with a room-temperature clear bore of 400mm and a 25T polyhelix-type water-cooled magnet(25TWM) with a 30mm bore(Fig. 2). The 40THM system has a 15MW dc power supply(the maximum of 430V and 35kA) and a 15MW water-cooling system(an inlet temperatue of 10 ℃, a flow rate of 700 t/hr and a pressure loss of 147MPa), and a helium liquefier(150 l/hr or 450W at 4.4K).

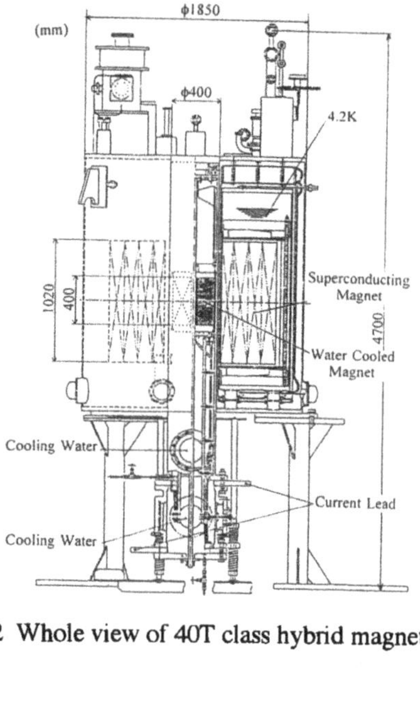

Fig. 2 Whole view of 40T class hybrid magnet

Table 2 Parameters for 15T superconducting magnet of 40T hybrid magnet

Central Field(T)	15			
Operation Temperature(K)	4.2			
Operation Current(A)	1476.1			
Spacer Ratio between Turns(%)	30			
Spacer Ratio between Pancakes(%)	50			
Inductance(H)	58.17			
Stored Energy(MJ)	63.37			
Net Weight of Winding Part(ton)	9.42			
No. of Pancakes	116			
Conductor Grade	G1	G2	G3	G4
Inner Diameter(mm)	471.56	669.80	922.28	1152.96
Outer Diameter(mm)	663.50	918.20	1150.16	1391.00
Winding Height(mm)	1020.80	1020.80	1020.80	1020.80
No. of Turns	16	23	27	32
Superconducting Material	$(Nb,Ti)_3Sn$	$(Nb,Ti)_3Sn$	Nb-Ti	Nb-Ti
Conductor Type	Cu-Housing	Cu-Housing	Monolith	Monolith
Copper Ratio	4.39	6.87	5.70	9.60
Coil Current Density(MA/m^2)	27.96	31.06	39.75	45.09
Maximum Field(T)	15.52	12.63	8.55	6.47
Joule Heat Flux(kW/m^2)	3.13	2.58	3.01	2.80

The 15TSM-HM is composed of 58 double pancake coils; each pancake is made of four kinds of superconductors(G1, G2, G3 and G4). The specifications of the 15TSM-HM and the four superconductors are listed in Table 2. The magnet is cryogenically stabilized to operate at 4.2K under the maximum current of 1476.1A and stored energy of 63.37MJ. We fabricated a new structural Cu-housing $(Nb,Ti)_3Sn$ conductors used in high field region(G1 and G2), as shown in Fig 3, in order to avoid the stress-induced Ic degradation of the conductor due to a large compressive transverse stress of 50-70MPa loaded to the magnet in a full operation. A clearance without solder between the Cu-housing and the $(Nb,Ti)_3Sn$ monolith reduced the stress to about 35 MPa and restrained the Ic degradation to only 18% at innermost part. The magnet was cooled down for about 145hr from room temperature to 4.2K[5].

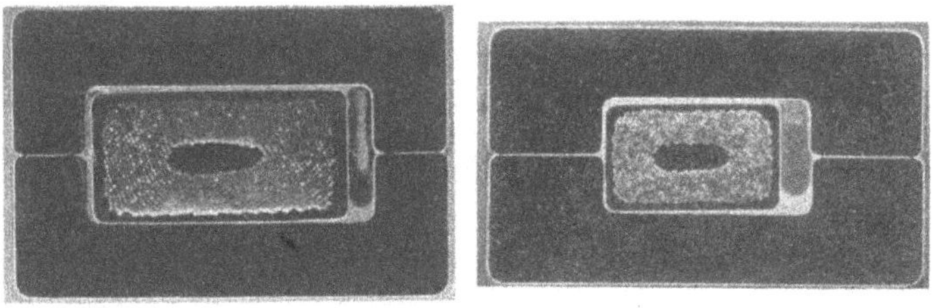

G1 G2

Fig. 3 Cross sections of Cu-housing $(Nb,Ti)_3Sn$ conductors; G1(5.4x8.2mm)and G2(4.8x8.2mm) conductors used for 15T superconducting magnet

The 25TWM is composed of 18 coaxial helical coils with a 30mm clear bore. The Specifications of the polyhelix coils are listed in Table 3. The inner 9 coils were fabricated by cutting from large Cu-0.73wt%Al_2O_3 alloy blocks, and the outer 9 coils by winding and machining cold-worked Cu-0.5wt%Al_2O_3 alloy rectangular strips. The magnet is expected to generate 25T at an operating current of 33.3kA, comsuming a power of 13.3MW, in a back-up field of 15T. The temperature and the hoop stress in any helix are designed to be lower than 100 C and 340MPa, much less than a yield strength of 420 MPa, respectively. The magnet is cooled directly by flowing deionized water at a rate of 700 t/hr, by three turbo-refrigerators, each with a cooling power of 5.2MW. We have also another 20T polyhelix-type water-cooled magnet with a 50mm bore(20TWM), composed of 15 polyhelix coils, which are fabricated by spark-cutting large Cu-Cr alloy blocks(Table 3).

Table 3 Parameters for polihelix coils of 40T hybrid magnets

Coil	25TWM	20TWM
Material	Cu-Al_2O_3	Cu-Cr
Total Field(T)	40	35
Incremental Field(T)	25	20
Clear Bore Diameter(mm)	30	50
Outer Coil Diameter(mm)	340	340
Coil Height(mm)	490	490
No. of Helix	18	15
Maximum Current(kA)	33.3	31.7
Maximum Voltage(V)	13.3	10.8
Flow Rate of Water(ton/hr)	646	700
Flow Velocity of Water(m/s)	18.5	17.9

4. Pulsed Magnets

We have been constructing two types of pulsed magnets, in order to achieve our final target to generate a field up to 80T in long pulsed field: One is a small magnet with a pulse duration of 5-10ms, matching to a small 10mF capacitor bank of 125kJ/5kV. The other is a large magnet with a 100-200ms duration for a large 128mF capacitor bank of 1.6MJ/5kV. Recently, we have developed a new Cu-Ag alloy wire with high strength(>1GPa) and high conductivity (>80%IACS), sufficiently useful for pulsed magnets[8]. The wire having a fine filamentary composite structure is heavily cold drawn(a reduction of area of 99%) with intermediate heat treatment at about 450 C.

Using 2 x 3mm rectangular Cu-16%Ag alloy wires we have fabricated several small multi-layered solenoid-type pulsed magnets(SPM) with a pulse duration of 5 - 10ms Internally reinforcements by glass fiber windings are carried out between the layers. Schematic view and specifications of the pulsed coils are shown in Fig. 4 and Table 4, respectively. The maximum field achieved so far is 73.4T for the SPM in a 10mm bore, as shown in Fig. 5[6]. For long pulsed magnets, on the other hand, using a large size, 4 x 6 mm a rectangular Cu-Ag wire, we have fabricated a large multi-layered pulsed magnet(LPM) with a 16mm bore(Table 4). The magnet generates 64.5T in a pulse duration of 100ms without destruction by the 1.6MJ bank(Fig. 5). The upper limit of generating field is mainly determined by the strength and conductivity of the conductor.

Since a higher tensile strength of 1.1-1.2GPa have been already realized in a laboratory-scale short sample, higher generating field will be expected in the near future.

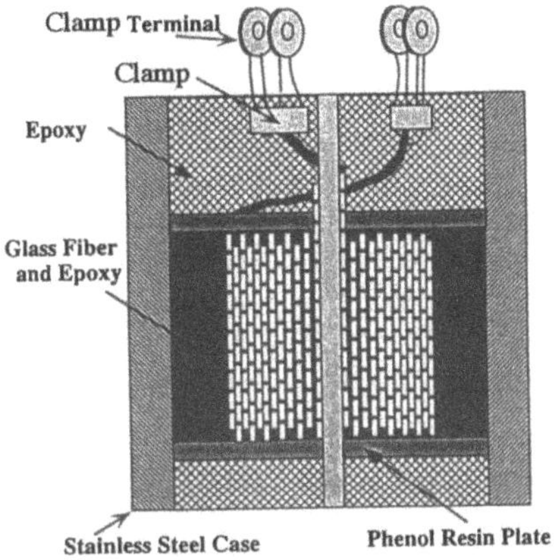

Fig.4 Structure of multilayered pulsed coil wound by Cu-Ag alloy wire

Table 4 Parameters of Cu-Ag multilayered pulsed coils

Coils	SPM	LPM
Inner Winding Diameter(mm)	10	16
Outer/Inner Diameter Ratio	5.68	9.34
Winding Length(mm)	55	130
No. of Turns/Layer	16	19
No. of Layers	8	12
Inter Layer Material	Stycast	Epoxy

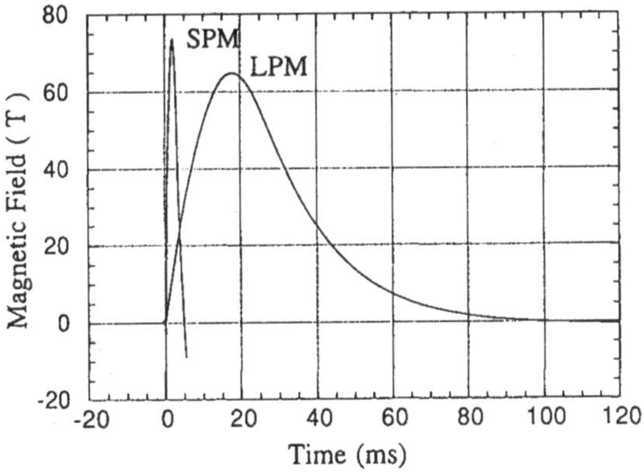

Fig.5 Pulsed magnetic field generated by Cu-Ag multilayered coils

5. Future Plan

In addition to the magnets described above, the TML have five compact superconducting magnets with clear bores of 30-70mm and a 12T split magnet with a 20mm gap. The present status of TML, however, seems to be not enough for accepting many near-future users. We have a plan to install some superconducting magnets including a 15T split magnet with a 35mm clear gap, a 1GHz-NMR(24T) magnet, etc., and several 20 - 27T water-cooled resistive magnets. For the water-cooled magnets the Cu-Ag alloy sheets, which we have developed recently, will be used. The sheets were testified to be useful for a Bitter-type water-cooled magnet: Last spring the Francis Bitter National Magnet Laboratory-MIT succeeded in generating 35.25T in the Hybrid III, the innner most part of which were made of our Cu-Ag sheets[9]. To develop the 1GHz-NMR superconducting magnet we will have to develop high performance high-T_C oxide coils to generate about 4T in a back-up field of 20T, for instance, using Bi-2212 wire.

6. References

1. K.Inoue, T.Takeuchi, T.Kiyoshi, K.Itoh, H.Wada, H.Maeda, T.Fujioka, S.Murase, Y.Wachi, S.Hanai, and T.Sasaki, *IEEE Trans. Magn.* **28**(1992)493
2. T.Kiyoshi, K.Inoue, K.Itoh, T.Takeuchi, H.Wada, H.Maeda, K.Kuroishi, F.Suzuki, T.Takizawa, N.Tada, and H.Mori, *IEEE Trans. Appl. Supercond.* **3**(1993)78
3. M.Oshikiri, K.Inoue, T.Kiyoshi, T.Takeuchi, K.Itoh, K.Takehana, M.Kosuge, Y.Iijima, and H.Maeda, to be published in *Physica B*(1994)
4. N.Tomita, M.Arai, E.Yanagisawa, T.Morimoto, H.Fujii, H.Kitaguchi, H.Kumakura, K.Inoue, K.Togano, H.Maeda, and K.Nomura, to be published in *Appl. Phys. Lett.*(1994)
5. K.Inoue, T.Takeuchi, K.Takehana, K.Itoh, T.Kiyoshi, H.Maeda, T.Fujioka, S.Murase, Y.Wachi, S.Hanai, Y.Dozono, and T.Kojo, to be published in *Cryogenics*(1995)
6. T.Asano, Y.Sakai, M.Oshikiri, K.Inoue, H.Maeda, G.Heremans, L.V.Bockstal, L.Li, and F.Herlach, to be published in *Physica* **B**(1994)
7. K.Takeuchi, H.Sekine, and Y.Iijima, Japan Patent No.1378791(1987)
8. Y.Sakai, K.Inoue, T.Asano, H.Wada, and H.Maeda, *Appl. Phys. Lett.* **59**(1991)2965
9. Y. Iwasa, M.J.Leupold, R.J.Weggel, and J.E.C.Williams, *IEEE Trans. Appl. Supercond.* **3**(1993)58, and Y. Iwasa, private communication(June, 1994)

Contributed Papers

1. Vortex Structure and Dynamics

CURRENT-VOLTAGE CHARACTERISTICS AND VORTEX-GLASS TRANSITION IN $Bi_2Sr_2Ca_2Cu_3O_x$ SUPERCONDUCTORS

H. Yamasaki, K. Endo, Y. Mawatari, S. Kosaka, M. Umeda, S. Yoshida & K. Kajimura
Electrotechnical Laboratory, 1-1-4 Umezono, Tsukuba-shi, Ibaraki 305, Japan

ABSTRACT

The current-voltage (*I-V*) characteristics have been investigated by transport measurement in epitaxial $Bi_2Sr_2Ca_2Cu_3O_x$ (Bi-2223) thin films and Ag-sheathed Bi-2223 tapes in magnetic fields applied parallel to the *c*-axis. Scaling analysis of the nonlinear *I-V* characteristics demonstrates the existence of a quasi-two-dimensional vortex-glass–liquid transition, in which the vortex-glass correlation length in the *a-b* plane scales with $|T - T_g|$ but the correlation along the *c*-axis is restricted only to adjacent a few CuO_2 planes, where T_g is the vortex-glass transition temperature. In the vortex-liquid phase above T_g there is a finite linear resistivity $\rho_{lin} = (E/J)_{J\to 0}$ = constant >0, while in the vortex-glass phase the linear resistivity seems to vanish.

1. Introduction

Bi-oxide cuprate superconductors are technologically very important because relatively high transport critical current densities, $J_c \sim 10^8$–10^9 A/m^2, can be obtained in polycrystalline samples such as Ag-sheathed tapes.[1] Investigations on how J_c is controlled or how dissipation occurs in the Bi-oxides are of crucial importance for the prevalent use of this material. Fisher, Fisher and Huse proposed a second-order phase transition from vortex liquid to vortex glass in disordered superconductors having random pinning centers.[2] In their theory the vortex-glass phase is regarded as a "true" superconductor having zero linear resistivity: $\rho_{lin} = (E/J)_{J\to 0} = 0$. A number of studies have been made on the vortex-glass–liquid transition of $YBa_2Cu_3O_7$ (YBCO),[3–6] and today it is accepted that the transition to a correlated glass phase with vanishing ρ_{lin} generally occurs in YBCO. Recently there have been several reports which suggest the existence of the vortex-glass–liquid transition in the highly anisotropic Bi-oxides.[7–9]

We have prepared high-quality, epitaxial $Bi_2Sr_2Ca_2Cu_3O_x$ (Bi-2223) thin films showing high critical temperatures $T_c(R{=}0)$ = 92–97 K,[10] and high critical current densities J_c = 10^{10}–10^{11} A/m^2 at 30–40 K.[11] Very recently we have investigated the current-voltage (*I-V*) characteristics in these high-quality Bi-2223 films in magnetic fields applied parallel to the *c*-axis, and a scaling analysis of the nonlinear *I-V* characteristics has demonstrated the existence of a quasi-two-dimensional (quasi-2D) vortex-glass–liquid transition.[12] In this paper we review the above work on the vortex-glass–liquid transition and present new results on Ag-sheathed Bi-2223 tapes.

2. Experimental

The details of film preparation were reported elsewhere.[10] The 60–100-nm thick films deposited on $LaAlO_3(100)$ single-crystal substrates were characterized by X-ray diffraction to be single-phase Bi-2223 films with *c*-axis oriented perpendicular to the film surface. High-resolution SEM and RHEED observations confirmed that they were epitaxially grown without apparent grain boundaries. The Ag-sheathed $Bi_{1.8}Pb_{0.4}Sr_2Ca_{2.2}Cu_3O_x$ tapes were prepared using a powder-in-tube process. The tapes, 4 mm wide and 0.25 mm thick, have 37 oxide filaments, and the volume fraction of the oxide is about 1/4.

The *I-V* characteristics were precisely measured with an error of several nanovolts by a four-probe dc method for small strip lines (50 μm × 20–50 μm) of thin films and for Ag-sheathed tapes with voltage taps 5 mm apart. We measured the resistance of the Ag sheath alone, R_n, and the supercurrent flowing in the oxide I_s is calculated by $I_s = I - V/R_n$, assuming the Ag-sheathed tape to be a parallel circuit of the oxide and the Ag sheath.

3. Quasi-2D vortex-glass transition in epitaxial films and Ag-sheathed tapes

We investigate the vortex-glass–liquid phase transition by measuring the electric field *E* vs current density *J* relation for a wide temperature range in our epitaxial Bi-2223 film (Fig. 1). The *E-J* curves shown in Fig. 1 are typical of all films and at all fields investigated (0.5–5 T). According to the vortex-glass–liquid transition theory by Fisher, Fisher and Huse,[2, 3] a power-law *E-J* relation is expected at the transition temperature T_g:

$$E(J; T=T_g) \sim J^{(z+1)/(D-1)}, \tag{1}$$

where z is the dynamical critical exponent and D is the dimension number. In the log*E* vs log*J* plots of Fig. 1 we define the vortex-glass transition temperature T_g to be the highest temperature at which a power-law *E-J* relation is observed in our electric-field window ($\sim 5 \times 10^{-1}$–10^4 μV/cm). In Fig. 1 we observe (1) in the vortex-liquid regime above T_g

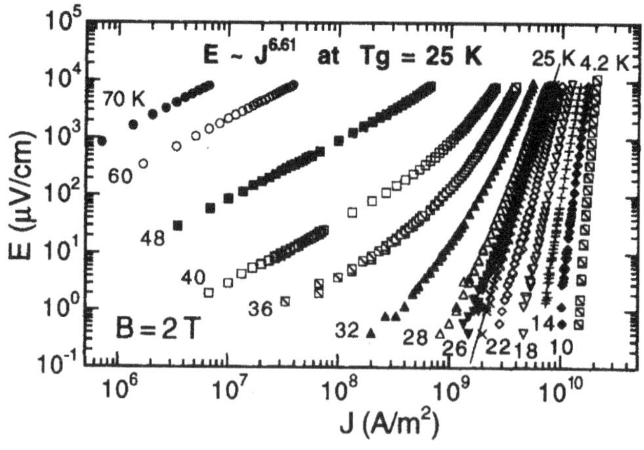

Fig. 1 The electric field *E* vs current density *J* relationship measured at $B = 2$ T at various temperatures in a 60-nm-thick Bi-2223 film.

there is a finite linear resistivity ρ_{lin} = constant >0 for small J, and the log$E$-log$J$ curve shows upward curvature; (2) in the vortex-glass regime below T_g the log$E$-log$J$ curve is linear or shows downward curvature corresponding to vanishing ρ_{lin}.

Figure 2 shows linear resistivity ρ_{lin} data above T_g plotted against $T - T_g$. Except for high temperature data there is a clear linear relationship between logρ_{lin} and log$(T - T_g)$, showing the validity of the critical scaling relation:

$$\rho_{lin} \sim (T - T_g)^{\nu(z+2-D)}, \tag{2}$$

where ν is the static critical exponent.[2, 3]

We assume a quasi-2D vortex-glass transition with $D = 2$ in Eqs. (1) and (2) for highly anisotropic Bi-oxides, in contrast to $D = 3$ in YBCO.[3–6] This means that the vortex-glass-transition correlation length in the a-b plane only scales with $|T - T_g|$ but the correlation length along the c-axis does not scale with $|T - T_g|$, remaining nearly constant.[12]

Figure 3 shows the critical scaling collapses of the vortex-glass transition.[2, 3] All the E-J curves shown in Fig. 1 and the data at $B = 4$ T collapse into two curves, corresponding to the vortex-liquid phase (upper curve) and to the vortex-glass phase (lower curve), respectively. In high fields, 2 T $\leq B \leq$ 5 T, similar scaling collapses have been observed for all films with constant (within experimental error) critical exponents $z \approx 5.5$ and $\nu \approx 1.4$,[12] which are comparable to the YBCO results.[3–6] This supports our proposal that the vortex-glass–liquid transition of Bi-oxides in high fields is a quasi-2D transition.

We have also investigated the I-V characteristics in Ag-sheathed Bi-2223 tapes, and observed similar quasi-2D vortex-glass–liquid transition.[13] In high fields ($B \geq 2$ T) we have observed in the log$E$-log$J$ plots a clear transition from the upward curvature to the

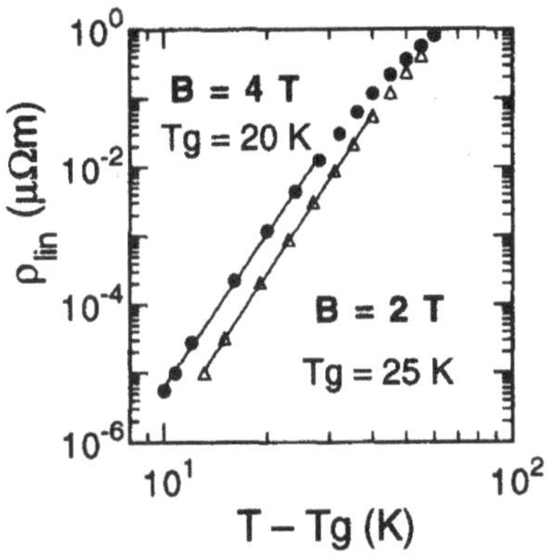

Fig. 2 ρ_{lin} vs $(T - T_g)$ plots at B = 2 T and 4 T.

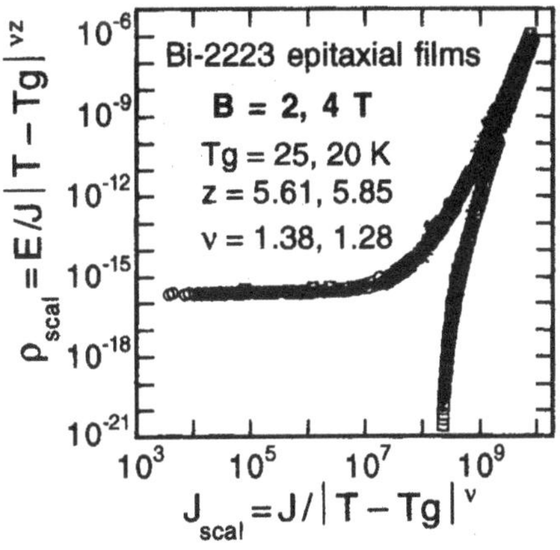

Fig. 3 Critical scaling collapses of the vortex-glass transition observed for the scaled J and the scaled ρ, for the E-J data at B = 2 T and 4 T.

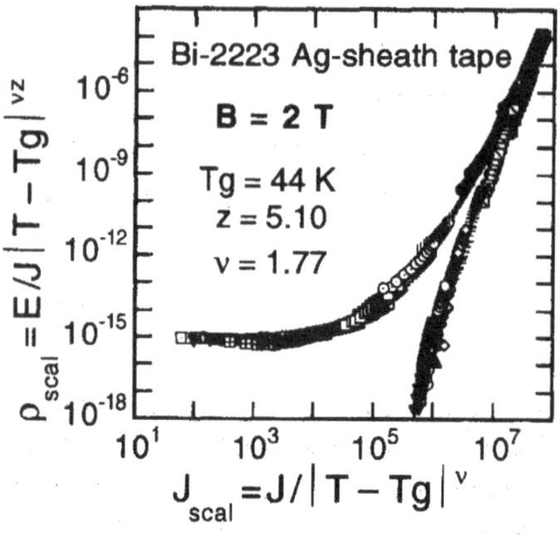

Fig. 4 Critical scaling collapse of the vortex-glass transition observed for a Ag-sheathed Bi-2223 tape.

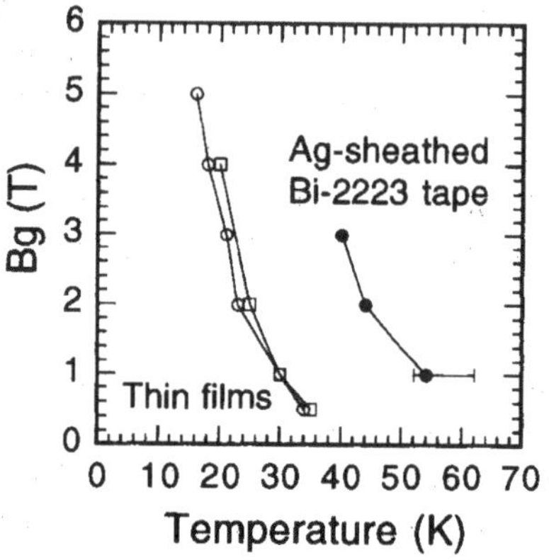

Fig. 5 Temperature dependence of the B_g of our Bi-2223 films and the Ag-sheathed Bi-2223 tape.

linear behavior or the downward curvature, and clear scaling collapses with similar z and ν which are obtained by Eqs. (1) and (2) with $D = 2$ (Fig. 4).[13] In Fig. 5 it is seen that the vortex-glass transition field B_g of the Ag-sheathed Bi-2223 tape is much higher than that of epitaxial films. One reason for this is the much higher critical temperature of the Ag-sheathed Bi-2223 tape: $T_c(R{=}0) = 108$ K, than those of epitaxial films: $T_c(R{=}0) \approx 90$ K.

4. Acknowledgement

The authors would like to thank Fujikura Co. Ltd. for providing their Ag-sheathed tapes.

5. References

1. K. Sato *et al.*, *J. Appl. Phys.* **70** (1991) 6484.
2. M. P. A. Fisher, *Phys. Rev. Lett.* **62** (1989) 1415; D. S. Fisher, M. P. A. Fisher and D. A. Huse, *Phys. Rev. B* **43** (1991) 130.
3. R. H. Koch *et al.*, *Phys. Rev. Lett.* **63** (1989) 1511; *ibid.* **64** (1990) 2586.
4. P. L. Gammel, L. F. Schneemeyer and D. J. Bishop, *Phys. Rev. Lett.* **66** (1991) 953.
5. C. Dekker, W. Eidelloth and R. H. Koch, *Phys. Rev. Lett.* **68** (1992) 3347.
6. D. G. Xenikos, J.-T. Kim and T. R. Lemberger, *Phys. Rev. B* **48** (1993) 7742.
7. H. Safar *et al.*, *Phys. Rev. Lett.* **68** (1992) 2672.
8. K. Shibutani *et al.*, *Appl. Phys. Lett.* **63** (1993) 3515; *ibid.* **64** (1994) 924.
9. Q. Li *et al.*, *Phys. Rev. B* **50** (1994) 4256.
10. K. Endo *et al.*, *Nature* **355** (1992) 327.
11. H. Yamasaki *et al.*, *IEEE Trans. Appl. Supercond.* **3** (1993) 1536; *ibid.*, *Phys. Rev. Lett.* **70** (1993) 3331.
12. H. Yamasaki *et al.*, *Phys. Rev. B* **50** (1994) in press.
13. Y. Mawatari, H. Yamasaki, S. Kosaka and M. Umeda (unpublished).

THE DIMENSIONALITY OF PINNED FLUXOIDS IN HIGH-T_c SUPERCONDUCTORS WITH LAYERED CRYSTAL STRUCTURES

KAORU YAMAFUJI
Depertment of Electornics, Kyushu University 36, 6-10-1 Hakozaki, Higashi-ku, Fukuoka 812, Japan

and

TAKANORI FUJIYOSHI
Depertment of Electrical Engineering and Computer Science, Kumamoto University, 2-39-1 Kurokami, Kumamoto 860, Japan

ABSTRACT

When an external magnetic field, $\boldsymbol{B}$, is applied in the direction parallel to the c-axis of a single crystal of layered superconductor, it is shown theoretically that the strength of the correlation of thermally vibrating fluxoids along the c-axis between neighboring layers begins to decrease as $(B/B_{cr})^{1/2}$ above the crossover field, B_{cr}, compared with the correlation of fluxoids belonging to the same layer. However, it is also shown that the medium-range correlation of fluxoids along the c-axis due to flux-pinning remains even at $B > B_{cr}$ with the almost same order as in isotropic low-T_c superconductors. Then the pinned fluxoids can be regarded to have nearly three-dimensional ($3D$) character even at $B > B_{cr}$, while they begin to have a tendency of psuedo-two-dimensional ("$2D$") character at $B > B_{cr}$.

1. Inroduction

For a single crystal of high-T_c superconductor with a layered crystal structure, the correlation between fluxoids belonging to different layers is expected to be weak owing to a weak superconducting coupling along the c-axis. In the presence of an applied magnetic field, $\boldsymbol{B}$, parallel to the c-axis, therefore, it has been suggested [1] that active thermal fluctuation brings about the disappearance of the correlation of a fluxoids along the c-axis between neighboring layers above a crossover field, B_{cr}, given by

$$b_{cr} \equiv B_{cr} / B_{c2}^{\parallel} \sim \eta \xi_{ab}^{2} / \gamma^2 d^2 , \qquad (1)$$

where $B_{c2}^{//}$ is the upper critical flux density parallel to the c-axis, ξ_{ab} is the coherence length in the (a, b) plane, d is the thickness of each crystal layer, $\gamma = \lambda_c / \lambda_{ab}$ is the anisotropy factor of the penetration depth, and η is a weak function of the magnetic field, B, and the temperature, T, with the order of magnitude of unity.

In this paper, the behavior of the correlation function of thermally vibrating fluxoids along the c-axis is investigated theoretically to obtain information on the dimensionality of pinned fluxoids below and above the crossover field.

2. Correlation Function Below and Above the Crossover Field

The correlation function between the thermal displacement of fluxoid segment at z=0 and the one at z=z is given by [2]

$$G(z; B, T) = \langle \boldsymbol{u}_{th}(z) \cdot \boldsymbol{u}_{th}(0) \rangle$$

$$= \frac{k_B T}{(2\pi)^3} \int_0^{K_M} 2\pi K_\perp dK_\perp \int_{-k_M}^{k_M} dk_z \frac{\exp(ik_z z)}{C_{66}K_\perp^2 + C_{44}(K_\perp, k_z)k_z^2 + \alpha_s}, \qquad (2)$$

where the z-axis is chosen in the derection of the c-axis, C_{66} and C_{44} are the shear and tilt moduli of fluxoids, α_s is the Labusch modified pinning parameter [3], k_B is the Boltzmann constant, and k_M and K_M are defined respectively as $k_M = \pi/d$ and $K_M = (4\pi B / \Phi_0)^{1/2} \sim 2\pi / a_f$ with a_f denoting the fluxoid spacing.

Let us define the relative strength of the correlation, $g(d; B, T)$, which is given by the ratio of the correlation between the thermal displacements of fluxoids in a given layer at z=0 and those in a neighboring crystal layer at z=d, *i.e.*, $G(d; B, T)$, to the auto-correlation of fluxoids belonging to the same layer, *i.e.*, $G(0; B, T)$:

$$g(d; B, T) = \frac{G(d; B, T)}{G(0; B, T)}. \qquad (3)$$

When the fluxoids have the pseudo-three-demensional ("3D") character, the correlation of a fluxoid along the c-axis between neighboring layers should be as strong as the auto-correlation of fluxoids inside the same layer. Since the integration of Eq.(2) leads to the result of $g(d; B, T) \sim 1$ at $B < B_{cr}$ [2], we can regard that the pinned fluxoids have

"$3D$ " character at $B < B_{cr}$

If we get the result of $g(d; B, T) \ll 1$ at $B > B_{cr}$, on the other hand, the correlation of fluxoids along the c-axis between neighboring layers can be regarded as almost disappeared at $B > B_{cr}$, and we can insist that the pinned fluxoids have pseudo-two-dimensional ("$2D$") character at $B > B_{cr}$. However, the intergration of Eq.(2) merely results in $g(d; B, T) \sim (B_{cr}/B)^{1/2}$ at $B > B_{cr}$ [2], and hence $g(d; B, T)$ does not decrease sharply across B_{cr}. Then it is reasonable to regard that B_{cr} is the crossover field above which the fluxoids ***begin*** to have a ***tendency*** of "$2D$" character.

At $B > B_{cr}$, furthermore, we can easily show by integrating Eq.(2) that the medium-range ($z \gg d$) correlation of pinned fluxoids along the c-axis is given by [2]

$$G(z; B, T) \sim \frac{k_B T}{4\pi}\left(\frac{B_{c2}^{/\!/ 2}}{\mu_0}\right)^{-1}\frac{\kappa^2}{\delta_{66}}\frac{1}{b}\frac{1}{|z|}\exp\left[-\left(\frac{\mu_0 \alpha_s}{B_{c2}^{/\!/ 2} b^2}\right)^{1/2}|z|\right], \tag{4}$$

where κ is the Ginzburg-Landan parameter and $\delta_{66} = (1-b)^2(1 - 0.58b + 0.29b^2)/4$ is a weakly decreasing function of b. Since neither the anisotropy factor, γ, nor any substancial constant characterizing the Josephson-like coupling between successive crystal layers is appearing in Eq.(4), it is clear that the same expression for $G(z; B, T)$ as Eq.(4) can be obtained for low-T_c isotropic superconductors except for a minor difference of $B_{c2}^{/\!/}$ from B_{c2}. It is worth reminding that the pinned fluxoids in low-T_c isotropic bulk superconductors have been regarded as $3D$-fluxoids due to the existence of the medium-range correlation of fluxoids with this order of magnitude. Even in the layered bulk superconductors, therefore, we can regard that the pinned fluxoids have $3D$-character, so far as the medium-range correlation of $|z| \gg d$ along the c-axis is concerned.

3. Discussion

Two kinds of the correlation along the fluxoid direction appear for the *pinned* fluxoids in the *layered* superconductors: one is the medium-range correlation of $|z| \sim l_{44}^{pin} = (C_{44}(0, 0)/\alpha_s)^{1/2} = (B_{c2}^{/\!/ 2} b^2/\mu_0\alpha_s)^{1/2} \gg d$ due to the flux-pinning, and the another is the short-range correlation of $|z| \sim l_{44}^{layer} = (\eta\xi_{ab}^2/\gamma b^2)^{1/2}$ due to the existence of layered structure.

In the isotropic superconductors with $\gamma \sim 1$ including the metallic superconductors, the pinned fluxoids have been defined to have the three-dimensional ($3D$) character when

l_{44}^{pin} is smaller than the sample thichness, D_s, while they are regarded as having the two-dimesional (2D) character when l_{44}^{pin} becomes larger than D_s. As can be seen from Eq.(4), the same correlation length as l_{44}^{pin} also appears for the medium-range correlation of *pinned* fluxoids in the *layered* superconductors. This result originates from the fact that the softening of $C_{44}(K_\perp, k_z)$ is not remarkable for small wave numbers of $k_z \sim 1/l_{44}^{pin}$, even in the layered superconductors with $\gamma >> 1$. Thus the definition of the 3D or 2D fluxoids due to flux-pinning is also applicable even in the layered superconductors with $\gamma >> 1$, where each fluxoid can be discontinuous locally for the scale of $|z| \sim d$ at $B > B_{cr}$.

Only in the layered superconductors with $\gamma >> 1$, on the other hand, l_{44}^{layer} can become smaller than the layer thickness, d, at $B > B_{cr}$, and even the *pinned* fluxoids begin to have the tendency of pseudo-two-dimensional ("2D") character. This is the reflection of the fact that $C_{44}(K_M, k_z)k_z^2$ can become smaller than $C_{66}(K_\perp)^2_{max} = C_{66}K_M^2$ at $B > B_{cr}$ due to a remarkable softening of $C_{44}(K_M, k_z)$ at $k_z \sim 1/d$. In other words, the "2D" character of fluxoids does not appear if a remarkable softening of $C_{44}(K_M, k_z)$ for $k_z \sim 1/d$ does not occur.

The definition of the 3D or 2D fluxoids is related to the correlation with the scale of $|z| \sim 1/l_{44}^{pin} >> d$, while the another definition of the "3D" or "2D" fluxoids is related to the correlation of $|z| \sim d$. Futhermore, the correlation becomes stronger by the crossover of 3D → 2D and "2D" → "3D", respectively. Since both kinds of correlations do co-exist for the pinned fluxoids in layered superconductors, the interpretation on the dimensionality of *pinned* fluxoids in *layered* superconductors should be made very carefully by taking account of the spatial scale of the observed phenomena. In conclusion, a more proper rearrangement of the concept of fluoxid dimension is desirable for a further detailed understanding of the behavior of *pinned* fluluxoids in *layered* superconductors.

4. References

1. L.I.Glazman and A.E.Koshelev, *Phys.Rev.* **B43** (1991) 2835.
2. K.Yamafuji, T.Wakuda, T.Fujiyoshi and K.Funaki, *Physica* **C**, to be published.
3. K.Yamafuji, T.Fujiyoshi, K.Toko, T.Matuno, K.Kishio and T.Matsushita, *Physica* **C212** (1993) 424.
4. R.Labusch, *Crystal Lattice Defects*, **1** (1969) 1.
5. A.I.Larkin and Yu.N.Ovchinnikov, *J.Low Temp.Phys.* **34** (1978) 409.

THE MIXED STATE IN HIGH-T_c SUPERCONDUCTORS BY THREE DIMENSIONAL NEUTRON DEPOLARIZATION

W. ROEST and M.Th. REKVELDT
Interfacultair Reactor Instituut, Mekelweg 15,
2629 JB Delft, The Netherlands

ABSTRACT

In studying the vortex behaviour in the high-T_c superconductors, mainly surface techniques and magnetization measurements are used up to now, giving no information about the internal vortex distribution. Three Dimensional Neutron Depolarization (3DND) does give this information about the local magnetization together with the average flux bundle size and the average direction of the flux bundles. Hence, 3DND is most sensitive in the glass and liquid region of the phase diagram of the vortices and can be of help in solving the problems that are still present in this part of the vortex phase diagram. The present 3DND measurements are done in low magnetic fields or in the remanent state only because the sensitivity is the highest in these field conditions. A review of the main results as a function of field, time, temperature and position will be presented, showing a strong correlation between the vortex distribution and the sample characteristics.

1. Introduction

A review of an extended study of vortex distributions in several types of high-T_c superconductors by Three Dimensional Neutron Depolarization[1] (3DND), will be given. The advantage of using this method instead of more available methods as common magnetization measurements, is that 3DND opens the possibility of studying the flux distribution inside a bulk sample on a micrometer scale.

Moreover, 3DND has proven to be an easy and accurate method for determining simultaneously the correlation length of local magnetic structures and the magnetization in bulk samples, statically[2] as well as dynamically[3]. It will be shown that 3DND is a powerful method especially in the glass and liquid region of the vortex distribution phase diagram by a selection of our recent experimental results. Neutron depolarization experiments have been done before[4–10] on vortices in superconductors usually in one dimension and not as well understood as the results obtained here.

2. Three Dimensional Neutron Depolarization (3DND)

The 3DND technique is based on Larmor precession of a polarized neutron beam in a magnetic field. When a polarized neutron beam is passing through a sample, the change from the initial to the final polarization of the beam can be expressed as:

$$\vec{p}_f = \tilde{D}\vec{p}_0, \tag{1}$$

where the depolarization matrix $\tilde{D}$ is a combination of a Larmor precession matrix $\tilde{R}$ and a diagonal depolarization matrix $\tilde{D}_p$. The depolarization matrix $\tilde{D}$ is the quantity that will be measured by 3DND.

The first effect in the depolarization matrix $\tilde{D}$ is Larmor precession of the polarization vector over an angle ϕ around the mean magnetic induction. This Larmor precession angle is given by: $\phi = c\lambda L\langle\vec{B}\rangle$. Here, the typical neutron constant is given by $c = 4.67 \times 10^{14}$ ($T^{-1}m^{-2}$), λ is the neutron wavelength, $\langle\vec{B}\rangle$ is the average induction and L is the transmission length through the sample.

The second effect contained in the depolarization matrix, is depolarization defined as shortening of the polarization vector. It can directly be related to the correlation functions $\alpha_i = \xi\langle\Delta B_i^2\rangle$, $i = x, y, z$ of the local fluctuations in the induction compared to the average induction. The correlation length ξ is defined as the characteristic length over which the local magnetic fluctuation $\Delta\vec{B}$ can change or in other words, the size of the local induction fluctuation. The fluctuation $\Delta\vec{B}$ is defined by $\Delta\vec{B} = \vec{B}(\vec{r}) - \langle\vec{B}\rangle$. Generally it is not unreasonable to assume that $\langle\Delta B^2\rangle \sim \langle B\rangle^2$, allowing the definition of a field independent, reduced correlation function $\delta \equiv \frac{\sum_i \alpha_i}{\langle B\rangle^2}$ which has the dimension of a length.

3. Vortices in Sintered $YBa_2Cu_3O_7$

The sample used in this experiment is a sintered Y(123) sample consisting of grains of ~ 1 μm size clustered in strongly correlated poly-crystalline clusters of 10 μm to 100 μm diameter[11]. The sample had a size of $5 \times 10 \times 20$ mm^3 and was clamped in a magnetic yoke to prevent demagnetization fields outside the superconductor and to simulate infinitely long fluxlines. The temperature dependence of the remanent induction after pulsed field input is measured to study the thermal excitation of vortices in this kind of material. The field was pulsed into the sample after ZFC with a pulse-width of 500 μs and a peak value of 0.7 T.

After the field pulse, the trapped vortex distribution is studied by 3DND. This trapped vortex distribution has no measurable time relaxation over 24 hours showing that time effects are not important on the time scale of the experiment. Therefore, the sample could be heated and meanwhile the depolarization matrix could be measured without worrying about time effects.

Because the field was applied in the z-direction, the trapped induction is the largest parallel to this direction resulting in a large precession angle around the z-axis. This large precession angle destroys the possibility to separate the fluctuations in the induction in the x-direction from those in the y-direction. Therefore, α_x and α_y had to be averaged, defining the perpendicular correlation function $\alpha_p = (\alpha_x + \alpha_y)/2 = \xi\langle\Delta B_p^2\rangle$, where ΔB_p is the induction fluctuation perpendicular to the z-direction.

The depolarization results are given in fig. 1 where both the remanent induction in fig. 1a and the correlation functions in fig. 1b can be explained by the following model for the temperature dependent vortex behaviour in a granular superconductor[12]. At 4.2 K, right after the pulse, the vortices are pinned in a 3D vortex glass in the grain-clusters causing correlated vortex bundles and an $\alpha_p > 0$. Heating the sample, at first causes vortices to straighten out reducing α_p and increasing $\alpha_{||}$. Above 20 K, all vortices are straight and a 2D vortex glass is formed. Up to 35 K, thermal excitations make the vortex density fluctuations disappear and a lattice is created. The lattice is the lowest energy state and will persist up to T_c where vortices cease to exist.

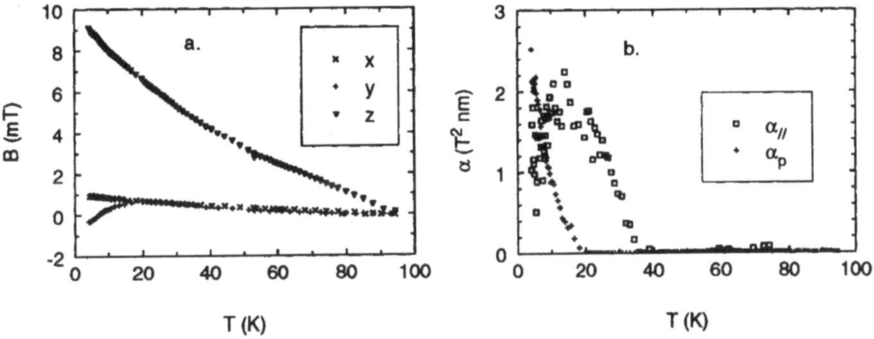

Figure 1: a.) Remanent magnetic induction; b.) Depolarization parameters $\alpha_{||} \equiv \alpha_z$ and $\alpha_p \equiv (\alpha_x + \alpha_y)/2$; all as a function of temperature after ZFC and a magnetic field pulse of 0.7 T during 500 μs at 4.2 K

It is possible to derive an estimate for the size of the vortex bundles by calculating the reduced correlation function δ. This estimate starts at 80 μm at 4.2 K and decreases about linearly with increasing temperature to zero at 40 K. The result for δ is of the order of the grain-cluster size in the sample (which is between 10 μm and 100 μm) supporting a picture of vortices pinned in the clusters and bending with them.

4. Reorientation of Vortices in Textured $Bi_2Sr_2CaCu_2O_4$

The textured Bi(2212) sample used in these experiments had dimensions $6 \times 6 \times 20$ mm^3 with the b-axis oriented along the longest sample side. It consists of many plate like single crystals with the c-axes assumed to be random in the plane perpendicular to the longest sample side. In the experiments, the sample was clamped in a magnetic yoke to prevent demagnetizing fields outside the sample from depolarizing the beam and to simulate infinitely long flux lines. The sample orientation was with the longest side (b-axis) along the z-direction which was also the applied field direction. The x-axis was defined as the transmission direction of the neutron beam.

On this textured sample, first, a measurement of trapped flux as a function of the applied field after ZFC down to 69 K, was done. Before every measured point, starting at zero field, the field is switched on during 3 s after which the measurement is performed in zero field.

In fig. 2a, it is observed that in the applied field direction (z) the induction increases with increasing applied field until a certain saturation is reached. Because of the sample texture, this remanence is accounted to Josephson vortices, trapped between the plate like crystals[13]. At an applied field of approximately 30 mT, perpendicular (x and y) magnetization components start to appear. At this field, the critical field value H_{c1} for flux penetration parallel to the c-axis is reached and vortices can penetrate the crystals. After penetration in field and switching off the field, the vortices reorient to a direction parallel to the c axis for energy reasons (see ref.[14]).

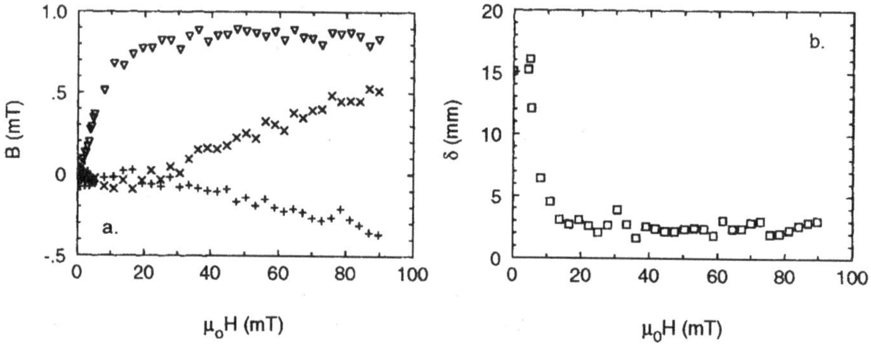

Figure 2: Trapped flux properties as a function of applied field after ZFC at 69 K: a.) Remanent induction, b.) Correlation function. The different symbols denote: ($\times$) x-direction; ($+$) y-direction and (∇) z-direction

The reduced correlation function δ, plotted as a function of field in fig. 2b, decreases quickly in low fields and becomes constant above 30 mT where the reorientation starts to occur. When reorientation has happened, the correlation length of the vortices will be determined by the crystal sizes and therefore it will be a constant as a function of field.

To study the relaxation in time of an initiated vortex distribution, a dynamic 3DND experiment was done after a short field pulse and ZFC. A field pulse of 0.9 T was applied during 1 ms and time channels were filled with neutron counts during, and after the field pulse. The procedure is repeated until reasonable counting statistics is obtained[3]. It was checked that every pulse gave the same time-effect and so it was made certain that the obtained result was not an addition of different contributions of each pulse.

The time relaxation of the vortex distribution can, in general, be described by an interpolation between the Anderson-Kim model[15] and the magnetization as a function of time of Fisher et al.[16], given by Geshkenbein et al.[17]. If the system is in the limit of single vortex creep, this interpolation formula can be reduced to[18]:

$$M(t) = M_0 \left(\frac{t}{\tau_0} + 1\right)^{-\frac{T}{U_c}} \tag{2}$$

with τ_0 the characteristic time, M_0 the magnetization at $t = 0$ and U_c the energy barrier in the critical state.

The time dependence of the induction is shown in fig. 3a. There, a very fast increase of the x-induction component after the field pulse is observed. This finite perpendicular induction component is explained by the fact that in the experiment the crystals can not exactly be aligned with the field because they do not have the same orientation throughout the sample. Therefore, vortices penetrate the crystals in the pulse. When the external field is removed the vortices reorient to a direction parallel to the c-axis.

Because of the sample texture the remanence along the z-axis was accounted to Josephson vortices trapped between the plate like crystals and so the time relaxation can be fitted by eq. (2), leading to the following fit parameters: M_0=2.75 mT, U_c = 355 K, and τ_0 = 1.14 ms.

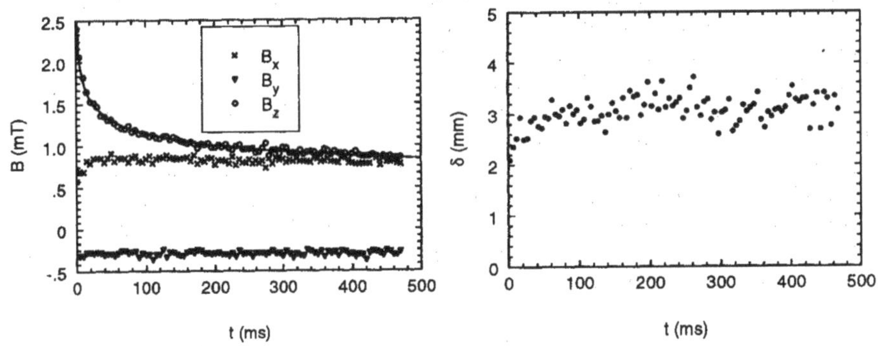

Figure 3: Time relaxation at 69 K after a field pulse of 0.9 T during 1 ms of: a. left) The average induction, where the drawn line is a fit made by eq. (2); b. right) The reduced correlation function

Furthermore, the total reduced correlation function δ is plotted as a function of time in fig. 3b. There, the induction independent time development of the correlation length of the local induction fluctuations is given. A fast time relaxation shortly after the field pulse, caused by the proposed reorientation of vortices, is observed. Clearly, after the reorientation, the reduced correlation function δ is constant within the accuracy of the measurement consistent with the theory that two coexisting vortex distributions are present, one with Josephson vortices parallel to the z-axis and one with Abrikosov vortices parallel to the crystal c-axis. The size of the correlated bundle is the same as in the field dependent experiment and is of the order of the sample size, supporting the reorientation model.

5. Position Dependence of a Vortex State in Single Crystalline La(214)

A single crystalline $La_{1.85}Sr_{0.15}CuO_4$[19] sample of optimal Sr doping and a mosaic spread of 0.5 degree is used to study the vortex distribution in a single crystal. The sample used in the experiment had a T_c of 35 K and was of the dimensions $2.4 \times 3.3 \times 4.1\ mm^3$ corresponding to the c-, a- and b-axis respectively. It was field cooled (FC) down to 4.2 K in a field of 8.9 mT which was applied parallel to the neutron beam transmission direction and the c-axis. At 4.2 K the field was removed and a 3DND scan was made. To study the vortex distribution locally, the crystal, including its environment, was scanned in the direction perpendicular ($= y$) to the beam and the field by a well collimated beam of $1.6 \times 0.2\ mm^2$. The method of scanning neutron depolarization was first applied on a sintered high T_c superconductor by Akselrod et al[20].

The remanent induction and the correlation functions are given in fig. 4. There it is observed that, after FC, in the remanent state, vortices are trapped in the applied field direction and a non-zero induction gradient towards the sample edge is present. The trapped vortex density is going to zero at the edge as expected. The gradient is not consistent with the Bean model because it is not a constant as a function of position. Furthermore, the trapped induction has a maximum that is apparently 50 % larger than the originally applied field. This can be explained by the fact that 3DND measures the line integral along the neutron path. From the Maxwell equations it can be derived that for this line integral the demagnetization fields of the magnetization in the sample, are canceled out. Therefore, 3DND only measures the magnetization multiplied by the transmission length ML. Because of the large demagnetizing factor $N_d \approx 0.5$ of the sample geometry used and the fact that the remanent induction $B \approx \mu_0 H_{app}$ for FC, the magnetization $M = B/\mu_0 = H_{app}/(1 - N_d)$ can be larger than the applied field H_{app}.

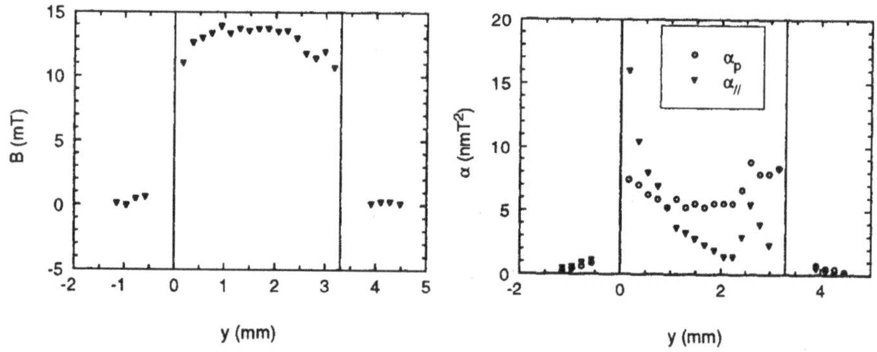

Figure 4: Scanning 3DND on a single crystalline, optimal Sr doped, La(214) sample in remanence after FC in an 8.9 mT field parallel to the c-axis, down to 4.2 K: a. left) Trapped induction as a function of position. b. right) The correlation functions parallel and perpendicular to the original field direction as a function of position

Vortex density fluctuations can be characterized by looking at the correlation function parallel to the applied field. The results are plotted in fig. 4b. In this configuration where the vortices are parallel to the beam, it is possible to calculate a length of straightness of the vortices[12]. This length is about ± 1 μm in the left edge increasing to about 10 μm in the middle. The asymmetry observed is caused by geometry effects. The crystal is not a perfect cube leading to an asymmetric correlation function.

In fig. 4b, also the fluctuations perpendicular to the applied field show a position dependence. First of all it is clear that the vortices are not straight but meander through the superconductor every where as can be concluded from the fact that $\alpha_p > 0$. Furthermore, the curving of vortices increases slightly towards the edge, despite the decrease in the induction (see fig. 4a). This increase could be explained by demagnetizing fields outside the superconductor (which will have an effect on the depolarization). By calculating the reduced correlation function for the perpendicular fluctuations it is possible to obtain an estimate for the vortex bundle size. The reduced correlation function is almost the same for all positions $\delta \approx 30\mu$m.

6. Discussion

By means of a couple of illustrative 3DND results, the wide range of applications of the technique has been shown. It is useful to compare these 3DND results with other techniques for the investigation of vortices inside bulk samples. The first important method is Small Angle Neutron Scattering (SANS) on vortices in high T_c superconductors well shown by Yethiraj et al.[21]. In that paper it is shown that SANS is sensitive to a vortex lattice, but the signal disappears when melting of the lattice takes place or when a glass state is present. Furthermore, the sensitive field range for SANS is limited for contrast reasons.

A second technique is Muon Spin Resonance (μSR) sensitive for both the vortex itself and for vortex density fluctuations[22]. When vortex density fluctuations are present it is almost impossible to separate the single vortex contribution from the fluctuation contribution.

3DND is only sensitive to vortex distribution fluctuations and so is most sensitive in the glass and liquid region of the vortex phase diagram as shown in this paper. Therefore 3DND is complementary to SANS and μSR. Furthermore, it is very easy to measure as a function of field, time and temperature, something that is not possible with the other techniques.

Acknowledgements

The Authors would like to acknowledge D.H.A. Blank of the University of Technology Twente, V.A.M. Brabers of the University of Technology Eindhoven and V.H.M. Duijn of the University of Amsterdam for the preparation of the samples. Furthermore, we would like to acknowledge P.F.M. Verhoeven of the University of Technology Delft for his contribution in the measurements.

References

1. M.Th. Rekveldt, *Text. Microstruct.* **11**, (1989) 127.
2. R. Rosman and M.Th. Rekveldt, *J. of Magn. and Magn. Mat.* **95**, (1991) 319.
3. F.J. van Schaik, J.W. Burgmeyer and M.Th. Rekveldt, *J. Appl. Phys.* **52**, (1981) 352.
4. H.W. Weber, *J. of Low Temp. Phys.* **17**, (1974) 49.
5. D. Petitgrand, B. Dillon, I. Mirebeau, G. Parette and G. Collin, *Proceedings ICTPS'90*, Rio de Janeiro (Brazil), (1990) 175.
6. L.A. Akselrod, G.P. Gordeev, V.N. Zabenkin, I.M. Lazebnik, V.I. Sbitnev and B.P. Toperverg, *Physica B* **174**, (1991) 354.
7. R.J. Papoular and G. Collin, *Phys. Rev.* **38**, (1988) 768.
8. M.L. Crow, R.J. Goyette, A.C. Nunes, S.J. Pickart, T.R. McGuire, S. Shinde and T.M. Shaw, *J. Appl. Phys.* **67**, (1990) 4542.
9. V.V. Aksenov, E.B. Dokukin, Yu.V. Nikintenko, A.V. Petrenko and S.A. Sergeenkov, *Phys. Scripta* **T49**, (1993) 650.
10. P.A. Miles, S.J. Kennedy, K.N.R. Taylor and G.D. Gu, *proc. of the ICM'94*, (Warsaw, Poland, 1994) to be published.
11. D.H.A. Blank, H. Kruidhof and J. Flokstra, *J. Phys. D* **21**, (1988) 226.
12. W.Roest and M.Th. Rekveldt, *Phys. Rev. B* **48**, (1993) 6420.
13. W.Roest and M.Th. Rekveldt, *Proc. of the* 7th *IWCC*, (Alpbach, Austria, 1994) to be published.
14. P.H. Kes, *Nature* **350**, (1991) 192 and references in there.
15. P.W. Anderson and Y.B. Kim, *Rev. Mod. Phys.* **36**, (1964) 39.
16. K.H. Fischer and T. Nattermann, *Phys. Rev. B* **43**, (1991) 10372.
17. V.B. Geshkenbein, A.I. Larkin, M.V. Feigel'man and V.M. Vinokur, *Physica C* **162-164**, (1989) 239.
18. M.V. Feigel'man, V.B. Geshkenbein and V.M. Vinokur, *Phys. Rev. B* **43**, (1991) 6263.
19. V.H.M. Duijn, N.T. Hien, A.A. Menovsky and J.J.M. Franse, to be published.
20. L.A. Akselrod, G.P. Gordeev, V.N. Zabenkin, M.R. Kolkhidashvili, I.M. Lazebnik, E.L. Ploshchanskaya, V.I. Sbitnev, *Physica C* **221**, (1994) 219.
21. M. Yethiraj, H.A. Mook, G.D. Wignall, R. Cubitt, E.M. Forgan, D.M. Paul and T. Armstrong, *Phys. Rev. Lett.* **70**, (1993) 857.
22. B. Pümpin, H. Keller, W. Kündig, W. Odermatt, I.M. Sacić, J.W. Schneider, H. Simmler, P. Zimmermann, E.Kaldis, S Rusiecki, Y. Maeno and C. Rossel, *Phys. Rev. B* **42**, (1990) 8019.

MAGNETIC FLUX DISTRIBUTIONS IN SINTERED $YBa_2Cu_3O_7$

W. ROEST and M.Th. REKVELDT
Interfacultair Reactor Instituut, Mekelweg 15,
2629 JB Delft, The Netherlands

ABSTRACT

The past few years, many attempts have been made to determine the vortex distribution in high T_c superconductors with only limited successes. Three Dimensional Neutron Depolarization (3DND) is a good tool to study these vortex distributions because it provides information about the internal magnetic structure in a bulk sample. By scanning a neutron beam through the sample, the magnetization and the depolarization, related to the size of a correlated flux area and the induction fluctuation in this area, are determined as a function of the beam transmission position. In low field measurements it was observed that a constant fraction of the applied field penetrated freely. This constant fraction indicates that certain parts of the sample have an H_{c1} smaller than the lowest applied field (1.8 mT) which can be understood from the granular sample structure. At somewhat higher fields, a small induction gradient is observed near the edges, suggesting pinning between the grains. Furthermore, an asymmetry with respect to the edges of the sample, is observed for the vortex-bundle size and the flux direction.

For the equilibrium state of trapped vortices, usually the Bean model[1] is used to describe the internal vortex distribution. A good experimental confirmation of the validity of this model has not been published yet. Techniques like scanning hall probe and Kerr effect measurements have been used to get this experimental confirmation, but those methods can only provide information about the surface distribution and nothing about the internal vortex behaviour. Three Dimensional Neutron Depolarization[2] (3DND), the technique used here, opens the possibility of studying the flux distribution inside a bulk sample on a micrometer scale. Therefore, by scanning 3DND, a good impression of the flux distribution and the local flux orientation can be obtained[3]. In this paper, an example of scanning 3DND on vortices in sintered $YBa_2Cu_3O_7$ will be given and the results will be compared with the Bean model.

The 3DND technique is based on Larmor precession of a polarized neutron beam in a magnetic field. When a polarized neutron beam is passing through a sample, the change from the initial to the final polarization of the beam can be expressed as:

$$\vec{p}_f = \tilde{D}\vec{p}_0, \tag{1}$$

where the depolarization matrix $\tilde{D}$ is a combination of a Larmor precession matrix $\tilde{R}$ and a diagonal depolarization matrix $\tilde{D}_p$. The depolarization matrix is the quantity that will be measured by 3DND.

The first effect, contained in the depolarization matrix $\tilde{D}$, is Larmor precession of the polarization vector over an angle ϕ around the mean magnetic induction in the sample. This Larmor precession angle is given by: $\phi = c\lambda L\langle\vec{B}\rangle$. Here, the typical neutron constant is given by $c = 4.67 \times 10^{14}$ ($T^{-1}m^{-2}$), λ is the neutron wavelength, $\langle\vec{B}\rangle$ is the average induction and L is the transmission length through the sample.

The second effect, contained in the depolarization matrix, is depolarization defined as shortening of the polarization vector. This depolarization can directly be related to the, so called, correlation functions $\alpha_i = \xi\langle\Delta B_i^2\rangle$, $i = x, y, z$ of the local fluctuations in the induction compared to the average induction. The correlation length ξ is defined as the characteristic length over which the local magnetic fluctuation $\Delta\vec{B}$ can change or, in other words, ξ is equivalent to the average size of the local induction fluctuation. The induction fluctuation $\Delta\vec{B}$ is defined by $\Delta\vec{B} = \vec{B}(\vec{r}) - \langle\vec{B}\rangle$. In the case where the average induction is large so that $\phi > \pi/2$, it is no longer possible to separate the three correlation functions and it is necessary to average the correlation functions perpendicular to the applied field. In the experiment described here, the applied field direction is x and we define the perpendicular correlation function: $\alpha_p \equiv (\alpha_y + \alpha_z)/2$ and the parallel correlation function $\alpha_{\|} \equiv \alpha_x$.

From vortex depolarization theory[4] it is known that single vortices give the main contribution to the parallel correlation function when the beam is parallel to the applied field. This can be seen from the expression for $\alpha_{\|}$ in that case[5]:

$$\alpha_{\|} = \frac{\langle B\rangle\phi_0 L}{8\pi\lambda_{eff}t\Delta\theta}. \tag{2}$$

Here, $\langle B\rangle$ is the average induction, t is the transmission length of the neutron beam through the vortex or can also be considered as the average length of straightness of the vortices, $\phi_0 = 2.07 \times 10^{-15}$ Wb is the magnetic flux quantum, λ_{eff} is the effective penetration depth and $\Delta\theta$ is the angular spread of the vortices. All the quantities in Eq. (2) are known from either the other depolarization results or from literature and so the length of straightness times the angular spread $t\Delta\theta$ can be calculated.

The sample used in this experiment is a sintered $YBa_2Cu_3O_7$ sample consisting of grains of ~ 1 μm size clustered in strongly correlated poly-crystalline clusters of 10 μm to 100 μm diameter[6]. In between these clusters only weak superconductivity can occur in the form of Josephson effects. The volume fill fraction was 82% of the theoretical prediction, meaning that holes are present inside. The size was $5 \times 10 \times 20$ mm^3 and the critical temperature was 92 K. In the experiment, the field was applied along the 10 mm sample side parallel to the neutron beam transmission (x-)direction. The scan (y-)direction was along the 5 mm side perpendicular to the neutron beam and two cadmium diaphragms of 0.2×15.0 mm^2 placed in front and behind the sample, collimated the scanning beam.

After cooling the sample in zero field (ZFC), vortices had to be put into the sample. This was done by means of a copper coil wound around the sample with a volume much larger than the sample volume to ensure the field homogeneity over the sample. Heating by the coil resulted in an upper- continues field limit, at 4.2 K, of about 20 mT. Therefore, three external field values (1.83 mT, 5.3 mT and 10.3 mT) well below this maximum, were chosen to be used in the scanning measurements.

The relevant depolarization parameters, as they were measured and the length of straightness of the vortices as a function of position are given in figs. 1a and 1b. Computer simulations of the experiments showed that the perpendicular correlation functions (α_p) are mainly caused by demagnetization fields outside the sample and not by superconductivity effects. For this reason and because $\alpha_p \ll \alpha_{||}$, the perpendicular correlation functions are not considered relevant and are not given in a figure.

A couple of remarkable features are observed in the magnetization picture (fig. 1a). The first is that at all field values, so even at the very low fields, a fraction of the applied field penetrates homogeneously. This penetration is almost free and is taking place in the holes that are present in between the clusters. Because of the weak superconductivity in the holes, this penetration will take place even at very low field values.

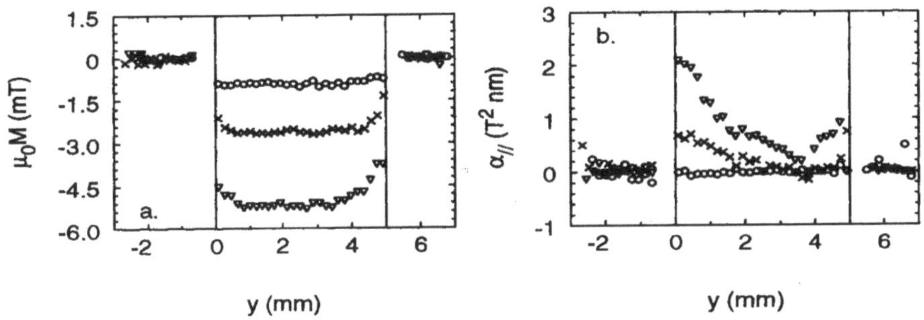

Figure 1: a.) The magnetization and b.) the correlation functions parallel to the applied field in external fields H with $\mu_0 H =$: (o) 1.83 mT, (×) 5.3 mT and (▽) 10.5 mT

At 1.83 mT the penetrated induction is homogeneous because only the, weak superconducting, inter-cluster holes can be penetrated by field. At the other two fields, 5.3 mT and 10.5 mT, an induction gradient is observed near the edge. The Bean model, and so the assumption that the induction gradient is linear, can provide an estimate of the critical current density for both the highest applied fields. This fairly crude estimate of the critical current density has the same outcome for both the fields: $j_c = 8 \times 10^5$ Am^{-2}. This result is rather low compared to other critical current measurements in similar samples[7]. However, the result is not conflicting with the results in ref.[7] because of the strong field- and grain-boundary dependence of the critical currents.

Furthermore, it seems that the field penetration at the right side of the sample is a little bit larger than the field penetration at the left side for both the highest applied fields. The most probable reason for this asymmetry is that during sintering, the oxygen could not penetrate the sample on both sides just as well, giving different critical currents.

The position dependence of the induction shows that in a sintered sample, a distinction between inter– and intra–cluster vortices is necessary to understand the scanning 3DND results. The free penetration is caused by penetration of magnetic flux in between the clusters. The observed induction gradient in the higher external fields is caused by penetration into the strongly correlated clusters, starting from the sample edge.

From the position dependence of the depolarization the relevant correlation functions are derived and plotted as a function of position in fig. 1b. An asymmetry is observed between the left and the right side of the sample in $\alpha_{||}$. This asymmetry is caused by the same effect of different oxygen contents at the left and the right side of the sample as was argued for the difference in the field penetration.

It is possible to calculate the length of straightness of the vortices by substitution of the penetration depth and the average trapped induction in Eq. (2). The penetration depth for sintered $YBa_2Cu_3O_7$ has been measured by μSR quite accurately[8], giving $\lambda_{eff} \approx 150$ nm. The resulting length of straightness is of the order of 200 μm in the highest applied fields.

At 1.83 mT field, $\alpha_{||}$ is immeasurably small because the Josephson vortices are more smeared out than Abrikosov vortices giving negligible depolarization. The order of magnitude of the vortex straightness in the higher fields is in agreement with the cluster size, confirming the picture of pinning in the clusters.

To conclude, the authors would like to acknowledge D.H.A. Blank of the University of Technology Twente for the preparation of the sample.

References

1. C.P. Bean, *Phys. Rev. Lett.* **8**, (1962) 250.
2. M.Th. Rekveldt, *Text. Microstruct.* **11**, (1989) 127.
3. L.A. Akselrod, G.P. Gordeev, V.N. Zabenkin, M.R. Kolkhidashvili, I.M. Lazebnik, E.L. Ploshchanskaya, V.I. Sbitnev, *Physica C* **221**, (1994) 219.
4. W.Roest and M.Th. Rekveldt, *Phys. Rev. B* **48**, (1993) 6420.
5. L.A. Akselrod, G.P. Gordeev, V.N. Zabenkin, I.M. Lazebnik, V.I. Sbitnev, B.P. Toperverg, *Physica B* **174**, (1991) 354.
6. D.H.A. Blank, H. Kruidhof, and J. Flokstra, *J. Phys. D* **21**, (1988) 226.
7. D. Dimos P. chaudhari and J. Mannhart, *Phys. Rev. B* **41**, (1990) 4038.
8. B. Pümpin, H. Keller, W. Kündig, W. Odermatt, I.M. Sacić, J.W. Schneider, H. Simmler, P. Zimmermann, E. Kaldis, S Rusiecki, Y. Maeno, and C. Rossel, *Phys. Rev. B* **42**, (1990) 8019.

FLUX DYNAMICS AT THE GRAIN BOUNDARIES OF BULK Y-Ba-Cu-O SUPERCONDUCTORS

M. Nikolo and M. J. Missey
Physics Department, Saint Louis University
St. Louis, Missouri 63103

ABSTRACT

A critical state model for granular superconductors is applied to model frequency dependence of ac susceptibility at the grain boundaries of a bulk $YBa_2Cu_3O_7$ superconductor in small ac magnetic fields. We obtain an excellent fit to our experimental data. This fit is used then to obtain frequency dependent Josephson magnetic flux profiles for different temperatures. A clear separation of the 10 Hz and 1000 Hz flux fronts is observed. The separation of the flux creep and no flux creep fronts inside the sample then leads to the Josephson flux creep velocities and the flux creep hopping rates at the grain boundaries as a function of temperature. Depending on the temperature, the Josephson flux creep velocity varies between 10^{-4} to about 1 m/s and the hopping rate between 10 to 10^5 Hz, respectively, in an applied ac field of 1 Oe and in the temperature range just a few degrees below the transition temperature.

I. Introduction and Calculation Outline

The frequency shift in temperature of the intergranular component of ac susceptibility in small magnetic fields for the granular bulk oxides was first reported by the author in collaboration with Goldfarb[1]. The shift was attributed to flux creep at the grain boundaries. We examine the frequency effects in small applied magnetic fields to study the flux dynamics at the grain boundaries of our Y-Ba-Cu-O bulk superconductor. We obtain frequency dependent flux profiles which are then used to obtain the flux creep hopping rates at the grain boundaries as a function of temperature.

One can calculate the real and imaginary parts, χ', and χ'', of the ac susceptibility by evaluating the lowest Fourier component of $\langle B\rangle(t)$, where $\langle B\rangle$ is the flux density which is averaged over the cross-sectional area of the sample[2,3].

$$\chi' = -1 + \frac{2f}{\mu_0 H_{a0}} \int_0^{1/f} \langle B\rangle(t)\cos(2\pi f t)\,dt \tag{1}$$

and

$$\chi'' = \frac{2f}{\mu_0 H_{a0}} \int_0^{1/f} \langle B\rangle(t)\sin(2\pi f t)\,dt \tag{2}$$

Here H_{a0} is the amplitude of the applied ac field and f is the frequency.

The average flux density must be separated into the intergranular and intragranular contributions $\langle B\rangle = \langle B_j\rangle + \langle B_g\rangle$. For a long, cylindrical geometry of the sample where R is the radius of the sample the grains are approximated by superconducting cylinders of grain radius R_g aligned along the direction parallel to the applied field. When the fields are greater than the lower critical field of the grains, H_{c1j}, then the Josephson vortices start to penetrate the grains and the spatially averaged flux density $\langle B_j\rangle$ is given by

$$\langle B_j\rangle = \mu_{eff}\frac{2\mu_0}{R^2}\int_0^R H_j(r)\, r\, dr \tag{3}$$

where $H_j(r)$ is the intergranular magnetic field distribution and μ_{eff} is the effective permeability[4].

The spatially averaged intragranular flux density $\langle B_g\rangle$ is then given by

$$<B_g> = (1-\mu_{eff})\frac{4\mu_0}{R^2R_g^2}\int_0^R\int_0^{R_g} H_g(r,r')\, r\, r'\, dr'\, dr \qquad (4)$$

where r' is the radial coordinate inside a grain that is located at r inside the sample and $H_g(r,r')$ is the intragranular magnetic field distribution[2,3].

The vortex behavior is defined by the critical state equations subject to the boundary conditions. The Ampere's law is applied to find $H_j(r,t)$ and $H_g(r,r',t)$

$$\frac{dH_J(r,t)}{dr} = \pm j_{cJ}(r,t) \qquad (5)$$

where

$$j_{cJ}(r,t) = \frac{F_J(T)}{\mu_{eff}(T)\mu_0}\frac{1}{|H_J(r,t)|+H_{0J}} \qquad (6)$$

and

$$\frac{dH_g(r,r',t)}{dr} = \pm j_{cg}(r,r',t) \qquad (7)$$

where

$$j_{cg}(r,r',t) = \frac{F_g(T)}{\mu_0}\frac{1}{|H_g(r,r',t)|+H_{0g}} \qquad (8)$$

$F_j(T)$ and $F_g(T)$ are the temperature dependent inter- and intragranular pinning force densities.

The thermally activated flux bundles will hop in the direction of the Lorentz force, F_L, at a rate of

$$\nu = \nu_0[\exp(-(U - F_L V_b d)/kT) - \exp(-(U + F_L V_b d)/kT)] \qquad (9)$$

$$F_p = \frac{kT}{V_b r_p}\,\mathrm{arcsinh}\left[\frac{\nu}{\nu_0}\exp\frac{U(T,H)}{kT}\right] \qquad (10)$$

Typically, the backward hopping term in Equation (9) can be neglected since the larger Lorentz force density makes the second term in Equation (9) much smaller. Then one can rewrite Equation (10) for the effective pinning force density F_{eff} as

$$F_{eff} = (1/(r_p V_b))\,(U + kT\ln(\nu/\nu_0)) = F_{J0}(T) + kT/(r_p V_b)\ln(\nu/\nu_0) \qquad (11)$$

The flux bundle volume can be approximated by $V_b = (2R_g)(\Phi_0/|B|)$. Equation (5) then takes the form[3,5,6]

$$dH(r)/dr = (\alpha H_0/(|H| + H_0) + \beta) \qquad (12)$$

where $\alpha = F_{J0}(T)/(\mu_{eff}\mu_0 H_0)$ and $\beta = kT/2R_g{}^2\Phi_0 \ln(\nu/\nu_0)$. Equation (12) is integrated.

$$H(r) - H_{a0} - \frac{\alpha H_{a0}}{\beta}\ln\left(\frac{\beta[H(r)+H_0]+\alpha H_0}{\beta[H_{a0}+H_0]+\alpha H_0}\right) + \beta r = 0 \qquad (13)$$

This equation determines H(r) from the cylinder surface at r=R, where $H(R) = H_{a0}$, to r=0, where R is the radius of the cylinder. Here $H_a=H_{a0}\cos 2\pi ft$ is the external field.

We speak of the critical state, but there is no precise "critical state" since flux creep continues at any value of the current density and the magnetic field. Nevertheless, one may define the critical state as that at which the rate falls below a practically observable limit. In this case, if the sample is probed by an ac field frequency f, the minimal observable flux creep rate, ν_{min}, can be set $\nu_{min}=\eta f$ [3,5,6]. Equation (11) then becomes

$$F_{eff} = F_{J0}(T) + \frac{kT}{2R_g^2\phi_0}\ln\left(\frac{\eta f}{\nu_0}\right) \qquad (14)$$

where $F_{J0}(T)= F_{J0}(0)(1-T/T_c)^{\alpha}$.

The spatially averaged intergranular and intragranular flux densities $<B_j>$ and $<B_g>$ defined in Equations (3) and (4) are then used to calculate χ'' as defined in Equations (1) and (2), as a function of the ac field amplitude, H_{a0}, and frequency, f. The intergranular pinning force density $F_{J0}(0)$, and the exponent α are chosen such that $T_p(H_{a0})$ at f = 100 Hz agrees best with the experimental data.

II. Results and Conclusion

The experimental data of ac susceptibility as a function of temperature for an amplitude of 1 Oe (rms) at frequencies f = 10 Hz and 1000 Hz were fitted. Here α=2 and the other material parameters are R = 2.8 mm, f_n=0.5, $F_{J0}(0)$=5600 TA/m^2, η/v_0=0.001 and $F_g(0) = 10^6$ TA/m^2. The above pinning force densities are equivalent to inter- and intra- granular critical current densities at zero temperature of 1.5×10^4 and 10^8 A/cm^2, respectively. The intragranular pinning force density, is assumed to have a linear temperature dependence, $F_g(T)=F_g(0)(1-T/T_c)$. There was a very good agreement. We emphasize that all fitting parameters with the exception of the pinning force densities and the η/v_0 term (very small effect) were experimentally determined. This χ' and χ'' fit allows us to obtain the corresponding Josephson frequency dependent flux profiles $H_J(r,t,T)$ inside the cylindrical sample. We plot the flux profiles for two different frequencies, 10 and 1000 Hz, at different temperature points on the χ vs. T curve. Fig. 1 shows the respective profile at a temperature of 86.0 K.

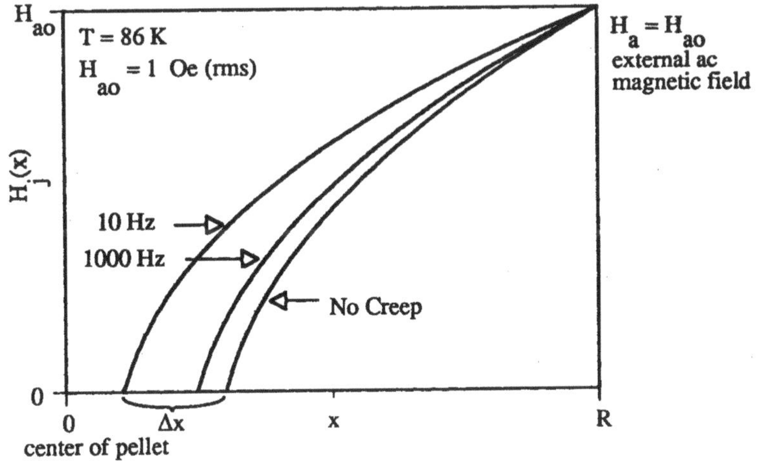

Figure 1. Separation Δx of the flux creep and no flux creep Josephson magnetic fronts $H_j(x)$ inside the sample at 86 K and 1 Oe (rms).

From the separation of the flux creep and no flux creep flux fronts inside the sample, Δx, we obtain the Josephson flux creep velocities at the grain boundaries as a function of temperature and correlate it to different points on the susceptibility curve. See Fig. 2a-b. The dashed lines in Figure 2a-b, although picked fairly arbitrarily, are meant to separate the region of flux creep from the onset of flux flow at least for some grain boundaries. Beyond that region, the backward flux hopping cannot be neglected and we are no longer strictly in a flux creep regime.

Furthermore, setting the intergranular hopping distance to the dimension of the grain, R_g, we also show the intergranular hopping rate at the grain boundaries as a function of temperature. See Figure 2b. Depending on the temperature, the Josephson flux creep

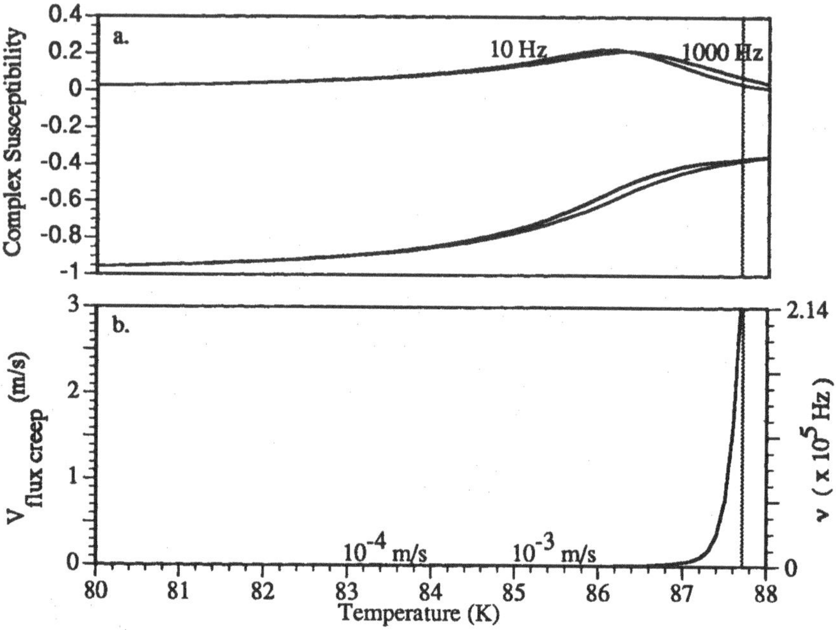

Figure 2a-b. Calculated ac susceptibility vs. temperature at f=10 Hz and 1000 Hz for an ac field of 1 Oe (rms) and the corresponding flux creep velocities and hopping rates at the grain boundaries.

velocity varies between 10^{-4} to about 1 m/s and the hopping rate between 10 to 10^5 Hz, respectively, in the temperature range just a few degrees below the transition temperature.

In conclusion, we apply the critical state model to examine the frequency shift of the intergranular part of ac susceptibility and to study the flux dynamics at the grain boundaries of a Y-Ba-Cu-O bulk superconductor. We show frequency dependent flux profiles inside our sample obtained by matching our experimental data of $\chi(f)$ in temperature with the critical state model calculation. The profiles are then used to obtain the flux creep hopping rate at the grain boundaries as a function of temperature.

Acknowledgments

The authors wish to acknowledge useful discussions with K. H. Müller.

References

1. M. Nikolo and R. B. Goldfarb, Phys. Rev. B **39**, 6615 (1989).
2. K.-H. Müller, Physica C **159**, 717 (1989).
3 M. Nikolo, Ph. D. thesis, U. Colorado, 1991.
4. J. R. Clem, Physica C **153-155**, 50 (1988).
5. K.-H. Müller, M. Nikolo, and R. Driver, Phys. Rev. B **43**, 7976 (1991).
6. K.-H. Müller, M. Nikolo, N. Savvides, and R. Driver, Proc. 3rd Inter. Symp. on Superc. (ISS '90), Nov. 6-9, 1990 Sendai, Japan, Springer-Verlag Tokyo (1992).

VORTEX PINNING IN MULTILAYERED SUPERCONDUCTING STRUCTURES WITH TUNNEL MAGNETIC BARRIERS

S. V. KUPLEVAKHSKY

Department of Physics, Kharkov State University, 310077 Kharkov, Ukraine

and

I. I. FAL'KO

Physico-Technical Department, Kharkov Polytechnical University, 310002 Kharkov, Ukraine

ABSTRACT

Within the framework of a microscopic theoretical approach, we consider the problem of an Abrikosov vortex pinning in superconducting structures with tunnel ferromagnetic and paramagnetic barriers. We show that tunneling with a spin-flip suppresses Cooper pairing at the interfaces between superconducting banks and a barrier and leads to the reduction of the dc Josephson critical current. For temperatures close to the superconducting transition temperature, we make self-consistent calculations of the arising long-range spatial variations of the pair potential and derive the boundary conditions for the Ginzburg-Landau equations. Using these results, we determine the force of interaction between an Abrikosov vortex and a tunnel magnetic barrier and estimate the relevant pinning force.

1. Introduction

The problem of pinning by planar defects has always been of considerable interest because of its role in determining the critical current density in high-field superconductors[1]. Depending upon the physical nature of these defects, the mechanisms of the pinning force are often very different. In recent years much theoretical work has been done to bring understanding into this field. (See, for example, the discussion of intrinsic pinning in high-T_c oxide superconductors[2], pinning by twin boundaries[3-5], normal metallic layers[6], and pinning in Josephson systems[7,8].)

In this paper, we study theoretically pinning of an Abrikosov vortex by a parallel magnetic tunnel barrier. We assume the barrier to be sufficiently thin, so that we can neglect the component due to core pinning and consider only the effect of the image force[8]. For *nonmagnetic* tunnel barriers, image-force pinning is decreased by Josephson currents when compared to its value at a superconductor/vacuum interface[9]. It was shown elsewhere[10] that magnetic atoms localized within the tunnel barrier cause destruction of Cooper pairs in processes of tunneling and reflection with a spin flip. Physically, we expect the reduction of the dc Josephson critical current and a depression of the superconducting order parameter, $\Delta(\mathbf{r})$, in the vicinity of the barrier. Thus, the presence of magnetic atoms facilitates pinning in two different ways. Firstly, the negative influence of Josephson currents is diminished. Secondly, a completely new positive contribution arises owing to the local depression of Δ. These two effects are discussed in detail below.

2. Microscopic Model

Consider two semi-infinite superconductors with s-wave pairing in the clean limit separated by a magnetic tunnel barrier in the yz-plane. In second quantized form, the system can be described by the Hamiltonian

$$H = H_0 + H_b, \tag{1}$$

$$H_0 = \int d^3\mathbf{r}\psi_\alpha^+(\mathbf{r})\left\{\frac{-1}{2m}\left(\nabla - \frac{ie}{c}\mathbf{A}\right)^2 - \mu\right\}\psi_\alpha(\mathbf{r}) - \frac{g}{2}\int d^3\mathbf{r}\psi_\alpha^+(\mathbf{r})\psi_{-\alpha}^+(\mathbf{r})\psi_{-\alpha}(\mathbf{r})\psi_\alpha(\mathbf{r}) + \int d^3\mathbf{r}\frac{[\nabla\times\mathbf{A}]^2}{8\pi}$$

Here the notations are standard, and a summation over electron spin is implied. For the Hamiltonian of the tunnel magnetic barrier, we take

$$H_b = \int d^3\mathbf{r}\psi_\alpha^+(\mathbf{r})\left\{\left[V\delta_{\alpha\beta} + J\sum_j\left(\mathbf{S}_j\boldsymbol{\sigma}_{\alpha\beta}\right)\delta\left(\boldsymbol{\rho} - \boldsymbol{\rho}_j\right)\right]\delta(x)\right\}\psi_\beta(r), \tag{2}$$

where V is the nonexchange part of the potential, $\boldsymbol{\rho} = (y,z)$, J is the exchange integral, and $\mathbf{S}_j$ is the spin of the magnetic atom at the point $\mathbf{r}_j = (0,\boldsymbol{\rho}_j)$. Ignoring the quantum nature of the spin $\mathbf{S}_j$, we are treating it as a classical vector.

For the model Eqs. (1) and (2), the Josephson critical current ($\mathbf{A} = 0$) is given by[10,11]:

$$j_J^c = \frac{e\pi N(0)v_0\Delta_\infty}{2}\int_0^1 dt t\left[T(t) - 2T_S(t)\right]\tanh\left(\frac{\Delta_\infty}{2T}\right). \tag{3}$$

Here Δ_∞ is the gap parameter in the interior of superconductors; $N(0)$ is the density of free electron states at the Fermi surface; v_0 is the Fermi velocity; $T_S(t)$ is the exchange part of the tunneling probability which depends on the cosine of the angle of incidence on the barrier, t; $T(t)$ is the total tunneling probability. It was already pointed out[10] that the typical difference $T(t) - 2T_S(t)$ in Eq. (3) can be interpreted as the tunneling probability for a Cooper pair, while the emerging term $2T_S(t)$ characterizes pair-breaking in the process of tunneling. Explicit expressions for $T_S(t)$ and $T(t)$ in the ferromagnetic and paramagnetic limits are obtained in Ref. 10 and Ref. 11, respectively. Note that Eq. (3) applies only if $T(1) \geq 2T_S(1)$. In principle, one can imagine a situation[12] when $T(1) < 2T_S(1)$. In this case[10], $j_J^c \equiv 0$.

The problem of the determination of $\Delta(\mathbf{r}) \equiv g\langle\psi_\uparrow(\mathbf{r})\psi_\downarrow(\mathbf{r})\rangle$ can be resolved analytically on the basis of Eqs. (1), (2) for temperatures near T_c. At distances from the barrier $\leq \xi_0 \equiv v_0/2\pi T_c$ ($\mathbf{A} = 0$), Δ is ruled by a linear integral equation first derived in

Ref. 10. We use this equation in Sec. 3 to determine the boundary conditions for the Ginzburg-Landau equations.

3. Ginzburg-Landau Equations

Confining our attention to temperatures in the vicinity of T_c, we consider a single vortex line parallel to the barrier. At distances from the barrier $|x| >> \xi_0$ the Ginzburg-Landau (GL) equations come into play. In the limit $\kappa >> 1$ ($\kappa \equiv \lambda/\xi$), we have:

$$-\xi^2\left(\nabla - \frac{2\pi i}{\phi_0}\mathbf{A}\right)^2 f - f + f|f|^2 = 0, \tag{4}$$

$$\nabla \times \nabla \times \mathbf{A} = \frac{|f|^2}{\lambda^2}\left(\frac{\phi_0}{2\pi}\nabla\theta - \mathbf{A}\right). \tag{5}$$

Here we have introduced the dimensionless GL order parameter, $f \equiv \Delta_{GL}/\Delta_\infty$ ($f = |f| \exp(i\theta)$); ϕ_0 is the flux quantum. Note that the phase θ has a singularity on the line $\mathbf{r}_0 = (x_0 > 0, 0, z)$ (vortex core).

Eqs. (4), (5) have to be complemented by boundary conditions on f and $\mathbf{A}$. For example, in the case of a nontransmitting barrier the boundary conditions must match the physical demand of the absence of the x-component of the supercurrent at $x = +0$. In the general case, the appropriate boundary conditions can be derived microscopically[13]. For $\mathbf{A} = 0$ (f is real), we obtain:

$$f(x) = \tanh\left[\frac{|x|}{\sqrt{2}\,\xi} + C\right], \tag{6}$$

where $C = \frac{1}{2}\,\mathrm{Arcsinh}\left(\sqrt{2}\,b/\xi\right)$. The extrapolation length[13], b, is given by

$$b = \frac{7\zeta(3)\xi_0}{3\pi^2}\left[\int_0^1 dt t T_S(t)\right]^{-1} \quad (2T_S(1) \leq T(1),\ T_S(1) << 1);$$

$$b = \frac{14\zeta(3)\xi_0}{3\pi^2}\left\{\int_0^1 dt t[T(t) + 2T_S(t)]\right\}^{-1} \quad (2T_S(1) > T(1),\ T_S(1) << 1). \tag{7}$$

4. Pinning Force

For $T(1) << 1$, the force of interaction between the vortex and the barrier can be represented as $F^i = F_0^i + F_1^i + F_2^i$, where F_0^i is the surface image force[9], F_1^0 is the component due to the Josephson currents[8], and F_2^i is the contribution of the spatial variations of $|f|$. Note that for $x_0 << \lambda$, $F_0^i \sim x_0^{-1}$, and $F_1^i \sim -j_J^c F_0^i$.

To calculate F_2^i, we first derive on the basis of Eqs. (4)-(7) the correction to the energy of the vortex[13] arising due to the local depression of $|f|$ (per unit length):

$$\varepsilon_2^i(x_0) = -\frac{2\sqrt{2}}{16\pi^3}\frac{\phi_0^2}{\lambda^2\kappa}\left(1 - \tanh C\right)\int_0^\infty \frac{dy K_1\left(\sqrt{\left(x_0/\lambda\right)^2 + y^2}\right)}{\left(x_0/\lambda\right)^2 + y^2}, \tag{8}$$

where $K_1(u)$ is the modified Bessel function of order one. Differentiating Eq. (8) with respect to x_0 yields:

$$F_2^i(x_0) = -\frac{\phi_0^2\xi}{16\pi^3\lambda^2}\left(1 - \tanh C\right)\frac{1}{x_0^2}, \qquad \xi \le x_0 << \lambda . \tag{9}$$

Evaluating the three conthibutions, F_0^i, F_1^i, F_2^i, at the cutoff range, ξ, we finally obtain an order of magnitude estimate for the pinning force:

$$F^{pin} \approx -\frac{\phi_0^2}{\lambda^2\xi} + \Theta\left(T(1) - 2T_S(1)\right)\int_0^1 dtt\left[T(t) - T_S(t)\right]\frac{\phi_0^2}{\lambda^2\xi} - \left(1 - \tanh C\right)\frac{\phi_0^2}{\lambda^2\xi}. \tag{10}$$

Acknowledgements

The research described in this publication was made possible in part by Grant No. U28000 from the International Science Foundation.

References

1. A. M. Campbell and J. E. Evetts, *Adv. Phys.* **54** (1972) 199.
2. M. Tachiki and S. Takahashi, *Solid State Commun.* **70** (1989) 291.
3. A. A. Abrikosov, A. I. Buzdin, M. L. Kulic, and D. A. Kuptsov, *Sov. Phys. JETP* **68** (1989) 210.
4. G. Blatter, J. Rhyner, and V. M. Vinokur, *Phys. Rev.* **B43** (1991) 7826.
5. D. Agassi and J. R. Cullen, *Physica* **C195** (1992) 277.
6. D. Davidovic and L. Dobrosavljevic-Grujic, *Phys. Rev.* **B43** (1991) 2809.
7. A. Gurevich, *Phys. Rev.* **B43** (1992) 3187.
8. L. D. Cooley, A. Gurevich, and D. C. Larbalestier, *Proc. 7th Int. Workshop on Critical Currents in Superconductors* (January 1994, Alpbach, Austria).
9. C. P. Bean and J. D. Livingston, *Phys. Rev. Lett.* **12** (1964) 14.
10. S. V. Kuplevakhsky and I. I. Fal'ko, Sov. *J. Low Temp. Phys.* **10** (1984) 361; *Theor. Mat. Phys.* (Moscow) **67** (1986) 483.
11. S. V. Kuplevakhsky and I. I. Fal'ko, this Conference.
12. L. N. Bulaevskii and V. V. Kuzii, Sov. *J. Low Temp. Phys.* **3** (1977) 352.
13. P. G. DeGennes, *Superconductivity of Metals and Alloys* (Benjamin, New York, 1966).

2. Critical State Phenomena

ANISOTROPY OF THE "SECOND PEAK" IN MAGNETIZATION CURVES OF $YBa_2Cu_3O_y$ SINGLE CRYSTALS

N. KOBAYASHI, K. HIRANO*, T. NISHIZAKI, H. IWASAKI**, S. AWAJI, K.WATANABE,
Institute for Materials Research, Tohoku University,
2-1-1 Katahira, Aoba-ku, Sendai 980-77 Japan,

H. ASAOKA
Japan Atomic Energy Research Institute
2-4 Shirakata, Tokai-mura, Ibaraki 319-11 Japan

and

H. TAKEI
Institute for Solid State Physics, University of Tokyo,
7-22-1 Roppongi, Minato-ku, Tokyo 106 Japan.

ABSTRACT

The angular dependence of the second peak in magnetization curves have been investigated as well as the irreversibility field of $YBa_2Cu_3O_y$ single crystals with oxygen content $6.6 \leq y \leq 6.9$. The magnetic field H_p at which the second peak appears depends linearly on $(1-T/T_c)$. The angular dependence of H_p is well described by the Ginzburg-Landau effective mass model. For the sample with y=6.9, the effective mass ratio m_c/m_{ab} $(=\gamma^2)$ is estimated to be about 36, which is in agreement with that obtained from a scaling of the angular dependence of the resistivity. The γ value increases with decreasing oxygen content.

1. Introduction

A field-induced pinning behavior called "second peak" or "fish-tail" in magnetization curves has been observed in the mixed state of high-T_c oxide superconductors [1-3]. As for the $YBa_2Cu_3O_y$ system, Daeumling et al. [4] proposed that oxygen-deficient regions having lower T_c and H_{c2} act as significant pinning centers as the magnetic field was applied. On the other hand, it has also been reported [5] that the anomalous magnetization arises from a steep relaxation of the magnetization at lower field due to a collective creep which is pointed out in recent flux creep theory [6].

From both viewpoints of practical application and basic research, it is important to know what kind of pinning mechanism contributes to the field-induced magnetization. In the present work, we have studied the anisotropic properties of the field-induced magnetization and discuss the angular dependence on the basis of the effective mass expression.

* Present address: Hitachi Cable, Ltd.
** Present address: Japan Advanced Institute of Science and Technology, Hokuriku.

2. Experimental

$YBa_2Cu_3O_y$ single crystals were grown by slow cooling of a solid-melt mixture under the peritectic condition. The as-grown crystals were annealed at appropriate temperature for 1-2 weeks in a flowing oxygen and then quenched into room or liquid nitrogen temperatures. All samples used in this study showed sharp superconducting transitions in both resistive and magnetic measurements. The oxygen content of the samples was estimated from the annealing condition and the transition temperature. The T_c value and the oxygen content are listed in Table I. Sample dimensions are typically $1.2 \times 1.2 \times 1 mm^3$. Detailed technique of the crystal growth has been described elsewhere [7].

The magnetization measurements were performed by using a commercial SQUID magnetometers (Quantum Design). The sample was rotated around an axis perpendicular to the magnetic field and parallel to the ab-plane. The rotated angle θ was defined by the angle between the c-axis and the magnetic field. The magnetization was measured after zero-field cooling.

3. Experimental results and discussion

Figures 1(a) and 1(b) show magnetization curves at various temperatures for a fully oxidized sample and a reduced one, respectively. The magnetization curves exhibit a large hysteresis indicating a strong pinning of flux lines. As for the fully oxidized sample, the second peak is observed in the narrow temperature region above 70 K. As the temperature decreases, the magnitude of the magnetization of the second peak first increases and after taking a maximum at about 84 K, then decreases. The background hysteresis grows up with decreasing temperature and the second peak is hidden by the background below 70 K. On the other hand, the second peak of the reduced sample increases monotonically with decreasing temperature. The magnetization curves at each temperature exhibit a similar behavior in shape. The magnetic field H_p at which the second peak appears varies linearly with temperature except near T_c.

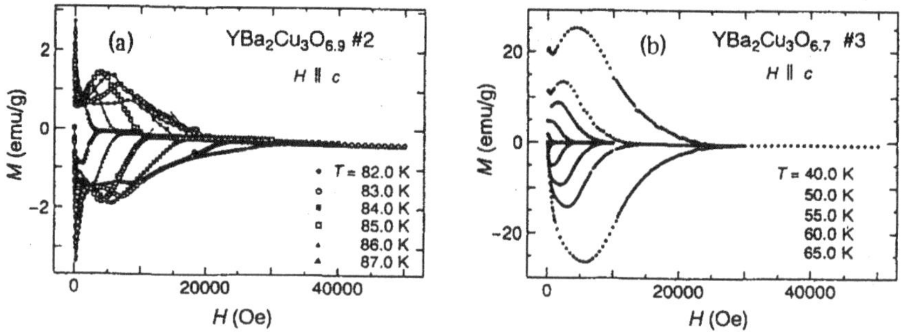

Fig. 1 Magnetization hysteresis curves for the samples with oxygen content 6.9 (a) and 6.7 (b).

This result is different from the temperature dependence of the irreversibility field which is proportional to $(1-T/T_c)^{1.5}$.

The field dependence of the magnetization of the fully oxidized sample at 85.5 K is shown in Fig. 2 for various tilt angles. The value of H_p increases with tilt angle, but the magnitude of the magnetization at the second peak decreases with increasing angle and the second peak almost disappear for the magnetic field perpendicular to the c-axis. Similar tendency is also found for the reduced samples. The angular dependence of H_p for all samples studied are shown in Fig. 3 as a function of tilt angle. Observed behavior is very similar to the angular dependence of the upper critical field H_{c2} for anisotropic superconductors with a uniaxial symmetry. Solid lines are fitting curves by the Ginzburg-Landau (GL) effective mass model,

$$H_p(\theta)=H_p(0)/(\cos^2\theta+\gamma^{-2}\sin^2\theta)^{1/2}, \tag{1}$$

where γ^2 is the effective mass ratio m_c/m_{ab}. The experimental results are well fitted by the effective mass model. The γ value is obtained to be 6.0 for the fully oxidized sample and increases with decreasing oxygen content as shown in Table I.

We also measured the angular dependence of the resistivity in the superconducting transition region for the fully oxidized sample. The angular dependence of the resistivity at various fields is well scaled by a reduced field $H(\cos^2\theta+\gamma^{-2}\sin^2\theta)^{1/2}$ with $\gamma=5.0\pm0.5$, which is in good agreement with the γ value for H_p.

Table I. Transition temperature T_c, oxygen content y and anisotropic parameter γ.

sample	T_c(K)	y	γ
#2	90	6.9	6.0
#1	89	6.85	6.3
#3	68	6.7	18
#4	58	6.6	26

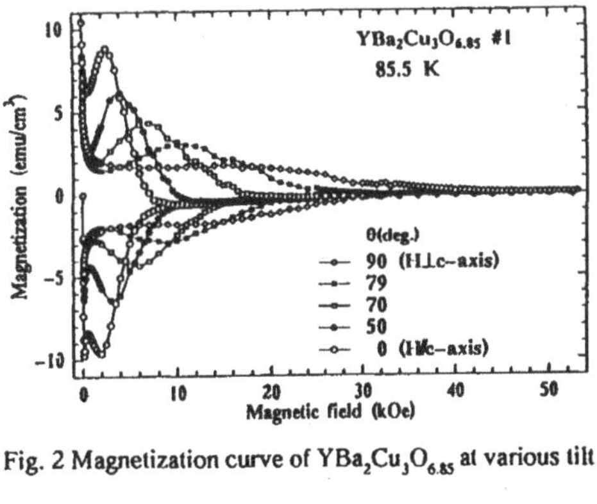

Fig. 2 Magnetization curve of $YBa_2Cu_3O_{6.85}$ at various tilt angles.

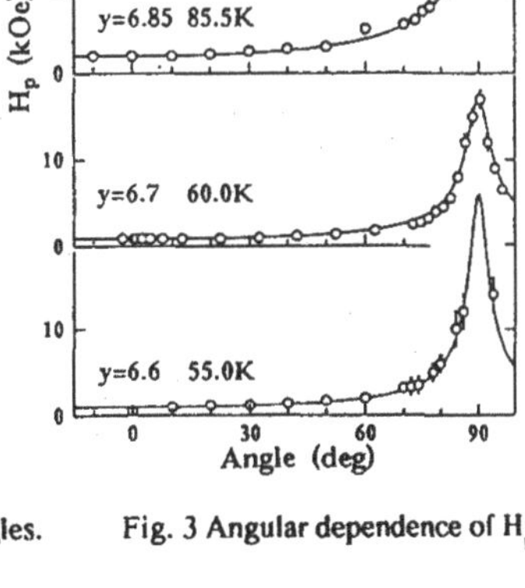

Fig. 3 Angular dependence of H_p.

In the case that isolated flux lines are individually pinned by strong pinning centers, the anisotropy of the magnetization hysteresis would be influenced by the anisotropic property peculiar to the pinning center. On the other hand, in the case of single-vortex weak-collective pinning or strong interaction between flux lines, the anisotropy in the pinning character is expected to arise from the vortex structure determined by the mass anisotropy of superconducting electrons. Thus, the angular dependence of H_p must be attributed to the anisotropic vortex structure intrinsic to the material. Moreover, the linear temperature dependence of H_p reminds that the value of $\lambda^2(T)$ which means a interacting area between flux lines may relates to the temperature dependence of H_p.

The irreversibility field H_{irr} is defined as the lowest field where the magnetization is reversible. The angular dependence of H_{irr}, however, cannot be described by the effective mass model, because it shows a steeper change at high tilt angles than the effective mass model. Our experimental results are well fitted by a relation $H_{irr} \ln H_{irr} \propto 1/(\cos^2\theta+\gamma^{-2}\sin^2\theta)^{1/2}$, which is explained by taking account of the influence of the flux creep on the irreversibility field.

4. Conclusion

We measured the anisotropy of the field-induced pinning effect observed as the second peak in the magnetization curve. For all samples studied, the angular dependence of the magnetic field at which the second peak appears was described by the effective mass expression. Our experimental results suggest that the field-induced pinning effect is attributed to a collective pinning due to a number of weak-pinning and/or a flux line interaction.

Acknowledgements

This work was supported by a Grant-in-Aid for Scientific Research on Priority Areas, "Science of High-T_c Superconductivity" from the ministry of Education, Science and Culture, Japan.

References

1. H. Küpfer et al., Cryogenics 28 (1988) 650.
2. T. Kimura et al., Physica C192 (1992) 247.
3. K. Kadowaki and T. Mochiku, Physica C195 (1992) 127.
4. M. Daeumling et al., Nature 346 (1990) 332.
5. L. Krusin-Elbaum et al., Phys. Rev. Lett. 69 (1992) 2280.
6. M. V. Feigel'man and V. M. Vinokur, Phys. Rev. B41 (1990) 8986.
7. H. Takei et al., Jpn. J. Appl. Phys. 30 (1991) L1102.

CRITICAL CURRENT DENSITY ESTIMATION OF YBCO BY FLUX DISTRIBUTION MEASUREMENT USING A HALL PROBE

T. Higuchi, N. Sakai, M. Murakami
Superconductivity Research Laboratory, ISTEC, 1-16-25 Shibaura, Minato-ku, Tokyo 105, Japan

and

M. Hashimoto
Polytechnic University, 4-1-1 Hashimotodai, Sagamihara-shi, Kanagawa 229, Japan

ABSTRACT

We have developed a system for visualizing flux distribution on a superconductor measured by a Hall probe into a color graphics. The sample was placed in the center of a Cu electromagnetic solenoid and cooled in liquid nitrogen. Under an applied field, flux distributions on the sample surface were mapped by scanning the Hall probe fixed to an XY-stage. Then the obtained flux distribution was visualized into a color contour map. First, in the measurement using large bulk samples (40mm×40mm×2mm) after zero-field-cooling, the overall shape of the distribution under the applied field up to 490G reflected the grain morphology. Since the flux penetrated into the sample through the weak-link regions such as grain boundaries or microcracks, we could easily distinguish weak-link regions from strongly coupled regions. Second, we measured a trapped field using a single-domained YBCO sample (10mm×10mm×1mm). By a comparison between the measured and calculated profiles, the intragranular J_c was estimated to be $2.6\times10^4 A/cm^2$ (77K, 0T). Finally, this method enabled us to evaluate the sample quality and to estimate J_c without destroying the sample.

1. Introduction

The measurement of the magnetic field using a Hall probe is very useful for the macroscopic observation of flux distribution on a superconductor. This method enables us to measure the field distribution accurately even in a large area and under a high field. Moreover, a quantitative estimation of Jc inside the sample can be performed by a comparison between the measured and calculated field distributions. Much work has been done on the numerical calculation of the field distribution for a thin-disk sample [1-3] or a small rectangular sample [4,5]. However, almost no work has been done on the field distribution using a large rectangular bulk.

In the present study, we have developed a system for visualizing flux distribution on a large bulk superconductor measured by a Hall probe into a two-dimensional color graphics. First, we performed the measurement of flux distributions using large YBCO bulk samples prepared by the MPMG process after zero-field-cooling. Second, intragranular Jc of a single-domained YBCO was estimated by a comparison between the measured and simulated field distributions.

2. Experimental

Fig. 1 shows a schematic illustration of the measurement system for magnetic field distributions. This system essentially consists of a Cu electromagnetic solenoid applying a field and a Hall probe measuring the field strength. The Hall probe (BHT-921/Bell) having a round active area of $0.196mm^2$ and a magnetic sensitivity constant of 0.8mV/G is fixed to an XY-stage. Since the Hall probe and the XY-stage are interfaced by a computer, the scanning and the data acquisition can be automatically executed.

First, we used two 40mmx40mmx2mm rectangular YBCO bulk samples prepared by the MPMG process (sample A and B). The sample A melt-grown with a two-dimensional temperature gradient was composed of several large grains. On the other hand, the sample B melt-grown without temperature gradient was composed of many small grains. The measurement procedure was as follows. The sample was placed in the center of the solenoid and zero-field-cooled in liquid nitrogen. Under the field up to 490G, magnetic flux distributions in the area of 48mmx52mm were mapped by scanning the Hall probe on the sample surface at the interval of 1.5mm.

Second, for the J_c estimation, we performed the measurement of a trapped field using a 10mm×10mm×1mm rectangular YBCO sample (sample C). Since the sample C was cut off from a large-domained bulk material prepared by the MPMG method, we could obtain a single-domained sample with the c-axis perpendicular to the 10mm×10mm square surface. The sample C was field-cooled in liquid nitrogen under the field of 4000G applied perpendicular to the 10mm×10mm square surface. After the applied field was turned off, a field distribution in the area of 10mm×10mm was mapped at the interval of 0.5mm. The active area inside the Hall probe was 1.2mm above the sample surface.

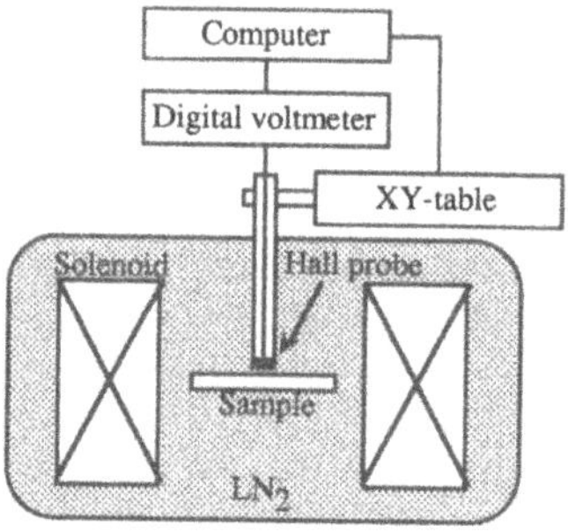

Fig. 1. Schematic illustration of the measurement system for magnetic field distributions. This system essentially consists of a Cu electromagnet solenoid applying the field and a Hall probe measuring the field strength.

3. Results and Discussion

Fig. 2 shows visualized two-dimensional contour maps of flux distributions on the sample A under the applied field of 98G (a) and 294G (b). The magnitude of a field is represented by a color scale normalized by a maximum in each measurement. Under the field of 98G, almost all the magnetic flux was excluded from the sample. Under the 294G field, however, the distribution was not homogeneous because the flux began to penetrate into the sample. Therefore, the shielding effect was considered to become small with increasing the field strength. On the other hand, the sample B showed almost no shielding effect even under 98G. Therefore, the shielding effect was considered to become large with increasing the grain size. In both the cases, compared with the grain morphology, it is clearly seen that the flux penetrated into the sample through the weak-link regions such as the grain boundaries or the microcracks parallel to the ab-plane. Consequently, this method was very useful for the evaluation of the sample quality.

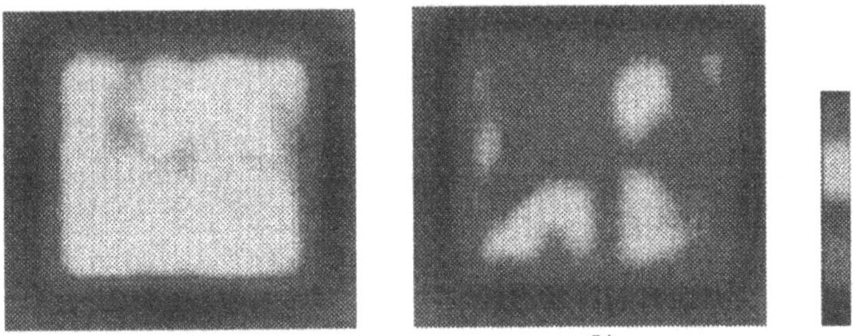

(a) Scale: -131 to 131G

(b) Scale: -366 to 366G

Fig. 2. Visualized two-dimensional contour maps of flux distributions on the sample A under the applied field 98G (a) and 294G (b). The magnitude of a field is represented by a color scale normalized by the maximum in each measurement.

Fig. 3 shows a contour map of a trapped field distribution on the sample C without an applied field. The trapped field distribution showed the maximum of 1.36×10^3G at the center. The trapped field decreased gradually toward the edges and formed ridges along the diagonals. The overall shape of the field distribution had a four-fold symmetry reflected by the sample geometry. Consequently, we see that the sample was homogeneous without weak-link regions.

With this trapped field distribution, we estimated an intragranular Jc . For the estimation, the experimental field distribution was compared with theoretical field distributions. In the calculation of the theoretical field distribution, some simple model assuming current paths inside the superconductor in the critical state is needed. In our numerical calculation, the whole sample was represented as an array of 10×10 co-axial square loops of a field-independent current. The magnetic field B generated by the current element vector Ids is represented by the Biot-Savart law

$$dB = \frac{\mu_0 I}{4\pi}\frac{ds \times r}{r^3}, \qquad (1)$$

where μ_0 is the permeability of free space and r is the location of the current element vector to the observation point. In this case, the z-component of the trapped field B_z was calculated by the sum of the z-component of B in Eq. (1), generated by each current loop.

Fig. 4 shows a comparison between the measured and calculated B_z profiles along the line crossing the center and parallel to the edge. The measured and calculated profiles are represented by open circles and solid lines, respectively. There was no difference between both the profiles except in the high field region near the maximum. Therefore, the critical state model assuming square current loops inside the sample and a field-independent Jc was applicable in low fields. However, field dependence of Jc shoud be taken into consideration in high fields. Fitting the measured profile to the calculated profile, the intragranular Jc was estimated to be 2.6×10^4A/cm^2 (77K,0T).

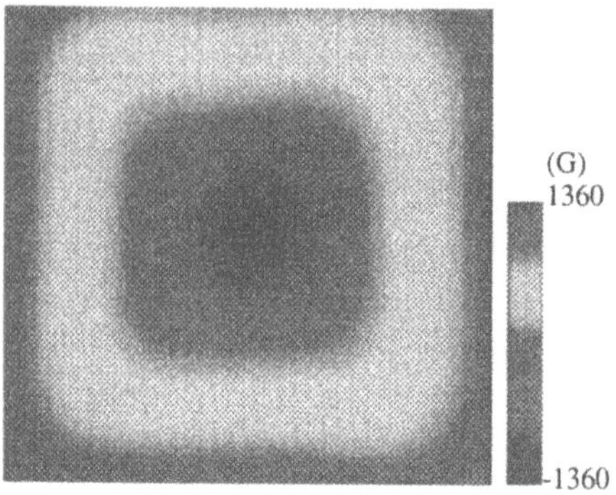

Fig. 3. Visualized two-dimensional contour map of a trapped field distribution on the sample C without an applied field. The trapped field distribution showed the maximum of 1.36×10^3G in the center.

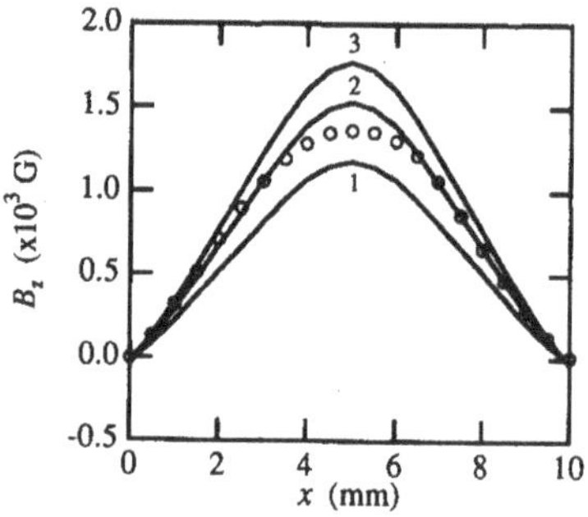

Fig. 4. Comparison between the measured and calculated B_z profiles along the line crossing the center and parallel to the edge. The measured profile, the calculated profiles using Jc value of 2.0×10^4, 2.6×10^4 and 3.0×10^4A/cm^2 are represented by open circles, solid lines of 1, 2 and 3, respectively.

4. Conclusion

The magnetic flux distribution on the surface of MPMG-processed YBCO bulk materials (40mm×40mm×2mm) was measured by a Hall probe and its distribution was visualized into a two-dimensional color graphics. Since the flux penetrated into the sample through the weak-link regions such as grain boundaries or microcracks, we could easily distinguish weak-link regions from strongly coupled regions. Furthermore, an intragranular J_c of a single-domained YBCO bulk sample (10mm ×10mm×1mm) was estimated by a comparison between the measured and calculated field distributions. The critical state model assuming square current loops inside the sample and a field-independent J_c can be adopted in low fields. The intragranular J_c was estimated to be 2.6×10^4A/cm^2 (77K, 0T). Finally, it is found that the measurement of flux distribution using a Hall probe is very useful for the evaluation of the sample quality and the estimation of an intragranular J_c without destroying the sample.

5. Acknowledgements

The authors would like to thank Dr. S. I. Yoo at SRL-ISTEC for helpful discussions, M.Deguchi at SRL-ISTEC for providing the sample and K.Fukuoka at Polytechnic University for his assistance in the experiment. This work was partially supported by NEDO for the R&D of industrial Science and Technology Frontier Program.

6. References

1. D. J. Frankel, *J. Appl. Phys.* **50** (1979) 5402.
2. M. Daumling and D. C. Larbalestier, *Phys. Rev. B* **40** (1989) 9350.
3. L. P. Conner and A. P. Malozemoff, *Phys. Rev. B* **43** (1991) 402.
4. T. Tamegai, L. Krusin-Elbaum, L. Civale, P. Santhanam, M. J. Brady, W. T. Masselink and F. Holtzbelg, *Phys. Rev. B* **45** (1992) 8201.
5. T. Tamegai, Y. Iye, I. Oguro and K. Kishio, *Physica C* **213** (1993) 33.

DETERMINATION OF IRREVERSIBILITY LINES AND CRITICAL CURRENT DENSITIES FROM AC AND DC EXPERIMENTS

M. C. FRISCHHERZ*, F. M. SAUERZOPF*, H. W. WEBER*,
G. A. EMEL'CHENKO** AND M. MURAKAMI***

*Atominstitut der Österreichischen Universitäten, A-1020 Vienna, Austria
**Institute of Solid State Physics, Chernogolovka, 142432 Russia
***International Superconductivity Technology Center, Superconductivity Research Laboratory, Tokyo 135, Japan

ABSTRACT

Critical currents and irreversibility lines of HTS were determined by dc magnetization measurements and by Campbell's ac flux profile technique. The critical currents are evaluated in terms of a critical state model, which cannot fully describe relaxation effects. Therefore, the complementary resistive skin effect model is discussed and its validity confirmed at high temperatures.

1. Introduction

The critical current density J_c of high temperature superconductors is the most interesting feature of these materials. Contactless magnetic measurement techniques prevail because of the large currents that are needed for transport measurements in bulk superconductors. In our experiments we wanted to compare two of these methods, namely SQUID magnetization and ac susceptibility measurements.

The experiments were made on two samples, a MPMG YBCO sample[1] with 30 mole% Y_2BaCuO_5 inclusions, a size of 2x2x1 mm^3 with a T_c of 92 K, and a big (3.3x3.0x1.2 mm^3) YBCO single crystal[2] with a T_c of 90.7 K. The dc magnetization was measured in a non-commercial SQUID magnetometer based on a SHE VTS in fields up to 8 T and at temperatures from 5 K to T_c. The critical currents were evaluated from magnetization loops at constant temperature by a standard Bean formalism because demagnetization corrections were proven to yield only minor corrections. The relevant time scale of this experiment is set by the waiting time of 10 minutes after each field change. Irreversibility lines were measured by temperature sweeps at constant field. The onset of distortions in the SQUID's response curves resulting from the sample movement in the slightly inhomogeneous field, was used to define the irreversibility temperature[3]. This criterion is equivalent to a shielding current density of about 10^4 A/m^2. The time scale of the experiment is typically 50 seconds, which corresponds to the periodicity of the sample movement.

The ac experiments were made in a set of two externally compensated pickup coils, with the sample mounted in one of them. They were compensated in the Meissner state, which resulted in a slope of $S_n = \chi'_n/h_0$ in the normal conducting state. The ac field amplitude h_0

ranged from 0.05 to 15 mT. Most critical current measurements were done at a frequency of 9 Hz and evaluated using Campbell's flux profile technique[4]. Here, the factor α in the quadratic term of the dependence of χ' on h_0, is related to the critical current by $J_c = S_n (a+b) / \mu_0 \alpha a b$, where a and b are the sample dimensions perpendicular to the field. The experiments were made in dc fields up to 15 T. While critical currents were measured at temperatures between 5 K and 77 K, irreversibility lines were determined from temperature sweeps (5 K/h) at fixed fields. Previous experiments showed good agreement between the evaluation of IL's from the maximum in χ'' or from higher harmonics[5]. Therefore, we chose the first method for the definition of the IL in our experiments. This leads to a shielding current density criterion of typically 10^5 A/m^2 at $\mu_0 h_0 = 0.1$ mT, a factor of 10 larger than in the SQUID experiments.

2. Critical Current Densities

J_c values were determined for both field orientations, H_a parallel to the c-axis and parallel to the ab-planes, resulting in the current densities $J_c^{ab;H\|c}$ and $J_c^{c;H\|ab}$. At very high J_c-values the SQUID experiment failed because the maximum measurable moment was exceeded, and the ac method was severely limited in resolution because of the resulting small ac penetration depth. Therefore, at low temperatures only $J_c^{c;H\|ab}$ was measured. The critical current densities for this field orientation (single crystal Big2) are shown in figure 1 on a logarithmic scale. We notice good agreement at 4 K, but rather large differences between the two methods at higher temperatures. The SQUID data are lower by a factor of up to 4, which is consistent with the much longer relaxation times occurring in this experimental technique.

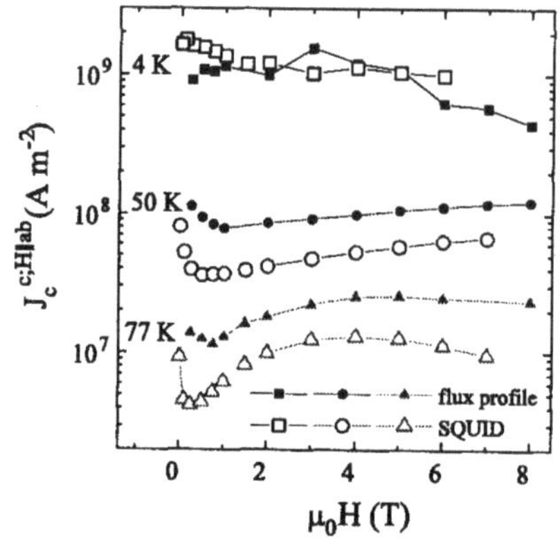

Figure 1: Critical current densities of the crystal Big2 for various temperatures and $H_a \| ab$.

Such good agreement with regard to the time scales of the experiments is not generally observed for $H_a \| c$, cf. figure 2. The SQUID data are higher than the corresponding ac values over a wide range of dc fields and temperatures. Two immediate explanations for

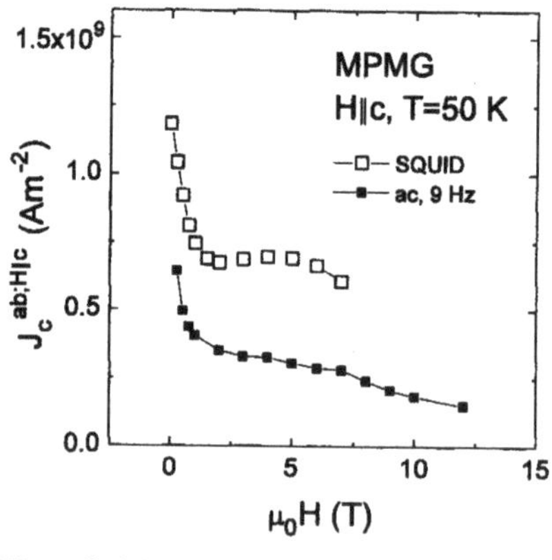

Figure 2: Critical current densities of the MPMG sample (#4) at 50 K and for $H_a \| c$.

this effect can be offered, both of which contribute to the experimental results. Firstly, the ac experiment is only scanning a surface sheath of the superconductor, while the SQUID averages over the whole sample volume. Secondly, and probably even more importantly, the interpretation of data in terms of a critical state is only a crude approximation. While for the SQUID measurements an interpretation in terms of relaxing shielding currents is rather straightforward, the flux profile measurement may be influenced in various ways by the presence of strong relaxation. In this case, the critical state model has to be replaced by an interpretation in terms of a skin effect with a resistivity ρ_{TAFF} due to thermal activation of the flux lines. An excellent review of these aspects is given in[6], where the non-trivial interactions between the various mechanisms are emphasized, which render simple approaches to solve the ambiguity so difficult.

3. Irreversibility Lines

The extreme case of relaxation limited shielding currents is described by the irreversibility line. An interpretation of these data faces the same ambiguity as discussed above. The peak in χ'' can be due to either one of two mechanisms, and is explained either by a critical state expression, with the peak position at the value $h_0 = J_c.r$, or as a consequence of the skin effect, with $\rho_{TAFF} = \mu_0.\pi.\nu.r^2$, where r is the sample radius and ν the measuring frequency. In the first case, the peak position should be independent of *frequency*, while in the second case it should be independent of the driving field *amplitude*. In figure 3 the irreversibility temperatures for a dc field of 1 T are shown for ac experiments as a function of frequency and amplitude. A pronounced dependence of T_{irr} on both the frequency ν and the driving field amplitude h_0 is observed. T_{irr} evaluated from the SQUID measurements is also included. The frequency (0.02 Hz) is derived from the periodicity of the movement and a "theoretical" amplitude is evaluated from

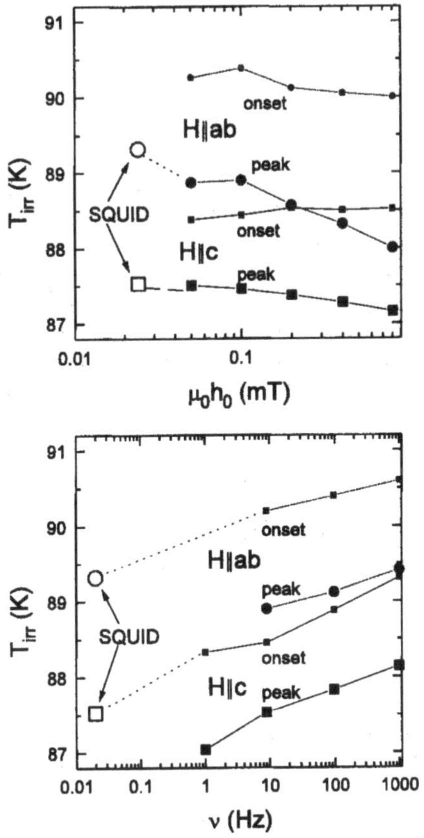

Figure 3: Irreversibility temperatures of sample MPMG at H_a=1 T of both main field orientations. Ac data were evaluated from the peak and the onset position of χ''.

the sensitivity criterion of $J_c = 10^4$ Am^{-2} according to the Bean model. Again, the data can be understood from two extreme positions of interpretation. The SQUID data are in good agreement with the ac values derived from the peak in χ'', if the amplitude dependence is considered, indicating that the critical state approach is valid to some extent, even at these high temperatures. It should be noted that the amplitude dependence is much less pronounced and even opposite (for $H_a \| c$), if the onsets of χ'' are considered. The frequency dependence, on the other hand, shows some disagreement between the ac-peak values and the SQUID data, which can be attributed to the influence of the different resolution limits of the two experiments. This is in turn supported by the good correlation of the ac-onset values with the SQUID data. At the higher "onset-irreversibility" temperatures the critical state model is no longer valid, the amplitude dependence vanishes and the frequency dependence is the determining quantity.

4. Summary

We have presented an experimental comparison between two magnetic methods to determine critical currents and irreversibility lines of high T_c superconductors. Data from SQUID magnetization measurements with a relatively long time constant were compared to relatively fast ac experiments. The critical currents were evaluated from the SQUID data using the standard Bean model and extracted from the ac data at various driving fields by the flux profile technique. The IL was determined from the onset of distortions in the SQUIDs response curves, and from the peak and the onset of the ac out-of-phase signal. The data are consistent with the time scales and the resolution of the experiments.

5. Acknowledgements

This work was supported by Fonds zur Förderung der Wissenschaftlichen Forschung, Vienna, under contract no. 8849.

6. References

1. M. Murakami, S. Gotoh, H. Fujimoto, K. Yamaguchi, N. Koshizuka and S. Tanaka, Supercond. Sci. Technol. **4** (1991) 43.
2. A. A. Zhokov and G. A. Emel'chenko, J. Crystal Growth **129** (1993) 786.
3. H. Wiesinger, Ph. D. thesis, Technical University Vienna (1992).
4. A. M. Campbell, J. Physics C **2** (1969) 1492.
5. M. Wacenovsky, Ph. D. thesis, Technical University Vienna (1993).
6. J. R. Clem, *Magnetic Susceptibility of Superconductors and other Spin Systems* edited by R. A. Hein (Plenum Press, New York 1992).

INFLUENCE OF SPECIMEN ASPECT RATIO ON LOW FIELD MAXIMUM IN MAGNETIC HYSTERESIS CURVES

M.DÄUMLING[a], E.WALKER[b], G.GRASSO[b,c], and R.FLÜKIGER[a,b]

[a] *Groupe de Physique Appliquée,* [b] *Département de Physique de la Matière Condensée, Université de Genève, CH-1211 Genève 4, Switzerland*

([c] on leave from Consorzio INFM, Università di Genova, Italy)

ABSTRACT

In this work we measure the magnetization of $YBa_2Cu_3O_{7-\delta}$ and $(Bi,Pb)_2Sr_2Ca_2$-Cu_3O_x superconductors with different specimen shapes. The focus here is the low field peak of the irreversible magnetization upon decreasing the external field. We find that for a rectangle (approaching slab geometry) this peak is at a significant negative field, whereas for plate shaped specimens (the field is applied parallel to the thin direction) this peak occurs at zero, or even a positive field. Possible reasons for this behavior include the demagnetization effects occuring in disks in perpendicular magnetic field, as well as the anisotropy in the j_c-H dependence due to transversal magnetic fields in the disk geometry.

1. Introduction

The critical state[1] in thin superconducting disks in perpendicular magnetic fields has been of considerable interest lately. [2-5] The particular features that occur in this case due to the vortex curvature are a radial magnetic field (perpendicular to the applied magnetic field) on the specimen surface, as well as a history dependent demagnetization field at the specimen edge, opposed to the trapped field in the center. One of the particular features noted in the irreversible magnetization of $YBa_2Cu_3O_{7-\delta}$ zone melted slabs is that the peak which occurs for low magnetic fields is virtually centered around zero for a disk, but occurs at a significant negative field for a slab-like specimen. [6] We have now expanded on this work and included more specimens of different aspect ratio.

2. Experimental

The specimens of $YBa_2Cu_3O_{7-\delta}$ were obtained by pulling a rod of cold isostatically pressed $YBa_2Cu_3O_{7-\delta}$ through an optical image furnace. The rod melted locally and resolidified in a quasi-singlecrystalline manner. Single grained specimens were then cut from the rod and oxygenated by slow-cooling them in an oxygen atmosphere. Critical temperatures of 92.5K and transitions width ΔT_c of less than 0.5K were obtained.

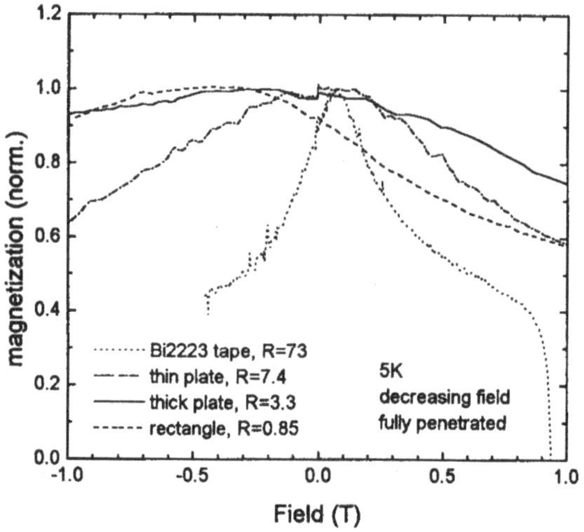

Fig. 1: *Magnetization normalized to the value at the peak in decreasing external magnetic field applied parallel to the c-axis for specimens of different aspect ratio R=(dimension ⊥ H)/ (dimension || H) .*

The Bi2223 tape was prepared by pressing. [7] The critical current density in self field is of order 30 kA/cm^2 at 77K, and 200 kA/cm^2 at 5K.

The magnetization was measured using a vibrating sample magnetometer (VSM) in a superconducting solenoid.

3. Results and Discussion

In figure 1 our results are summarized. Depicted are the reverse legs of the magnetization curves for all specimens (always H || c) in a fully penetrated critical state. The full magnetization curves for two of the specimens are shown in ref. [6] Full penetration fields for the $YBa_2Cu_3O_{7-\delta}$ specimens are of order 1 to 2 Tesla, and about 0.2 to 0.3T for the Bi2223 tape. As the aspect ratio of the specimens increases we note a definite trend: the position of the maximum in the magnetization moves from negative field first towards zero, and then even to positive field for the thinnest specimens.

The position of the maximum in a slab shaped superconductor is determined by the value of the full penetration field, and the fact that the critical current density j_c is field dependent. The maximum appears at negative fields because in the center of the superconductor the magnetic field is positive (in decreasing field), and a negative applied field is necessary to decrease this field to zero. The maximum occurs at the negative field where there rests some 'effective' minimum field in the specimen, leading to maximum shielding and thus a peak in the magnetization.

Due to the demagnetization effects present in a disk this mechanism does not work as simply as described above. Results of calculations in fact show that a peak at or close to zero field is obtained. [8,9] No account of anisotropy is made in these calculations. However, anisotropic current flow significantly changes the current flow patterns. [10] This is the case because the field dependence of the critical current density is different for H $\parallel$ c and H $\perp$ tape, thus causing shifts of the shielding current due to the radial magnetic surface field. Quantitative calculations taking into account anisotropic current flow are needed to clarify the question whether a magnetization peak at positive fields can be explained this way.

4. Acknowledgement

This work has been supported by the Swiss National Funds (NFP30).

5. References

[1]C.P.Bean. *Rev. Mod. Phys.* **36** (1964) 31.
[2]M.Däumling and D.C.Larbalestier. *Phys. Rev. B* **40** (1989) 9350.
[3]E.H.Brandt. *Phys.Rev.B* **48** (1993) 6699.
[4]M.V.Indenbom, Th.Schuster, M.R.Koblischka, A.Forkl, H.Kronmüller, L.A.Dorosinskii, V.K.Vlasko-Vlasov, A.A.Polyanskii, R.L.Prozorov, and V.I.Nikitenko. *Physica C* **209** (1993) 259.
[5]M.Oussena, P.A.J.de Groot, R.Gagnon, and L.Taillefer. *Phys.Rev.B* **49** (1994) 9222.
[6]M.Däumling, E.Walker, and R.Flükiger. *Phys. Rev. B*, accepted, (1994).
[7]G.Grasso, A.Perin, B.Hensel, and R.Flükiger. *Physica C* **217** (1993) 335.
[8]K.V.Bhagwat and P.Chaddah. *Physica C* **224** (1994) 155.
[9]E.Zeldov et al. *Phys. Rev. Lett.* **73** (1994) 1428.
[10]L.Conner and A.P.Malozemoff. *Phys. Rev. B* **43** (1991) 402.

CURRENT-INDUCED QUASI-BOUND STATES IN A JOSEPHSON JUNCTION WITH A WEAKLY INHOMOGENEOUS TUNNEL BARRIER

S. V. KUPLEVAKHSKY
Department of Physics, Kharkov State University, 310077 Kharkov, Ukraine

and

I. I. FAL'KO
Physico-Technical Department, Kharkov Polytechnical University, 310002 Kharkov, Ukraine

ABSTRACT

Starting from a self-consistent microscopic model, we study theoretically the influence of random structural inhomogeneities of the tunnel barrier on the newly predicted bound states induced by the supercurrent in a Josephson junction. We show that tunneling with the non-conservation of the momentum component parallel to the barrier, κ, leads to a shift and damping of the corresponding energy levels. In the limit of small fluctuations of the potential of the barrier, we derive a general formula for the energy of the resulting quasi-bound states in terms of the total tunneling probability and the tunneling probability with the non-conservation of κ.

1. Introduction

Recent theoretical studies reveal that supercurrents in Josephson structures with tunnel barriers induce quasiparticle states within the superconducting gap. The energies of these states are parameterized by the coherent phase difference, ϕ, and the cosine of the angle of incidence on the barrier, $\cos\theta$. For example, in an isolated junction with an ideally flat tunnel barrier a single discrete level[1,2] occurs:

$$E_0(t;\phi) = \Delta_\infty\left[1 - \tfrac{1}{2}T(t)\sin^2\tfrac{\phi}{2}\right], \tag{1}$$

$$|t| >> \left(\Delta_\infty/E_F\right)^{1/2}.$$

Here Δ_∞ is the energy gap in the bulk of superconductors, $t \equiv \cos\phi$, $T(t)$ is the tunneling probability for given t ($T(1) << 1$), E_F is the Fermi energy. In a superlattice, the level Eq. (1) is transformed into a Bloch-type band[3]. The possibility of resonance excitation of the level[1,2] by an electromagnetic field was studied in Ref. 4.

The consideration of the papers[1-3] is based essentially on the assumption of translational ivariance in the directions parallel to the interfaces. This assumption guarantees the conservation of the corresponding momentum components, yielding a quasi-one-dimentional problem to solve. On the other hand, it is quite clear that owing to the inevitable presence of structural inhomogeneities (at least on the atomic scale) real tunnel barriers are not translationally invariant. In this case, one should expect a shift and damping of the level Eq. (1). Below we give an explicit mathematical derivation of these results on the basis of a simple but rather general microscopic model of a Josephson junction.

2. Formalism

Consider two semi-infinite superconductors in the clean limit separated by a thin tunnel barrier in the yz-plane. In terms of the retardered Green's function, the problem is formulated as follows:

$$\left[E\tau_0+\left(\tfrac{1}{2m}\nabla^2+E_F\right)\tau_3-\tfrac{1}{2}\left(\tau_1+i\tau_2\right)\Delta(\mathbf{r})-\tfrac{1}{2}\left(\tau_1-i\tau_2\right)\Delta^*(\mathbf{r})-\hat{W}(\mathbf{r})\right]$$

$$\times\tilde{G}_E\left(\mathbf{r},\mathbf{r}'\right)=\delta\left(\mathbf{r}-\mathbf{r}'\right)\tau_0. \tag{2}$$

Here E is the energy parameter with an infinitesimal positive imaginary part; m is the mass of an electron; τ_k $(k=1,3)$ is a Pauli matrix in Gor'kov-Nambu space, and τ_0 is the corresponding unit matrix. The potential of the barrier is described by

$$\hat{W}(\mathbf{r})=\left[V+U(\boldsymbol{\rho})\right]\delta(x)\tau_3, \tag{3}$$

where V is a positiv constant, $\boldsymbol{\rho}=(y,z)$, $U(\boldsymbol{\rho})$ is a random function of coordinates with the properties

$$\langle U(\boldsymbol{\rho})\rangle=0, \tag{4}$$

$$\langle U(\boldsymbol{\rho})U(\boldsymbol{\rho}')\rangle=\frac{\bar{u}^2}{p_0^2}\delta\left(\boldsymbol{\rho}-\boldsymbol{\rho}'\right), \tag{5}$$

where $p_0=mv_0$ (v_o is the Fermi velocity), $\bar{u}^2$ is a constant, and $\langle\ldots\rangle$ means averaging. Discarding the self-consistency condition, we adopt for the pair potential the approximation

$$\Delta(\mathbf{r})=\Delta_\infty\exp\left[i\,\mathrm{sgn}(x)\phi\right]. \tag{6}$$

This approximation is valid for the evaluation of the subgap states in the limit of a low barrier transmission[2] ($V>>v_0$). Additionally, we restrict ourselves to the limit of small structural fluctuations:

$$\varepsilon \equiv \bar{u}^2/V^2 << 1. \tag{7}$$

We are interested in the solution of Eqs. (2), (6) averaged over the structural fluctuations. Making use of Eqs. (4), (5), (7) and mathematical methods developed for the treatment of superconducting alloys[5], we derive

$$\langle \tilde{G}_E \rangle (x,x';\mathbf{\kappa}) = G_E(x,x';\mathbf{\kappa}) + G_E(x,0;\mathbf{\kappa})\,\Sigma\left[\tau_0 - G_E(0,0;\mathbf{\kappa})\,\Sigma\right]^{-1} G_E(0,x';\mathbf{\kappa}), \tag{8}$$

where we have taken Fourier transforms in the coordinates x and y, with $\mathbf{\kappa}$ being the momentum component parallel to the barrier. The Green's function $G_E(x,x;\mathbf{\kappa})$ gives us the solution of Eqs. (2), (6) for $U(\mathbf{\rho}) \equiv 0$. Diagonal elements of G_E with one or both coordinates in the plane of the barrier have a pole at the energy of the current-carrying level Eq. (1). The self-energy part, Σ, satisfies the self-consistent equation

$$\Sigma = \frac{\varepsilon V^2}{p_o}\tau_3 \int \frac{d^2\mathbf{\kappa}}{(2\pi)^2}\left[G_E^{-1}(0,0;\mathbf{\kappa}) - \Sigma\right]^{-1}\tau_3, \tag{9}$$

The solution of Eq. (9) is expressed in terms of the matrices τ_0, τ_1, τ_3 with complex coefficients. Note that the emergence of ImE indicates damping caused by tunneling with the non-conservation of $\kappa \equiv |\mathbf{\kappa}| = p_0|\sin\theta|$.

3. Excitation Spectrum

The subgap part of the energy spectrum of our system is determined by the poles of diagonal elements of $\langle \tilde{G}_E \rangle (0,0;\mathbf{\kappa})$, which is equivalent to the condition

$$\det\left[G_E^{-1}(o,o;\mathbf{\kappa}) - \Sigma\right] = 0. \tag{10}$$

Solving Eqs. (9), (10) in the lowest order in tunneling probabilities yields

$$E_0(t;\phi) = \Delta_\infty\left\{1 - \tfrac{1}{2}[T(t) + T_\kappa(t)]\sin^2\tfrac{\phi}{2} - i3\pi T_\kappa(t)|t|^3\sin^2\tfrac{\phi}{2}\right\}, \tag{11}$$

where

$$T(t) = [1 + \varepsilon/2\pi]v_0^2 t^2/V^2 + T_\kappa(t) \tag{12}$$

is the averaged total tunneling probability, while

$$T_\kappa(t) = \varepsilon v_0^2 |t| / V^2$$

is the averaged probability of tunneling with the non-conservation of κ; $|t| >> \max\{(\Delta_\infty/E_F)^{1/2}, \varepsilon\}$. (See Appendix.) As expected, in the absence of structural inhomogeneities Eq. (11) reduces to Eq. (1). Physically, $T(t)$ determines the resistance of the junction in the normal state[6]. It is interesting to observe that $T_\kappa(t)$ not only gives rise to damping, but also contributes to the shift of ReE_o.

Acknowledgements

The authors thank A. N. Omelyanchuk for fruitful discussions.

The research described in this publication was made possible in part by Grant No. U28000 from the International Science Foundation.

Appendix

We calculate the transmission probability for electrons at the Fermi surface using the results[6] for the potential Eq. (3) in the limit $V >> v_0$, $U(\boldsymbol{\rho})$ being an *arbitrary deterministic* function:

$$\tilde{T}(t) = \frac{v_0|t|}{m\Omega V^2} \int \frac{d^2\kappa'}{(2\pi)^2} \sqrt{p_0^2 - \kappa'^2} \left| \int \frac{d^2\rho \exp\left[i\rho(\kappa - \kappa')\right]}{1 + U(\rho)/V} \right|^2, \tag{A1}$$

where Ω is the area of the barrier. After expanding Eq. (A1) in U/V to second order and averaging with the help of Eq. (4) and its Fourier transform we obtain

$$T(t) \equiv \langle \tilde{T} \rangle(t) = \frac{v_0^2 t^2}{V^2} + \frac{\varepsilon v_0 |t|}{4\pi m V^2} \int d^2\kappa' \sqrt{p_o^2 - \kappa'^2} \left[\delta(\kappa - \kappa') + 1 \right]. \tag{A2}$$

Performing the last integration in Eq. (A2) finally yields Eq. (12).

References

1. A. Furusaki and M. Tsukada, *Physica* **B165-166** (1990) 967; *Phys. Rev.* **B43** (1991) 10164.
2. S. V. Kuplevakhsky and I. I. Fal'ko, *Sov. J. Low Temp. Phys.* **17** (1991) 501.
3. S. V. Kuplevakhsky and I. I. Fal'ko, *Sov. J. Low Temp. Phys.* **18** (1992) 776.
4. V. S. Shumeiko, G. Wendin, and E. N. Bratus, *Phys. Rev.* **B48** (1993) 13129.
4. A. A. Abrikosov, L. P. Gorkov, and I. E. Dzyaloshinski, *Methods of Quantum Field Theory in Statistical Physics,* Englewood Cliffs, 1963.
5. I. O. Kulik, Yu. N. Mitsay and A. N. Omelyanchuk, *Zh. Eksp. Teor. Fiz.* **66** (1974) 1051.

JOSEPHSON CURRENT AND QUASI-DISCRETE ENERGY LEVELS IN A SUPERCONDUCTING JUNCTION WITH A TUNNEL PARAMAGNETIC BARRIER

S. V. KUPLEVAKHSKY
Department of Physics, Kharkov State University, 310077 Kharkov, Ukraine

and

I. I. FAL'KO
Physico-Technical Department, Kharkov Polytechnical University, 310002 Kharkov, Ukraine

ABSTRACT

On the basis of a self-consistent microscopic model, we make theoretical calculations of the dc Josephson critical current and the energy spectrum of a clean superconducting junction with a tunnel paramagnetic barrier. We show that tunneling with a spin-flip induces formation of quasi-discrete energy levels of a finite width. The damping of the levels occurs due to the non-conservation of the momentum component parallel to the barrier. In the absence of the Josephson current, the quasi-discrete levels exist only for large angles of incidence on the barrier. This condition is relaxed in the current-carrying state.

1. Introduction

The superconductivity of multilayered structures with magnetic barriers is a subject of continuing and long-standing interest. In superconducting junctions and superlattices with *ferromagnetic* tunnel barriers, tunneling with a spin flip results in the reduction of the Josephson critical current[1] and the formation of polarized quasiparticle states at energies within the superconducting gap[2,3]. We predict that analogs of these two effects occur also in superconducting junctions with *paramagnetic* tunnel barriers, but the subgap part of the energy spectrum is characterized by strong damping.

2. Formalism

Consider two semi-infinite superconductors with s-wave pairing in the clean limit separated by a thin paramagnetic tunnel barrier in the yz-plane. In terms of the retardered Green's function, the problem can be formulated as follows:

$$\left[E\tau_0\sigma_0+\left(\tfrac{1}{2m}\nabla^2+E_F\right)\tau_3\sigma_0-\tfrac{1}{2}\left(\tau_1+i\tau_2\right)i\sigma_2\Delta(\mathbf{r})-\tfrac{1}{2}\left(\tau_1-i\tau_2\right)i\sigma_2\Delta^*(\mathbf{r})-\hat{W}(\mathbf{r})\right]$$

$$\times\widetilde{G}_E\left(\mathbf{r},\mathbf{r}'\right)=\delta\left(\mathbf{r}-\mathbf{r}'\right)\tau_0\sigma_0. \qquad (1)$$

Here E is the energy parameter with an infinitesimal positive imaginary part; m is the mass of an electron; $\tau_k\sigma_l$ is the direct product of Pauli matrices in Gor'kov-Nambu space (τ_k) and in spin space (σ_l), and τ_0 and σ_0 are the corresponding unit matrices. The potential of the barrier is described by

$$\hat{W}(\mathbf{r}) = \left[V\tau_3\sigma_0 + \sum_i \hat{J}_{\mathbf{S}_i}\delta\left(\boldsymbol{\rho} - \boldsymbol{\rho}_i\right) \right]\delta(x),$$

$$\hat{J}_{\mathbf{S}_i} = \frac{J}{2}\left[\left(\tau_0 + \tau_3\right)\left(\boldsymbol{\sigma}\mathbf{S}_i\right) - \left(\tau_0 - \tau_3\right)\left(\boldsymbol{\sigma}^*\mathbf{S}_i\right) \right]. \tag{2}$$

where V is the nonexchange part of the potential, $\boldsymbol{\rho} = (y,z)$, J is the exchange integral, and $\mathbf{S}_i$ is the spin of the magnetic atom at the point $\mathbf{r}_i = (0,\boldsymbol{\rho}_i)$. Ignoring the quantum nature of the spin $\mathbf{S}_i$, we are treating it as a classical vector. For simplicity we assume that the magnetic atoms are distributed at random in the plane of the barrier.

Discarding the self-consistency condition, we adopt for the pair potential the approximation

$$\Delta(\mathbf{r}) = \Delta_\infty \exp\left[i\,\mathrm{sgn}(x)\phi\right], \tag{3}$$

where Δ_∞ is the energy gap in the bulk of superconductors, and ϕ is the coherent phase difference. This approximation is valid in the limit of a low barrier transmission

$$V >> v_0, JS,$$

where v_0 is the Fermi velocity, and $S = |\mathbf{S}_i|$.

We are interested in the solution of Eqs. (1)-(3) averaged over orientations of the spins of magnetic atoms and their positions in the plane of the barrier $x = 0$. Making use of mathematical methods developed for the treatment of superconducting alloys[4], we derive

$$\left\langle \tilde{G}_E \right\rangle\left(x,x';\boldsymbol{\kappa}\right) = G_E\left(x,x';\boldsymbol{\kappa}\right) + G_E\left(x,0;\boldsymbol{\kappa}\right)\Sigma\left[\tau_0\sigma_0 - G_E\left(0,0;\boldsymbol{\kappa}\right)\Sigma\right]^{-1}$$

$$\times G_E\left(0,x';\boldsymbol{\kappa}\right), \tag{4}$$

where we have taken Fourier transforms in the coordinates x and y, with $\boldsymbol{\kappa}$ being the momentum component parallel to the barrier. The Green's function $G_E\left(x,x;\boldsymbol{\kappa}\right)$ gives us the solution of Eqs. (1)-(3) at $J = 0$. Note that diagonal elements of G_E with one or both coordinates in the plane of the barrier have poles at the energies of supercurrent-induced bound states[5,6] (for $\phi \neq 0$). The self-energy part, Σ, satisfies the self-consistent equation

$$\Sigma = \left(JS\right)^2 c \int \frac{d^2\kappa}{(2\pi)^2}\left[G_E^{-1}\left(0,0;\kappa\right) - \Sigma\right]^{-1}, \tag{5}$$

where c is the concentration of magnetic atoms in the plane of the barrier.

3. Josephson Current

By analytic continuation $E \to i\omega_n$ (ω_n is the Matsubara frequency) of Eqs. (4)-(5) we obtain the temperature Green's function $\langle \tilde{G}_{\omega_n}\rangle\left(x,x';\kappa\right)$. It is straightforward to evaluate the Josephson current, j_J, with the help of $\langle \tilde{G}_{\omega_n}\rangle$:

$$j_J = \frac{ie}{4m} T \sum_{\omega_n}\left\{\left(\frac{\partial}{\partial x'} - \frac{\partial}{\partial x}\right)\int \frac{d^2\kappa}{(2\pi)^2} \mathrm{Tr}\left[\left(\tau_0 - \tau_3\right)\langle \tilde{G}_{\omega_n}\rangle\left(x,x';\kappa\right)\right]\right\}_{x'=x=0}$$

$$= \frac{e\pi N(o) v_o \Delta_\infty}{2} \int_0^1 dt t\left[T(t) - 2T_S(t)\right] \tanh\left(\frac{\Delta_\infty}{2T}\right) \sin\phi. \tag{6}$$

The general form and the interpretation of Eq. (6) is the same as in the case of a junction with a ferromagnetic tunnel barrier[1]. Thus $T(t) = T_N(t) + T_S(t)$ is the averaged total tunneling probability for a given angle of incidence, θ ($t \equiv \cos\theta = \left(1 - \kappa^2/p_0^2\right)^{1/2}$, $p_0 = v_0 m$), with

$$T_N(t) = \frac{v_0^2 t^2}{V^2} + \frac{c p_0^2 v_0^2 J^2 S^2 t^2}{4\pi V^4}$$

and

$$T_S(t) = \frac{c p_0^2 v_0^2 J^2 S^2 |t|}{6\pi V^4}$$

being the non-spin-flip part and the spin-flip part, respectively. The quantity $T(t)$ determines the resistance of the junction in the normal state, while the negative term $-2T_S(t)$ describes pair-breaking. For a paramagnetic barrier, $T_S(t)$ also characterizes tunneling with the non-conservation of $|\kappa|$. (Notice that the expression of the type Eq. (6) was first obtained in Ref. 7 by the tunneling Hamiltonian method.)

4. Excitation Spectrum

Diagonal elements of $\langle \tilde{G}_E\rangle\left(0,0;\kappa\right)$ have poles at complex energies, with $\mathrm{Re}E < \Delta_\infty$. The emergence of $\mathrm{Im}E$ indicates damping caused by tunneling with non-conservation of $|\boldsymbol{\kappa}|$. This is a completely new feature in comparison with a junction with a ferromagnetic tunnel barrier[3]. Supposing the damping rate to be small, one obtains

$$E_{0\pm}\left(t;\phi\right) = \Delta_\infty\left\{1 - 2T_S(t) - \tfrac{1}{2}\left[T(t) - 3T_S(t)\right]\sin^2\tfrac{\phi}{2} - i6\pi T_S(t)t^2\gamma(t)\left(1 + \sin^2\tfrac{\phi}{2}\right)\right\}, \quad (7)$$

where $\gamma(t) = \left\{\frac{\partial}{\partial|t|}\ln\left[\Delta_\infty - \mathrm{Re}E_{0\pm}(t;\phi)\right]\right\}^{-1} < 1$ and $|t| >> T_S(1)/T(1)$. As a result of damping, the existence of quasi-discrete levels Eq. (7) is restricted by the condition $|\mathrm{Im}E_{0\pm}| << |\mathrm{Re}E_{0\pm}|$. The lower indices $\pm$ on the left hand side of Eq. (7) refer to the orientation of the quasiparticle spin. Physically, the second term in the figure brackets on the right hand side of Eq. (7) describes the formation of the quasi-bound states due to tunneling with a spin flip. It was first derived in Ref. 8. The second term is the contribution of the Josephson current. Here again $-3T_S(t)$ characterizes pair-breaking in all the processes of tunneling. For $\phi = 0$, the quasi-bound states occur only if $T_S(1)/T(1) << |t| << (3\pi)^{-1/3}$. The Josephson current facilitates the formation of the subgap levels. In the absence of magnetic atoms, Eq. (7) reduces to the well-known expression[5,6] for the supercurrent-induced bound states.

Acknowledgements

The authors thank A. N. Omelyanchuk for fruitful discussions.

The research described in this publication was made possible in part by Grant No. U28000 from the International Science Foundation.

References

1. S. V. Kuplevakhsky and I. I. Fal'ko, *Sov. J. Low Temp. Phys.* **10** (1984) 361; *Theor. Mat. Phys.* (Moscow) **67** (1986) 483.
2. M. J. DeWeert and G. B. Arnold, *Phys.Rev. Lett.* **55** (1985) 1522; *Phys. Rev.* **B39** (1989) 11307.
3. S. V. Kuplevakhsky and I. I. Fal'ko, *Sov. Phys. Solid State* **34** (1992) 96; *JETP Lett.* **55** (1992) 387.
4. A. A. Abrikosov, L. P. Gorkov, and I. E. Dzyaloshinski, *Methods of Quantum Field Theory in Statistical Physics,* Englewood Cliffs, 1963.
5. A. Furusaki and M. Tsukada, *Physica* **B165-166** (1990) 967; *Phys. Rev.* **B43** (1991) 10164.
6. S. V. Kuplevakhsky and I. I. Fal'ko, *Sov. J. Low Temp. Phys.* **17** (1991) 501.
7. I. O. Kulik, *Zh. Eksp. Teor. Fiz.* **49** (1965) 1211.
8. S. V. Kuplevakhsky and I. I. Fal'ko, *JETP Lett.* **58** (1993) 195.

MACROSCOPIC EQUATIONS FOR THE DESCRIPTION OF NON-REVERSIBLE SUPERCONDUCTORS OF ARBITRARY SHAPE

R.Labusch

Institut für Angewandte Physik, Technische Universität Clausthal, 3992 Clausthal-Zellerfeld, Germany.

and

T.B.Doyle

Department of Physics, University of Natal, Durban 4001, South Africa.

ABSTRACT

Macroscopic equations, in terms of the constitutive Abrikosov $B[H_{rev}]$ relation and the bulk pinning force density are obtained for the description of the critical state in a singly-connected Type II superconductor of arbitrary shape.

1. Introduction

The detailed magnetic behaviour of any magnetic medium, in a sample geometry with non-zero demagnetisation field and spatially non-uniform magnetisation, requires a self-consistent calculation of the local demagnetisation tensor or current distribution. The treatment given here is based [1,2] on a rigorous analysis of the free energy of an Abrikosov vortex element in the presence of a fixed but arbitrary distribution of other vortices in a conventional type II superconducting specimen of arbitrary shape.

2. Theory

We consider a singly-connected specimen of minimum macroscopic dimension D such that quantities of order λ/D and ξ/D can be neglected. The Gibbs free energy ΔG required to create a vortex in a specimen of arbitrary shape with a fixed but arbitrary distribution of other vortices is given [3] by

$$\Delta G = \int\left[\Delta\boldsymbol{B}(\boldsymbol{r})\cdot\boldsymbol{H}(\boldsymbol{r}) + \frac{1}{2}\Delta\boldsymbol{B}(\boldsymbol{r})\cdot\Delta\boldsymbol{H}(\boldsymbol{r})\right]d^3\boldsymbol{r} + \int\left[\Delta E_k(\boldsymbol{r}) + \Delta E_c(\boldsymbol{r})\right]d^3\boldsymbol{r} = I_1 + I_2 \quad (1)$$

where I_1, E_k and E_c are the magnetic field, kinetic and the condensation energy contributions respectively. $\Delta\boldsymbol{H}(\boldsymbol{r})$ and $\Delta\boldsymbol{B}(\boldsymbol{r})$ are the changes in $\boldsymbol{H}(\boldsymbol{r})$ and $\boldsymbol{B}(\boldsymbol{r})$ on the introduction of the vortex. Inside the superconductor $\boldsymbol{H}(\boldsymbol{r})$, $\Delta\boldsymbol{B}(\boldsymbol{r})$, $\Delta\boldsymbol{H}(\boldsymbol{r})$, ΔE_k and ΔE_c are restricted to a "microscopic" channel. The integral I_1, which is taken over all space, is separable into three component parts; namely: $I_1 = I_i + I_e + I_\delta$ where

$$I_i = \int_{sc}\left[\Delta\boldsymbol{B}\cdot\boldsymbol{H} + \frac{1}{2}\Delta\boldsymbol{B}\cdot\Delta\boldsymbol{H}\right]d^3\boldsymbol{r} \quad (2a)$$

is an integral over the superconductor volume,

$$I_e = \int_{ext}\Delta\boldsymbol{B}\cdot\boldsymbol{H}\,d^3\boldsymbol{r} \quad \text{and} \quad I_\delta = \frac{1}{2}\int_{ext}\Delta\boldsymbol{B}\cdot\Delta\boldsymbol{H}\,d^3\boldsymbol{r} \quad (2b)$$

are integrals over the remaining space. Since ΔB and ΔH, in I_i, and ΔE_k and ΔE_c, in I_2, are restricted to a microscopic tube we can cut out a narrow, but nevertheless "macroscopic" tube around each line element of the newly created vortex and replace the surrounding material, with its other vortices, by some parallel field without changing I_i or I_2. To be in equilibrium with the local flux density this field must be equal to $H_{rev}(B)$, the thermodynamic reversible field in equilibrium with B. In this situation ΔG is known [4] to be given by

$$\Delta G = \phi_o \int \boldsymbol{H}_{rev}(\overline{B}) \cdot d\boldsymbol{l} = I_i + I_2 \tag{3}$$

where $\phi_o = h/2e$ is the flux quantum and $\overline{B}$ is an average induction over a small but macroscopic volume. For $B = 0$, H_{rev} is indeterminate, but clearly $H_{rev} = H_{c1}$ in Eq.3 because $I_i + I_2$ is simply the self-energy of an isolated vortex in this limit.

For the calculation of I_e we note that; (i) $\boldsymbol{H}(\boldsymbol{r})$ can be expressed in terms of the applied field intensity and the potential of magnetic charges at the surface of the superconductor: $\boldsymbol{H}(\boldsymbol{r}) = \boldsymbol{H}_o - \nabla\varphi(\boldsymbol{r})$, (ii) $\nabla.\boldsymbol{B} = 0$ and (iii) far from the specimen the surface integral $\oint_\infty \Delta\boldsymbol{B}\varphi(\boldsymbol{r})\, d^2\boldsymbol{s} \to 0$. With these relations I_e can be transformed to

$$I_e = -\int_{ext} \Delta\boldsymbol{B}\cdot\nabla\varphi\, d^3\boldsymbol{r} = -\int_{ext} \nabla(\Delta\boldsymbol{B}.\varphi)\, d^3\boldsymbol{r} = \oint_{sc} (\Delta\boldsymbol{B}.\varphi) d^2\boldsymbol{s} = -\int_{sc} \Delta\boldsymbol{B}\cdot\boldsymbol{H}_e(r) d^3\boldsymbol{r}. \tag{4}$$

The field $\boldsymbol{H}_e(\boldsymbol{r})$ can be constructed according to well known principles and is conveniently given by

$$\boldsymbol{H}_e(\boldsymbol{r}) = \boldsymbol{H}_o + \frac{1}{4\pi}\int_{sc} \frac{[\boldsymbol{M}(\boldsymbol{r}')\cdot\boldsymbol{n}](\boldsymbol{r}-\boldsymbol{r}')}{|\boldsymbol{r}-\boldsymbol{r}'|^3} d^2\boldsymbol{s} \tag{5}$$

where $\boldsymbol{M}(\boldsymbol{r}) = \frac{1}{\mu_o}\boldsymbol{B}(\boldsymbol{r}) - \boldsymbol{H}_e(\boldsymbol{r})$ is the local average magnetisation, $\boldsymbol{n}$ is a unit vector normal to the surface and the integral is taken over the surface of the superconductor. Using the quantisation condition $\boldsymbol{n}\int \Delta\boldsymbol{B}\, d^2\boldsymbol{r} = \phi_o$, we can transform Eq.4 into a line integral along the vortex, and obtain

$$I_e = -\phi_o \int \boldsymbol{H}_e(\boldsymbol{r})\cdot d\boldsymbol{l}. \tag{6}$$

The contributions from I_δ and surface penetration terms in I_i can be shown to be insignificant [2] and will be neglected here. Thus finally, collecting the significant terms from Eqs.3 and 6, we obtain

$$\Delta G = \phi_o \int (\boldsymbol{H}_{rev} - \boldsymbol{H}_e)\cdot d\boldsymbol{l}. \tag{7}$$

In equilibrium ΔG must be zero when a vortex, parallel to the surface of the superconductor, is nucleated into the surface (providing that there is no special surface barrier) and we have therefore the boundary condition $H_{rev\parallel} = H_{e\parallel}$ at the surface. Inside a reversible superconductor $\Delta G = 0$ also and therefore $H_{rev} = H_e$. This is a well known and accepted result [5] although it has until now been derived only intuitively.

In the non-equilibrium situation we calculate the energy difference for the vortex at two positions $\boldsymbol{r}$ and $\boldsymbol{r} + \delta\boldsymbol{r}$ (see Fig.1) and obtain

$$\delta G = \phi_o \int [\boldsymbol{H}_{rev}(\boldsymbol{r}+\delta\boldsymbol{r}) - \boldsymbol{H}_{rev}(\boldsymbol{r})]\cdot d\boldsymbol{l} - \phi_o \int [\boldsymbol{H}_e(\boldsymbol{r}+\delta\boldsymbol{r}) - \boldsymbol{H}_e(\boldsymbol{r})]\cdot d\boldsymbol{l} = \delta I_i + \delta I_e \quad (8)$$

where δI_i can be represented, according to Fig. 1, as a line integral over a closed loop with line elements perpendicular to $\boldsymbol{H}_{rev}$ at the ends of the vortex plus the two contributions $\boldsymbol{H}_{rev}.d\boldsymbol{l}_1$ and $\boldsymbol{H}_{rev}.d\boldsymbol{l}_2$, from the vortex-end segments, which result from the change in the length of the vortex with the displacement $\delta\boldsymbol{r}$.

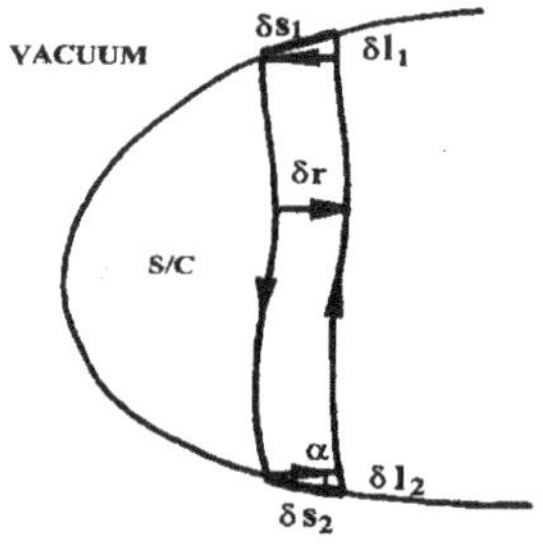

Fig.1 Displacement of a flux tube by $\delta\boldsymbol{r}$. The difference between the line along the displaced and the un-displaced positions of the flux tube is represented by a closed loop integral, as indicated by arrows, and by the contributions from $d\boldsymbol{l}_1$ and $d\boldsymbol{l}_2$

The integral over the closed loop, using Stokes' theorem, is equal to an integral over the area of the loop; $\phi_o \int [\nabla \times \boldsymbol{H}_{rev}]\cdot d\boldsymbol{A}$. Since $d\boldsymbol{A} = d\boldsymbol{l}\times\delta\boldsymbol{r}$ and $[\nabla\times\boldsymbol{H}_{rev}]\cdot[d\boldsymbol{l}\times\delta\boldsymbol{r}] = \delta\boldsymbol{r}\cdot[\nabla\times\boldsymbol{H}_{rev}]\times d\boldsymbol{l}$, it may be written as a single line integral along the vortex; $\phi_o \int \delta\boldsymbol{r}\cdot[\nabla\times\boldsymbol{H}_{rev}]\times d\boldsymbol{l}$. The contribution of an end segment is approximately $\phi_o H_{rev}\delta r.\sin\alpha/\cos\alpha$ where α is the angle between the surface and the flux direction. This is conveniently expressed as a line integral $\phi_o \int \delta(\boldsymbol{r}-\boldsymbol{s})\delta\boldsymbol{r}\cdot[\boldsymbol{n}\times\boldsymbol{H}_{rev}]\times d\boldsymbol{l}$ where $\delta(\boldsymbol{r} - \boldsymbol{s})$ is a Dirac delta function at the surface of the specimen and $\boldsymbol{n}$ is again a unit vector normal to the surface. (Notice that this integral represents the contributions from both ends of the vortex.)

For the evaluation of δI_e it is more convenient to go back to the alternative form given in Eq. 4; namely: $I_e = \oint_{sc} (\Delta\boldsymbol{B}.\varphi)d^2\boldsymbol{s} = \phi_o[\varphi(\boldsymbol{s}_1) - \varphi(\boldsymbol{s}_2)]$ where $\boldsymbol{s}_1$ and $\boldsymbol{s}_2$ are the points where the flux tube intersects the surface. It follows immediately that $\delta I_e = -\phi_o[\delta s_1\boldsymbol{H}_e(\boldsymbol{s}_1) - \delta s_2\boldsymbol{H}_e(\boldsymbol{s}_2)]$ where we have defined $\delta\boldsymbol{s}_1$ and $\delta\boldsymbol{s}_2$ as the displacements of the vortex along the surfaces [given by $\delta s = \delta r/\cos\alpha$ where α is again the angle between the surface and the flux direction (see Figure 1)]. Putting this in the line-integral form (as for δI_1 above) we obtain

$$\delta I_e = -\phi_o \int \delta(\boldsymbol{r}-\boldsymbol{s})\delta\boldsymbol{r}\cdot[\boldsymbol{n}\times\boldsymbol{H}_e]\times d\boldsymbol{l}. \quad (9)$$

Finally, collecting all the contributions to δG and defining a force vector $\boldsymbol{f}$ per unit length of vortex and using the relation $\delta G = \int (\delta \boldsymbol{r} \cdot \boldsymbol{f}) dl$ gives

$$\boldsymbol{f} = \phi_o \{ [\nabla \times \boldsymbol{H}_{rev}] + \delta(\boldsymbol{r} - \boldsymbol{s}) [\boldsymbol{n} \times (\boldsymbol{H}_{rev} - \boldsymbol{H}_e] \} \times \boldsymbol{e} \tag{10}$$

where $\boldsymbol{e}$ is a unit vector along the flux direction.

If there is no special surface barrier the δ-function at the surface will always produce flux motion unless $H_{rev\|} = H_{e\|}$. This is identical to the boundary condition that was obtained from the free energy expression. Inside the volume of the specimen we are concerned only with the first term of this expression. Now the volume force acting on the vortex assembly is $\boldsymbol{f}$ times the local area density of the vortices or flux tubes; namely: B/ϕ_o. In the presence of a volume pinning force $\boldsymbol{F}_p$ the critical state is therefore given by

$$\boldsymbol{F}_p = [\nabla \times \boldsymbol{H}_{rev}] \times \boldsymbol{B}. \tag{11}$$

The relation between the pinning force and H_{rev} inside the specimen is therefore independent of specimen shape. The field $\boldsymbol{H}_e(\boldsymbol{r})$, which depends on geometry, however imposes two boundary conditions; namely: $H_{rev\|} = H_{e\|}$ and $B_\perp(H_{rev}) = \mu_o H_e{}^o{}_\perp$ at the surface. Here $H_e{}^o{}_\perp$ is the perpendicular component of the outside field. The latter boundary condition derives from $\nabla \cdot \boldsymbol{B} = 0$. $\boldsymbol{H}_e(\boldsymbol{r})$ is obtained from Eq.5 or, alternatively from the differential equation $\nabla \cdot \boldsymbol{H}_e = \delta(\boldsymbol{r} - \boldsymbol{s})\, \sigma(\boldsymbol{s})$ with $\sigma = H_e{}^o{}_\perp - H_e{}^i{}_\perp$ and the boundary condition $H_e \rightarrow H_o$ for $r \rightarrow \infty$, where $H_e{}^i{}_\perp$ is the perpendicular component of the internal field at the surface. Thus finally Equations 5, 11, the two surface boundary conditions and the Abrikosov constitutive relation $|\boldsymbol{B}| = \boldsymbol{B}(|H_{rev}|)$, as obtained for a long cylindrical specimen of reversible material, constitute the basic equations by which the system is described.

3. Conclusions

Macroscopic equations for the description of the critical state in a singly-connected Type II superconductor of arbitrary shape have been derived. The analysis gives two terms for the driving force on a vortex; the usual bulk term which does not depend on specimen shape and a term which allows for the spatially dependent demagnetising effects in the specimen.

4. References

1. R. Labusch, Private communication 1974.
2. R. Labusch and T.B. Doyle, to be submitted to *Phys. Rev. Lett.*
3. See for example "*Landau and Lifschitz*" *Theoretical Physics* Vol. **8.**
4. A.M. Campbell, and J.E Evetts, *Adv. Phys.* **21** (1972) See, in particular, p222.
5. A.L. Fetter, and P.C. Hohenberg, "*Superconductivity*" Vol. **2**, p817. Edited by R.D.Parks, Marcel Dekker Inc., N.Y. (1969).

THE MAGNETISATION BEHAVIOUR OF AN IDEAL TYPE II SUPERCONDUCTING DISC SPECIMEN IN TRANSVERSE FIELD

T.B.Doyle

Department of Physics, University of Natal, Durban, 4001, South Africa.

and

R.Labusch

Institut für Angewandte Physik, Technische Universität Clausthal, 3992 Clausthal-Zellerfeld, Germany.

ABSTRACT

A model analysis, which gives the essential features of the $M(H_o)$ behaviour in terms of the Abrikosov constitutive relation $B_e[H_{rev}]$ for an ideal Type II superconducting disc in transverse applied field H_o is compared with experiment on ultra-pure, defect 'free', niobium single-crystal specimens.

1. Introduction

This work is based on an early investigation on a conventional Type II superconductor [1-3] but has new relevance to the problem of the magnetic behaviour in disc and platelet-shaped crystals of the high-T_c ceramic-oxide systems. Until recently publications on conventional and high-T_c systems have, with the exception of that due to Clem et al [4] on Type-I superconducting strips, usually assumed the validity of the use of a single scalar demagnetising coefficient which in situations of large demagnetising field and non-uniform magnetisation leads to significant error. The problem of the magnetic field profile and critical current distribution in thin-disc and platelet high-T_c specimens in transverse applied field has been the subject of several recent publications [see 5 and references therein]. Although it is now recognised that the use of a single scalar demagnetising factor is incorrect [6] the problem of the spatially dependent magnetic component of the energy of a vortex segment and the applied field of initial vortex penetration has not previously been explicitly treated.

2. Experiment

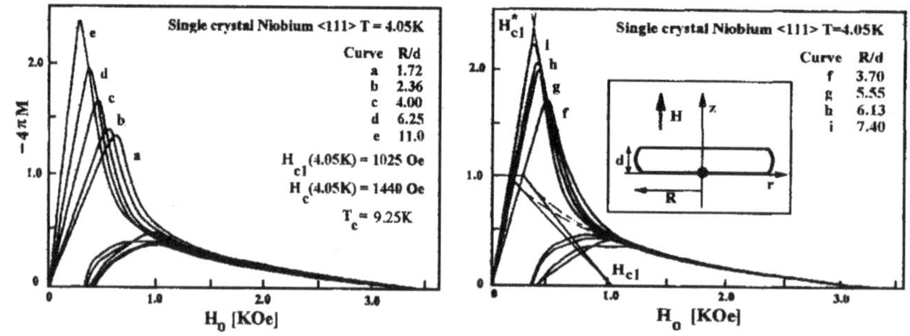

Fig.1 M-H_o curves for Nb single-crystal disc-specimens of various R/d ratio at $T = 4.05$K (with $R \cong 2.9$mm in Fig. 1(a) and $d \cong 0.37$mm in Fig.1(b)). The inset in Fig. 1(b) shows the specimen-field geometry.

Experimental M-H_o data for niobium single-crystal disc-shaped specimens at $T = 4.05$K are reproduced from Ref. [1] in Figs.1(a) and (b). Fig.3 gives a plot of the gradient $-4\pi dM/dH_o$, in the Meissner regime, against $(1-D_m)^{-1}$, where D_m is the demagnetising factor for a right-cylinder [7], and demonstrates the validity of the use of this factor in this regime where M is spatially uniform.

3. Theory and discussion

The basic theory has been given in the paper preceeding this one [8]. Consider now a suitably 'barrelled' disc-shaped specimen, of radius R and thickness d, in a transverse magnetic field H_o, for which the field intensity just outside of the barrelled surface is everywhere parallel to it (see insert Fig.1(b)). In cylindrical co-ordinates, we obtain [3], for the z-component of the reduced thermodynamic field

$$\Phi(\boldsymbol{r},z) \equiv \frac{H_e(\boldsymbol{r},z)}{H_o} = \left\{1 + \frac{1}{2}\int \varphi_o(\boldsymbol{r}')\left[\gamma(|\boldsymbol{r}-\boldsymbol{r}'|, d-z) + \gamma(|\boldsymbol{r}-\boldsymbol{r}'|, z)\right] d^2\boldsymbol{r}'\right\} \qquad (1)$$

where $\gamma(r,z) = z/2\pi(r^2+z^2)^{3/2}$ and $\varphi_o(\boldsymbol{r})$ is a normalised surface charge density which satisfies the integral equation $\varphi_o(\boldsymbol{r}) - \int \varphi_o(\boldsymbol{r}')\gamma(|\boldsymbol{r}-\boldsymbol{r}'|, d)\, d^2\boldsymbol{r}' = 2.$

Numerical solutions [3] for $\Phi(\rho,0) = \Phi(\rho,d)$, $\Phi(\rho,d/2)$, and $\Phi_m(\rho)$ as a function of $\rho = r/R$, for various values of the ratio R/d, are shown in Fig.2(a) where it is evident that the free energy of an isolated vortex is a increasing function of ρ_o.

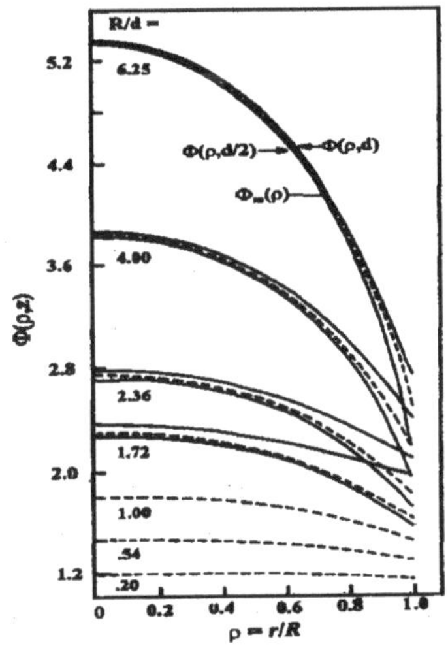

Fig.2(a). Reduced fields $\Phi(\rho, d)$, $\Phi(\rho, d/2)$ and $\Phi_m(\rho)$ as a function of $\rho \equiv r/R$ for various R/d.

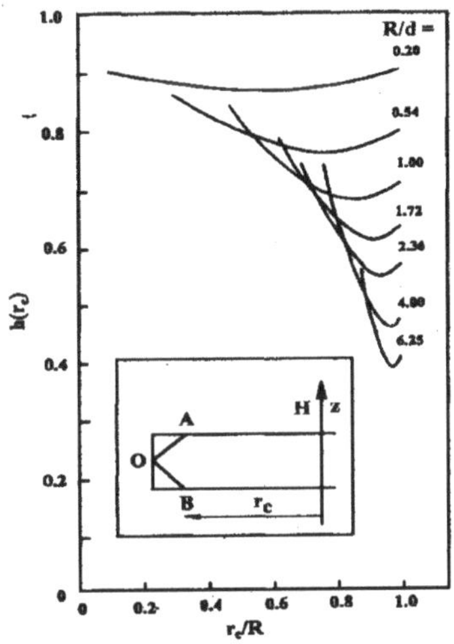

Fig.2(b). Reduced field of initial vortex penetration $h(r_c)$ as a function of r_c/R for various R/d.

We consider now the transverse reduced field $h^* \equiv H_{c1}^*/H_{c1}$ for initial vortex penetration. Thus, with reference to Fig.2(b)(insert), as H_o increases from zero, vortex segments begin to penetrate the rims of the disc. At a critical field $H_o = H_{c1}^*$ the leading vortex segments will lie along AOB and will become free to join, straighten and penetrate, under the influence of the potential gradient, into the bulk of the disc. The transition to the mixed state at H_{c1}^* is expected, and is indeed found (see Figs. 1(a) and (b)) to be very sharp. The Gibbs energy increment for a vortex along AOB is given by

$$\delta G = H_{c1}\phi_o d\left\{\left[1+(2\Delta r_c/d)^2\right]^{1/2} - h\Phi_m(r_c)\right\} \tag{2}$$

where $\Delta r_c \equiv R - r_c$. Bulk penetration occurs for $\delta G = 0$ when h is a minimum; that is at

$$h = h^* = \left\{\left[1+(2\Delta r_c/d)^2\right]/\Phi_m(r_c)\right\}_{\min}. \tag{3}$$

This equation can be solved either by numerical or by graphical method as in Fig.2(b). The results for h^* are compared with experiment and with the empirical expressions $h^* = (1 + R/d)^{-1/2}$ and $h^* = 0.03 + 0.94(1 + R/d)^{-1/2}$ in Fig. 4.

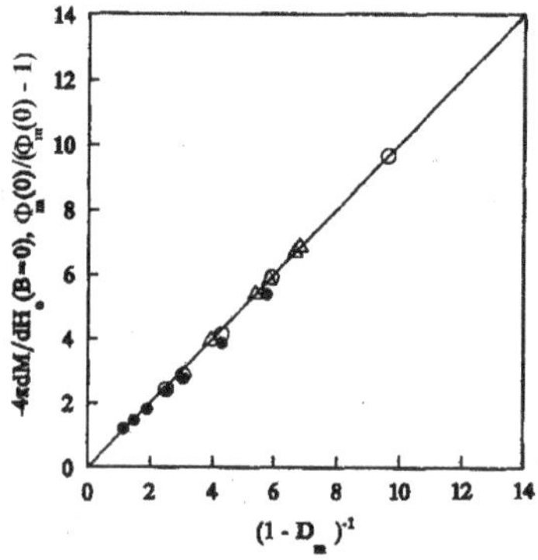

Fig.3. Experimental results for $-4\pi dM/dH_o$ in the Meissner regime (o and Δ for the disc-series $R \cong$ const. and $d \cong$ const. respectively) and the mean field $\Phi_m(0)$ (•) as function of $(1-D_m)^{-1}$.

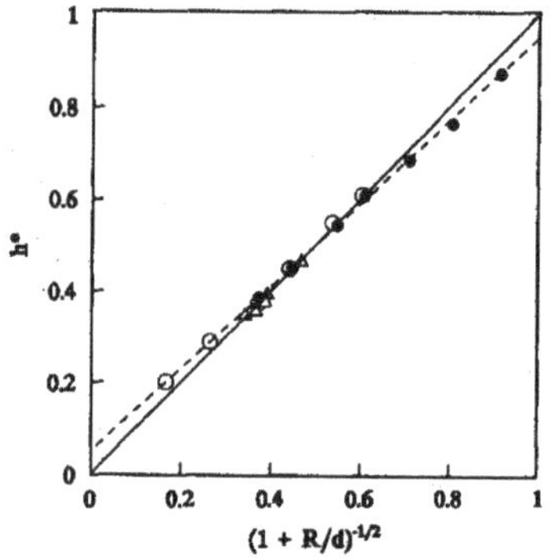

Fig.4. Experimental results for the field for initial vortex penetration $h^* \equiv H_{c1}^*/H_{c1}$ (o and Δ for the disc-series $R \cong$ const and $d \cong$ const. resp) as a function of $(1 + R/d)^{-1/2}$. Theory (•). Dashed line: $h^* = 0.03 + 0.94(1 + R/d)^{-1/2}$.

For h increasing from zero; the system, initially in the Meissner regime, is clearly in a non-equilibrium state for $(1-D_m) < h < h^*$. For $h > h^*$ the penetration of vortices will, in the absence of pinning forces, be limited by dissipative viscous forces and by a long-range back-stress due to the already penetrated vortices which will be confined to a "pool", of radius $0 < r_p < R$, concentric with the disc axis. A detailed treatment of this regime will be given elsewhere [3]. We consider here only the case $r_p \to R$ for which the total Gibbs free energy of the vortices, assumed straight and to be of uniform density, is given by

$$G \approx (\pi R^2 d)\Big[BH_{c1} + \alpha(B) - BH_o\Psi + (B^2/2\mu_o)(\Psi - 1)\Big] \tag{4}$$

where $\Psi \approx \frac{1}{\pi R^2 d}\int_0^R \Phi_m(r)d^2r$ and $\alpha(B)$ is the usual internal short-range vortex interaction [9]. The reduced field h_R at which $r_p \rightarrow R$ is determined by the condition that the vortex magnetic energy is exactly compensated by the repulsive long-range back-stress and internal short-range interaction terms, that is by:

$BH_R(\Psi - 1) = (B^2/2\mu_o)(\Psi - 1) + \alpha(B)$. On substitution into Eq.4 with $G = 0$ we obtain finally $h_R = H_R/H_{c1} = 1$ in agreement with experiment and intuition.

For $r_p = R$ and $h > h_R$ the vortices near the perimeter of the disc will tend to curve along the field direction and to move reversibly in and out of the specimen and the system will be in equilibrium in accordance with the constitutive $B[H_{rev}]$ relation for the material. In decreasing h, from above h_R to $h = 0$, $r_p = R$ for all h. The rims of the disc however present a barrier to vortex escape and bevelling tends to restore equilibrium $M(H)$ behaviour to lower fields where eventually bulk vortex pinning apparently becomes significant [3]. For B to remain in equilibrium requires that Eq.4, with $G = 0$, is satisfied for all H_o. In the limit $B \rightarrow 0$ (from above); $\alpha(B) \rightarrow 0$ and $\Psi \rightarrow \Phi_m(0)$ and we obtain $H_o(B \rightarrow 0) = H_{c1}/\Phi_m(0)$. This is an intuitively satisfying result because it implies $\Phi_m(0) = 1/D_m$. This is confirmed, within experimental and analytical error, in Fig.3 which includes a plot of $\Phi_m(0)/[\Phi_m(0) - 1]$ against $(1 - D_m)$. It also suggests the possibility that an ideal disc-specimen can trap a single vortex at the origin.

4. References

1. T.B. Doyle, PhD Thesis, University of the Witwatersrand, 1975.
2. R. Labusch, Private communication 1974.
3. T.B. Doyle and R. Labusch, to be submitted to *Phys. Rev. B*.
4. J.R. Clem , R.P. Huebener and D.E. Gallus, *J. Low Temp. Physics* **12** (1973) 449.
5. Th. Schuster, M.V. Indenbom, H. Kuhn and E.H. Brandt, *Phys. Rev. Lett.* **73** (1994) 1424.
6. E.H. Brandt and M. Indenbom, *Phys. Rev. B* **48** (1993) 12893.
7. W.F. Brown, *Magnetostatic Principles in Ferromagnetism,* North Holland Pub.Co., Amsterdam, 1962.
8. R. Labusch and T.B. Doyle, present Conference Proceedings.
9. P.G. DeGennes, *Superconductivity in Metals and Alloys*, W.A. Benjamin, Inc. New York, Amsterdam (1966).

3. Microstructure and Critical Current

MAGNETO OPTICAL IMAGING OF THE LOCAL CURRENT PATHS IN Ag-SHEATHED $Bi_2Sr_2CaCu_2O_x$ HIGH T_c SUPERCONDUCTOR TAPE

A.A. Polyanskii*, A.E. Pashitski, A. Gurevich, D. C. Larbalestier

Applied Superconductivity Center, University of Wisconsin, Madison, WI 53706 USA

V. K. Vlasko-Vlasov, V. L Nikitenko

Institute of Solid State Physics, 142432 Chernogolovka, Moscow District, Russia.

ABSTRACT

We have utilized in-plane magnetized YIG magneto optical (MO) indicator films to analyze the flux density distribution above a Ag-sheathed $Bi_2Sr_2CaCu_2O_x$ (BSCCO-2212) composite having a high transport critical current density (J_{ct}) of $1.2x10^5$ A/cm^2 (50 mT, 4K). Although the tape has a generally good microstructure consisting of well aligned grain blocks with a low second phase content, the sample is markedly non-uniform in its resistance to external applied fields. Some regions strongly resist flux penetration and the transport current is preferentially carried in these regions. Other regions, particularly in the tape center, permit easy flux entry and the current paths are there confined to local loops which have length scales of order a few tape thicknesses. MO images explicitly show that the current does not flow uniformly throughout the tape and suggest that the measured transport critical current density is an average determined by percolation through the well connected high J_c regions.

1. Introduction

Underpinning all efforts to develop power applications from high temperature superconductors (HTS) is the fabrication of conductors carrying high transport current densities (J_{ct}) in magnetic fields. Rapid progress is being made with Ag-sheathed tapes of the 2212 and 2223 Bi-Sr-Ca-Cu-O (BSCCO) compounds. Such tapes can be made in km lengths with J_{ct} values ~10^4 A/cm^2 (77 K, 0 T) [1,2]. There is much effort aimed at raising J_{ct} to >10^5 A/cm^2. In fact the best local or short length J_{ct} values in such tapes exceed $7.5x10^4$ A/cm^2 (77 K, 0 T) [3] while the best epitaxial thin films have achieved >10^6A/cm^2 [4,5]. Since the microstructure of all presently made tapes is highly variable, one interpretation of the lower J_{ct} values of long length polycrystalline examples of these compounds is that current flow is percolative, the effective current-carrying cross-section being less than the whole and variable from place to place. Independent support for this view comes from the study of microslices cut from different parts of a BSCCO-2223 tape in which J_{ct} varied by factors of 3-5 [3]. We here show that MO imaging reveals the local percolative current path on a scale of ~ 5 μm, thus permitting a direct visualization of how defects in the microstructure control current flow.

2. Magneto Optical Imaging of BSCCO tapes

We have imaged both BSCCO-2212 and BSCCO-2223 tapes and found them to display highly non-uniform flux penetration properties [6,7]. We here report on a high J_c monofilament BSCCO-2212 tape which was rolled to dimensions ~ 2 mm by 50 μm and specially processed to produce well-aligned grains [8]. The imaged sample was cut down its middle into a slab ~1 mm high x 50 μm thick x 2 mm long so as to expose the mid-thickness of the longitudinal section. Fields up to 120 mT could be applied parallel to the slab, thus inducing go-and-return currents to flow longitudinally along the tape in the ab planes of the BSCCO grains. The MO images were obtained by placing a ferrimagnetic Bi-doped Y-iron-garnet film directly on the polished surface of the tape, the film and tape then being cooled to ~ 10 K in a continuous flow cryostat [9]. The images were recorded on videotape and film and then imported into a computer for subsequent processing [7].

Fig. 1 presents a montage of a) the MO image, b) the polarized light microstructural image and c) the field contours extracted from MO image over one representative section ~0.5 mm long in the tape. Prior to cutting the sample its transport J_{ct} was measured to be 1.2 x 10^5 A/cm^2 , defined at 1 μV/cm at 50 mT, 4 K. This high J_c value is consistent with its generally low (5-10 vol. %) second phase content (principally $Bi_2(Sr,Ca)_4O_x$ and $(Sr,Ca)_{14}Cu_{24}O_x$) and well aligned grain structure. However, Figs. 1.a) and 1.b) show that the residual microstructural non-uniformities in the tape produce a markedly variable flux penetration. Although the 2212 grains exhibit generally good alignment, in fact the grains form into colonies or blocks which often have significant misalignments at the block boundaries. Other defects are associated with undulations of the Ag-BSCCO interface. The crack which is evident at the left end of the tape in Fig. 1.b) was formed on cooling below room temperature, apparently in response to the major grain block misalignment which occurs in its vicinity. Such defects are common in 2212 and 2223-BSCCO tapes.

The influence of the crack is very obvious in the MO image in that it provides an easy path for flux entry. Significant regions of the tape show strong flux penetration along the center line. Indeed the tape appears to divide itself into two parallel bands which strongly exclude flux, these regions being separated by a region of easy flux penetration which occupies the center line of the tape. This picture can be made more quantitative by converting the MO image into field contours after calibrating the image for the local light intensity, polarizer/analyzer settings and the characteristics of the film [7], as is shown in Fig. 1.c). Regions of high J_c in this image correspond to densely spaced contours ($J_c = \nabla H$). In this representation the contour lines also correspond to the streamlines of current and these streamlines clearly show that the current flows preferentially in those regions of the filament which exclude flux and which appear dark in the MO image. By contrast, the center of the tape is preferentially occupied by current loops having dimensions in the range ≈50-500 μm. Thus only a portion of the

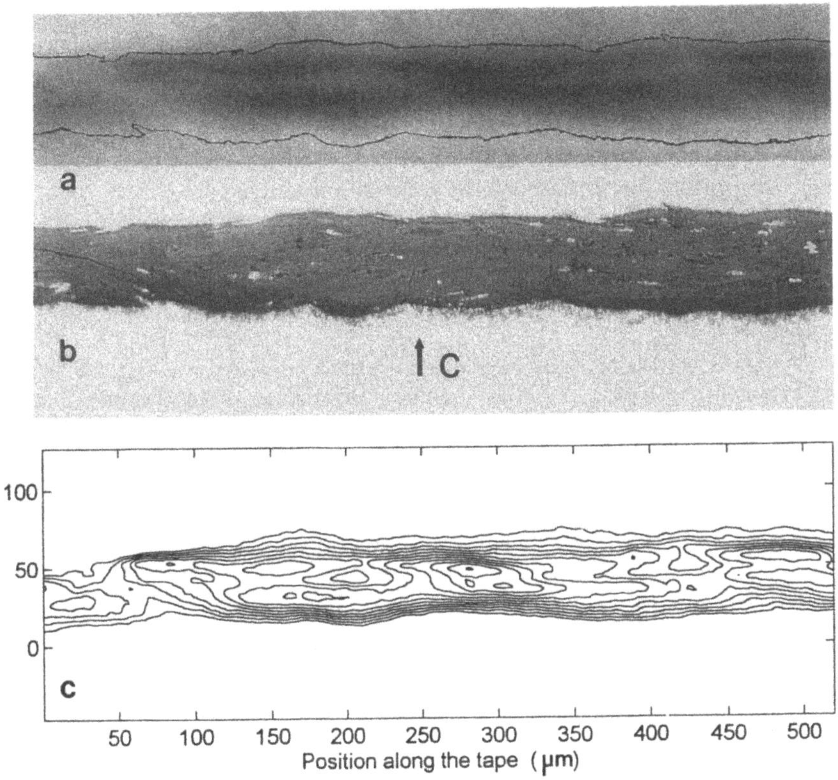

Fig. 1. a) MO image of the tape as viewed in a perpendicular field of 120 mT after it was zero field cooled to 15 K. Darker regions are those resisting flux entry. Preferential flux penetration at a crack associated with the misaligned grain blocks at left in Fig. 1.b) is seen in the MO image, as is the preferential flux penetration at the center of the tape along its whole length. b) is the polarized light microstructural image which shows the generally well aligned grain structure and small second phase content of the tape. c) shows contours of the perpendicular component of the field at intervals of 8 mT. The contours are also current streamlines. The long range, continuous nature of the current path in the outer, better aligned sections of the tape is in strong contrast to the shorter range current loops found in the tape center. Misalignments between blocks of aligned grains appear to produce barriers to current flow which cause many of these loops.

tape cross-section is active in carrying transport current over long distances and the current flow is percolative. In this case about half the tape cross-section is carrying shorter range magnetization currents and about half the (long range) transport current.

3. Summary

We have shown by direct imaging that the flux penetration into a high transport critical current density BSCCO-2212 tape is non-uniform and that the current paths are percolative. It was demonstrated that the regions along the center line of the tape easily admit magnetic flux, while the regions near the Ag-BSCCO interface strongly exclude it. The center of the tape is shown to be preferentially occupied by current loops whose length is only a few filament thicknesses. The transport current follows paths which are closer to the Ag interface regions of the tape. In this high J_c tape the transport current appears to occupy about half of the total cross-section of the BSCCO core.

Acknowledgments

This work benefited greatly from trial experiments performed with the help of H. Edelman (UW), F. Gayle, D. L. Kaiser and M. Turchinskaya of NIST-Gaithersburg. We are also grateful to Wei Zhang and E. Hellstrom for the sample studied and to W. Starch and A. Squitieri for much assistance in setting up the experiment. The work was supported by ARPA (N00014-90-J-4115) and EPRI(RP8065-6).

*On leave from Institute of Solid State Physics, 142432 Chernogolovka, Moscow District, Russia

References

1. G. N. Riley, D.R. Parker, C.J. Christopherson, P.K. Miles, J.J. Pickett, S.E. Hughson, J.D. Schreiber, A.A. Polyanskii, A. Pashitski, D.C. Larbalestier, Proc of the M^2S conference Grenoble, August 1994, *Physica C,* to appear.
2. K. Ohkura, H. Mukai, T. Hikata, M. Ueyama, T. Kato, J. Fujikami, K. Muranaka, and K. Sato, *Proc. 6th Int. Symposium on Superconductivity, Hiroshima, Japan* (Nov. 1993), to appear.
3. D.C. Larbalestier, X.Y. Cai, Y. Feng, H. Edelman, A. Umezawa, G.N. Riley Jr., and W.L. Carter, *Physica C* **221**, 299-303 (1994).
4. V.V. Metlushko, G. Guntherodt, P. Wagner, H. Adrian, I.N. Goncharov, V.V. Moshchalkov, and Y. Bruynseraede, *Appl. Phys. Lett.* **63**, 2821-2823 (1993).
5. Y. Hakuraku, and Z. Mori, *J. Appl. Phys.* **73**, 309-315 (1993).
6. D.C. Larbalestier, X.Y. Cai, H. Edelman, Y. Feng, A. Gurevich, J. Parrell, A. Pashitski, A. Polyanskii, *J. of Metals,* To appear **46 (12)** December 1994.
7. A. Pashitski, A. Polyanskii, A. Gurevich, J. Parrell, D.C. Larbalestier, V.I. Nikitenko, Submitted to *Physica C.*
8. W. Zhang, and E.E. Hellstrom, *Physica C* **218**, 141-152 (1993).
9. L.A.Dorosinskii, M.V. Indenbom, V.I. Nikitenko, Yu.A. Ossip'yan, A.A. Polyanskii, and V.K. Vlasko-Vlasov, *Physica C* **203**, 149-156 (1992).

PINNING ENERGIES AND DEFORMATION INDUCED EFFECTS ON CRITICAL CURRENTS IN Bi(Pb)2223 SILVER SHEATHED TAPES

M.DÄUMLING[a], F.MARTI[b], G.GRASSO[b,c], and R.FLÜKIGER[a,b]

[a] *Groupe de Physique Appliquée,* [b] *Département de Physique de la Matière Condensée, Université de Genève, CH-1211 Genève 4, Switzerland*

([c] on leave from Consorzio INFM, Università di Genova, Italy)

ABSTRACT

In specially aligned tapes with transport j_c=0 (no intergrain contributions) pinning activation energies were determined using the inversion scheme of Schnack et al.[1] Best fitting results are obtained using a 2D local pinning approach, and a maximum activation energy of about 80 meV (at T=0, j=0, B=0.2T, H || c) is obtained. We also have examined the effect of mechanical deformation on the flux pinning properties of the aligned $(Bi,Pb)_2Sr_2Ca_2Cu_3O_x$ phase inside a silver sheath, using both resistive and magnetic methods. In cold deformed tapes an increase of the critical current density inside the grains by a factor of about two was obtained. In tapes deformed at high temperature significant improvements in the transport current density j_{ct} for the H ⊥ tape orientation are observed. They amount to up to more than a factor of two at 77K and 0.4T, while only small changes are apparent in the j_c-H characteristic for the other orientation (H || tape) and in self-field. Thus intra-grain flux pinning rather than weak links appears to be controlling j_{ct} for H ⊥ tape.

1. Introduction

Cold deformation is an important step in the fabrication of silver sheathed $(Bi,Pb)_2$-$Sr_2Ca_2Cu_3O_x$ tapes. [2] This processing step densifies the tape between anneals, and thus also enhances reaction rates and phase formation velocity. However, it also breaks the grains and the connections between grains and thus lowers the transport current density. In ductile low T_c superconductors deformation is used to increase the critical current density (j_c). [3,4] This enhancement of flux pinning by cold deformation appears also to be possible in Bi(Pb)2223. [5] However, this method will never serve as a means of increasing the transport j_c since the brittle ceramic also breaks during the deformation. It will be shown that the deformation at high temperature apparently not only just avoids the creating of cracks, but even aids in the creation of flux pinning centers.

2. Experimental

The special tapes which were used to investigate the intra-grain currents were produced as described previously. [6] Due to the special nature of these tapes (transport

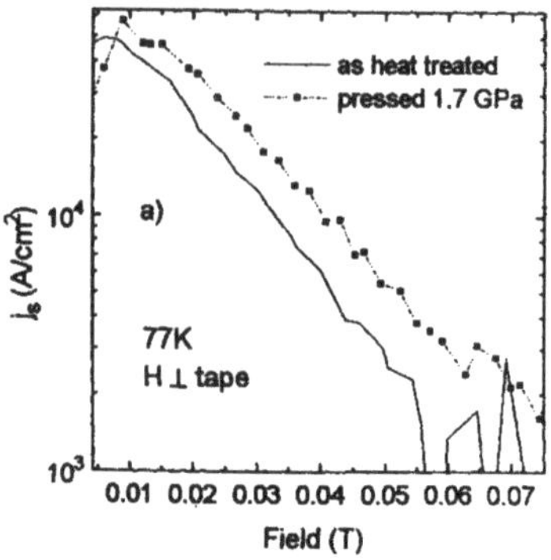

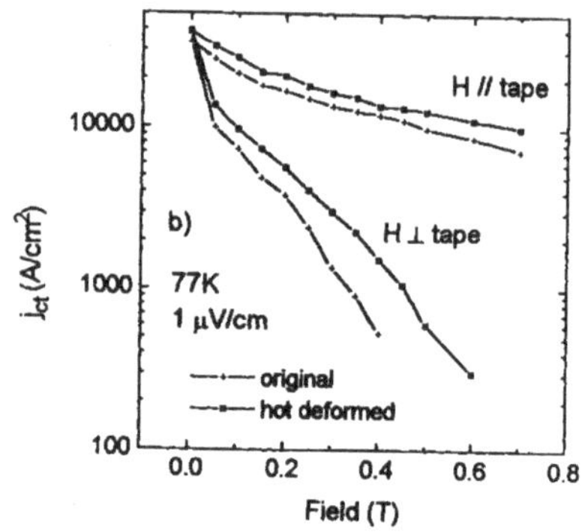

Fig. 1: *a) Intra-grain critical current densities determined magnetically in cold pressed tapes , and b) transport j_c in hot deformed tape at a temperature of 77K.*

j_c=0) the critical current density in the grains can be measured directly magnetically[5] with a vibrating sample magnetometer (VSM). For the conversion of the hysteresis into a current density the critical state model was used. [7] The regular tapes which were used in the hot deformation experiments were prepared by the standard pressing technique. [2] The resistive measurements of the transport current density j_{ct} were carried out in a four point probe configuration in liquid nitrogen, using a criterion of 1 $\mu V/cm$.

3. Results and Discussion

3.1. Deformation Effects on j_c

The results of the cold[8] and hot deformation have been summarized in figure 1. Here only results at 77K are shown. The complete results will be published elsewhere. It is evident that cold deformation raises the critical current density in the grains by about 70 % at all magnetic fields at 77K for H ⊥ tape, but only slightly changes the field dependence of j_c. Hot deformation seems to have a slightly different effect. It should be noted that deformation at high temperature is necessary. An identical heat treatment without deformation leaves j_c unchanged. The critical current density j_{ct} in self field increases (for the particular tape shown) by about 15 %. For H || tape the critical currents increase somewhat but the field dependence of j_{ct} is unaltered by the deformation. The situation is different when H ⊥ tape, where the increase of j_c reaches a factor of more than two for fields of order 0.5T. Thus the field dependence of j_{ct} when H ⊥ tape is quite different after the deformation.

This effect is reminiscent of the data of Kummeth et al.[9] who irradiated their tapes with heavy ions. The irradiation drastically increased intra-grain flux pinning, and

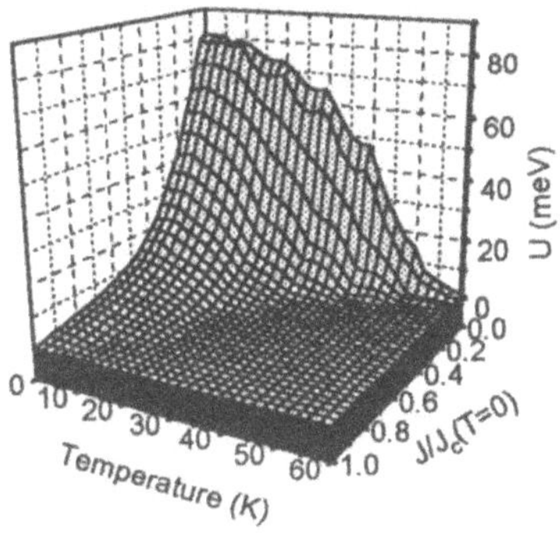

Fig. 2: *Pinning activation energy as function of current and temperature in a magnetic field of 0.2T perpendicular to the undeformed tape.*

also weakened the field dependence of the transport j_c for fields larger than about 0.2T, leading to significantly enhanced in-field j_c values. Thus we conclude that increasing the flux pinning inside the grains (see also fig. 1a) increases the transport j_c significantly for H $\perp$ tape. j_{ct} does not seem to be weak link limited. Is H $\|$ tape the critical currents are significantly higher and possibly the j_{ct} is limited by a weak link mechanism. However, it is not to be excluded that the kind of pinning center created by the deformation is not the correct one to pin flux in the H $\|$ tape field orientation.

3.2. Pinning Activation Energy

The activation energy for flux pinning was determined from magnetic measurements of the dynamic relaxation rate with different field ramp speeds at temperatures from 5K to T_c using the inversion scheme of Schnack et al.. [1] The result of the fitting procedure is shown in figure 2. The maximum pinning energy U at 0.2T for the magnetic field perpendicular to the tape is about 80 meV (1100K, or $10kT_c$). The best fits result in p=0, where U$\propto$ $(j_c)^p$. Thus the flux pinning appears to be in the 2D limit (flux pancakes) and the pancakes appear to be individually pinned. We now have to compare this value to those in the literature. Yamasaki et al.[10] found values of U_0 (using a resistive method) of about 180eV (0.2T) in thin films of Bi(Pb)2223 with a very high current density, almost one order of magnitude above ours. Thus their activation energies should not be directly compared with ours. Ries et al.[11] measured activation energies for bulk two layer Bi2212 material, where they find values of order 200 to 300 meV (at 0.2T, H $\|$ c), determined resistively and magnetically. The fact that our value is smaller might be an indication that the anisotropy in the three layer material is greater than the one for the two layer phase. Increasing anisotropy increases the isolation of the superconducting layers in the structure and decreases the activation energy, as was shown recently using thin film multilayers. [12]

4. Conclusion

We have determined pinning activation energies U_0 in undeformed Bi(Pb)2223 tapes using an inversion scheme[1] that uses magnetization data from 5K to T_c. We find U_0=80meV (0.2T, H $\parallel$ c).

The influence of deformation on the critical currents was investigated. Deformation at room temperature raises the intra-grain j_c by about a factor of two, but breaks grains as well, and is thus not useful for raising transport current densities j_{ct}. However, deformation at high temperature appears to avoid this problem. The j_{ct} in self field and for H $\parallel$ tape is only slightly enhanced, but the j_{ct} for H $\perp$ tape is increased more than twofold at 0.4T, 77K. It is concluded that intra-grain flux pinning controls j_{ct} in this case.

5. Acknowledgement

This work has been supported by the Swiss National Funds (NFP30).

6. References

[1]H.G.Schnack, R.Griessen, J.G.Lensick, and Wen Hai-Hu. *Phys. Rev. B* **48** (1993) 13178.

[2]G.Grasso, A.Perin, B.Hensel, and R.Flükiger. *Physica C* **217** (1993) 335.

[3]A.M.Campbell and J.E.Evetts. *Adv. Phys.* **21** (1972) 199.

[4]D.C.Larbalestier. *in: Superconductor Materials Science.* S.Foner and B.B.Schwartz (eds.), Plenum 1981.

[5]M.Däumling, J.-C.Grivel, B.Hensel, and R.Flükiger. *Physica C* **219** (1994) 429.

[6]M.Däumling, A.Jeremie, and R.Flükiger. *Supercond. Sci. Techn.* **6** (1993) 721.

[7]C.P.Bean. *Rev. Mod. Phys.* **36** (1964) 31.

[8]F.Marti, M.Däumling, and R.Flükiger. *IEEE Trans. Magn.*, to be published.

[9]P.Kummeth, C.Struller, H.-W.Neumüller, G.Ries, M.Kraus, M.Kraus, G.Wirth, J.Wiesner, O.Eibl, and G.Saemann-Ischenko. *Proceedings of European Conference on Applied Superconductivity* (1994) 685.

[10]H.Yamasaki, K.Endo, M.Umeda, S.Kosaka, S.Yoshida, and K.Kajimura. *Phys. Rev. Lett.* **70** (1993) 3331.

[11]G.Ries, H.-W.Neumüller, R.Busch, P.Kummeth, M.Leghissa, P.Schmitt, and G.Saemann-Ischenko. *Proceedings of Fall Meeting of European Materials Research Society*, Strasbourg, November 1992.

[12]J.M.Triscone, P.Fivat, M.Andersson, M.Decroux, and Ø.Fisher. *Phys. Rev. B* **50** (1994) 1229.

THE REAL STRUCTURE OF COLUMNAR PINNING CENTERS IN HEAVY-ION-IRRADIATED CUPRATE SUPERCONDUCTORS

D. O. WELCH and YIMEI ZHU
Department of Applied Science, Brookhaven National Laboratory
Upton, NY 11973, USA

and

R. C. BUDHANI
Department of Physics, Indian Institute of Technology Kanpur
Kanpur, 208016 India

ABSTRACT

There has been considerable recent interest in the use of columnar defects produced by irradiation with energetic heavy ions to raise the irreversibility line and improve the critical current density of cuprate superconductors. In the interpretation and theoretical modeling of the flux-pinning characteristics of heavy-ion tracks, it is generally assumed that they are simply columns of non-superconducting material. In this paper we present a more realistic description, based both on resistivity measurements and on detailed, quantitative transmission electron microscope methods (both imaging and analytical studies), of the nature of heavy-ion damage, including defects, disorder, strain fields, and oxygen deficiencies in the matrix of the superconductor surrounding the amorphous columns. The presence of such disorder appears to be a consequence of the mechanism of track formation, which involves partial epitaxial regrowth of a molten region which follows the passage of sufficiently energetic ions.

1. Introduction

There is widespread recognition that columnar tracks, produced in cuprate superconductors by the passage through them of heavy ions accelerated to of a few hundred MeV, cause substantial increases in critical current-densities and raise the irreversibility line. (See, e.g., the work of Budhani et al.,[1,2] among many others.) Recently, the use of GeV protons to induce fission of bismuth ions in Bi-cuprates, thereby creating columnar fission tracks, has been reported[3] and promises to provide a method for improving the critical current density in more massive structures than the thin films, or even tape conductors, used in most studies of the effects of heavy-ion damage. Theoretical models and discussion of the pinning of magnetic flux by heavy-ion tracks generally treat them as simple cylindrical columns, a few nanometers in diameter, of non-superconducting material embedded in a matrix of undamaged superconductor. This, however, is a considerable oversimplification because the damage process by which the tracks are created results in the creation of defects, disorder, and strain fields in the matrix of the surrounding superconductor, as shown by detailed studies with electron microscopy and the measurement of normal-state resistivity and critical temperature.[4-6]

2. The Damage Process

Ions moving with MeV energies in crystals transfer their energy to the crystal primarily by electronic excitation, and the rate of energy deposition Q per unit of path length can be calculated with reasonable accuracy by well-known methods.[7] This energy is rapidly transferred to the ions by inelastic scattering processes, and the result is a spike of thermal energy in a very localized region along the path of the irradiating ion. The rapidity with which energy is transferred from the hot electrons to the lattice of ions depends on whether the material is metallic, semiconducting, or ionic, in increasing order of rapidity.[8] (This accounts for variations in the diameter of ion-induced columnar defects observed in $Bi_2Sr_2Ca_2Cu_3O_x$ and $YBa_2Cu_3O_{7-\delta}$ with varying degrees of oxygenation.[5]) The energy thus deposited locally in the crystal lattice then spreads out by conduction, in the process heating the region around the ion track to temperatures in excess of the melting point.

Such a "thermal-spike" model[8] of the damage process can be constructed for the case of heavy-ion damage to cuprates and provides a consistent and reasonably quantitative description of several features of the observed damage.[5] If the dissipation by the diffusion of heat through the lattice is approximately described by Fourier's linear law of heat conduction, neglecting the effects of the latent heat of melting and freezing, one finds that the diameter d_{max} of the circular region around an ion track parallel to the crystallographic c axis of the cuprate in which the temperature has reached or exceeded the melting point is:

$$d_{max} \simeq (4Q/\pi e C \theta_{mp})^{1/2} \tag{1}$$

where e is the base of natural logarithms, C is the heat capacity per unit volume, and θ_{mp} is the difference between the melting point and the ambient temperature. When the ion travels parallel to the a or b crystal axes, the ion track is elliptical, with major and minor diameters of the region which has exceeded the melting point given by:

$$d_{ab} = d_{max}(D_{ab}/D_c)^{1/4} \quad , \qquad d_c = d_{max}(D_c/D_{ab})^{1/4} \tag{2}$$

where D_{ab} and D_c are the thermal diffusivities for directions in the a-b plane and along the c axis. The experimentally-observed size and ellipticity of ion tracks in cuprates are in qualitative agreement with these predictions, but detailed comparison strongly suggests that the entire region which has reached or exceeded the melting point does not remain amorphous after cooling, and that partial epitaxial regrowth occurs. A consistent description of the experimental results is obtained if it is assumed that there is a critical length of time Δt_c, in the range 10-40 ps, which a given region of material must remain above the melting point in order to be retained in the amorphous state. Based on this assumption, numerical calculations with the thermal spike model show that a good approximation to the diameter of a c-axis track which remains amorphous after partial epitaxial regrowth is given by:

$$d \simeq d_{max}\,[1-(16D_{ab}\Delta t_c/ed^2_{max})^2]^{1/2} \qquad (3)$$

This expression leads to a threshold value of the energy deposition rate Q, below which amorphous tracks do not register. For track formation:

$$Q > Q_{threshold} \simeq 4\pi D_{ab} C\theta_{mp}\Delta t_c \qquad (4)$$

Note that the predicted threshold rises with increasing thermal diffusivity, in accord with the experimental observations.

3. Damage to the Matrix Surrounding Columnar

The transmission electron microscope (TEM) studies,[4-6] show that the material within the columnar defects is amorphous and that there is an elastic strain field in the surrounding matrix. For an ion track along the crystal c-axis, the displacement field u(r) which gives rise to the strain field, which is pure shear in character, is given by:

$$u(r) = \varepsilon_{eff}\,(R^2/r) \qquad (5)$$

where R is the radius of track and

$$\varepsilon_{eff} \equiv [(C_{11} + C_{12})\,\varepsilon_{ab} + C_{13}\,\varepsilon_c]/2C_{11} \qquad (6)$$

where the C_{ij} are elastic constants and ε_{ab}, ε_c are the fractional strains of the cylindrical column which result upon its amorphization. The TEM studies show that $\varepsilon_{eff} \simeq 0.02$ in $YBa_2Cu_3O_7$. Nanoprobe (~2 nm) electron energy loss spectroscopy (EELS) shows that the electronic hole density is severely depleted up to 5-10 nm from the edge of the amorphous region in $YBa_2Cu_3O_7$, due to a reduction in the oxygen content, disorder in the crystal lattice, or perhaps to the elastic strain field. In addition, the creation of the columnar defects in $YBa_2Cu_3O_{7-\delta}$, especially for ion tracks parallel to the a-b planes, often results the formation of wing-like planar faults attached to the columns and extending out to about the calculated dimension of the original molten zone. Nanoprobe chemical analysis, high-resolution imaging, and image simulation showed that the majority of faults consist of one or two extra CuO planes, bounded by sessile dislocations. These faults are consistent with a mechanism of partial epitaxial regrowth of the molten region which is induced by the passing heavy ion.

The creation of defects and strain fields in the matrix surrounding the columnar defects is consistent with the observation in $YBa_2Cu_3O_{7-\delta}$ thin films of a depression in the superconducting critical temperature T_c and an increase in the normal-state resistivity.

A universal linear increase in resistivity and suppression of T_c is seen with the increase in the fractional area of amorphization, irrespective of the irradiating ion type. T_c decreases by approximately 0.4% per % area amorphized, while the normal-state resistivity increases by about 6% per % area amorphized.

The discussion above indicates that the "real structure" of the damage created by heavy-ions is somewhat more complex than the formation of simple cylindrical columns of damaged material, as is often assumed. Quantitative theoretical models of the effects of heavy-ion damage on flux pinning should take this into account.

4. Acknowledgments

This research was supported by the U.S. Department of Energy, Division of Materials Sciences, Office of Basic Energy Sciences under Contract No. DE-AC02-76CH00016.

5. References

1. R. C. Budhani, Y. Zhu, and M. Suenaga, *Appl. Phys. Lett.* **61** (1992) 985.
2. R. C. Budhani, M. Suenaga, and S. H. Liou, *Phys. Rev. Lett.* **69** (1992) 3816.
3. Martin Maley, Private communication; L. Krusin-Elbaum et al., *Appl. Phys. Lett.* (in press).
4. Yimei Zhu, R. C. Budhani, Z. X. Cai, D. O, Welch, M. Suenaga, R. Yoshizaki, and H. Ikeda, *Philos. Mag. Lett.* **67** (1993) 125.
5. Yimei Zhu, Z. X. Cai,, R. C. Budhani, M. Suenaga, and D. O. Welch, *Phys. Rev.* **B48** (1993) 6436.
6. Yimei Zhu, Hong Zhang, M. Suenaga, and D. O, Welch, *Philos. Mag. A* **68** (1993) 1079.
7. J. F. Ziegler, *Handbook of Stopping Cross-Sections For Energetic Ions in All Elements* (Pergamon, New York, 1980).
8. L. T. Chadderton and I. M. Torrens, *Fission Damage in Crystals.* (Methuen, London, 1969) Chaps. 2 and 12.

CRITICAL STATES IN PROTON IRRADIATED YBCO SLABS

R.GERBALDO, G.GHIGO, L.GOZZELINO, E.MEZZETTI, B.MINETTI
Physics Department, Politecnico di Torino, c.so Duca degli Abruzzi 24, 10129 Torino, Italy

and

R.CHERUBINI
INFN-Laboratori Nazionali di Legnaro, via Romea 4, 35020 Legnaro (PD), Italy

ABSTRACT

The effect of monoenergetic proton irradiation of good quality sintered YBCO samples on both the intragrain and the intergrain critical state has been studied. The characteristic role of implantation in modifying critical current density in both regimes is discussed. In intergrain critical state a significant enhancement of critical current densities has been found for suitable field and temperature ranges.

1. Introduction

Sintered HTC superconductors consist of anisotropic grains. The grains are separated by grain boundaries, which act as Josephson junctions with different coupling energies [1]. For strong magnetic fields Abrikosov-like vortices form inside grains. For weak magnetic fields the response of granular superconductors is determined by the diamagnetic response of the grains and the pinning of Josephson vortices which form along grain boundaries. In both regimes the concept of critical state can be applied: the existence of two regimes, intragranular critical state and intergranular critical state, each of them operating in a specific range of magnetic fields, can be assumed [2]. However in sintered materials the various inhomogeneities of the specimens make it extremely difficult to calculate the total losses and as a consequence the comparison with the experimental data is very puzzling. If we introduce other inhomogeneities, e.g. by means of irradiation, the problem is further complicated by the possible interplay of various loss mechanisms, acting in different parts of the specimens [3].

However, from the experimental point of view, proton induced defects act on the intragranular as well as on intergranular critical state in a very peculiar way. Far to have deeply understood all the underlying mechanisms, we present here a set of intragrain as well as intergrain characterisation in which we compare experimental results concerning the critical state before and after monoenergetic proton irradiation. The results were analysed on the basis of the Bean model, that reveals to be the most reliable in order to bring out the main effects.

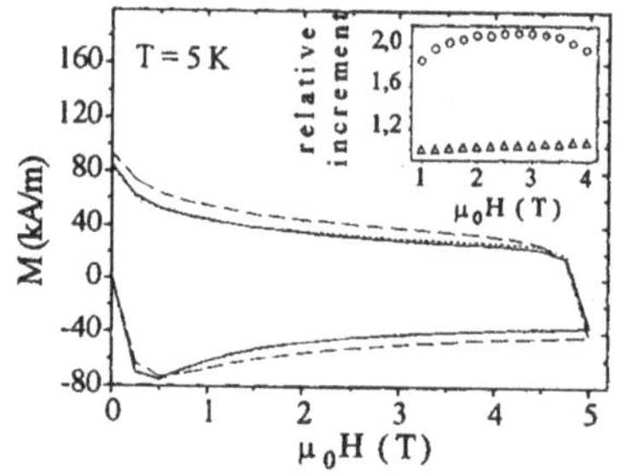

Fig.1 - Hysteresis cycles performed before irradiation (solid lines), after the 6.5MeV irradiation (dotted) and after the 3.5MeV irradiation (dashed) at T=5K. The inset shows corrected relative enhancements: $\Delta M^*(6.5MeV)/\Delta M(unirr)$ (triangle); $\Delta M^*(3.5MeV)/\Delta M^*(6.5MeV)$ (circle).

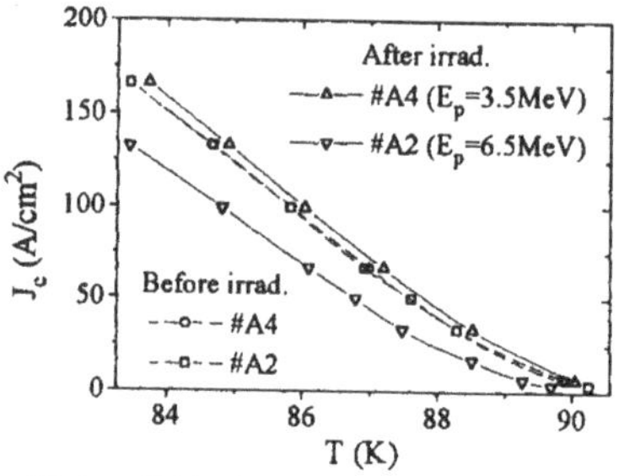

Fig.2 - Critical current density as a function of temperature, as deduced from a.c. susceptibilty measurements in zero d.c. field.

2. Experimental details

Irradiation experiments at room temperature were performed in vacuum at the APT scattering chamber of the 7MV Van de Graaff CN accelerator of INFN Laboratori Nazionali di Legnaro. Different sets of twin slab-shaped sintered YBCO slices were used, labelled in the text as T-, A-, S-samples. By "twin" we mean specimens cut from a single pellet with the same accurately measured shape. Proton energies will be specified case by case.

3. Proton damage: energy loss and implantation

Studying proton irradiation effects we deal with defects due to either energy loss by elastic scattering or proton implantation. A preliminary task of our research was to distinguish as clearly as possible the effects of these kinds of defects on the superconductive properties of our samples. For this purpose we calculated the implantation distribution as a function of depth, by means of the TRIM 92 code, for the measured density of our sintered YBCO slabs [4]. Then we chose suitable proton energies in order to irradiate the samples (labelled T-samples) without (E_p=6.5MeV) and subsequently with proton implantation (E_p=3.5MeV). A comparison between the intragrain J_c values obtained before irradiation, after the first irradiation and after the second one, allows to distinguish the effects of the two damage mechanisms. J_c values were deduced from d.c. hysteresis loops for applied field up to 4000kA/m. In fig.1 the cycles at 5K are reported. The inset shows the relative enhancement of the cycle width for the 6.5 MeV loop with respect to the not irradiated one, and for the 3.5 MeV loop with respect to the 6.5 MeV one. The width values were corrected to take into account the partial sample irradiation [4]. As shown in fig.1, at 5K the magnetic properties of the materials were not significantly changed by 6.5MeV protons, crossing the sample without implanting. On the contrary the subsequent 3.5MeV irradiation causes a clear enhancement of the intragranular critical currents (proportional to the width of the loops, as assumed in the Bean model). It should be noted that the total energy lost by elastic scattering within the sample is about the same for both irradiations. At higher temperatures also the first 6.5MeV irradiation causes width increments, but the second 3.5MeV irradiation produces enhancements significantly higher [4]. On the basis of these results, excluding strong non linear fluence dependence, we deduce that also proton implantation damage plays a positive role in enhancing critical currents.

We were also interested in studying the effects of proton implantation on intergrain critical currents. Aiming at this, we cut some twin sample from a different batch ("A-samples"). The intergrain characterisation of these specimens consisted in a.c. susceptibility measurements. In fig.2 we report the J_c dependence on temperature deduced by means of the Bean model. The sample irradiated without implantation (sample #A2; E_p=6.5MeV) shows a zero field J_c damage, while the sample irradiated with implantation (#A4; E_p=3.5MeV) shows a weak J_c enhancement. It should be noted that the effects of implantation are probably underestimated because they are localised in a thin layer of the specimen [4].

What emerges from these measurements is that both energy loss by elastic scattering and proton implantation are effective in enhancing intragrain critical currents, but implantation becomes crucial in enhancing intergrain critical currents.

4. Effects of proton irradiation on intergrain critical currents

A confirmation of the results on the sample #A4 comes from the intergrain susceptometric characterisation of some YBCO samples, cut from a different batch and labelled "S-samples". These samples were irradiated at the same energy (6.5MeV) with different doses ((0, 1.8, 3.8, 8.6) 10^{16} p/cm^2) in order to study the changes of intergrain critical currents as a function of the irradiation fluence. Due to the sample thickness, at this beam energy protons implant in the specimens. Fig.3 and fig.4 show J_c dependence on temperature, deduced from a.c. susceptibility measurements in zero and 4kA/m d.c. applied field, respectively. All the irradiated samples show a J_c enhancement and this enhancement decreases with the irradiation fluence, indicating the existence of an optimum dose. The enhancement is more clear as a d.c. field is applied.

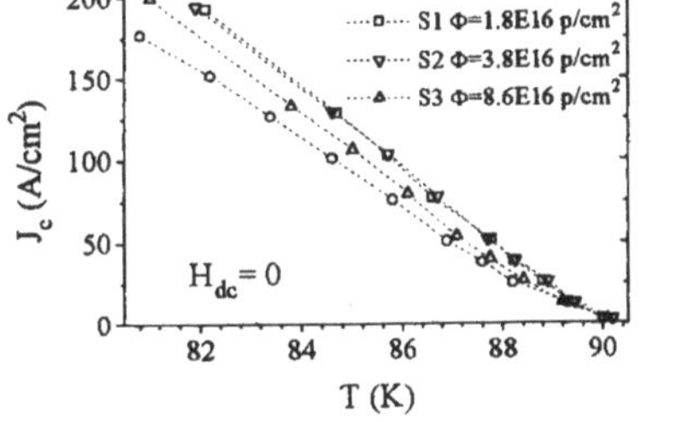

Fig.3 - Critical current density vs. temperature, derived from a.c. susceptibility measurements in zero d.c. magnetic field.

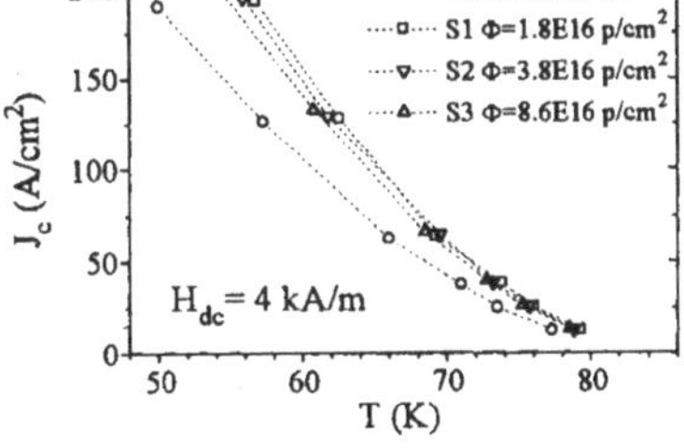

Fig.4 - Critical current density vs. temperature, derived from a.c. susceptibility measurements in 4kA/m d.c. magnetic field.

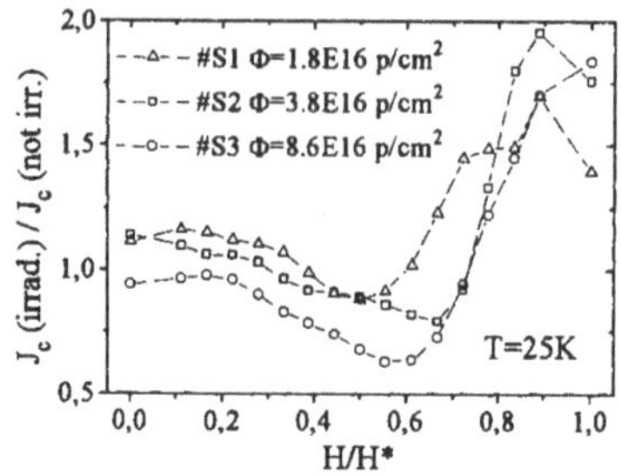

Fig.5 - Relative intergrain enhancements vs. normalised field for the irradiated samples with respect to the not irradiated one.

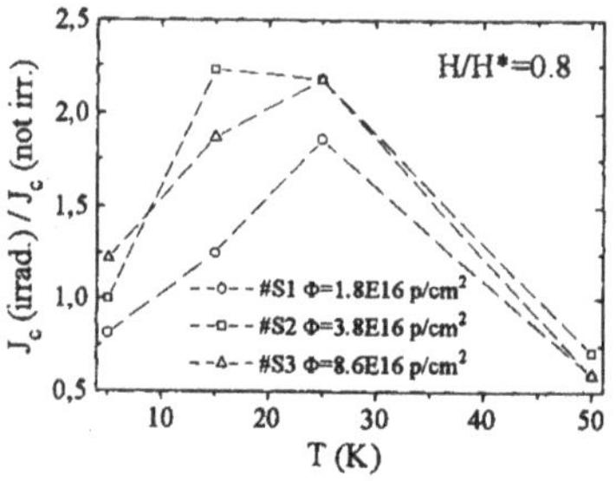

Fig.6 - Relative intergrain enhancements vs. temperature for the irradiated samples with respect to the not irradiated one.

In order to study the intergrain critical currents of these specimens also by means of d.c. methods, we measured d.c. hysteresis cycles in low fields, i.e. fields lower than H_{c1} for grains. The ratio between irradiated sample J_c and not irradiated one was studied as a function of magnetic field and temperature [4]. The results obtained as a function of magnetic field at T=25K and as a function of temperature for H/H*=0.8 are plotted in fig.5 and fig.6. (H* is the field at which the cycle closes of 90%, i.e. a evaluation of intergrain H_{c2}).

What emerges from the d.c. measurements is a sort of proton induced modulation of J_c values as a function of field and temperature. This picture can be justified on the basis of collective pinning models in Josephson junction networks [5], as our samples can be modelled

This trend is essentially confirmed by electric transport data, published elsewhere [4].

5. Conclusions

Proton implantation on sintered materials deeply affects the intragranular as well as intergranular critical state. Intragranular critical currents present enhancements increasing with the irradiation fluence. About intergranular critical currents, the existance of a range of fields and temperatures in which the J_c response can be optimised makes it possible a modulation of this response by means of proton implantation. This effect decreases with the fluence, suggesting the existence of an optimal irradiation dose.

Future experiments would concern the study of structures with a suitable distribution of implanted defects, optimised in relation to the working temperature and field.

6. References

1. J.Jung, I.Isaac, M.A.K.Mohamed, *Phys. Rev. B* **48** (1993) 7526
2. K.H.Muller, *Magnetic Susceptibility of Superconductors and Other Spin Systems*, ed. R.Hein, T.Francavilla, D.Liebenberg (Plenum Press, New York, 1991) p.229
3. J.Clem, *ibid.*, p.177
4. E.Mezzetti et al., *J. Supercond.*, in press
5. R.Fehrenbacher, V.B.Geshkenbein, G.Blatter, *Phys. Rev. B* **45**, 5450 (1992)

PROCESSING AND PROPERTIES OF SEEDED DIRECTIONAL SOLIDIFIED YBCO SUPERCONDUCTORS FOR HIGH CURRENT APPLICATIONS

D. F. LEE†, R. G. PRESSWOOD, Jr. and K. SALAMA
Texas Center for Superconductivity, University of Houston, Houston, Texas 77204-4792 USA

T. L. FRANCAVILLA
Naval Research Laboratory, Washington DC 20375-5000 USA

M. EISTERER and H. W. WEBER
Atominstitut der Österreichischen Universitäten, Wien, Austria

and

S. ZANNELLA
CISE Spa, Milan, Italy

ABSTRACT

In order for melt-textured YBCO to be utilized in high current conductor applications, the size of the oriented domains and the degree of a-b plane alignment need to be improved. A seeded directional solidification process has been developed to accomplish these tasks. Due to controlled movement of the growth front and the presence of a seed, superconductors fabricated by this method are found to be consisted of one single domain with grains aligned along the sample axis. The superconducting characteristics of these long samples have been determined by transport and magnetic means, and are found to be similar to those of well textured short specimens.

1. Introduction

Melt-texturing has been shown to be capable of increasing the critical current density (J_c) of bulk YBCO high temperature superconductors (HTS). Critical current densities as high as 10^5 A/cm^2 at 77 K have been achieved for currents flowing within a single oriented domain [1]. When the current has to transverse domain boundaries, however, the J_c is found to decrease drastically. Therefore, the size of the domains has to be increased and the number of domain boundaries need to be minimized if the HTS is to be utilized in conductor applications. Typically, these features can be obtained through directional solidification. However, increase in domain size is not sufficient to obtain high current capacity in these highly anisotropic HTS. In addition, the strongly superconducting a-b planes need to be aligned along the sample axis.

Due to the existence of lateral temperature gradients, large cross-sectional area YBCO fabricated by conventional directional solidification methods consisted of multiple domains with locally oriented grains that are misaligned with respect to the sample axis [2]. In order to improve the domain size as well as the grain alignment, we have developed a seeded directional solidification process which combines the beneficial features of seeding with the modified Bridgman method [3]. Samples fabricated by this process are found to contain one domain with oriented grains aligned along the sample axis. In this paper, the processsng method and the superconducting characteristics of these long HTS will be presented and compared with those of short melt-textured specimens.

---

† Present address: Metals & Ceramics Div., Oak Ridge National Laboratory, Oak Ridge, TN 37921-6116 USA

2. Experimental

Seeded directional solidification is briefly described in the following. A Detailed description of the process can be found elsewhere [3]. In this process, SmBCO seeds were obtained by the Liquid Phase Processing method where slow cooling was carried out at 0.5 - 1°C/h between 1065 and 1000°C. Rectangular seeds were cut from single domains with the surfaces and sides parallel to the a-b plane and <110> directions, respectively. For seeded directional solidification, a seed was inserted into a slot in a sintered YBCO bar such that the a-b plane of SmBCO was parallel to the sample axis. Solidification was carried out in a modified Bridgman furnace with a longitudinal thermal gradient and a sample traveling rate of 55°C/cm and 0.8 - 1 mm/h, respectively [4].

Transport critical current (I_c) measurements were performed on large cross-sectional area samples as well as small cross section specimens cut along the bulk sample axis. Silver wires were attached to the HTS with silver paste, and the HTS were heated to 900°C in flowing oxygen and held for 2 h to anneal the contacts. Then, the specimens were cooled to 500°C with a holding time of 120 h for oxygen uptake. Self-field I_c measurements were performed with a pulsed current source of 1,000 A capability at 77 K and a pulse width of 2 ms. In the determination of field dependency of J_c, samples of cross-sectional areas of ≈ 1.2 x 0.9 mm^2 were measured using continuous dc current and the 1 μm/cm criterion.

In addition to transport dc experiments, magnetization measurements were also performed on a specimen of 3.08 x 2.94 x 3.08 (c-axis) mm^3 which was subjected to a similar oxygenation treatment. Moreover, AC losses of a 1.45 x 1.27 x 38 mm^3 HTS carrying a sinusoidal current were measured at 77 K by an electrical method where the instantaneous values of the voltage and current signals were amplified and digitally multiplied to give the power dissipation during each cycle [5].

3. Results and Discussion

Samples of 4 x 8 mm^2 cross-sectional area and up to 65 mm long have been fabricated by seeded directional solidification. Optical microscopy observations revealed that these HTS essentially consisted of a single oriented domain. During the initial stage of solidification, i.e. before the growth front has reached the seed, the textured material is composed of multiple misoriented domains. Once the growth front has reached the seed, however, YBCO nucleation occurrs at the SmBCO/Y211 + Liquid interface and grows outward such that a single domain microstructure is developed. Due to difficulties involved in cutting the seed exactly parallel to the a-b planes and maintaining the seed along the sample axis in the semi-solid state, final grain misorientation of up to 10° with respect to the sample axis can result. Cross-polarized microscopy of the twin traces revealed that the <110> direction of the solidified YBCO is also aligned along the sample axis, thereby confirming that seeded growth has occurred. Optical tracking of the orthogonal twin traces showed that the a- or b-axis misorientation (c-twist) along an entire sample length is less than 5° [3].

Because of the single domain nature of the samples together with the preferred grain alignment with respect to the direction of current flow, the I_c of these HTS are much higher than the 1,000 A capability of our current source. Transport measurements performed on small cross section specimens revealed a self-field J_c of 47,000 A/cm^2 at 77 K, which would give a bulk sample I_c of approximately 15,000 A. This I_c value is well in excess of most practical conductor applications, and makes the seeded YBCO an excellent candidate for these usages. The J_c field dependency (H||c) of seeded HTS was determined on 2specimens of roughly 40 mm long, and is shown in Fig. 1 together with that of a short

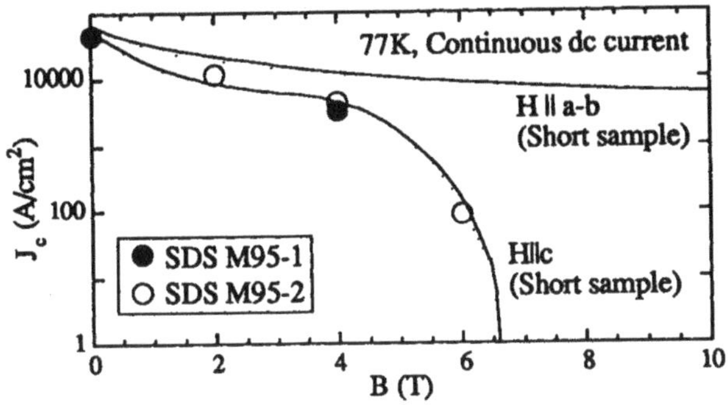

Fig. 1. Field dependency of J_c of long seeded YBCO and short textured specimen.

melt-textured specimen [6]. It can be seen from the figure that the field dependency of J_c of long seeded YBCO is in excellent agreement with that of short well textured specimens, and demonstrates the single domain nature and absence of weak links in these seeded HTS.

From low-frequency ac susceptibility measurements, the irreversibility field (H||c) of a seeded YBCO specimen (SY-1) at 77 K was determined to be 7.7 T [7], and is similar to the value of high quality HTS produced by other texturing methods. The field dependency of magnetization J_c of seeded YBCO, which was determined by the flux profile technique is shown in Fig. 2 together with that of a specimen fabricated by the melt-powder-melt-growth (MPMG) process [7]. It can be seen from this figure that the J_c of seeded YBCO as determined by this technique is similar in value to the well textured MPMG specimen (SY-2). While the J_c values are similar, the fish-tail pattern, which is usually seen in textured YBCO, is absent in the seeded HTS indicating that the peak pinning mechanism is not operative. Also included in the figure are the J_c characteristics of the same specimens after neutron irradiation to a fluence of 4×10^{21} m^{-2}. It can be seen from this figure that the seeded specimen responds well to irradiation, with a J_c enhancement factor of 21.5 at

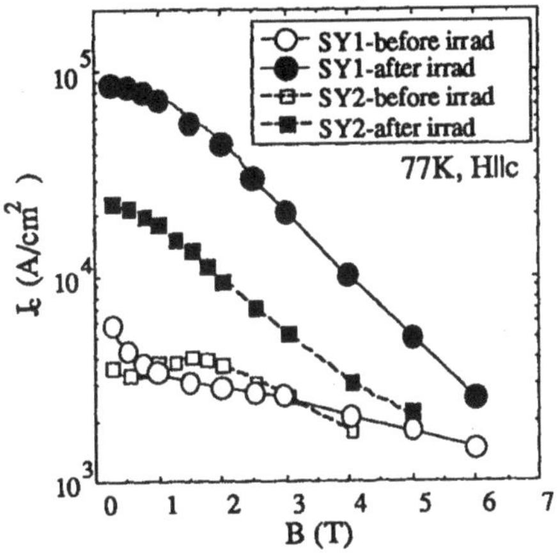

Fig. 2. Magnetizaton J_c of Seeded and MPMG YBCO before and after neutron irradiaton.

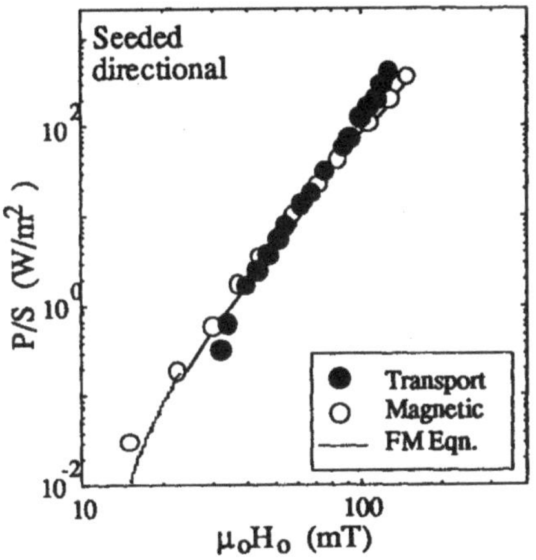

Fig. 3. AC power losses of seeded YBCO.

1 T as compared to a factor of 4.7 in the MPMG specimen. One possible reason for the favorable response in the seeded HTS may be due to the fact that no special effort was taken to improve pinning during the fabrication stage (e.g. Y211 refinement or addition) such that a greater leeway is available for the introduction of pinning defects.

In the 50 Hz AC power losses measurements, self-field losses varied as $I^{4.67}$ up to 310 A where they deviated upward due to resistive transition during part of the current cycle [8], which corresponds to a J_c of 56,000 A/cm^2. The AC losses per unit surface of the seeded YBCO as obtained by transport and magnetization means is shown in Fig. 3. Also included in the figure is the theoretical curve predicted by the Fournet and Mailfert equation [9]. The good agreement between experimental and theoretical losses indicates that seeded HTS are well described by the critical state model if a field dependent J_c (Kim's model) is assumed.

4. Summary

Large cross-sectional area single domain YBCO with preferred grain alignment along the sample axis have been fabricated by a seeded directional solidification process. Due to the absence of weak links and grain boundaries along the current path, these HTS behave similarly to short specimens, and possess J_c of 47,000 A/cm^2 at 77 K. Irradiation experiments indicate that substantial enhancement in pinning can still be introduced into these HTS, and they are found to be well described by the critical state model from AC losses measurements. All of these results indicate that seeded YBCO possesses well behaved properties, and can be used in high current conductor applications.

5. Acknowledgments

This work is supported by ARPA and the state of Texas. DFL would like to acknowledge partial support through ORAU under prime contract No. DE-AC05-76OR00033. HWW wishes to acknowledge partial financial support by the Austrian Science Foundation, Vienna, under contract No. 8849.

6. References

1. K. Salama, V. Selvamanickam and D. F. Lee, in *Processing and Properties of High-T_c Superconductors - Vol. I Bulk Materials*, ed. S. Jin (World Scientific, Singapore, 1993), p. 155.
2. For example, S. Pinol, F. Sandiumenge, V. Gomis, B. Martinez, J. Fontcuberta, X. Obradors, E. Snoeck and C. Roucau, in *Applied Superconductivity*, ed. H. C. Freyhardt (DGM, Oberursel, 1993), p. 365.
3. D. F. Lee, C. S. Partsinevelos, R. G. Presswood, Jr. and K. Salama, *J. Appl. Phys.* **76** (1994) 603.
4. V. Selvamanickam, C. Partsinevelos, A. V. McGuire and K. Salama, *Appl. Phys. Lett.* **60** (1992) 3313.
5. S. Zannella, L. Jansak, M. Majoros, V. Selvamanickam and K. Salama, *Physica C* **205** (1993) 14.
6. J. W. Ekin, K. Salama and V. Selvamanickam, *Nature* **350** (1991) 26.
7. M. Eisterer, Diploma Thesis, Atominstitut der Österreichischen Universitäten, Wien (1994).
8. S. Zannella, M. Majoros, V. Ottoboni, D. F. Lee and K. Salama, in *Proc. 8th CIMTEC Conf.* (1994) in press.
9. G. Fournet and A. Mailfert, *J. de Physique* **31** (1970) 357.

THE INFLUENCE OF MELTING AND CRYSTALLIZATION TIMES ON THE STRUCTURE AND SUPERCONDUCTING PROPERTIES OF $YBa_2Cu_3O_{7-x}$

B.P. MIKHAILOV and A.R. KADYRBAEV

Baikov Institute of Metallurgy, Russian Academy of Sciences, Leninsky prospekt, 49, Moscow, 117334/Russia

ABSTRACT

The carring capacity of HTSC wires depends mainly on ceramic corn structure, phase composition, oxygen content, density, grain texture, insulating phases at the grain boundaries. The core structure can be improved in decisive degree by means of melting and subsequent crystallization. The report represents the results of melting HTSC $YBa_2Cu_3O_{7-x}$ compound wires with zirconium and silver-palladium covers completed by different methods: electric current, laser beam, infrared radiation. The heating temperature was varied from 1000 to 1100°C. The dependence of the core structure and superconducting properties on the melting and crystallization regimes has been investigated. The possibility of forming a practically nonporous single-phase structure with a critical current density over 10^4 A/cm^2 at 77 K has been shown.

The improvement of HTSC ceramics current-carrying capacity can be achieved by melting and subsequent directional solidification. Ceramic plates were treated in such a way [1,2]. In this paper the influence of heating of the $YBa_2Cu_3O_{7-x}$ ceramics up to the melting temperature on its structure and superconducting properties was studied for the wire samples.

In our work the melting and subsequent solidification of HTSC ceramic composite wire core in refractory metal shell was studied. The composite HTSC wire was fabricated by the "powder in tube" method and consists of a metallic shell with $YBa_2Cu_3O_{7-x}$ compound core. The composite HTSC wire heating to the ceramic core melting temperature was provided by three methods: 1) passage of current through the wire; 2) infrared irradiation; 3) laser beam.

The electric curret heating of zirconium shell HTSC wires took place in air. Both samples with intact shell and with half a shell along the wire moved away ("open shell") were heated. The heating time to the ceramic core melting temperature (1000 – 1100°C) was 5–6 s. Then the temperature was maintained at this level by current regulation until the sample failed to conduct current due to sample destruction or without visual external damage. At the first moment of current heating the maximum temperature was reached on about 70% of the wire length in its central part, but this zone progressively reduced in size until it transformed into a point (it is this point where the sample destruction took place). In the maximum heating zone

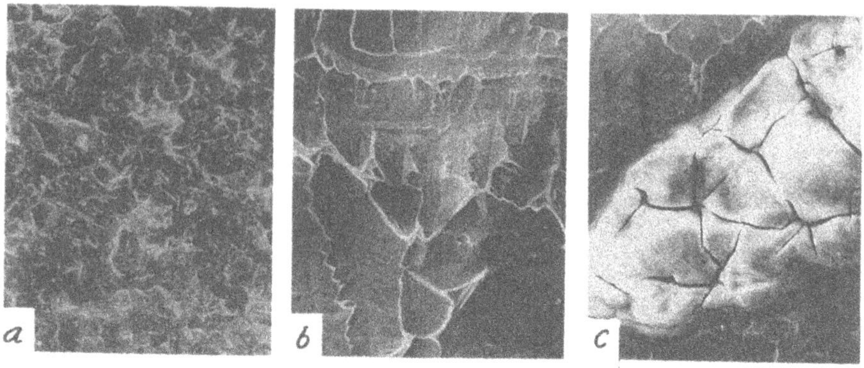

Figure 1. Microstructure of the $YBa_2Cu_3O_{7-x}$ wire core melted under different temperature-time regimes. (*a*) 1000°C - 10 s, (*b*) 1000°C - 10 s + 1050°C - 10 s, (*c*) 1050°C - 120 s. ×1000.

the ceramic core was in a wholly melted state. Due to the reduction of the heating zone during current passing the melted core solidification zone moves slowly from the ends of the sample to its middle. The temperature gradient appears along the wire axis. The final cooling of the samples took place in air.

Both $YBa_2Cu_3O_{7-x}$ ceramic pellets and HTSC wire samples with Pd-Ag alloy shell were heated by infrared radiation. The wire samples had an intact shell. The heating was carried out in 10 mPa vacuum. The heating consisted of two steps: first, the samples were heated to 900 – 930°C, cured at this temperature for 120 s, then the temperature was raised to 1100°C and maintained for 10 – 20 s. A slow cooling was provided by a gradual radiation power decrease.

The laser beam heating of HTSC wire samples with the zirconium shell was also carried out in air. Only open shell wire samples were heated. During the heating the laser radiation power and the beam movement speed along the sample were varied.

Melted samples were annealed in the oxygen atmosphere. The microstructure of melted and nonmelted parts of the sample was studied with an optical and scanning electron microscopes. The chemical composition of the phases was determined by an X-ray microprobe. The superconducting transition temperature and critical current were measured by an inductive method by inducing a.c. magnetic fields of with a frequency 122 Hz and various intensities (from 0.1 to 5 mT) in the sample.

The electric current melting and following crystallization of the ceramic core in the wire samples leads to a decrease of the core volume. This creates a hollow canal and the melted material settles on cover walls. The core structure straightly depends on the treatment conditions. After a short-term high-temperature heat treatment (10 s at 1000 and 1100°C) the core structure does not change (Fig. 1,*a*). At the time

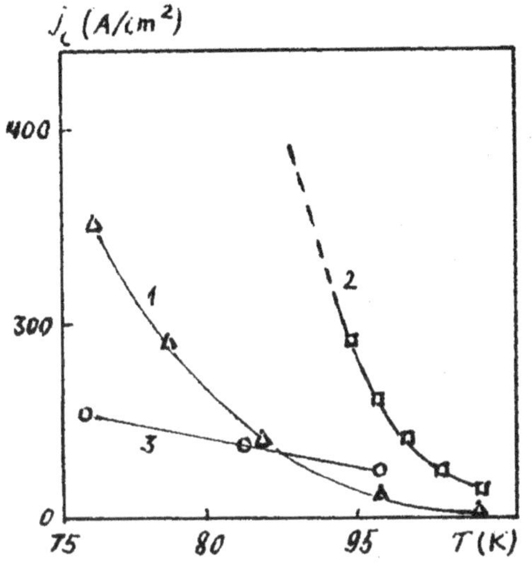

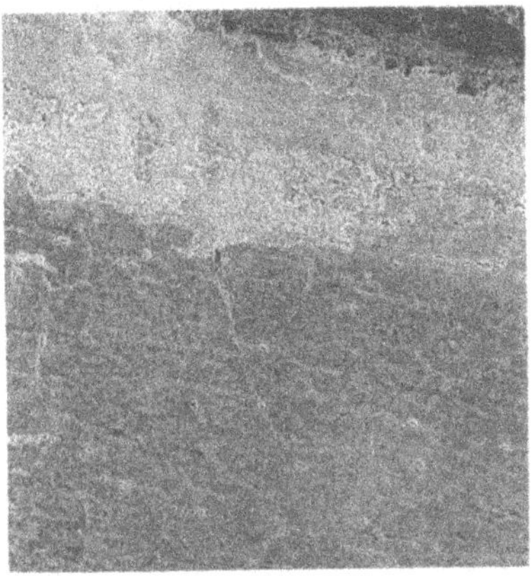

Figure 2. Critical current density versus measured temperature for $YBa_2Cu_3O_{7-x}$ wires heated by the electrric current. (*1*) nonmelted sample, (*2*) after heating at 1000°C - 10 s and annealing in the oxygen atmosphere at 900°C - 20 h, (*3*) after heating at 1050°C - 60 s and annealing in the oxygen atmosphere at 900°C - 20 h.

Figure 3. Microstructure of the $YBa_2Cu_3O_{7-x}$ pellet cross-section after the infrared radiation melting at 1000°C - 10 s. ×200.

increases (at 1050°C) up to 30–120 s the structure transforms. After 30 s at 1050°C core becomes nonporous small crystallites are dissolved and the boundaries along and perpendicular to the sample axis appear(Fig. 1,*b*). The material has a single-phase structure. The heat treatment at 1050°C during 60 s leads to local disintegration of the 123 phase with the Cu-rich phase precipitations. The second phase quantity increases with the heat treatment time growing up to 120 s (Fig. 1,*c*). Melting during 10–165 min results in an inhomogeneous microstructure.

The short-term melting (up to 60 s) does not change the element composition of the 123 phase. After melting during 120 s the Y concentration is increased in 123 phase (up to 19.46 %). This changed composition effects in superconducting properties. The short-term heating (10–30 s) by the electric current increases the wire critical current density by the factor of 5 (Fig. 2). A longer heat treatment (more than 30 s) leads to a sharp decrease of both critical temperature and critical current of the wires. This occurs because of oxygen losses by 123 phase and 123 phase transition from the cubic to tetragonal mode.

Figure 3 shows the cross sectional structure of the 123 pellet after the infrared heating at 1100°C during 10 s with a subsequent heat treatment in the oxygen atmosphere for 20 h at 900°C. On the irradiated surface there is 1/3-tablet-thickness

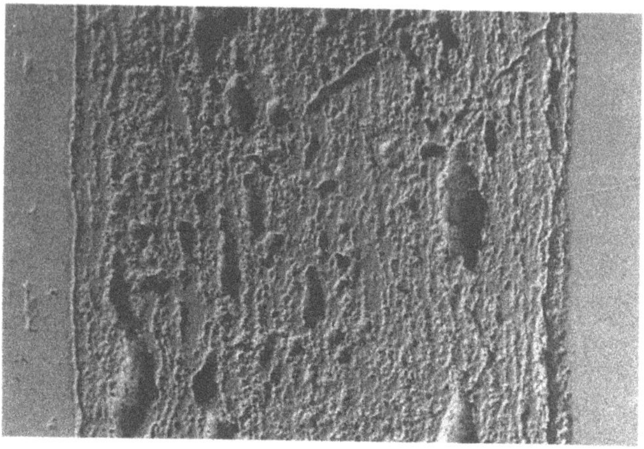

Figure 4. Microstructure of the $YBa_2Cu_3O_{7-x}$ wire with Pd-Ag alloy shell (longitudinal section) after the infrared radiation melting at 1000°C - 10 s. ×300.

layer with a more dense structure than other parts of the tablet. A high resolution investigation of this layer shows its inhomogeneity. Pores, cracks, single large blocks are absent in the melted layer. The sample melted by infrared heating has a greater increase of the critical current with the temperature decrease than nonmelted samples.

Infrared heating strongly changes the structure of the wire sample ceramic core (Fig. 4). After melting the grains orients along the wire axis contraction pits appear, the boundaries between core and cover is free of the intermediate phases.

The laser beam melting of the composite wire sample ceramic core leads to the creation of a dense nonporous ceramic layer. The critical current density of the laser-melted samples is several times higher, too.

Thus short-term melting of the 123 compound by the electric current, laser or infrared heating leads to perceptible compacting of ceramic material. This promotes grain knitting and preservation of the stoichiometric composition of the 123 phase. Marked structural transformation certainly favors the current-carrying ability of the composite wires and ceramic plates.

1. S. Jin, T.H. Tiefel, R.S. Sherwood at al, *Appl. Phys. Lett.* **51** (1987) 943.
2. K. Salama, V. Selvamanickam, L. Gao at al, *Appl. Phys. Lett.* **54** (1989) 2352.

Effect of Intergrain Ag on the Magnetic Properties of $YBa_2Cu_3O_7$ analyzed using the Generalized Critical State Model

P. Fournier* and M. Aubin**

* Edward L. Ginzton Laboratory, Stanford University, Stanford, CA 94305-4085
**Département de Physique and Centre de Recherche en Physique du Solide, Université de Sherbrooke, Sherbrooke, Québec (CANADA) J1K 2R1

We use the Generalized Critical State Model (GCSM) to analyze the effect of Ag-addition on the magnetic properties of polycrystalline $YBa_2Cu_3O_7$ with 5 % wt of intergrain silver (Ag). The GCSM parameters are obtained from the fits of the field penetration in a hollow cylinder as a function of the applied magnetic field for temperatures ranging from 20 to 50K. From 50K to T_c, difficulties in obtaining good fits indicate the contribution of the intergrain lower critical field (H_{c1j}). The parameters are compared with the ones previously determined with samples having no Ag. Other indications of the H_{c1j} contribution are deduced when comparing the transport critical current of 0% and 5% samples.

Introduction

To use high temperature superconductors in large scale applications, one has to improve the mechanical and electrical properties of the polycrystalline materials. Melt processing[1,2] (texturing) and Ag-cladding[3] (extruded wires) are the most promising avenues. The composite superconductors, in which grains are embedded in a metallic matrix (say, $YBa_2Cu_3O_7$ in Ag) have also been explored[4]. The interest for silver arises from the fact that it does not dope significantly the $YBa_2Cu_3O_7$ (YBCO) grains as indicated by T_c which is constant within a degree[5]. Silver is found to sit in the intergrain region and is a potential candidate for increasing the coupling between grains (resulting in an increase of the critical current density of the junctions). However, the transport critical current density in zero field for such a system is not significantly influenced when the amount of silver is varied from 0 to 20 wt % [5]. Moreover, the magnetic J_c extracted from the low field magnetization is depressed as the amount of Ag is increased[6]. The origin of this decreasing quality with silver addition can be clarified by extracting some information from the critical state model which describes properly the magnetic and transport properties of these polycrystalline samples.

In this work, we follow a procedure used previously to determine the Generalized Critical State Model [GCSM : $J_c(H) = J_{c0}\,(1+H/H_0)^{-n}$] parameters from two independent experiments[7,8] on the same sample, namely the critical current in thin slabs and the field penetration in a hollow cylinder as a function of magnetic field and temperature. We show that the low temperature field penetration in the hollow cylinder can be fitted exactly while neglecting the intergrain lower critical field H_{c1j}. However, as one approaches T_c, the fits are no longer possible. This is believed to come from an increasing contribution from the reversible currents. We finally present a theoretical simulation of the effect of reversible currents on the transport critical current.

Experimental Set-Up and Procedures

The samples are prepared using the usual solid state processing. Y_2O_3, $BaCO_3$ and CuO are initially mixed thoroughly and calcined at 900°C for 48 hours with an intermediate grinding. A portion of the resulting powder had been used previously to make the 0% materials[7,8] (pellets) while the rest is used here to make the 5% materials. This ensures that both materials present the same grain size distribution. To obtain the 5% materials, we mix the YBCO and AgO powders in the proper quantities to obtain 5% by weight of atomic Ag. The resulting powder is pressed into pellets and annealed at 940°C for 48 hours in ultra-pure oxygen followed by a further step at 410°C for another 48 hours. The resulting pellets are cut in the shape of thin slabs or drilled and sandblasted[7] to obtain the hollow cylinder. We find the YBCO density to be 4.96 g/cm^3 in the Ag-containing samples, a value smaller that the one obtained for the 0% pellets (5.23 g/cm^3).

The transport critical current measurements are made using a set-up described elsewhere[9]. This set-up allows one to reach temperatures in the 65 to 95K interval while the sample is in liquid nitrogen. The field penetration in a hollow cylinder is measured using a 10^{16} cm^{-3} n-type GaAs Hall probe[7] properly calibrated for the temperatures of interest.

Results and Discussion

The field penetration in hollow cylinders is shown in Figure 1(a) for both concentrations and for the same wall thickness (a = 4mm) at 50K. One can see that the trapped field has decreased approximately by a factor of 2 for the 5% cylinder. In Ref. 8, we describe a method to determine the full penetration field, H_p, for short hollow cylinders presenting a demagnetizing effect. This field is seen to be slightly different from the trapped field. Following this procedure, we determine H_p as a function of the temperature as shown in Fig. 1(b).

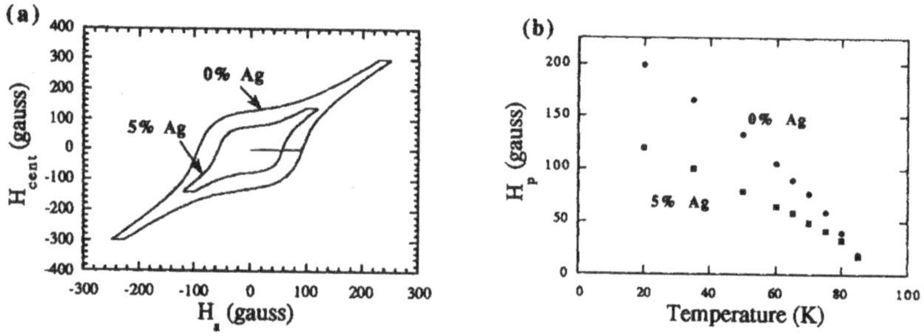

Figure 1 : (a) Field in the center of the cylinder (H_{cent}) as a function of the applied field (H_a) for both materials at 50K for a wall thickness of 4mm; (b) full penetration field as a function of temperature.

The results indicate that the addition of Ag has an effect on the strength of the flux pinning ($F_p \approx H_p \phi_0 / a$ from the Bean model). This can be shown by fitting the hysteresis loop of the 5% material introducing the GCSM parameters J_{c0}, H_0 and n , and parameters related to the demagnetizing effect, namely the demagnetization factor, N, and the shielded volume fraction of grains, f_g. This allows one to fit the loops for the whole range of temperature for the 0% sample, but it is seen to be possible only from 20 to 50K for the 5%

cylinder. Above 50K, the loops are impossible to reproduce within the approximations made initially in Ref. 7.

From 20 to 50K, we can compare the values of the GCSM parameters for the two hollow cylinders. A very important observation is that the exponent n remains equal to 2 for the two concentrations. This suggests that n does not depend on the superconducting properties, but, as suggested previously[8], on the microstructure, principally on the average angle of the junction interfaces relative to the internal field[10]. Figure 2(a) shows the values of J_{c0} as a function of temperature for both samples. J_{c0} is slightly changed by the addition of silver. In Fig. 2(b), we find that H_0 is depressed by a factor of approximately 2.

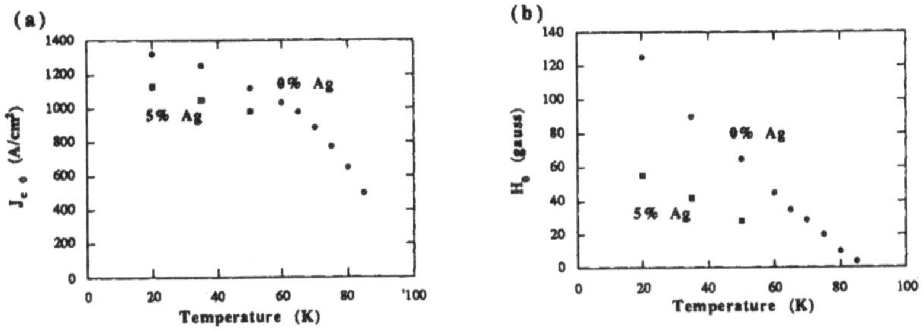

Figure 2 : (a) J_{c0} and (b) H_0 as a function of temperature for both materials.

As one increases the amount of Ag, the intergrain space increases (as seen with the measurement of the density of YBCO) and gives more room for the penetration of flux. As stated previously[8], a quantum of intergrain flux (ϕ_o) embeds several grains in the limit where λ_J , the intergrain penetration depth, is larger than the mean grain radius r_g. The grain cores are totally screened while the flux penetrates the surface over the London penetration depth λ_g . The cross-sectional area A_J penetrated by a quantum of flux is then the number of grains ($N_g \sim \lambda_J{}^2 / r_g{}^2$) multiplied by the penetrated area per grain ($\sim 8\, r_g\lambda_g$) giving $A_J \sim \lambda_J{}^2\lambda_g / r_g$. Keeping the grain radius constant, while changing the Ag content just increases λ_J and A_J (see Ref.11), thus decreasing H_0 ($= \phi_o / A_J$). Despite the increased non-superconducting volume, J_{c0} remains high (comparable to the undoped case) and can reveal an increased coupling between grains.

In Figure 3(a), we present the normalized transport critical current as a function of the applied magnetic field for the two concentrations at 85 K (close to T_c). Despite the fact that the zero-field transport critical current density (I_{c0}/S ; S : cross-section) is significantly changed by the Ag-addition[12] (see inset), the half-width of the field dependence is not seriously altered. This is in contradiction with the fact that H_0 is decreased by a factor of 2 at lower temperatures (supposing that it remains the case near T_c). To explain this inconsistency, we recall that the zero-field critical current is a summation of the irreversible currents ($2YH_p$: where Y is the width of the slab) and the reversible currents ($\sim 2YH_{c1j}$), usually neglected. This latter contribution can be added to our fitting procedure following the theory developed by LeBlanc[13], but it introduces two new free parameters. Supposing that the reversible currents are proportional to the reversible magnetization[13] (goes as $\ln(H_{c2j} / B)$) and reaches 0 at H_{c2j} (the intergrain upper critical field), we find, as shown in Figure 3(b), that the field dependence of the transport I_c can be modified by the reversible currents. This would explain the difficulties in fitting $I_c(H_a)$ from 65 to 85K and $H_{cent}(H_a)$ from 60 to 85K. Thus, the addition of Ag in the intergrain region increases H_{c1j}, through probably an increase of the mean critical current density of the junction, J_0. However the field dependence of $J_c(H)$ is badly affected by the dilution of the superconducting phase.

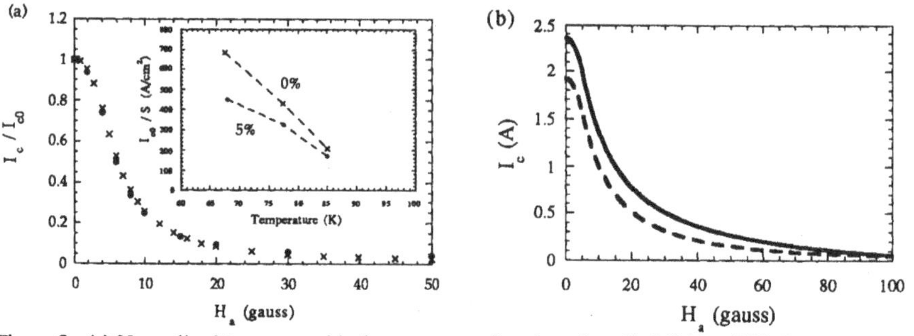

Figure 3 : (a) Normalized transport critical current as a function of applied field at 85K. Inset : transport critical current density as a function of temperature, (b) Critical current as a function of applied field with J_{c0} = 700 A/cm^2, H_0 = 15G, n = 2 , 2a = 0.2mm, Y = 2mm and H_{c2j} = 105G for H_{c1j} = 2G (solid line) and H_{c1j} = 0G (dashed line).

Acknowledgements

We are indebted to C.Julien for her technical support throughout this project and to M.A.R. LeBlanc for several stimulating discussions. The authors would like to thank NSERC (Canada), FCAR (Québec) and the Centre de Recherche en Physique du Solide for their financial support.

References

1- S.Jin, T.H.Tiefel, R.C.Sherwood, M.E.Davis, R.B.van Dover, G.W.Kammlott, R.A. Fastnacht and H.D.Keith, Appl. Phys. Lett. **52** (1988) 2074 .
2- M.Murakami, M.Morita and N.Koyama, Jpn.J.Appl.Phys. **28** (1989) L1125 .
3- See for example: K.E.Hautanen, M.Oussena and J.R.Cave, Cryogenics **33** (1993) 326.
4- G.G.Peterson, B.R.Weinberger, L.Lynds and H.A.Krasinski, J.Mat.Res. **3** (1988) 605.
5- N.Imanaka, F.Saito, H.Imai and G.Adachi, Jpn.J.Appl.Phys. **28** (1989) L580 .
6- J.Jung, M.A.-K.Mohamed, S.C.Cheng and J.P.Franck, Phys.Rev.B **42** (1990) 6181 .
7- P.Fournier and M.Aubin, Phys.Rev.B **49** (1994) 15976 .
8- P.Fournier and M.Aubin, Phys.Rev.B **50** (1994) Oct. 1st .
9- P.Fournier and M.Aubin, Rev.Sci.Instrum. **65** (1994) 788 .
10- R.-L.Peterson and J.W.Ekin, Physica C **157** (1989) 325 .
11- Z.Przelozny, T.R.Finlayson and T.F.Smith, in *Studies of High Temperature Superconductors*, Vol. **10**, Ed. A. Narlikar (Nova Science Publisher, New York), p.115.
12- This result is slightly different from the one of B. Dwir et al [Appl.Phys.Lett. **55** (1989) 399] who show a small initial increase of I_{c0} / S with the amount of Ag. This originates from different heat treatments giving different grain radii and the related behavior of the density of YBCO.
13- M.A.R.LeBlanc, Cryogenics **32** (1992) 813.

MAGNETIC FLUX PINNING OF METALLIC SUPERCONDUCTORS WITH ARTIFICIALLY DESIGNED PIN STRUCTURE

KANAME MATSUMOTO and YASUZO TANAKA
Metal Research Center, The Furukawa Electric Co.,Ltd.
500, Kiyotaki, Nikko 321-14, Japan

KAORU YAMAFUJI, KAZUO FUNAKI and MASATAKA IWAKUMA
Faculty of Engineering, Kyushu University
6-10-1 Hakozaki, Higashi-ku, Fukuoka 812, Japan

and

TERUO MATSUSHITA
Faculty of Computer Science and Systems Engineering, Kyushu Institute of Technology
680-4 Kawazu, Iizuka 820, Japan

ABSTRACT

High crtical current density (J_c), which exceeds the highest one in a conventional NbTi superconductor with α-Ti precipitates, has been attained by the introduction of artificial pinning centers into NbTi superconducting alloys. The pinning strength of the composites depends on both the designed pin structures and the materials for artificial pinning centers. The flux-pinning characteristics of artificial pinning centers are discussed on the basis of theoretical and experimental investigations.

1. Introduction

In conventional NbTi superconducting wires with high J_c, ribbon-like α-Ti precipitates are the dominant pinning centers. Since the microstructure of the pinning centers was elucidated in the 80's, the highest J_c values for the NbTi wires were largely increased by the development of the thermomechanical treatment.[1] However, record J_c values have not been improved in recent few years in spite of much effort, because it is difficult to introduce more adequate pin structures into NbTi only by means of conventional metallurgical methods.

On the other hand, a new technique for introducing artificial pinning centers into the NbTi alloys has been developed.[2-7] This technique makes it possible to obtain more higher J_c values. In this paper the flux pinning properties of artificial pinning centers are discussed.

2. Global Pinning Force of Artificial Pinning Centers

Several artificial pin structures have been presented for NbTi wires. These are fabricated by usual composite techniques, and are reduced to the nanometer scale keeping the designed morphology. We use the lamellar type structure[4,8,9] in order to control J_c because it is not deformed severely during reducing process and it is easy to increase the volume fraction of artificial pinning centers. According to the experimental results, the appropriate volume fraction is 25~35%.

The artificial pin structure is extended successively along the direction of the wire axis, so that the transport current does not cross the pin structure. Hence the density of the effective pinning centers (N_p) can be increased to a fairly high level without noticeably degrading the superconducting characteristics of the wires. If the distance between the effective pinning centers is expressed as d_{eff}, we have

$$N_p = 1/(a_f^2 d_{eff}); \qquad (a_f \gtrsim d_{eff})$$
$$= 1/(a_f d_{eff}^2); \qquad (a_f \lesssim d_{eff}) \qquad (1)$$

where a_f is the spacing of flux line lattice.

Nb is a typical material for artificial pinning centers, since it has a good workability and functions as strong pinning center above 0.5T at 4.2K. Using the elementary pinning force(f_p) and the pinning efficiency(η), the global pinning force(F_p) of artificial pinning centers is expressed by the linear sum of

$$F_p = J_c B = \eta f_p N_p \qquad (2)$$

It is known empirically that η approaches unity as f_p increases.

3. Experimental Results

In Fig.1, we show the cross sections of the artificial pin structure composed of Nb and Nb-50wt%Ti thin layers. The designed pin structures can be expressed by the parameters of d_p, d_w and d_s, where d_p, d_w and d_s are the pin layer thickness, width, and spacing.

The typical J_c vs. d_p curves due to artificial pinning centers are shown in Fig.2. As d_p decreases, the J_c values increase in proportion to d_p^{-1} and reach peak values in the range of d_p = 5~20nm. The maximum J_c values at 3T and 5T are three times and 1.4 times higher than those of the pracatical NbTi superconducting wires with α-Ti ribbon pins. The optimized

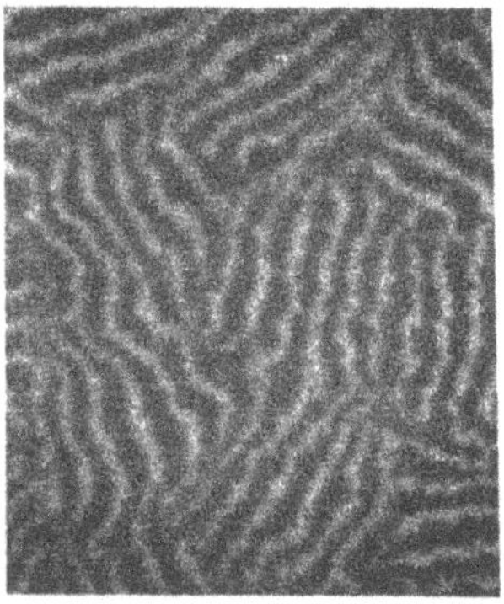

Fig.1 A cross section of the lamellar type artificial pin structure(d_p=200nm, Nb pin vol%=27%).

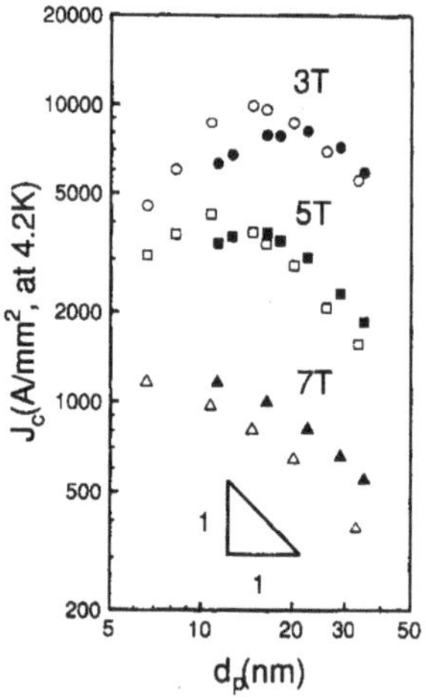

Fig.2 J_c vs. d_p curves of a typical Nb/NbTi artificial pin strucutre.

thickness of the Nb layers are 5~10 times thicker than those of the optimized α-Ti ribbon thickness (1~2nm).[1]

When $d_p \gtrsim 2\xi_{NbTi}$ and $a_f \lesssim d_{eff}$, we have the following formula[4]

$$J_c = F_p/B = C\eta(B_{c2}/\kappa)^2 B^{-1}(d_s+d_p)^{-1} b^{1/2}(1-b) \qquad (3)$$

where B_{c2} is the upper critical field, κ is the GL parameter, $b = B/B_{c2}$ and C is a numerical constant depending on the ratio of d_w/d_p. Eq.(3) holds in the vicinity of the peak values of J_c and the linear sum is achieved. In the high field range, however, the observed J_c values decrease and η becomes small.

4. Discussion

We can estimate a coherent volume, $V_c \simeq (2l_{66})^2 l_{44}$, from the view point of the collective pinning.[8] For strong pinning centers, the correlation lengths l_{66} and l_{44} are given by

$$l_{66} = (C_{66}a_f/4F_p)^{1/2},\ \ l_{44} = (C_{44}a_f/4F_p)^{1/2} \qquad (4)$$

where C_{44} and C_{66} are the elastic moduli of the fluxoid lattice. Using V_c we have

$$F_p = (N_p/V_c)^{1/2} f_p \qquad (5)$$

From Eqs.(2) and (5), η is expressed by

$$\eta = (N_p V_c)^{-1/2} = n_p^{-1/2} \qquad (6)$$

where $n_p = N_p V_c$ is the number of pinning centers in the coherent volume. If $l_{66} \lesssim a_f/2$, each fluxoid behaves independently as shown in Fig.3. For specimens having artificial pinning centers, we can easily see $l_{66} \lesssim a_f/2$ near the peaks of J_c vs. d_p curves in the low magnetic fields. Here we can expect $n_p \sim 1$, $\eta \sim 1$ and the linear sum. As the magnetic field increases, however,

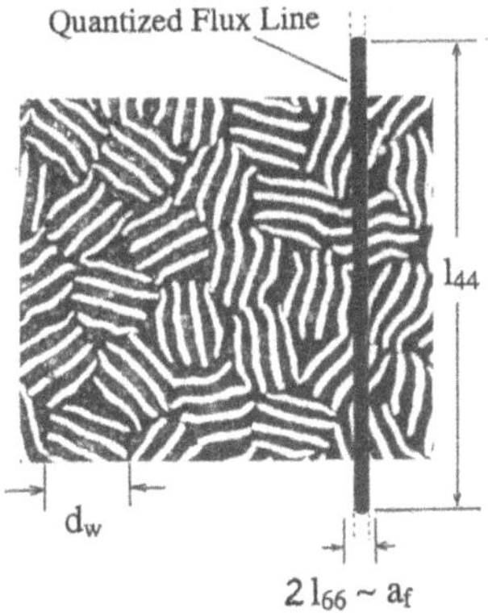

Fig.3 In strong pinning regime, $l_{66} \lesssim a_f/2$ is achieved and each fluxoid behaves independently. Then the coherent volume V_c becomes $a_f^2 l_{44}$.

l_{44} becomes longer. Then $n_p >> 1$ and η becomes small.

Furthermore, the proximity effect brings about the depression of B_{c2}.[4,9] The microstructure of the lamellar type pinning centers can be approximated by ramdomly oriented multilayered clusters composed of Nb and NbTi layers. Generally, multilayered superconductors have anisotropic properties of B_{c2}. The lower components of B_{c2} reduce F_p in high magnetic fields. When $d_p < 2\xi_{NbTi}$, f_p also decreases by the reduction of the overlapping volume between Nb pins and the core of the interacting fluxoid. The decrease of F_p in the range of $d_p < 2\xi_{NbTi}$ can be explained by both the reduction of f_p and the collective behavior of fluxoids.

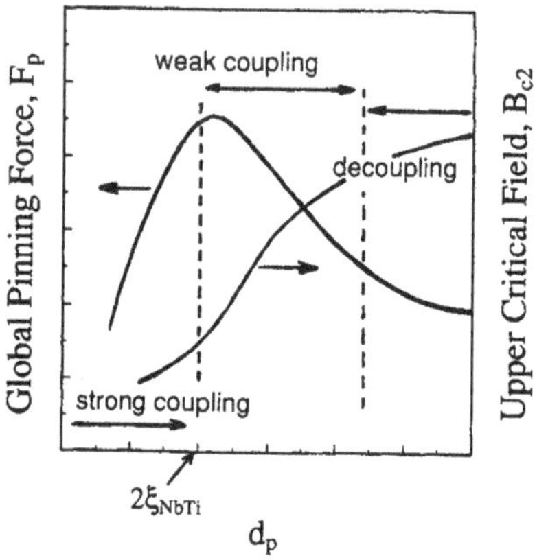

Fig.4 Variations of F_p and B_{c2} as a function of d_p. The peak value of F_p is obtained in the crossover region from the weak coupling to the strong coupling between NbTi layers.

In Fig.4, we summarize the variations of F_p and B_{c2} in the Nb/NbTi lamellar type structure as a function of d_p.[10] As d_p decreases, B_{c2} decreases gradually and reaches a minimum value when $d_p < 2\xi_{NbTi}$, where NbTi layers strongly couple through Nb layers by the proximity effect. On the other hand, F_p increases in proportion to d_p^{-1} and then decreases rapidly after F_p reaches a peak value around $d_p \sim 2\xi_{NbTi}$. Since the NbTi layers weakly couple through Nb layers when $d_p > 2\xi_{NbTi}$, the peak value of F_p can be obtained in the crossover region from weak coupling to strong coupling between NbTi layers.

References

1. C.Meingast et al., J.Appl.Phys. **66**,5962(1989).
2. L.R.Motowidlo et al., Adv.Cryog.Eng. **36A**,311(1990).
3. K.Yamafuji et al., Cryogenics **31**,421(1991).
4. K.Matsumoto et al., Supercond.Sci.Technol. **5**,684(1992).
5. R.M.Scanlan et al., IEEE Trans.Appl.Supercond. **3**,1358(1993).
6. O.Miura et al., Supercond.Sci.Technol. **6**,748(1993).
7. L.D.Cooley et al., Appl.Phys.Lett. **64**,1298(1994).
8. K.Yamafuji et al., Physica C **159**, 743(1989).
9. K.Matsumoto et al., IEEE Trans.Appl.Supercond. **3**,1362(1993).
10. K.Matsumoto et al., Appl.Phys.Lett. **64**,115(1994).

IRREVERSIBILITY LINE IN SUPERCONDUCTING Pb-Bi ALLOYS

TSUGIO HAMADA and TERUO MATSUSHITA
Department of Computer Science and Electronics,
Kyusyu Institute of Technology,680-4 Kawazu, Iizuka Fukuoka 820, Japan
and
BAORONG NI
Department of Electronic Materials Engineering,
Fukuoka Institute of Technology, 3-30-1 Wajirohigashi, Higashi-ku Fukuoka 811-02, Japan
and
TADAHIRO AKUNE and NOBUYOSHI SAKAMOTO
Department of Electrical Engineering,
Kyusyu Sangyo University,2-3-1 Matsugadai, Higashi-ku Fukuoka 813, Japan

ABSTRACT

Irreversibility line was studied for superconducting Pb-Bi alloys with different pinning strength. It was found that the magnitude and temperature dependence of the irreversibility field depend rather slightly on the flux pinning strength. The difference between the irreversibility field and the upper critical field remains finite even in specimens with very strong pinning force. The experimental result is compared with the flux creep theory .

1. Introduction

The critical current density, J_c, of high temperature superconductors suddenly decreases with increasing temperature or magnetic field. Such a poor property of critical current density is attributed mainly to weak pinning strength against large thermal agitation at high temperatures. It is desired, therefore, to improve the critical current property of high temperature superconductors by improving the pinning strength. For the purpose, the mechanism that determines the critical current property is required to be clarified. The poor critical current property in high temperature superconductors is reflected to the low irreversibility line. So, it is necessary to clarify quantitatively the dependence of the irreversibility line on the flux pinning. However, it is difficult to control the pinning force in high temperature superconductors. In this paper, therefore, superconducting Pb-Bi alloy specimens are used, since the pinning strength can be easily controlled. The experimental result is compared with the flux creep theory.

2. Experimental

2.1. Sample preparation

Superconducting Pb-Bi alloy specimens were prepared as follows : Lead and bismuth with the purity of 99.99% were mixed in a desired composition and melt at 400℃

Table 1. Parameters for two Pb-Bi specimens

	$B_{c2}(0)(\mathrm{T})$	$T_c(\mathrm{K})$	$AB_{c2}^m(0)(\mathrm{A/m^2})$	m	γ	δ
Bi28at%	1.687	8.35	8.51×10^7	1.62	0.53	2.00
Bi56at%	2.410	8.53	7.94×10^8	1.50	0.50	2.00

for 96 hours in a vacuum of 10^{-4} torr. In general, the Pb-Bi alloy consists of the superconducting epsilon phase and the normal bismuth-rich phase. We prepared two samples and the atomic ratio of bismuth was selected as 28% and 56% so as to obtain the volume fraction of normal precipitates of 0% and 20%, respectively. Superconducting parameters of these specimens are listed in Table 1.

2.2. Measurements

Magnetic property was measured by SQUID magnetometer. The upper critical field, B_{c2}, was obtained from the DC susceptibility measurement. The critical current density, J_c, and the irreversibility field, B_i, were estimated from the DC magnetization measurement. The Bean model was used for the calculation of J_c.

2.3. Results

The critical current densities of the two specimens at 4.5K and 0.2T were $4.3\times10^8\mathrm{A/m^2}$ and $2.7\times10^7\mathrm{A/m^2}$. Figure 1 shows the experimental result of temperature dependence of the irreversibility field normalized by the upper critical field at 0 K, $B_{c2}(0)$. The dashed line indicates the upper critical field approximated as $B_{c2}(T) = B_{c2}(0)[1-(T/T_c)^2]$. The difference between the upper critical field and the irreversibility field is smaller for the specimen with stronger flux pinning. Such a behavior agrees qualitatively with the prediction of the flux creep theory. However, the results are similar for the two specimens in spite of large difference in the flux pinning strength. That is, the difference between B_{c2} and B_i remains finite even for the specimen with the normal precipitates of 20% in volume fraction.

3. Discussion

In this section, we briefly explain the flux creep theory used in the numerical calculation. If we assume the empirically known scaling relation for J_{c0} as

$$J_{c0}(B,T) = AB_{c2}^m(0)\left[1-\left(\frac{T}{T_c}\right)^2\right]^m\left(1-\frac{B_i}{B_{c2}}\right)^\delta, \tag{1}$$

the irreversibility field is expressed as

$$B_i^{(3-2\gamma)/2} = \left(\frac{K}{T}\right)^2\left[1-\left(\frac{T}{T_c}\right)^2\right]^m\left(1-\frac{B_i}{B_{c2}}\right)^\delta, \tag{2}$$

where K is approximately a constant and is given by

$$K = \frac{0.835g^2[AB_{c2}^m(0)]^{\frac{1}{2}}}{\zeta^{\frac{3}{2}}\ln(Ba_f\nu/E_c)}. \tag{3}$$

In the above g^2 is a number of the fluxoids in the flux bundle, ζ is a constant dependent on the pinning center and a_f is the fluxoid spacing. g^2 is given by

$$g^2 = g_e^2 \left[\frac{5k_BT}{2U_e}\ln\left(\frac{Ba_f\nu}{E_c}\right)\right]^{4/3}, \tag{4}$$

with $g_e^2 = C_{66}a_f/\zeta J_{c0}B$, where C_{66} is the shear modulus of the fluxoid lattice, U_e is the pinning potential when g^2 is equal to g_e^2 and E_c is the electric field criterion for definition of the critical current density. The pinning parameters of the two specimens are also listed in Table 1. Since the theoretically obtained g^2 are smaller than 1, we used $g^2 = 1$ for the strongly pinned specimen.

The results obtained from Eq.(2) shows that B_i is very close to B_{c2} for the strongly pinned specimen. This is speculated to be caused by the theoretical definition that the irreversibility field is defined by the field at which the critical current density reaches completely zero. In experiments, on the other hand, B_i is defined in terms of some criterion for the critical current density, ΔJ_c. This condition is described as

$$U_0 = \frac{k_BT}{(1-\Delta J_c/J_{c0})}\ln\left(\frac{Ba_f\nu}{E_c}\right), \tag{5}$$

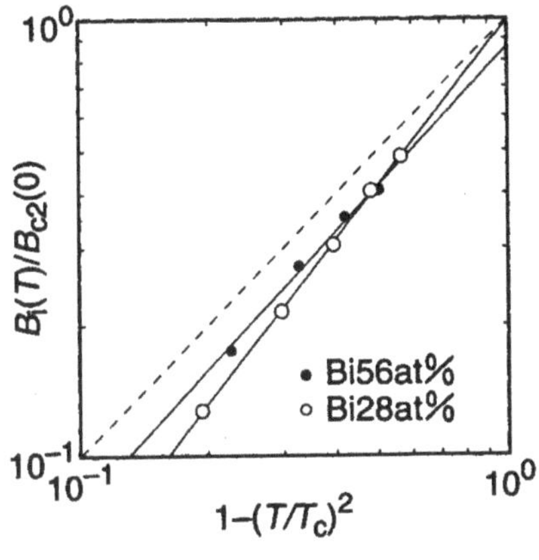

Fig. 1 Experimentally observed irreversibility fields of the two specimens.

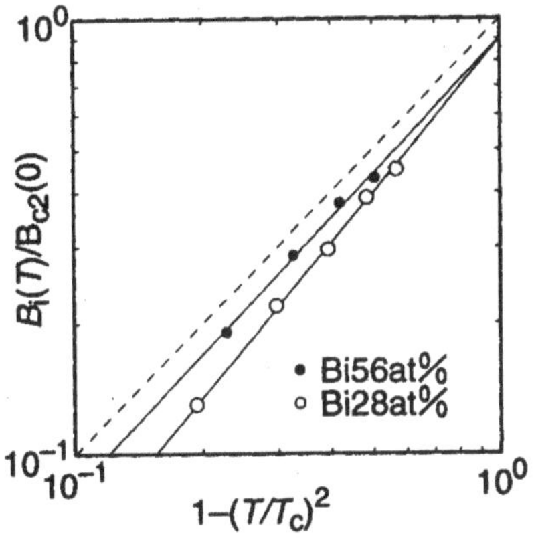

Fig. 2 Theoretically obtained irreversibility fields of the two specimens.

where U_0 is the pinning potential given by

$$U_0 = \frac{0.835 g^2 k_B J_{c0}^{1/2}}{\zeta^{3/2} B^{1/4}}. \tag{6}$$

Equation (6) leads to

$$B_i^{1/4} \ln(B_i a_f \nu / E_c) = \left(\frac{0.835 g^2 J_{c0}^{\frac{1}{2}}}{\zeta^{\frac{3}{2}} T} \right) \left(1 - \frac{\Delta J_c}{J_{c0}} \right). \tag{7}$$

The obtained theoretical results are shown in Fig. 2. It is found that the theoretical irreversibility field is also close to the upper critical field and does not depend largely on the flux pinning strength. This agrees with the experimental results shown in Fig. 1. It means that the flux creep theory explains well also the experimental results on metallic low temperature superconductors. The large difference between low and high temperature superconductors is the ratio of the pinning energy and the thermal energy. That is, the significant effect of the flux creep on the irreversibility field can be expected only for metallic superconductors with much weaker pinning strength than the Bi28at% specimen. However, further investigation is necessary for the irreversibility property in more weakly pinned superconductors in order to compare with the theoretical result in more detail.

4. Conclusion

The irreversibility lines in two superconducting Pb-Bi alloy specimens with largely different pinning strength are measured and compared with the result from the flux creep theory. The irreversibility field is found to be close to the upper critical field in the both specimens. This result can be explained by the theoretical result. For the further investigation it is necessary to measure the irreversibility field in more weakly pinned superconductors.

5. Acknowledgement

We would like to thank M. Kiuchi, K. Noda and N. Ihara for helpful discussion. We are also indebted to S. Yoshino for preparation of the manuscript.

6. References

1. T. Matsushita, T. Fujiyoshi, K. Toko and K. Yamafuji, *Apply. Phys. Lett.* **56** (1990) 2039.
2. T. Matsushita, *Proceeding of the 6th International Symposium on Superconductivity*, Hiroshima, 1993, p. 503.
3. T. Matsushita, A. Mastuda and K. Yanagi, *Physica C* **213** (1993) 477.
4. T. Matsushita, T. Nakatani, E. S. Otabe, B. Ni, T. Egawa, K. Kinouchi, S. Nozaki and S. Utsunomiya, *Cryogenics* **33** (1993) 251.

DEPENDENCE OF INTERACTION DISTANCE ON FLUX PINNING STRENGTH IN SUPERCONDUCTING Pb-Bi ALLOYS

Baorong NI

Department of Electronic Materials Engineering, Fukuoka Institute of Technology, Higashi-ku, Fukuoka 811-02, Japan

and

Tsugio HAMADA, Teruo MATSUSHITA

Department of Computer Science and Electronics, Kyushu Institute of Technology, Iizuka, Fukuoka 820, Japan

Tsuyoshi WAKUDA, Kazuo FUNAKI

Research Institute of Superconductivity, Faculty of Engineering, Kyushu University, Higashi-ku, Fukuoka 812, Japan

ABSTRACT

Ac inductive measurement on flux motion was carried out for two Pb-Bi alloy specimens with different critical current densities. The interaction distance, d_i, was evaluated based on the Campbell model. It was found that d_i depends more strongly on temperature in the specimen with lower critical current density, which is similar to that observed in high T_c superconductors. A discussion was given on the effect of flux creep on the flux pinning characteristics.

1. Introduction

It was recently found that the interaction distance, d_i, which gives a measure of the pinning potential radius, decreases with increasing termperature in Y-Ba-Cu-O superconductor[1]. This is deduced to result from the significant flux creep due to thermal agitation at high temperatures. Therefore, d_i is considered to depend on the flux pinning strength, i.e., stronger flux pinning will make a small dependence of d_i on temperature. In order to prove this consideration, measurements on a series of specimens with different pinning forces under the condition of similar superconducting characteristics are required. It is difficult to be realized for high-T_c superconductors, since the pinning characteristics cannot be controlled generally in fabrication process. In this study, we use several Pb-Bi alloy specimens, in which the critical current density, J_c, can be easily varied without large changes in other superconcuting characteristics. The magnetic field and temperature denpendences of J_c and d_i in Pb-Bi alloys with different flux pinning strength were estimated. The concept of flux creep due to thermal agitation was used for the explanation of d_i's temperature dependence.

2. Experimantal Detail

2.1 Specimen Preparation

Two specimens were prepared with a simple manner: Pb and Bi shots of about 2.0–5.0mm in diameter were mixed with the atomic composition of 28% Bi and 56% Bi, respectively. The mixtures were melted at 450℃ for 24 hours in a vaccumed glass tube, and then quenched in liquid nitrogen. The alloy with 28% Bi was annealed at about 100℃ for 7 days, while the alloy with 56% Bi was kept at the temperature of liquid nitrogen after the quench to prevent the room-temperature-aging. Slabs of about 0.5mm × 2.5mm × 10.0mm were cut out from the two alloys. The optical observation on specimen's microstructure confirmed the Pb-Bi phase diagram, i.e., the alloy with 28% Bi is homogeneous ϵ phase while the alloy with 56% Bi is an eutectic with fine and dense Bi precipitates in the matrix of ϵ phase.

2.2 Measurement and Numerical Calculation

The dependences of J_c and flux profile on magnetic field B and temperature T were measured using an ac inductive measurement (the Campbell method)[2,3], and the temperature dependence of upper critical field, $B_{c2}(T)$, was measured with a SQUID apparatus. A normal resistive method was employed for the mesurement of the flow resistivity, ρ_f.

The bulk J_c at 4.2K and 0.5T was higher than 10^8 A/m^2 in 56% Bi specimen, while it is about 10^6 A/m^2 in 28 % Bi specimen. Besides, the critical current density due to the surface pinning, J_{cs}, was found to be about 2 orders of magnitude larger than the bulk J_c in 28% Bi specimen. The Labusch parameter, α_L, which denotes the strength of flux pinning and the interaction distance, d_i, can be obtained from the experimental results. However, it is impossible to directly estimate the bulk d_i properly from the force-displacement characteristic, since the experimental infomation can be obtained only through the surface area and includes some contribution from the surface area necessarily. In order to eliminate the effect of surface area, we applied a numerical calculation of flux density distribution based on the Campbell model[3]. A differential equation is derived with this model as

$$\frac{d^2b}{dx^2} - \frac{b}{\lambda_0^2}\left(1 + \frac{1}{2\mu_0 J_c}\frac{db}{dx}\right) = 0, \tag{1}$$

where b is a variation in the flux density and λ_0 is the Campbell ac penatration depth, $\lambda_0 = (Bd_i/\mu_0 J_c)^{1/2}$. Figure 1 shows the flux distribution numerically obtained from Eq. (1). The effect of surface area of about 2.0 μm thick with large J_{cs} was taken into account. If we take $\zeta = a_f/d_i$, where a_f is fluxoid spacing, ζ can be treated as one of the adjustable parameters in solution of Eq. (1). Numerical flux profile and force-displacement characteristics are easily obtained by taking the spatial integral of $b(x)$. Figure 2 shows the temperature dependence of d_i which are specified by fitting the numerical flux profiles with the experimental ones.

3. Discussion

It is predicted theoretically that the value of ζ is about 4.0 – 2π, depending on the mechanism of the flux pinning, while experimental results for various materials give a wide range up to several tens. From this point of view, our result shows a reasonable values of ζ, since they vary from 8 to 14 for the 56% Bi specimen in the temperature range of 4.2K–7.2K and from 12 to 25 for 28% Bi specimen in 4.2K–6.5K, respectively. The difference between the two specimens seems to be caused by the different flux pinning mechanism. It is suggested by the fact that there are large amount of Bi precitipates in 56% Bi specimen, which may act as strong pinning centers.

Although d_i decreases with increasing temperature in both two specimens, the one with weaker flux pinning (28% Bi) shows a larger temperature dependence. This is

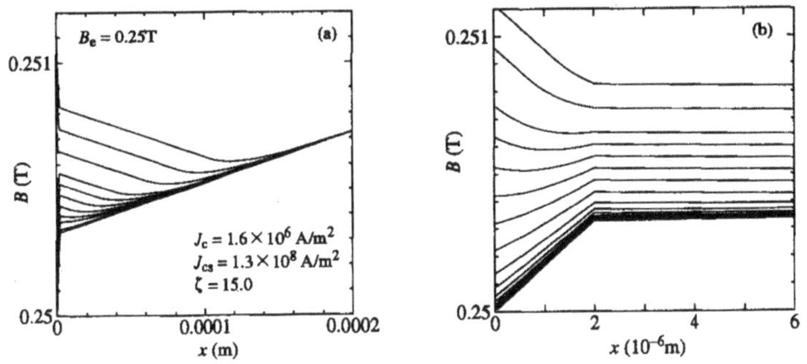

Figure 1: Flux distributions in superconducting slab obtaind from numerical calculation at $B = 0.25$T and $T = 5.6$K. The thickness of surface area was assumed as 2.0×10^{-6} m. The parameters of $J_c = 1.6 \times 10^6$ A/m^2, $J_{cs} = 1.3 \times 10^8$ A/m^2, $\zeta = 15.0$ were used. (a) is the overall figure and (b) the zoomed figure near surface area.

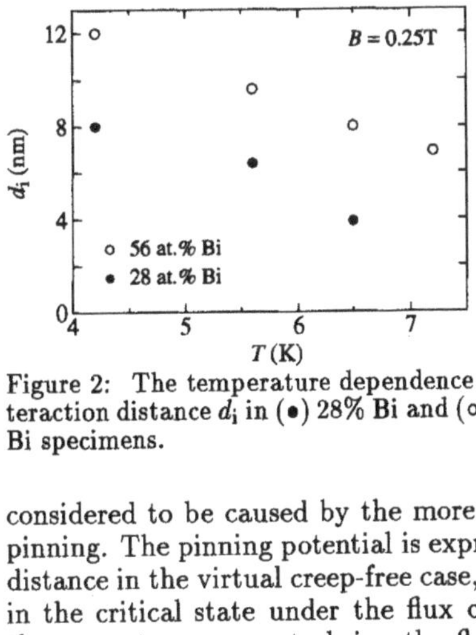

Figure 2: The temperature dependence of interaction distance d_i in (•) 28% Bi and (o) 56% Bi specimens.

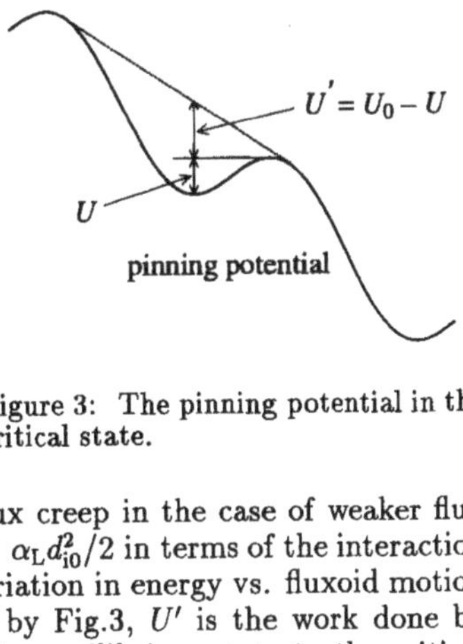

Figure 3: The pinning potential in the critical state.

considered to be caused by the more significant flux creep in the case of weaker flux pinning. The pinning potential is expressed as $U_0 = \alpha_L d_{i0}^2/2$ in terms of the interaction distance in the virtual creep-free case, d_{i0}. If the variation in energy vs. fluxoid motion in the critical state under the flux creep is given by Fig.3, U' is the work done by the current necessary to bring the fluxoids from the equilibrium state to the critical state. Hence, this quantity should be expressed as $\alpha_L d_i^2/2$ in terms of d_i observed experimentally. It is seen from Fig.3 that U' is given by $U_0 - U$, where U is the activation energy. If we assume that the pinning potential varies sinusoidally[4], then the equation of

$$U'(J_c) = U_0 \left[1 + j \cos^{-1} j - (1 - j^2)^{1/2}\right] = \frac{1}{2}\alpha_L d_i^2 \qquad (2)$$

is obtained, where $j = J_c/J_{c0}$ with J_{c0} denoting the critical current density in the creep-free case. J_{c0} is known to be given by some empirical scaling describing the

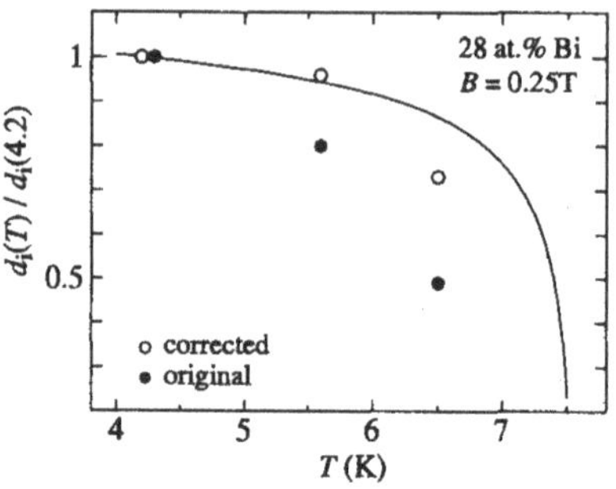

Figure 4: The temperature dependence of normalized d_i in 28% Bi specimen. The line shows the numerical result.

dependences on temperature and magnetic field. Then, we obtain $d_i(T)$ as

$$\frac{d_i(T)}{d_{i0}(T)} = \left[1 + j(T)\cos^{-1} j(T) - (1 - j^2(T))^{1/2}\right]^{1/2}. \tag{3}$$

Since $j(T)$ can be calculatated with the flux creep model[5] under the given J_{c0} and electric field criterion E_c, numerical calculation on temperature dependence of d_i can be carried out. The parameters of $T_c = 8.35$K, $B_{c2}(0) = 1.69$T, $\rho_n = 4.49 \times 10^{-7}\Omega$m, $E_c = 10^{-5}$V/m were assumed. Figure 4 shows the numerical calculation comparing with the experimental results denoted by solid symbols. We suppose that d_i in 56 % specimen is similar to d_{i0}, since the flux pinning is sufficient strong. This means that a primary temperature denpendence, as was shown in Fig.2 by open symbols, may appear in d_{i0}. If it is true, the temperature dependence of d_{i0} should be deducted from $d_i(T)$ of 28% Bi specimen in our estimation, since it seems to be disrelated to the flux creep. The corrected data are shown in Fig. 4 by open symbols, which show a closer agreement with our numerical results.

4. Conclusion

The interaction distances in two superconducting Pb-Bi specimens with different critical current densities were estimated by employing the Campbell model. It was found that d_i depends more strongly on temperature in the specimen with lower J_c, which is similar to that observed in high T_c superconductors. This is considered as the result of the significant flux creep. A numerical estimation on $d_i(T)$ was carried out. It turned out that our estimation is qualitatively consistent with the experimental results.

5. References

1. B. Ni, H. Kitagawa, T. Wakuda, T. Shigemitsu, K. Funaki, K. Yamafuji and T. Matsushita, *Jpn. J. Appl. Phys.* **32** (1993) L1215.
2. A. M. Campbell, *J. Phys.* C**2** (1969) 1492.
3. A. M. Campbell, *J. Phys.* C**4** (1971) 3186.
4. T. Matsushita and E. S. Otabe, *Jpn. J. Appl. Phys.* **31** (1992) L33.
5. P. W. Anderson and Y. B. Kim, *Rev. Mod. Phys.* **36**(1964) 39.

CRITICAL CURRENT DENSITY AND MICROSTRUCTURES OF V_3Si MULTIFILAMENTARY SUPERCONDUCTORS

T. Takeuchi, Y. Nemoto, Y. Iijima, M. Kosuge, K. Inoue and H. Maeda
Tsukuba Magnet Laboratories, National Research Institute for Metals
Tsukuba, Ibaraki 305, Japan

ABSTRACT

A flattened multifilamentary conductor was prepared, which induces a morphology and configuration change of the S-N interface, compared with that of a wire. Correlation between microstructures (grain boundary and S-N interface) and anisotropy in $F_{p,max}$ with respect to field direction indicates that the S-N interface acts as a dominant pinning center at least for composites with fine V filaments.

1. Introduction

Recently, a V_3Si multifilamentary superconductor with a large overall critical current density J_c (1.3×10^9 A/m^2 at 5 T), which is comparable to that of commercial bronze processed Nb_3Sn multifilamentary wires for ac use, has successfully been fabricated by a modified bronze process.[1-3] As for the flux pinning mechanism of V_3Si, fundamental studies have been intensively carried out on bulk samples which were irradiated with fast neutrons to induce various levels of point- and line-defect concentration.[4] In practical A15 superconducting wires, grain boundaries are, however, believed to be the major flux-pinning centers. From correlation of J_c with the reciprocal of grain size, Livingston indicated that grain boundary acts as the major pinning center, at least for the conventional bronze-processed V_3Si conductor with 20μm V cores.[5] However, J_c is much larger in the present V_3Si multifilamentary wires. Since the V filament size is one order of magnitude smaller than that of Livingston, there is a possibility of the enhancement of superconductor- normal metal (S-N) interface pinning, other than grain boundary pinning. The anisotropic feature in J_c with respect to the magnetic field direction for tape conductors gives us an information on flux pinning mechanism. The present study has thereby been made to check the S-N interface pinning contribution with anisotropy in J_c for flattened multifilamentary conductors.

2. Experimental

The samples used are the modified bronze processed V_3Si multifilamentary wire and tape. 82 short-pieces of a Cu-8.5at.%Si tube containing a V core were bundled, encased in a Ta tube, and drawn to a wire of 0.7 mm in diameter (V core diameter: 27 μm). Then, 134 pieces of the resulting single-stacked composite were assembled inside a Cu-8.5at.%Si alloy tube for wire drawing to the final V core diameter ranging from 1.2 to 7.4 μm. Two kinds of flattened multifilamentary conductors were prepared by rolling 0.7 mm diameter single-stacked and double-stacked composite wires with 27 and 1.9 μm V cores, respectively. Heat treatment was carried out at 700 - 850°C. Critical current I_c was defined as the current at which the second derivative of the V(I) transition reached a maximum. Overall J_c was defined as I_c/S, where S was the cross-sectional area of the composites excluding the two sheath materials at assembly. J_c was calculated by dividing I_c by the V_3Si cross-sectional area, which was estimated with backscattering electron microscope image, for samples with V core larger than 3.6 μm.

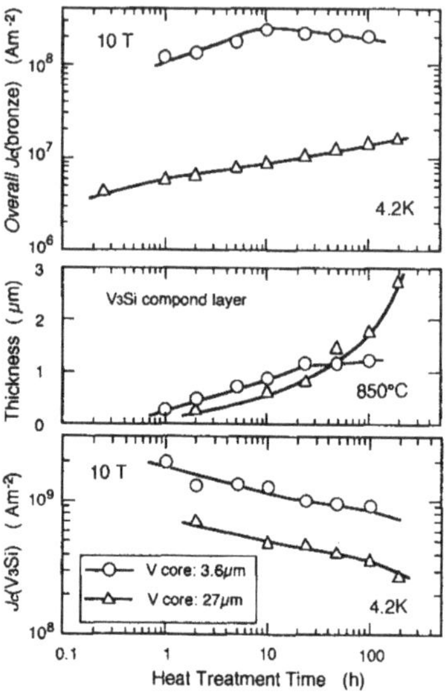

Fig. 1 The dependence of overall J_c at 10 T, thickness of V_3Si layer, and J_c at 10 T on heat treatment time at 850°C for composites with 3.6 and 27μm V filaments.

3. Result and Discussions

Fig. 1 shows the heat treatment time dependence of overall J_c, thickness of V_3Si layer and J_c. Reduction of filament size from 27 to 3.6 μm allows decomposition of V_5Si_3 and hence complete reaction to occur in much shorter times: the layer thickness of V_3Si

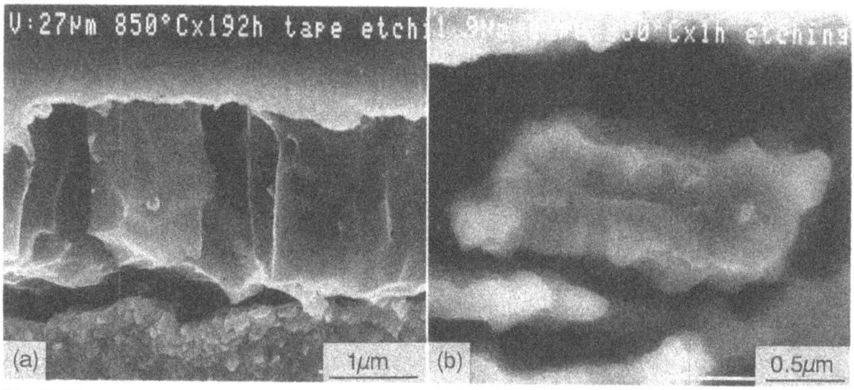

Fig. 2 The transverse cross-section for flattened (a) single-stacked (27μm V core) and (b) double-stacked (1.9μm V core) composites, heat treated at 850°C for 192 and 1 h, respectively.

is already saturated at 24 h. For 3.6 μm V core sample, overall J_c shows a peak against reaction time, despite the saturation of V_3Si layer thickness at long heat treatments. The discrepancy between their time dependence suggests that the grain boundary acts as a pinning center for fluxoids and the contribution of grain boundary pinning becomes small at long heat treatments due to grain growth. In actual, grain growth of V_3Si was confirmed with SEM, and J_c monotonously decreases with increasing reaction time accordingly.

There was a clear difference in the shape of the volume pinning force F_p vs. reduced field h curve between the large and small V core samples, suggesting the different pinning mechanism between them. Furthermore, F_p seems not to change appreciably in the high-field region, despite its remarkable enhancement in the low-field region caused by reducing the V core size. Although the saturation of F_p in high fields is reported for the bulk V_3Si with cascades, it is still unclear that the saturation occurs in the present V_3Si multifilamentary superconductors with different V core size.

In flattened multifilamentary conductors, V filaments were deformed into ribbon-like shapes and aligned parallel to the rolling plane. Fig. 2 shows the transverse cross-sectional structure after heat treatment at 850°C for a large and a small V core samples. Irrespective of V core size, columnar grains of V_3Si are grown perpendicularly to the rolling plane for both samples, as like for the conventional bronze processed Nb_3Sn tape conductor. For the large V core sample, [overall J_c x B] for magnetic fields perpendicular to the tape surface is actually larger than for field parallel to the surface as shown in Fig. 3(a), in agreement with the result for the Nb_3Sn tape where grain boundary acts as the major pinning center. However, anisotropy for the small V core sample is quite contrary to that for the large V core sample, as shown in Fig. 3(b). Reduction of V core size brings about an increase in S-N interface density and the drastic change in anisotropic feature of J_c. It is noted that reduction of V core size enhances not only overall J_c but also

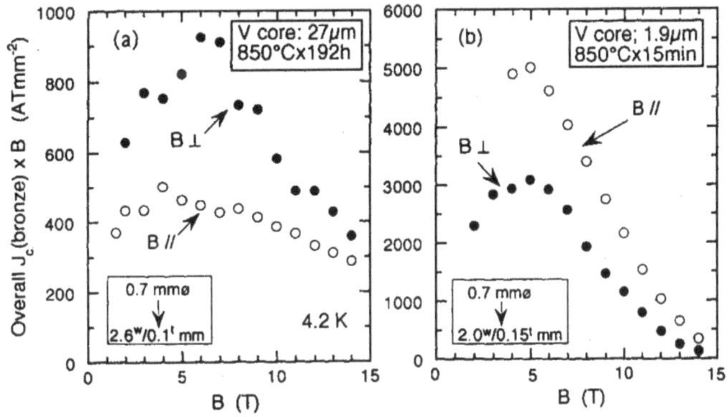

Fig. 3 The product of overall J_c by magnetic field as a function of magnetic field parallel or perpendicular to the conductor surface, for flattened (a) single-stacked (27μm V core) and (b) double-stacked (1.9μm V core) composites.

J_c (Fig. 1). Configuration of S-N interface is contrary to that of grain boundary: the density of S-N interface is higher in the direction perpendicular to the tape surface. Thus, if the contribution of the S-N interface pinning becomes larger than that of the grain boundary pinning, strongest pinning should occur when the fluxoids are parallel to the tape surface, in agreement with the result for the small V core sample.

Fig. 4 shows the dependence of anisotropy in $F_{p,max}$ on heat treatment time. Since the area of V_3Si layer has not been measured yet for tape conductors, [overall J_c x B]$_{max}$ (a maximum in the overall J_c x B vs. B curve) is plotted for reference. Anisotropy for the small-V-core sample decreases with increasing reaction time and approaches to unity, suggesting that the contribution of grain boundary pinning becomes large at long heat treatments. This is also the case with the large V core sample: at the early stage of the diffusion reaction, the large V core sample rather shows an anisotropy larger than unity, as well as the small V core sample.

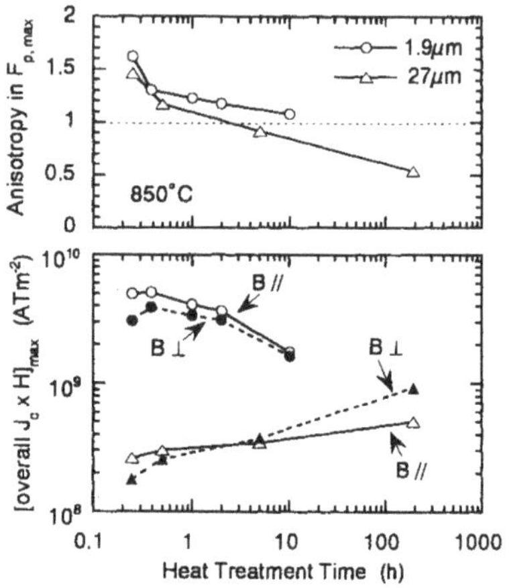

Fig. 4 Anisotropy in $F_{p,max}$ respective to magnetic field as a function of heat treatment time. A maximum in [overall J_c x H] vs. H curves is also plotted for reference.

In conclusion, an increment in S-N interface density is responsible for a remarkable increase in J_c and overall J_c, at least for the fine V filaments composites. Even for the large-V-core sample, the S-N interface was found to act as a major pinning center at the early stage of diffusion reaction, due to the extremely thin layer-thickness of V_3Si.

4. References

1. T. Takeuchi and K. Inoue, *J. Appl. Phys.* **74** (1993) 6454.
2. T. Takeuchi, K. Inoue, M. Kosuge, Y. Iijima and K. Watanabe, *Adv. Cryogenic Engine.* **40** (1994) 891.
3. T. Takeuchi, K. Itoh and K. Inoue, *Cryogenic Engine.* (Japanese) **29** (1994) 386.
4. R. Meier-Hirmer and H. Kupfer, *Phys. Rev.* **B 31** (1985) 183
5. J. D. Livingston, *Phys. Stat. Sol.* **(a) 44** (1977) 295

FABRICATION AND STRUCTURE OF HIGH-J_c Nb_3Sn WIRES WITH ALLOY ADDITIONS

B.P. MIKHAILOV and A.R. KADYRBAEV

Baikov Institute of Metallurgy, Russian Academy of Sciences, Leninsky prospekt, 49, Moscow, 117334/Russia

ABSTRACT

The addition of titanium or zirconium into niobium filaments results in a visible increase of Nb_3Sn layer thickness and keeps thin grain structure of the single and multifilament composite superconducting wires. The wires with only seven filaments showed the overall critical current density up to 10^4 A/cm^2 at 18 T and $5 \cdot 10^3$ A/cm^2 at 20 T. The rare metals (La, Ce) additions increased also the Nb_3Sn layer growth, decreased the grain size and improved the bronze deformability.

The greatest successes in the creation of multifilamentary Nb_3Sn wires have been achieved due to applying the "bronze technique" [1,2] based on the solid-state diffusion reaction between niobium and Cu-Sn bronze. The process allows one to create the wires with a very great amount of superconducting Nb_3Sn filaments (up to several dozens thousand filaments) with a diameter of about one micron and less. The critical current density in such wires per the total cross-sectional area achieved 10^5 A/cm^2 in a magnetic field of 5–10 T. However the modern level of the superconducting technology demands the further enhancement of the wire current capacity, especially in the magnetic field exceeding 15 T.

The wire current capacity depends on several factors: matrix-to-filament volume ratio; amount and dimension of filaments, their forms in the transverse and longitudinal cross sections; tin concentration in bronze matrix; modes and conditions of the heat treatment, etc. A promising way of the critical current density enhancement for Nb_3Sn wires is alloying of either niobium filaments or bronze matrix. The simultaneous alloying of both filaments and matrix is also promising .

The alloying has promoted achieving other important advantages: enhancement of the critical field and critical temperature of Nb_3Sn wires; enhancement of the Nb_3Sn layer growth rate; obtaining the Nb_3Sn layers with small grains; reduction of the sensitivity of superconducting properties to mechanical and other influences.

The set of alloying elements (zirconium, titanium, tantalum, hafnium, gallium at example) used up to now [3,4] is well known. However not all the problems concerning the optimization of the alloying element content, methods of introducing the alloying elements into Nb_3Sn, heat treatment regimes has been solved.

The aims of our work are: the study of the influence of zirconium, titanium, tantalum and rare earth additions on the Nb_3Sn layer growth kinetic in the wires with

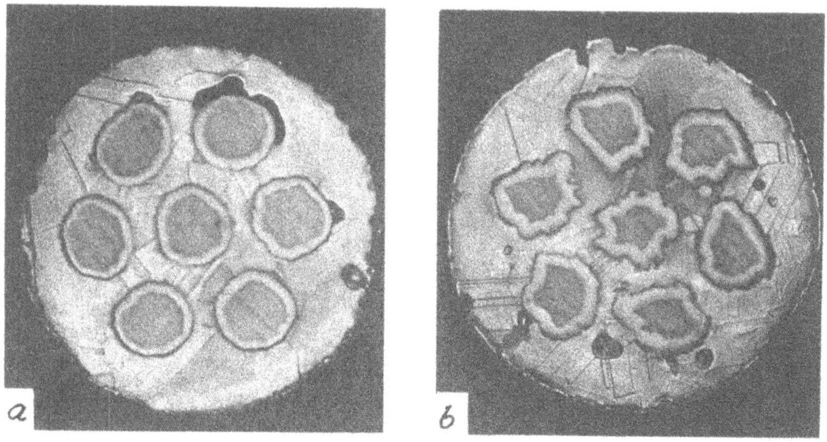

Figure 1. Cross-section of 7-filaments Nb_3Sn wires with outer diameter 0.4 mm. (*a*) non-doped wire, (*b*) 0.8 wt% Zr. ×160.

different filament amount (from 1 to 1615 filaments); the study of current capacity properties of Nb_3Sn wires in the magnetic field up to 20 T.

The selection of alloying elements is based on the fact that zirconium, titanium and tantalum enhance the niobium recrystallization temperature and do not decrease its critical temperature and increase the critical current in the magnetic field higher than 10 T. The lanthanum and cerium utilization has been connected with their strong modification effect on tin bronze and Nb_3Sn layers. The concentration limits of alloying additions into niobium were from 0.2 to 1.2 wt% Zr, from 0.4 to 2.0 wt% Ti and from 1.8 to 7.5 wt% Ta. The lanthanum and cerium were added in these dilution limits in tin bronze. The selection of mentioned concentrations was connected with the possibility of plastic deformation of composites.

The experiments has been performed on wires with diameters from 0.4 to 0.8 mm and different amounts of filaments. The heat treatment was performed in vacuum (10^{-5} millimeter of mercury) at the temperature from 680 to 750°C. The critical current measurements in the magnetic field (from 8 to 20 T) were performed in Kurchatov atomic energy institute, Moscow.

Microstructures of the cross-sections of 7-filament Nb_3Sn wires both unalloyed and alloyed with zirconium are shown in Figure 1. At the comparison it becomes evident that much more thicker Nb_3Sn layers were obtained on zirconium-doped filaments under the same conditions of heat treatment. The observation of cross sections of the wires (Fig. 2) shows that zirconium-doped wire grains of Nb_3Sn are smaller than those in non-doped wires. Alloying with titanium influenced the growth rate of Nb_3Sn

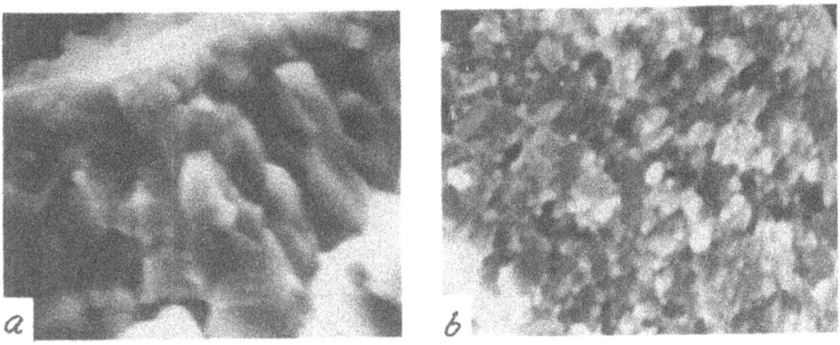

Figure 2. Microstructure of 7-filaments Nb_3Sn wire. (*a*) non-doped wire, (*b*) 0.8 wt% Zr. ×3000.

layers in the same way.

The results of measurements of the current flow characteristics of zirconium doped wires in strong magnetic fields (8–20 T) (Fig. 3) show that zirconium doped wires possess higher values of j_c. For instance, the critical current density of 7-filaments wire, containing 0.2 – 0.3 wt% Zr, in the 20 T magnetic field at 4.2 K was about $5 \cdot 10^3$ A/cm^2; for non-doped wire - $2 \cdot 10^2$ A/cm^2. Thus using a small quantity of filaments (7) zirconium-doped wires show the critical current density values quite sufficient for the application in high magnetic field systems (14–19 T).

Figure 4 shows the critical current density versus the external magnetic field for one-filament wires doped with different quantities of titanium and tantalum (j_c was counted on the Nb_3Sn layer volume). Comparing the curves one can conclude that in the relatively low fields (below 12 T) the critical current density of a non-doped wire is higher than that of the titanium-doped (0.97 wt% Ti) wire, but during the rise of the magnetic field (higher than 12 T) the critical current in the doped wire keeps at a high level in comparison with a low level for the non-doped sample.

Next, on the basis of the determined optimal titanium concentration in niobium filaments, 1615-filament wires with a diameter of 0.6–0.8 mm and length up to 300 m were produced in the collaboration with Electrotechnical Institute of Slovak Academy of Sciences, Bratislava. Beside of the superconductive filaments, the wires contain the pure copper stabilizing shell and the tantalum diffusion barrier. The shell and barrier occupy about 60% of the whole wire volume. The determination of the critical current in the 20-T magnetic field at 4.2 K showed the value of j_c up to $5 \cdot 10^3$ A/cm^2, and in the 18-T magnetic field - up to 10^4 A/cm^2.

In the bronze doped with La and Ce a considerable decrease of the bronze matrix grain size and the Nb_3Sn layer grain size (more than 10 times) is observed. The critical current measurements showed considerable increasing of j_c in the magnetic

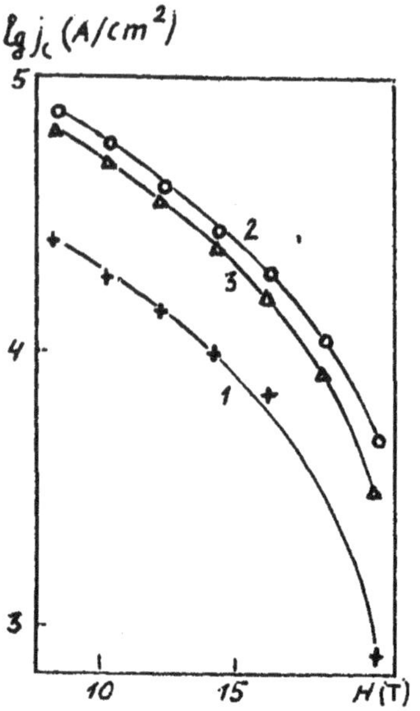

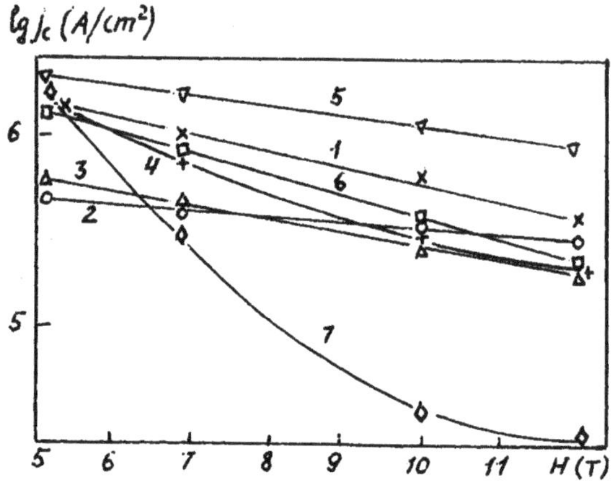

Figure 3. Overall critical current density versus external magnetic field for Nb_3Sn wires with diameter 0.4 mm. (*1*) non-doped wire, (*2*) 0.3 wt% Zr, (*3*) 0.8 wt% Zr.

Figure 4. Critical current density of Nb_3Sn layers versus outer magnetic field for Nb_3Sn wires. (*1*) non-doped wire, (*2*) 0.97 wt% Ti, (*3*) 1.2 wt% Ti, (*4*) 1.4 wt% Ti, (*5*) 1.97 wt% Ta, (*6*) 5.01 wt% Ta, (*7*) 6.69 wt% Ta.

fields up to 12 T, and after further magnetic field increasing the critical current falls as well as in the case of non-doped wires.

Thus, the conducted investigations have exhibited the fact that using doping it is possible to reach the constructive current density up to 10^4 A/cm^2 in 16–18-T magnetic fields. Our results confirm the possibility of considerable decreasing of the niobium filament amount in wires by using the dopands, that change the structure and growth kinetics of the diffusional Nb_3Sn layer.

1. G.A. Roberts and P.R. Roberts, Superconductors. UK patent No 1203291. (1966).
2. A.R. Kaufmann and I.I. Picket. *Bul. Amer. Soc.* **15** (1970) 839.
3. M. Suenaga, K. Tsuchija, N. Higuchi and K. Tachikawa, *Cryogenics* **25** (1985) 123.
4. B.P. Mikhailov, P. Kovach, P. Hutka et al, *Pisma v GTF* **16** (1990) 22.

FABRICATION OF Nb_3Al SUPERCONDUCTING WIRE BY OHMIC-HEATING REACTION METHOD

S.S. Oh, D.W. Ha, K.S. Ryu
Applied Superconductivity Lab. Korea Electro-technology Research Institute,
P.O.Box 20, Changwon 641-600, Korea
and
J.Y. Shim, K.W. Kim, H.K. Baik
Dept. of Metallurgical Engineering, Yonsei University, Shinchon-dong,
Seodaemun-ku, Seoul 120-749, Korea

ABSTRACT

Two kinds of composite wires have been fabricated for ohmic-heating reaction method. The one was powder processed composite from fine powders consisting of Nb_3Al and Nb_2Al which were obtained from plasma arc melting and pulverization. The other was tube processed one from Nb/Al-Mg solid composite and multi filamentary Nb/Al-Mg composites have been prepared for ohmic-heating reaction method. Ohmic-heating for high temperature heat treatment has been carried out by passing a d.c. current for short time. It was made clear that Nb_3Al phase can be formed during ohmic heating reaction and superconducting properties depend on the condition of supplied power.

1. Introduction

Nb_3Al superconducting compound has been considered to be a strict material for wire fabrication because formation of stoichiometric A 15 phase is difficult below the temperature of 1800°C and Cu reacts with Nb and Al, then forms non superconducting compound during heat treatment. So far, however, various fabrication processes such as Jelly Roll[1], P/M [2] and tube processes[3] using solid state reaction have been developed for Nb_3Al superconducting wire by many researchers. This was attributed to that the Nb_3Al phase can be formed at low temperature by extremely reducing a Nb/Al diffusion distance [4]. Therefore, high areal reduction in coldworking of composite, is required to obtain a high critical current density in these processes.

Ohmic heating reaction method has an advantage that a high reaction temperature can be easily obtained within short time by passing a current through a specimen. According to this, grain coarsening of Nb_3Al can be expected to be suppressed.

As a feasibility research, this study is intended as an investigation of Nb_3Al compound formation in ohmic heating reaction. When the ohmic heating conditions were varied for powder processed and tube processed composite, respectively, microstructure dependence of superconducting properties has been investigated.

2. Experimental

For powder process, the ingots prepared using plasma arc melting were roughly crushed into size of 1 ~ 3 mm and finely pulverized by high speed planetary ball mill in vacuum and grinding in alumina pestle. The obtained powders were compacted into a Nb tube, 10 mm in outer diameter and 7 mm in inner diameter. The powder compacted composites were swaged to 3 mm round rods and grooved rolled to about 1 mm square rods and then flat rolled to the tapes of various size. On the other hand, Al-10at%Mg alloy ingots were prepared using induction melting for tube process . The Al-Mg ingots were swaged to 5 mm round rods, and then inserted into Nb tube. The obtained composites were also deformed to round rods and groove rolled to square rods and bundled and inserted into Nb tube for stacking and this process was repeated two times for making a multi filamentary tape and finally rolled to tape specimen. Total number of filaments for tube processed specimen is 594. The final size of powder processed specimen was 0.4 mm in thickness and 3.5 mm in width and that of tube processed specimen was 0.36 mm in thickness and 2.1 mm in width. The areal reduction ratio of composites in cold working was 100.

After cold working, the composite tapes were ohmic heated with various conditions. The specimens of 70 mm in length were mounted on copper electrode clamps. After the reaction-chamber was pumped out up to ~ 10^{-6} torr vacuum , D.C. current began to flow firstly at higher level for 1 sec and dropped and maintained at some value as typically shown in figure 1. Since cooling water was flowing through Cu electrode continuously during operation, specimens just after reaction were rapidly cooled when the current transport was finished. The tube processed and powder processed specimens after ohmic heating were heat treated at 700 and 750°C, respectively for 72 hours in vacuum atmosphere.

T_c and I_c were measured with the definition of zero resistance and 1μV/cm, respectively by 4 terminal resistive method.

3. Result and Discussion

Observing with optical microscope, the starting powder size after final pulverization was confirmed to range from 0.5 to 5 μm.

Figure 2 shows the X-ray diffraction patterns of starting powder(a) for packing into Nb tube and the powder processed specimen(b) after ohmic heating , respectively. Ohmic heating was performed by transporting the current at 47 A for 19 sec and temperature of specimen was measured to about 1590°C. From the pattern(a), it can be confirmed that Nb_2Al, σ phase and Nb_3Al, A 15 phase are major phases for the specimen which was plasma arc melted. On the other hand, powder processed and ohmic heated specimen exhibits the larger intensity peaks from Nb_3Al and Nb phase and remarkably weakened intensity peaks from Nb_2Al. According to this, most of Nb_2Al existed in starting powder is considered to transform into Nb_3Al by ohmic heating reaction.

Figure 3 shows the microstructure of the powder processed(a) and tube processed specimen(b), respectively after ohmic heating . Compacted core was observed to be dense due to effect of sintering during ohmic heating reaction in the center part of powder processed specimen and this microstructure was differently observed every section along a

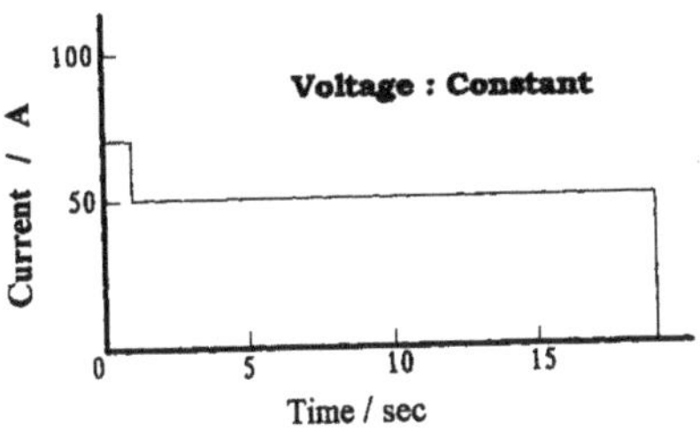

Fig.1 Input current pattern in ohmic heating process.

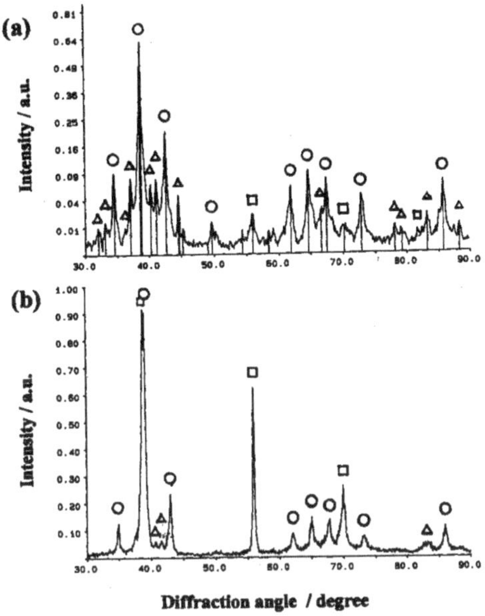

Fig.2 X-ray diffraction pattern of starting powder(a) and powder processed specimen(b) after ohmic heating. O, Δ and □ refer to Nb_3Al, Nb_2Al and Nb phase, respectively.

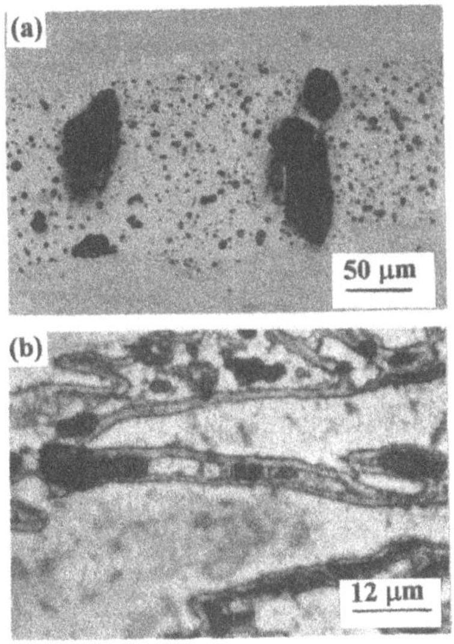

Fig.3 Optical micrographs of powder (a) and tube processed specimen(b), respectively after ohmic heating.

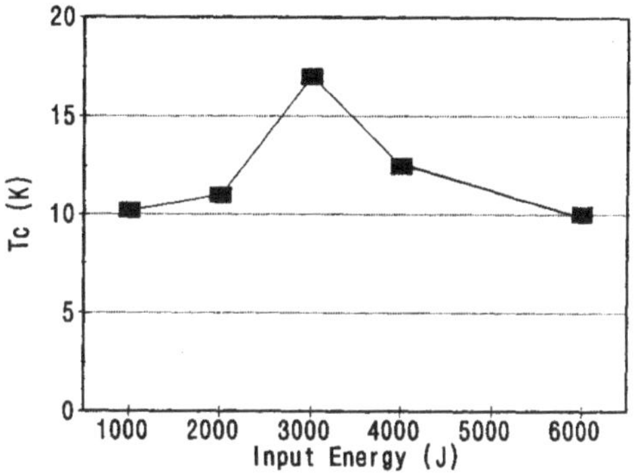

Fig.4 T_c vs input energy in ohmic heating for powder processed specimen.

specimen length. This is thought to be mainly resulted from temperature distribution. The temperature of specimen by ohmic heating increases with approaching the center part. Compound like layer, formed between Nb and Al-Mg layer, could be confirmed in tube processed specimen after ohmic heating and post heat treatment. In some tube processed specimens, the center part swelled out by over ohmic heating. Thus, optimization of input power was thought be important for tube processed specimen.

Figure 4 shows relationship between T_c and input energy in ohmic heating process for the specimens from powder process. T_c increases and reaches maximum at around 3.0 kJ and decreases with increasing input energy. T_c was confirmed to be improved within a narrow region of input energy. The powder processed specimen only ohmic heated with input energy of 3.4 kJ without post heat treatment had a T_c of 13.9 K. Therefore, T_c was confirmed to increase by post heat treatment. This is thought to be resulted from the improvement of stoichiometry of A 15 phase by heat treatment for long time.

Table 1 T_c and I_c (4.2 K) for the various ohmic heated specimens of powder(P) and Tube(T) process

Specimen	Input Energy(kJ)	T_c (K)	I_c(A) at 10 T
P1-N	3.57	12.5	110
P2-N	3.42	13.9	> 470
P3-C	1.0	10.2	> 470
T1-B	0.72	14.5	> 470
T2-B	4.48	< 10	—

Post heat treatment condition
N : No, B : 700°C x 72h, C : 750°C x 72h

As shown in table 1, I_c was exhibited with high value in the powder(P) processed specimen without post heat treatment. This is thought to be due to that a dense A 15 phase can be obtained from starting powders consisting of Nb_3Al and Nb_2Al phase by ohmic heating.

4. Conclusion

Ohmic heating reaction method was confirmed to be effective for Nb_3Al superconducting wire fabrication from composites prepared with a low areal reduction. A dense Nb_3Al phase can be formed within very short time for powder processed specimen and its I_c turned out to be very high. T_c could be improved by post heat treatment at 700~750°C for 72 h.

5. Acknowledgments

We are very grateful to Messes. M.H. Choi and T.H. Han for their help in specimen preparation and measurement. This work is part of the research performed by the support of the KEPCO(R.O.K)

References

[1] R. Bruzzese et al. : IEEE Trans on Mag, MAG-23(1987)653.
[2] R. Akihama et al. : IEEE Trans on Mag., MAG-17(1981)274.
[3] T. Takeuchi et al. : IEEE Trans on Mag, MAG-25(1989)2068.
[4] B. Annaratone et al. : IEEE Trans on MAG-17(1981)1000.

CRITICAL CURRENT DENSITY OF LIQUID-QUENCHED $YBa_2Cu_3O_{7-y}$

S. Fujino*, T. Ozaki*, K. Itoh*, K. Aoki*, A.Yazawa*, K. Oguri** and Y. Nishi**
*Student of Materials Science, Tokai University, 1117, Kitakaname,
Hiratuka, Kanagawa 259-12, Japan.
**Department of Materials Science, Tokai University,1117,Kitakaname,
Hiratuka, Kanagawa 259-12, Japan.

ABSTRACT

The temperature dependence of J_c just below offset temperature of superconducting transition is observed in weak magnetic filed. The J_c values of liquid-quenched sample annealed is higher than that of sintered sample just below offset temperature of superconducting transition.

1. INTRODUCTION

The J_c values of high T_c superconductors are limited by weak link at grain boundary near offset temperature of superconducting transition. In order to achieve high J_c values, many works have been reported[1-6]. The liquid-quenching enhances the density of sample, which may improve the superconducting performance. The liquid-quenching is one of the possible method to modify the weak link at grain boundary. Thus, we investigate the effects of liquid-quenching on J_c of annealed $YBa_2Cu_3O_{7-y}$ samples in weak magnetic fields.

2. EXPERIMENTAL PROCEDURE

Samples with nominal composition $YBa_2Cu_3O_{7-y}$ were prepared from high purity powders of Y_2O_3, $BaCO_3$ and CuO. The powders were mixed and reacted at 1210 K for 2 h and then cooled under pure oxygen. After crushing, a palletized tablet 1 mm thick and 13 mm diam, was sintered under pure oxygen at 1210 K for 8 h and furnace cooled. Liquid-quenched foils were prepared from chips cut from sintered tablets, using a twin-type piston-anvil apparatus shown in Fig. 1 [7-10].

The sample was annealed under pure oxygen at 1210K for 8h and furnace cooled. The measurement of J_c was performed using a standard 4-probe technique. The temperature was measured by Au∗Fe-normal silver (Ag-0.37at%Au) thermocouple attached to the specimen in a cryostat at equilibrium temperature. The structure of the samples was examined by means of the x- ray diffraction.

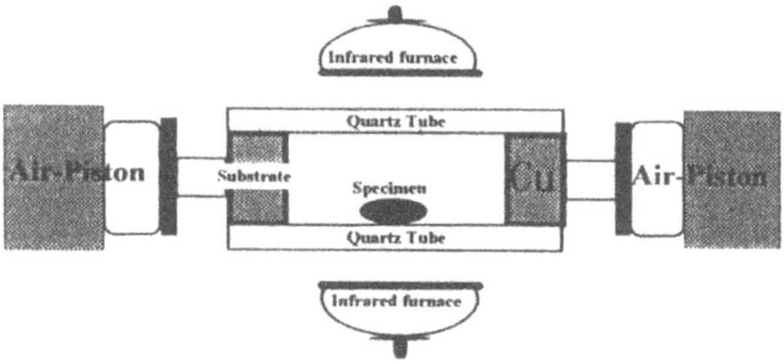

Fig 1. Schematic diagram of twin-piston-anvil apparatus

3. RESULTS AND DISCUSSIONS

Figure 2 shows the temperature dependence of J_c just below offset temperature of superconducting transition for liquid-quenched sample annealed and sintered sample in weak magnetic fields.

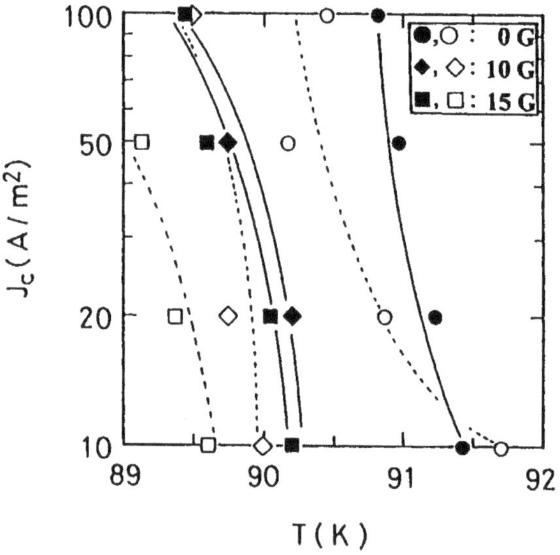

Fig 2. Temperature dependence on J_c of liquid-quenched sample annealed (●, ◆, ■) and sintered sample (○, ◇, □) in weak magnetic fields

The solid line is for the liquid-quenched sample annealed (●,◆, ■). The broken line is for the sintered sample (○, ◇, □). These samples show the temperature dependence of J_C under selected magnetic fields (0G, 10G, 15G). The J_C values of liquid-quenched sample annealed is higher than that of sintered sample just below offset temperature of superconducting transition in weak magnetic fields.

Figure 3 shows laser microscope photographs of surface for the liquid quenched sample annealed(a) and sintered (b) samples. Although, the contact area of the grain is small of the sintered sample, the intergrain zone is observed of the liquid-quenched sample annealed. Since the zone enhances the contact area, the liquid-quenching is one of the possible method to modify the weak link at grain boundary.

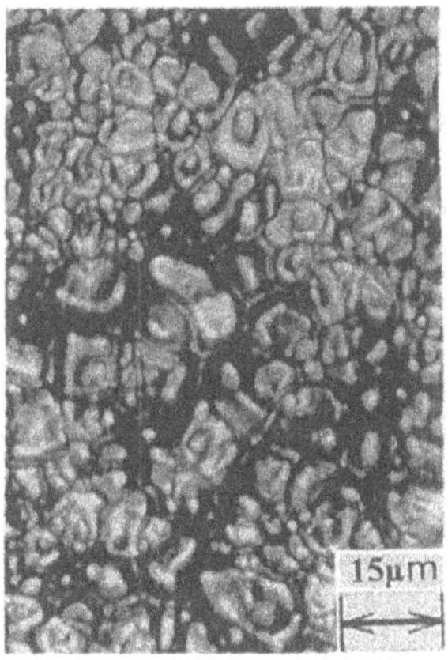

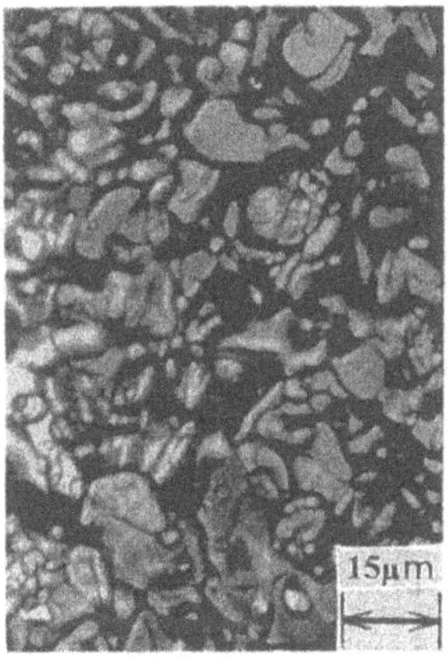

(a) Liquid-quenched sample annealed (b) sintered sample

Fig 3. Laser microscope photographs of surface for liquid-quenched sample annealed (a) and sintered sample (b)

4. CONCLUSION

In summary, the temperature dependence of J_C just below offset temperature of superconducting transition is observed in weak magnetic filed. The J_C values of liquid-quenched sample annealed is higher than that of sintered sample just below the offset temperature of superconducting transition in weak magnetic field.

5. ACKNOWLEDGMENTS

Authors would like to thank Ms. Megumi Kobori of Lasertec Co. Ltd. for taking the Laser microscope photographs.

6. REFERENCE

1. S. Jin, T. H. Tiefel, R. C. Sherwood, R. B. van Dover, M. E. Davis, G. W. Kammlott, and R. A. Fastnacht. *Phys. Rev. B* 37, (1988) 7875.
2. S. Jin, T. H. Tiefel, R. C. Sherwood, R. B. van Dover, M. E. Davis, G. W. Kammlott, and R. A. Fastnacht, and H. D. Keith.. *App. Phys. Lett.* 52(1988) 2074 .
3. K. Yamamoto, M. Murakami, H. Fujimoto. N. Koshizuka and S. Tanaka. *J. Advan. Sci.* 4 (1992) 54
4. T. Asaka, Y. Shiomi and K. Tachikawa, *J. Jpn. Inst. Met*, 56, (1992) 722.
5. J. W. Ekin and T. M.Larson. *Phsica. C.* 160. (1989) 489.
6. K. Tachikawa, T. Inoue and Y. Hikichi, *Supercond. Sci. Technol*, 5. (1991) 386.
7. Y.Nishi, Y. Kita, A. Igarashi and S, Tokunaga, *J. Appl. Phys*, 65, (1988) 3972.
8. Y.Nishi, K. Suzuki and T. Masumoto, *J. Jpn. Inst. Met*, 45, (1982) 818.
9. Y. Nishi, H. Harano, H. Ishizuki, M. Kawakami and E. Yajima, *Mater. Sci. Eng.* 98, (1988) 505.
10. Y. Nishi, S. Moriya and T. Manabe, *J.·Appl. Phys.* 65, (1989) 2389.

TEMPERATURE DEPENDENCE OF J_c IN HIGH T_c SINGLE PHASE

K.Aoki*, S.Ichimura*, S.Fujino*, N.Hirai*, K.Oguri** and Y.Nishi**.
*Student of Materials Science, Tokai University, 1117, Kitakaname, Hiratuka, Kanagawa, 259-12, Japan.
**Department of Materials Science, Tokai University, 1117, Kitakaname, Hiratuka, Kanagawa, 259-12, Japan.

ABSTRACT

The abnormal J_c increasing against temperature has been found from 50 to 80 K in $Bi_{1.6}Pb_{0.4}Sr_2Ca_2Cu_3O_x$. The sample has the low T_c phase, which converts to the normal state from 50 to 65 K. If fine particles of the low T_c phase exist, they act as additional pinning sites. Since the additional flux pinning sites enhance J_c above 50 K, the abnormal J_c change from 50 to 80 K can be explained. The small amount of Sb addition increases the volume of high T_c phase easily. In addition, the Sb addition enables to obtain nearly the high T_c single phase. The abnormal J_c value is not remarkably found in the high T_c single phase, although the high J_c value is found above 80 K for the high T_c single phase.

1. INTRODUCTION

T_c above 100 K has been reported for the Bi-Pb-Sr-Ca-Cu-O[1-3]. From an engineering point of view, it is important to investigate the superconducting critical current density (J_c). We have recently found that an abnormal J_c change with temperature has been found from 50 to 80 K in $Bi_{1.6}Pb_{0.4}Sr_2Ca_2Cu_3O_x$. A possible explanation is suggested. The sample has the low T_c phase, which converts to the normal state from 50 to 65 K. If fine particles of the low T_c phase exist, they act as additional pinning sites. Since the additional flux pinning sites enhance J_c above 50 K, the abnormal J_c change from 50 to 80 K can be explained. Although the J_c value is mainly dominated by the weak link model, the excess pinning model is one of the possible models to explain the abnormal J_c change[4]. The small amount of Sb addition obtains the high T_c phase easily. Thus, the temperature dependence of J_c is investigated for the Sb added Bi-Pb-Sr-Ca-Cu-O.

2. EXPERIMENTAL PROCEDURE

Samples with nominal composition were prepared by high purity powders of Bi_2O_3 (99.99 mass %), PbO (99.99 mass %), Sb_2O_3 (99.9 mass %), $SrCO_3$ (99.9 mass %), $CaCO_3$ (99.99 mass %), and CuO (99.9 mass %). The powders were mixed and sintered in air at 1023 K for 6 h and then air-cooled. After crushing, the sintered powders were resintered in air at 1073 K for 18 h and then air-cooled. After crushing, a pelletized tablet, 1 mm thick and 13 mm diameter, was sintered in air at 1123 or 1103 K for 100 h and then furnace-cooled. The deformation was performed by the stress (1.0 t/cm^2). The deformed sample was annealed for 1123 K for 100 h and then furnace-cooled.

The J_c was determined by an electrical resistivity measurement using a standard four-probe technique and a Keithley 182 nanovoletmater[5]. The superconducting critical current (I) was defined as the currents to achieve a voltage equivalent to 1 μ V/cm. The temperature was measured by Au*Fe-normal silver thermocouple attached to the specimen in a cryostat. Using the elasticity of the thermocouple, we attached the thermocouple to the sample surface.

The structure of the sample was examined by means of x-ray diffraction.

3. RESULTS AND DISCUSSIONS

Figure 1 shows the temperature dependence of J_c for the samples. The sample A (○) is sintered at 1123 K for 100 h and then the deformed sample is annealed for the $Bi_{1.6}Pb_{0.35}Sb_{0.05}Sr_2Ca_2Cu_3O_x$. The sample B (●) is sintered at 1123 K for 100 h for the $Bi_{1.6}Pb_{0.4}Sr_2Ca_2Cu_3O_x$. The sample C (△) is sintered at 1103 K for 100 h for the $Bi_{1.6}Pb_{0.4}Sr_2Ca_2Cu_3O_x$. The J_c values of all the samples decrease against temperature below 50 K, as usual. Although the J_c of the sample C can be observed below 60 K. The abnormal J_c change can be detected from 50 to 80 K for the sample B. The sample A decreases the abnormal J_c value from 50 to 80 K. The J_c value of the sample A above 80 K is larger than that of the sample B.

Figure 2 shows x-ray diffraction patterns of the samples. The sample A enables to obtain the high T_c single phase[6]. The sample B is a mixture of the low T_c phase[7] and the high T_c phase. The sample C enables to obtain the low T_c single phase.

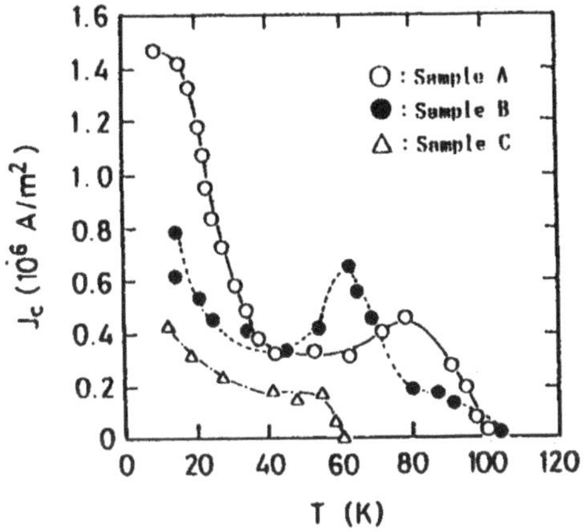

Figure. 1 Temperature dependence of J_c for samples. Sample A (○) is sintered at 1123 K for 100 h and then the deformed sample is annealed for $Bi_{1.6}Pb_{0.35}Sb_{0.05}Sr_2Ca_2Cu_3O_x$. Sample B (●) is sintered at 1123 K for 100 h for $Bi_{1.6}Pb_{0.4}Sr_2Ca_2Cu_3O_x$. Sample C (△) is sintered at 1103 K for 100 h for $Bi_{1.6}Pb_{0.4}Sr_2Ca_2Cu_3O_x$.

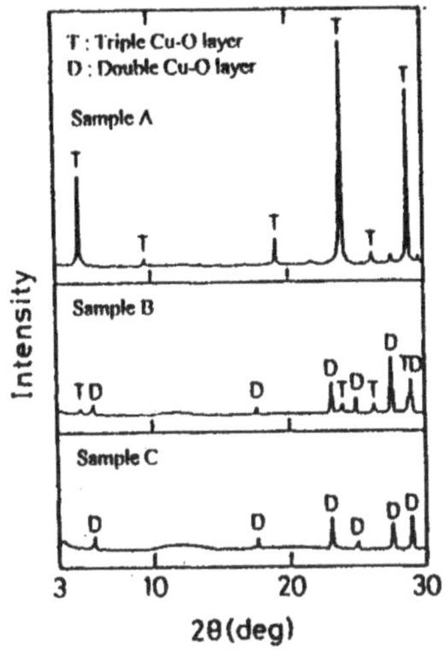

Figure 2 x-ray diffraction patterns of samples. Sample A enables to obtain high T_c single phase. Sample B is a mixtures of low T_c phase (double Cu-O layer) and high T_c phase(triple Cu-O layer). Sample C enables to obtain low T_c single phase.

Figure 3 shows schematic drawing of the phase transition for samples. The single phase samples (sample A and C) don't have the abnormal J_c change. The sample B is a mixture of the high T_c phase and the low T_c phase. The mixed sample (sample B) has the abnormal J_c change between 50 K and 80 K. The J_c value is mainly dominated by the weak link. The high T_c softens below T_c [8], whereas the low T_c phase doesn't soften above 60 K. If the high T_c phase is squeezed by the low T_c phase, the good contact condition can be formed. Namely, the abnormal high J_c value (sample B) can be explained by the modified weak link.

The presence of a magnetic field when an electric current having are due to Oersted (In 1820)[9]. Ampere, Biot and Savart presented the results of a series of experiments that may be generalized and expressed in modern mathematical language[9]. Using the current induced magnetic field (see *Figure* 3), we suggest another possible explanation.

The transition of low T_c phase is from 50K to 80K. If fine particles of the low T_c phase exist, they act as additional pinning sites. Since the additional flux pinning site enhances the J_c value between 50 K and 80 K, the abnormal J_c change from 50 to 80 K can be explained. Namely, it is the excess pinning model. Although the J_c value is mainly dominated by the weak link, we suggest that the excess pinning model is one of the possible models to explain the abnormal J_c change.

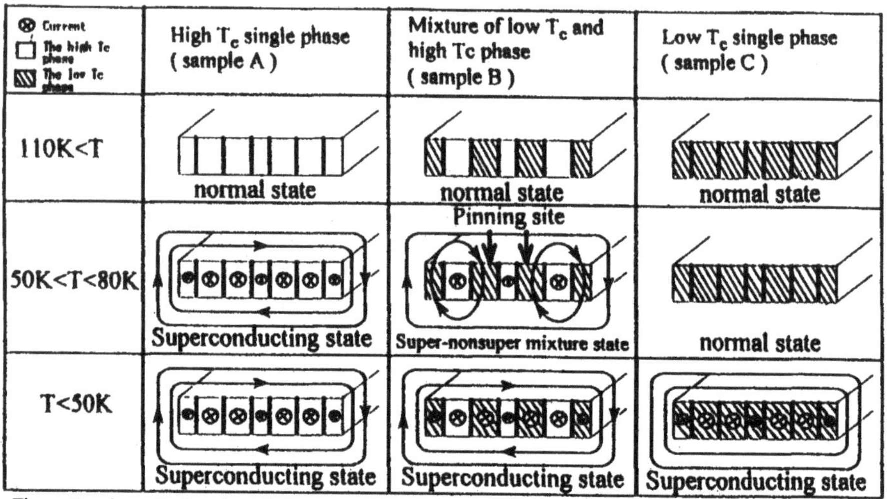

Figure 3 schematic drawing of phase transition for sample A (high T_c single phase), B (mixture) and C (low T_c single phase).

4. CONCULUTION

In summary, the temperature dependence of the superconducting critical current (J_c) is investigated of $Bi_{1.6}Pb_{0.35}Sb_{0.05}Sr_2Ca_2Cu_3O_x$. The small amount of Sb addition increases the volume of high T_c phase easily. In addition, the deformation and annealing enable to obtain nearly the high T_c single phase. Abnormal J_c change is not remarkable in the high T_c single phase sample, although the high J_c value is found above 80 K for the high T_c single phase.

5. Reference

[1]Y.Nishi, S.Moriya, and T.Manade, *J. Appl. Phys.* **65**, (1989) 2389.

[2]Y.Nishi, K.Oguri, H.Ohinata, K.Tanioka, Y.Kita, and N.Ninomiya, *Phys. Rev. B* **41**, (1990) 6520.

[3]Y.Nishi, Y.Kita, and K.Tanioka, *J. Mater. Sci.* **25**, (1990) 5105.

[4]Y.Nishi, K.Nozaki, and S.Ichimura, *J. Appl. Phys.* **71**, (1992) 5569.

[5]Y.Nishi, S.Moriya, and S.Tokunaga, *Phys. Lett. A* **126**, (1987) 55.

[6]U.Endo, S.Koyama, and T.Kawai, Jpn. *J. Appl. Phys.* **27** (1988) L1476.

[7]M.Onoda, A.Yamamoto, E.Takayama-Muromachi, and S.takekawa, *Jpn. J. Appl. Phys.* **27** (1988) L833.

[8]G.Cao, Y.Li and R.Ma *Physica C.* **205** (1993) 133.

[9]S.G.Starling,B.Sc.,A.R.C.Sc.,F.Inst.P.,"ELECTRICITY AND MAGNETISM" (LONGMANS, GREEN AND CO, London, 1925) p.51

CHANGES IN T_c AND J_c OF LIQUID-QUENCHED $Bi_{1.6}Pb_{0.4}Sr_2Ca_2Cu_3O_x$ AGAINST NUMBER OF TREATMENT TIMES OF COLD-PRESSING AND ANNEALING

Y. KAWAGUCHI, T. AIDA, K. SHIRAISHI, K. FUKUSHIMA and Y. NISHI*

Student of Materials Science, Tokai University, 1117 Kitakaname, Hiratsuka Kanagawa 259-12, Japan.

*Department of Materials Science, Tokai University, 1117 Kitakaname Hiratsuka, Kanagawa 259-12, Japan.

ABSTRACT

Superconducting properties may be affected by the sample density. The density of the liquid-quenched sample is obviously higher than that of the sintered sample. Furthermore, cold-pressing is also useful method to increase the density. Thus, influences of the cold pressing before final annealing on the T_c^{off} and J_c values are studied for the liquid-quenched $Bi_{1.6}Pb_{0.4}Sr_2Ca_2Cu_3O_x$ after the crystallization.

As a result, we found that the treatment (cold-pressing and annealing) was obviously effective process in increasing the T_c and J_c values. The 2nd treatment increases the T_c and J_c values.

1. INTRODUCTION

Since the melting points of high-T_c Bi oxide is low, it is easy to be melted and liquid-quenched. The density of the liquid-quenched sample is obviously higher than that of the sintered sample, which may improve their superconducting performance [1-3].

From an engineering point of view, it is important to know the influence of the cold-pressing on superconducting properties. Although the compressive stress improves the superconducting performance [4, 5], the influence of the cold-pressing process has not been studied, precisely. Thus, effects of the treatment (cold-pressing and annealing) on T_c and J_c are investigated for the liquid-quenched $Bi_{1.6}Pb_{0.4}Sr_2Ca_2Cu_3Ox$ crystallized.

2. EXPERIMENTAL PROCEDURE

Samples with nominal composition were prepared from high-purity powders of CuO (99.9 mass %), Bi_2O_3 (99.99 mass %), PbO (99.9 mass %), $SrCO_3$ (99.9 mass %) and $CaCO_3$ (99.99 mass %). The powders were mixed and reacted in air at 1023 K for 6 h and then air-cooled. The reacted powders were resintered in air at 1073 K for 18 h and then air-cooled. After crushing, a pelletized tablet, 1 mm thick and 13 mm in diameter, was sintered at 1123 K for 12 h and then furnace cooled in air. Liquid-quenched glassy foils were prepared from chips cut form the sintered tablets using a twin-type piston-anvil apparatus [6, 7]. The liquid-quenched sample was annealed at 1123 K for 26 h in air and then furnace cooled.

The cold-pressing was performed for the sample at 49 MPa. The cold-pressed sample was finally annealed at 1123 K for 2 h and then furnace cooled in air. The sample was performed by the treatment (cold-pressing and annealing).

The electrical resistivity was measured using a standard four-probe technique and a Keithley 181 nanovoltmeter. The temperature was measured by an Au*Fe/chromel thermocouple attached to the specimen in a cryostat at equilibrium temperature. The offset T_c of the transition is defined as the temperature where zero-resistivity is reached (below 10^{-9} Ω m at 2×10^3 A/m^2 current density). The J_c under earth field was determined by an electrical resistivity measurements using a standard four-probe technique and a Keithley 181 nanovoltmeter [8, 9]. The superconducting critical current (I) was defined as the currents to achieve a voltage equivalent to 1 μ V. We measured the J_c value at 102 K.

The structure of the sample was examined by means of x-ray diffraction (RU-200B, Rigaku Denki, Tokyo).

3. RESULTS AND DISCUSSIONS

Fig. 1 shows change in the T_c^{off} against number of the treatment (cold-pressing and annealing) times. The 2nd treatment increases the T_c^{off}. The T_c^{off} increase is approximately 5 K at 2×10^3 A/m^2. The treatments obviously improve the T_c^{off} value. We found the highest T_c^{off} value (107 K at 2×10^3 A/m^2) at the 2nd treatment.

Fig. 2 shows change in the superconducting critical current density (J_c) at 102 K against number of the treatment times. The treatments obviously improve the J_c value at high temperature of 102 K. Particularly, the 2nd treatment significantly increases J_c value. The J_c value of the 2nd treatment is about five times larger than that before the treatment.

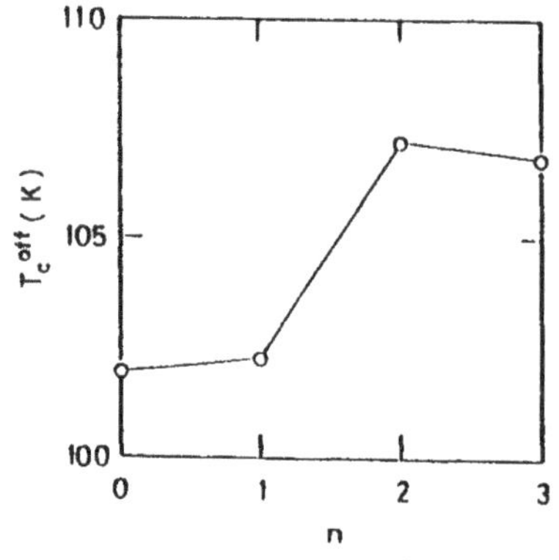

Fig. 1 Change in offset temperature (T_c^{off}) of superconducting transition temperature at 2×10^3 A/m^2 against number of treatment for $Bi_{1.6}Pb_{0.4}Sr_2Ca_2Cu_3O_x$.

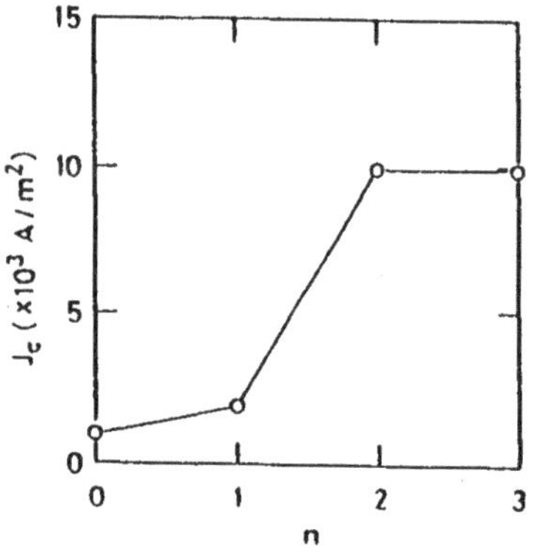

Fig. 2 Change in critical current density (J_c) against number of treatment for $Bi_{1.6}Pb_{0.4}Sr_2Ca_2Cu_3O_x$.

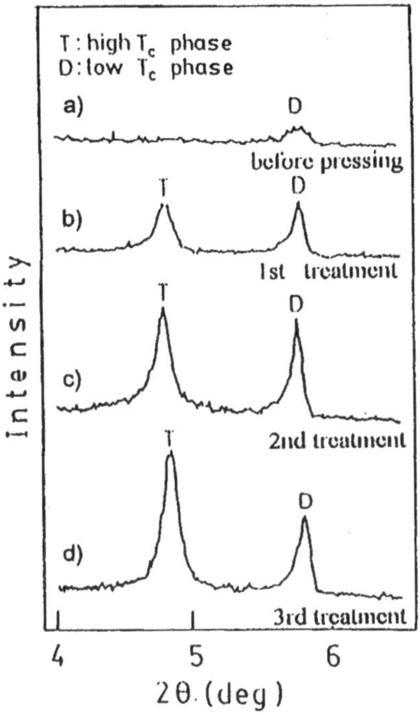

Fig.3 X-ray diffraction peaks of (002) plane for treated (cold-pressed and annealed) bulk samples. a), b), c) and d) are for before pressing, after first, second and third treatments, respectively.

Fig. 3 shows the x-ray diffraction patterns of (002) plane of the treated bulk samples. In alphabetical order, the samples are before pressing, after the first, the second and the third treatments, respectively. The samples are mixture of the high-T_c (triple Cu-O layer) and low T_c (double Cu-O layer) phases[10,11]. The treated sample has the large peak height of the high-T_c phase.

Fig. 4 shows the change in $\Delta 2\theta$ (FWHM : value of the full width at half maximum intensity) for the high-T_c (triple Cu-O layer) phase of (002) plane against number of treatment for $Bi_{1.6}Pb_{0.4}Sr_2Ca_2Cu_3Ox$. The high-$T_c$ sample shows minimum $\Delta 2\theta$ value of the peaks. It means that the 2nd treatment enhances the crystal perfection of the lattice ordering. The change in $\Delta 2\theta$ value approximately agrees with the change in T_c^{off} value (see Fig. 1). Namely, the high-T_c value can be maintained by the high crystal perfection of the crystal lattice ordering.

On the other hand, the small strain increases the defects density[12]. The defects improve the atom diffusivity[13]. Although the samples usually have heterogeneous internal stress among the ions, the 2nd treatment largely relaxes the residual stress in the sample. Therefore, it increases T_c because of the high crystal perfection. Since the 3rd treatment likely induces excess defects, it prevent to be the high crystal perfection because of high rate of nucleation of recrystallization. Thus, the high-T_c can be found at the 2nd treatment.

The FWHM is one of indicators to evaluate the grain boundary area. The superconducting current can partly path through the grain boundary. The rapid recrystallization decreases the boundary area and the defects density near the grain boundary. Thus, the J_c increase are found at the 2nd treatment (see Fig. 2).

Fig. 5 shows the change in the electrical resistivity R_e at 250 K against number of treatment times. The R_e value considerably increases at 3rd treatment. The higher the R_e values, the smaller the density of states becomes. It largely induces the low carrier concentration in the sample. Therefore, we explain the results that the T_c^{off} and J_c values don't largely increase at the 3rd treatment.

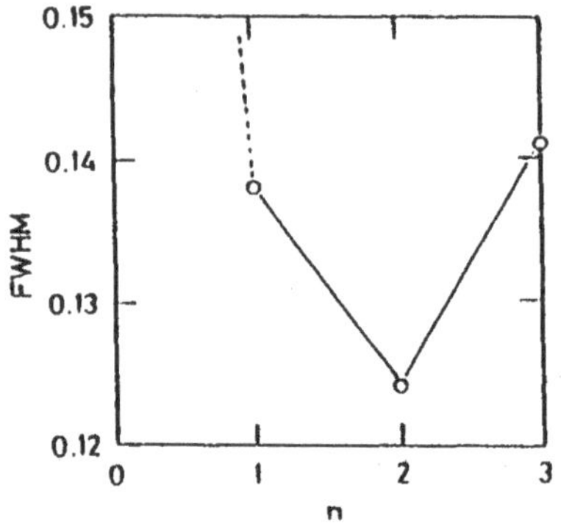

Fig. 4 Change in $\Delta 2\theta$ (value of the full width at half maximum intensity) for the high-T_c phase (triple Cu-O layer) of (002) plane against number of treatment for $Bi_{1.6}Pb_{0.4}Sr_2Ca_2Cu_3O_x$.

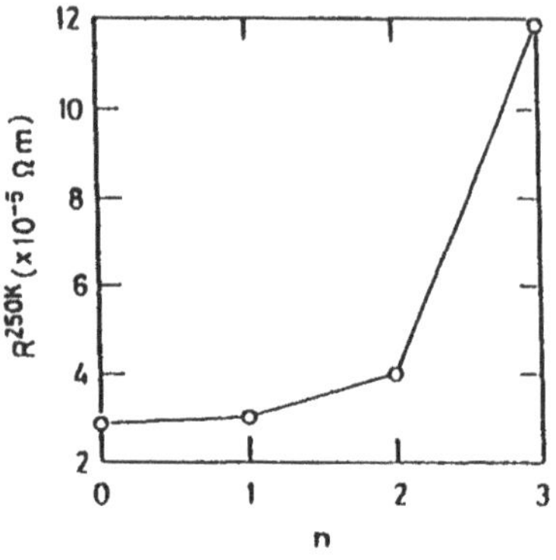

Fig. 5 Change in the electrical resist vity (Re) at 250 K against number of treatment for $Bi_{1.6}Pb_{0.4}Sr_2Ca_2Cu_3O_x$.

5. CONCLUSION

In summary, the effects of the treatments (cold-pressing and final annealing) on the T_c value are investigated for the liquid-quenched $Bi_{1.6}Pb_{0.4}Sr_2Ca_2Cu_3O_x$ which is crystallized. The 2nd treatment increases the T_c^{off} and J_c values. The high-T_c sample shows the minimum FWHM (the high crystal perfection) of (002) diffraction peak for high-T_c (triple Cu-O layer) phase.

6.REFERENCES

1. Y. Nishi, S. Moriya, and T. Manabe, *J. Appl. Phys.* **65**, (1989) 2389.
2. Y. Nishi, Y. Kita, and K. Tanioka, *J. Mater. Sci.* **25**, (1990) 5105.
3. Y. Nishi, S. Watanabe, T. Aida, K. Nozaki, K. Oguri, and T. Manabe, *Phys. Lett. A* **170**, (1992) 149 .
4. T. Asaka, Y. Okazawa, T. Hirayama, and K. Tachikawa, *J. Jpn. Appl. Phys.* **29**, (1990) 280.
5. T. Asaka, Y. Okazawa, and K. Tachikawa, J. *Jpn. Inst. Matl.* **56**, (1992) 715.
6. Y. Nishi, T. Kai, and K. Kitago, *Wear* **126**, (1988) 191.
7. Y. Nishi, and H. Harano, *J. Appl. Phys.* **63**, (1988) 1141.
8. Y. Nishi, K. Nozaki, and S. Ichimura, *J. Appl. Phys.* **71**, (1992) 11.
9. Y. Nishi, S. Moriya, and S. Tokunaga, *Phys. Lett. A* **126**, (1987) 55.
10. H. Nobusawε , K. Shimizu, Y. Kitano and T. Kawai, *J. Jpn. Appl. Phys.* **27**, (1988) L846.
11. H. Nobusawa, K. Shimizu, Y. Kitano and T. Kawai, *J. Jpn. Appl. Phys.* **27**, (1988) L1669.
12. Z. S. Basinski and S. J. Basinski, *Phil. Mag.* **9**, (1964) 51.
13. M. Cohen, Translated by K. Hirano, *J. Jpn. Inst. Metal*, **9** (1970) 271.

4. Fabrication of HTS Bulk and Wires with High J_c

FABRICATION AND PERFORMANCE TESTS OF Bi-2212 PANCAKE COILS

H. Kumakura, H. Kitaguchi, K. Togano and H. Maeda
National Research Institute for Metals, Tsukuba, Ibaraki 305 Japan

N. Tomita
Asahi Glass Co., Ltd., Yokohama, Kanagawa 221 Japan

K. Kato
Hitachi Cable, Ltd.,Tsuchiura, Ibaraki 300 Japan

ABSTRACT

We fabricated a double stacked small pancake coil using Bi-2212/Ag tapes prepared by the combination of continuous dip-coating process and melt-solidification method. The coil size was 13mmϕ(bore)x46.5mmϕ(outer diameter), and the tape length was 7.3m with a number of turns of 77 for each pancake. This coil was used as an insert magnet of a conventional superconducting magnet system. Critical current of the coil was 310A with a criterion of $10^{-13}\Omega m$ in a bias field of 20.9T at 1.8K, and the generated field by the coil was 0.9T. Hence we achieved the total field generation of 21.8T which was the highest fields obtained by the superconducting magnet system.

1. Introduction

Recently, a newly developed superconducting magnet system at National Research Institute for Metals accomplished the magnetic field generation of 21.1T at 1.8K[1]. However, it is very difficult to generate higher fields by conventional metallic superconducting wires, such as Nb_3Sn, because of its low critical current density J_c at ~20T. Grain oriented Bi-2212/Ag tapes, on the other hand, show excellent J_c values, higher than $10^5 A/cm^2$, even in magnetic fields above 20T at 4.2K[2]. Hence, it is expected that a coil made of this Bi-2212/Ag tape can be used as an insert magnet of the superconducting magnet system which can generate fields higher than 21T. We have been studying the fabrication of Bi-2212/Ag tapes and the application of the tapes to small superconducting coils[3,5]. In this paper, we summarize our recent results on the development of the Bi-2212/Ag coils

2. Fabrication of Bi-2212/Ag pancake coils

Bi-2212/Ag tapes were fabricated applying the combination continuous dip-coating process and slow cooling from a partially molten state[5]. The slurry for dip-coating was prepared by mixing calcined fine Bi-2212 powder, solvent binder and dispersant. Nominal composition of Bi-2212 was 2.00(Bi):2.00(Sr):0.96(Ca):2.00(Cu). Ag tapes with 30mm width and 80μm thickness were dipped into a slurry and the tapes were coated with slurry on both sides. Each dip-coated tape was wound into a pancake coil with small gap between each turn, and then heat treated. At first, the coils were heated at 810°C in air to remove organic materials and to decompose carbonates. Each pancake coil was, then, put on the powder mixture of $Bi_2Al_4O_9$ and Al_2O_3, covered with an alumina crucible and heat treated. During the heat treatment, bismuth was vaporized from $Bi_2Al_4O_9$. This heat treatment under bismuth atmosphere suppresses the vaporization of bismuth from oxide layers, which minimizes the precipitation of Bi-free impurity phases[6].

Melt-solidification was performed by heating up to 892°C and cooling down to 835°C at a rate of 5°C/hr. By this heat treatment, highly grain-oriented Bi-2212 superconducting layers of ~15μm thickness were formed on both sides of the Ag tape. Current leads and voltage taps were soldered to the Ag tape. 100μm thickness Ag plates and 20μm thickness Ag tapes were used for current leads and voltage taps, respectively. Organic films(Mylar tape) of 50μm thickness were inserted into the gaps between turns for insulation. The coil was tightly wound around a stainless-steel pipe of 14μm diameter which was inserted into the center of the coil, and then, encased in a stainless-steel frame of 46.5mm outer diameter. Finally, the coil was impregnated with epoxy resin cured with acid anhydride. In this way, two pancake coils were fabricated. These two pancakes were stacked coaxially and were connected electrically in series with Ag tapes. Figure 1 shows this double stacked Bi-2212/Ag coil. The parameters of the coil were listed in Table 1.

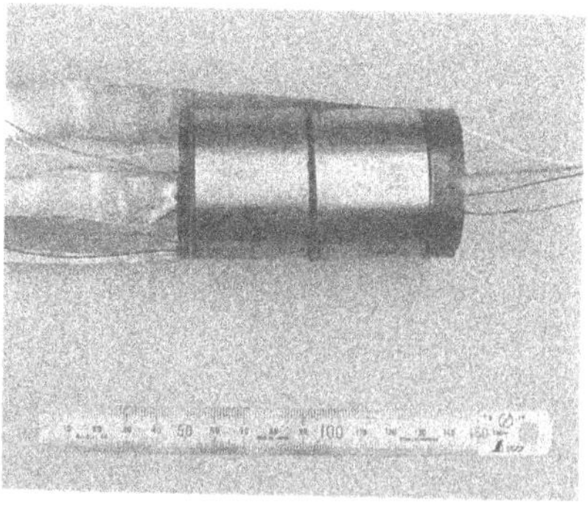

Figure 1. Double stacked Bi-2212/Ag coil.

Table1. Parameters of Bi-2212/Ag coil

Tape length	7.3mx2
Cross section (Bi-2212)	15μmx22mmx2
Number of turns	154
Inner bore	13mm
Outer diameter	46.5mm
Potting materials	Epoxy

3. Performance test

The small double stacked pancake coil was used as an insert magnet of a conventional superconducting magnet system. The magnet used in this experiment consisted of four types of Nb-Ti and Nb_3Sn superconducting coils[1]. The magnet system generated a magnetic field of 21.1T in 50mm bore by cooling with saturated superfluid helium. The performance test of the Bi-2212/Ag coil was also carried out at a saturated helium temperature under bias magnetic fields up to 20.9T. I-V characteristics of the coil were measured by a four probe resistive method. The generated field of the coil was measured by a Hall probe. Prior to the coil test, a generated field vs. applied current curve was obtained in zero bias field using a Hall probe set at the center of the coil. A generated magnetic field by the Bi-2212/Ag coil in a bias field was calculated from this current vs. field curve and measured I_c of the coil.

Figure 2 shows the I-V curve for one of the Bi-2212/Ag pancakes in the double stacked coil measured under a bias field of 20.9T at 1.8K. I_c of the coil was 310A with a criterion of $10^{-13}\Omega m$. Another pancake coil also showed similar I-V characteristic. This I_c of 310A corresponded to the field generation of 0.9T. Hence, the superconducting magnet system achieved generation of total magnetic field of 21.8T. The I-V curve shows a small slope in superconducting state. This slope increased with increasing bias magnetic field. This slope is probably due to micro cracks introduced in the oxide layers during the tightening process. Measured resistivity of a Ag tape was $\sim 5\times10^{-9}\Omega m$ in 20T at 4.2K. Because the temperature dependence of magnetoresistivity of Ag was small below 10K[7], the I_c criterion of $10^{-13}\Omega m$ was much smaller than the magnetoresistivity of Ag in 20.9T at 1.8K. Thus, current by-passing through the Ag tape was negligibly small for applied currents smaller than I_c.

J_c for the Bi-2212 layer was $4.7\times10^4 A/cm^2$ in 20.9T. This J_c was higher than that of the Bi-2212/Ag coil reported before[5], but was still smaller than the highest J_c of

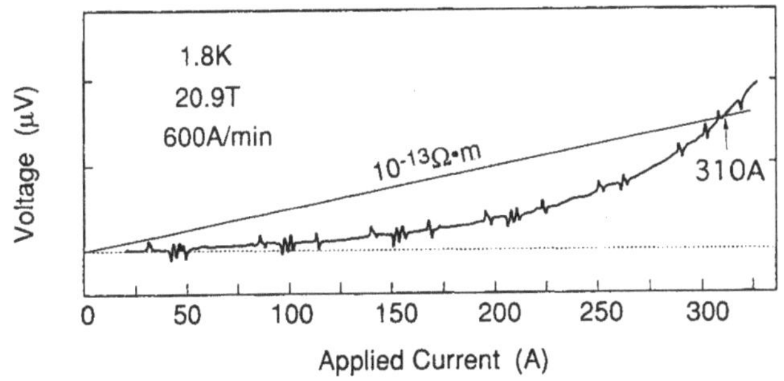

Figure 2 I-V curve of one of the Bi-2212/Ag pancake coils under bias field of 20.9T.

Bi-2212/Ag short samples, ~10^5A/cm^2. Furthermore, local I_c of the Bi-2212/Ag tape wound into the pancake was ~500A which was larger than that of the whole length of the tape. These results indicated that there was I_c scattering along the tape length due to the inhomogeneity of the superconducting oxide layer. Another reason for smaller I_c of the whole length is the micro cracks in the oxide layers. Because of the small curvature of the tape, and hence, large bending strain, it seems that micro cracks were introduced during the tightening process, especially at the area near the center of the coil. This suggests that larger I_c values and higher generated magnetic fields can be expected if the winding process of the coil is improved.

I_c of the coil increased with decreasing bias field. However, this increase of I_c with decreasing field was small. For example, in a field of 10T, I_c of the coil was 340A. This field dependence of I_c for the coil was much smaller than those of Bi-2212/Ag short samples. Because of the large I_c and high magneto-resistiviy of Ag, large amount of heat was generated at the Ag current leads and this heat diffused into the coil, increasing the temperature of the coil and suppressing the increase of I_c.

4. Conclusion

We have fabricated double stacked small Bi-2212/Ag pancake coil and tested this coil using a high field superconducting magnet. I_c and J_c values of practical level were obtained in bias fields of ~20T. The coil generated 0.9T in a bias field of 20.9T. Present results demonstrate that Bi-2212 can be used as the conductor of an inner magnet of a superconducting magnet system.

References

1. T. Kiyoshi, K. Inoue, K. Itoh, T. Takeuchi, H. Wada, H. Maeda, K. Kuroishi, F. Suzuki, T. Takizawa, N. Tada and H. Mori, Trans. Appl. Superconductivity **3**(1993)78.
2. H. Kumakura, K. Togano, H. Maeda, J. Kase and T. Morimoto, Appl. Phys. Lett. **58**(1991)2830.
3. H. Kumakura, H. Kitaguchi, K. Togano, H. Maeda, J. Shimoyama, T. Morimoto, K. Nomura and M. Seido, Cryogenics 32 ICMC Supplement(1992)489.
4. N. Tomita, J. Shimoyama, H. Kitaguchi, H. Kumakura, K. Togano, H. Maeda, H. Fujii and K. Nomura, Adv. Cryog. Eng. Materials **40A**(1994)297.
5. N. Tomita, M.Arai, E. Yanagisawa, T.Morimoto, H.Fujii, H.Kitaguchi. H.Kumakura, K. Inoue, K. Togano, H. Maeda, K. Nomura, Appl. Phys. Lett. **65**(1994)898.
6. J. Shimoyama, N. Tomita, T. Morimoto, H. Kitaguchi, H. Kumakura, K. Togano, H. Maeda, K. Nomura and M. Seido, Jpn. J. Appl. Phys. **31**(1992)L1328.
7. Y. Iwasa, E. J. McNiff, R.H. Bellis and K. Sato, Cryogenics **33**(1993)836.

Microstructure and J_c-B characteristics of Ag-sheathed $(BiPb)_2Sr_2Ca_2Cu_3O_y$ wires

T. Koizumi, Y. Hikichi, H. Kobayashi and T. Hasegawa
Showa Electric Wire & Cable Co. Ltd., Kawasaki, 210 Japan
H. Onoda and Y. Yamada
Toshiba Corporation, R&D Center, Kawasaki, 210 Japan

ABSTRACT

Intermediate compression is a technique to improve the orientation of superconducting grains for Bi-2223 superconductors. We prepared Ag-sheathed wires by three types of intermediate compression techniques. In a mono-filament wire, higher rolling forces in the intermediate rolling process formed smaller misalignment angles of superconducting grains and realized better J_c and J_c-B characteristics. In a multi-filament wire, high texture was realized in almost all the filaments, even if the wire was prepared by a conventional low rolling technique.

1. Introduction

Ag-sheathed $(BiPb)_2Sr_2Ca_2Cu_3O_y$ (Bi-2223) superconducting wire is one of the most promising candidates for a practical superconducting wire. Almost all Ag-sheathed wires employ the powder-in-tube technique and critical current density (J_c) of a short sample has come to be about $7x10^4$ A/cm^2 at 77 K[1]. J_c of a long wire (over 1 km in length) has been improved to $4x10^3$ A/cm^2 [2], but it is difficult for a long wire to realize the same superconducting properties as a short sample. This seems to be due to differences in the homogeneity of phase formation, misalignment of superconducting grains, the amount of impurity phases, the density of oxide cores and sausaging at the interface between the superconducting oxide core and the Ag-sheath. One of the factors which induces these differences is the intermediate compression process between heat treatments. In Bi-2223 superconductors, it is well known that doing several intermediate cold presses between heat treatments improves the alignment of superconducting grains and enhances superconducting properties. Therefore, the microstructure and misalignment of superconducting grains seem to be determined by the magnitude and direction of the compression forces in the intermediate compression process.

In this paper, we prepared mono- and multi-filament Ag-sheathed wires using three types of intermediate compression techniques (i.e. press and roll techniques with high and low rolling forces) and investigated the effects of intermediate compression on the microstructure and superconducting properties.

2. Experiment

The calcined powder with a cation ratio of Bi:Pb:Sr:Ca:Cu=1.85:0.35:2.0:2.0:3.0 was prepared by a conventional ceramic technique. The powder was packed into a silver tube and drawn into a round-shaped wire. The round-shaped wire was rolled into a mono-filament wire with 0.2-0.3 mm thickness and 3.0-3.5 mm width. This size was for a practical wire for a pancake coil. In a multi-filament wire, 55 round-shaped wires were bunched into a silver tube and drawn again. This was then rolled into a wire with the same size as the mono-filament wire. These wires were heat treated (HT1) at a predetermined temperature in 7.7 % oxygen partial pressure. After that, these wires were compressed using three types of techniques, which were pressing and rolling techniques with high and low rolling forces. Subsequently, heat treatment (HT2) was carried out in almost the same conditions as HT1. J_C was measured by the four-probe DC method with the criterion of 10^{-13} Ωm. The microstructure and crystal phases were investigated by scannning electron microscopy (SEM) and X-ray diffraction techniques, respectively. The density of the superconducting oxide core was calculated by a value of vickers hardness and graphic analysis of a SEM photograph.

3. Results and discussion

In order to approximate compression force in the rolling process to that of the pressing process, we applied the new rolling technique with high rolling forces to the intermediate compression process. Fig.1 shows J_C-B characteristics of mono-filament Ag-sheathed wires prepared by different intermediate compression techniques. The magnetic field was applied parallel to the tape surface. The J_C values of all samples decreased rapidly with increasing magnetic field at 77 K. In lower magnetic fields than 0.2 T, however, decrease of J_C depended on the intermediate compression technique. Higher compression forces resulted not only in higher J_C value at 0 T but also in slower decreases of J_C when magnetic fields were applied. A rapid decrease of J_C in low magnetic fields, such as lower than 0.2 T, is supposed to result from weak links. This result suggested that high rolling forces improved the connections between superconducting grains and reduced weak links.

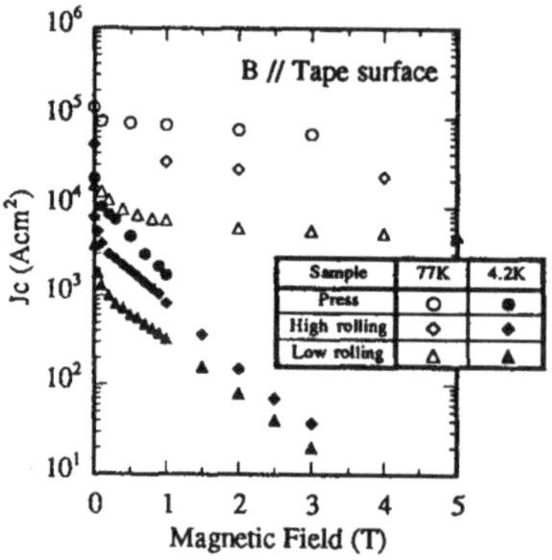

Fig. 1 J_C-B characteristics of multi-filament wire at 77 K and 4.2 K.

On the other hand, in higher magnetic fields than 0.2 T, decrease of J_C is said to be governed by the misalignment of superconducting grains. In order to physically evaluate the misalignment angles of superconducting grains of both rolling techniques, measurements of rocking curves and triple-axis maps were carried out for a surface from which the Ag sheath had been removed. Table 1 shows the full width at half maximum (FWHM) of the rocking curves and the off-axis angles ($\Delta\omega$)[3] of Bi-2223 crystals against the incident X-ray beam. These measurements were carried out on the (0 0 14) plane of the Bi-2223 phase. The values of FHWM for rolled samples were higher than that of the higher than that of the pressed sample reported by Hensel et. al[4]. But there was no difference in the FHWM values between the

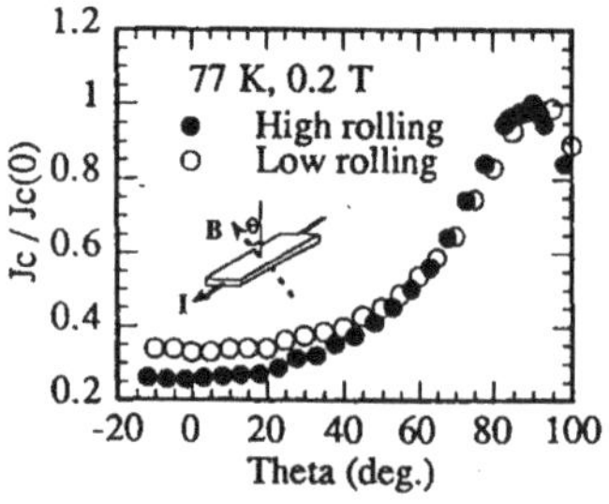

Fig. 2 Angular dependence of the relative J_C in magnetic field (0.2 T) at 77 K for Ag-sheathed wires.

high rolling and low rolling force techniques. In this measurement, these values reflected the orientation inside the superconducting oxide core, because the X-ray beam permeated about 20 μm and the highly textured part near the interface between the Ag-sheath and the supertconducting oxide core had been removed together with the Ag. These results suggeated that the orientation of superconducting grains inside the oxide core was not affected by the rolling force of the intermediate rolling process. It is thought that the superconducting properties are governed by the highly orientated and dense area at the interface between the oxide core and Ag sheath. The anguler dependence of J_C for wires rolled with high and low rolling force was measured at 77 K to estimate the practical misalignment of superconducting grains. Fig. 2 shows the angular the angular dependences of relative J_C in 0.2 T. The change of relative J_C value for the wire with high rolling force was more drastic in the higher θ area above 60°. According to the calculation method reported by Hensel et. al.[4] using J_C-θ characteristics, the misalignment angles for the high and low rolling force techniques were estimated to be 9° and 12°, respectively. These results indicated that the misalignment of the part of the oxide core which governed the superconducting properties of Ag-sheathed wire was affected by the rolling force of intermediate rolling. It is assumed that the orientation of superconducting grains near the Ag-sheath is improved by increasing the rolling force of the intermediate compression process but this force could not propagate to the inside of the superconducting oxide core.

Table 1 FHWM of rocking curves and offset angles of triple-axis maps (Δω)

Sample	FHWM (deg.)	Δω (deg.)
High rolling	16	17
Low rolling	17	16

Table 2 Vickers hardness and porosity of each sample

Sample	Vickers hardness (Hv)	Porosity (%)
Press	120 [5]	—
High rolling	80	19
Low rolling	72	23

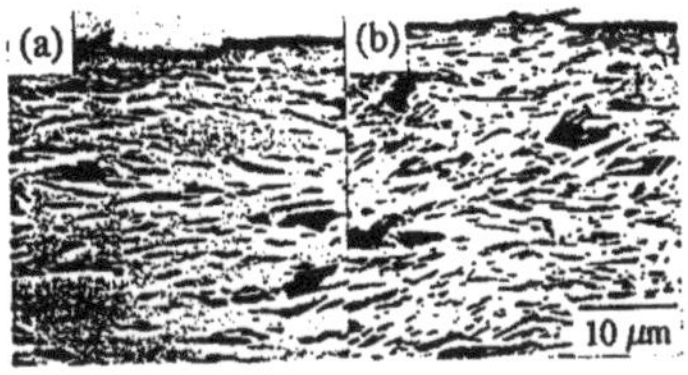

Fig. 3 Longitudinal cross-sectional SEM micrographs of Ag-sheathed wires rolled by high (a) and low (b) rolling forces.

The density of the oxide core is well known as a factor affecting J_C value. We measured vickers hardnesses for oxide cores rolled by high and low rolling forces in order to investigate the relationship between hardness and intermediate compression technique[5]. Porosities of oxide cores were measured by the graphic analysis method using SEM photographs shown in Fig.3. Table 2 is the summary of hardnesses and porosities of oxide cores. This table indicates that the density of oxide core increases in the order of pressing technique > high rolling force technique > low rolling force technique in the intermediate compression process. In comparison with the results of the X-ray diffraction analysis, the rolling force did not improve the orientation of Bi-2223 grains inside the oxide core but enhanced their density. Differences of J_C-B characteristic in low magnetic fields shown in Fig. 1 were considered to be due to this increase in the density of the oxide core.

Furthermore, we investigated the actual ability of multi-filament wire which was prepared by conventional rolling process and cut from a long wire of 100 m in length from the standpoint of the texture of superconducting grains. J_C-B and J_C-θ characteristics are shown in Fig. 4 and 5, respectively. Although the rolling force of the intermediate rolling process was quite lower than that of high rolling, J_C and the dependence of the magnetic field were similer to those of wire rolled by high rolling force. The misalignment angle calculated from the J_C-θ characteristic shown in Fig. 5 was 6°. This value was smaller than that of a wire rolled with high rolling force.

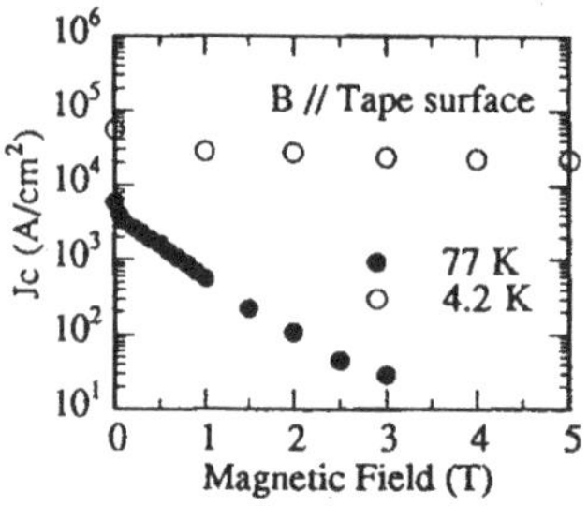

Fig. 4 J_C-B characteristics of multi-filament wire at 77 K and 4.2 K

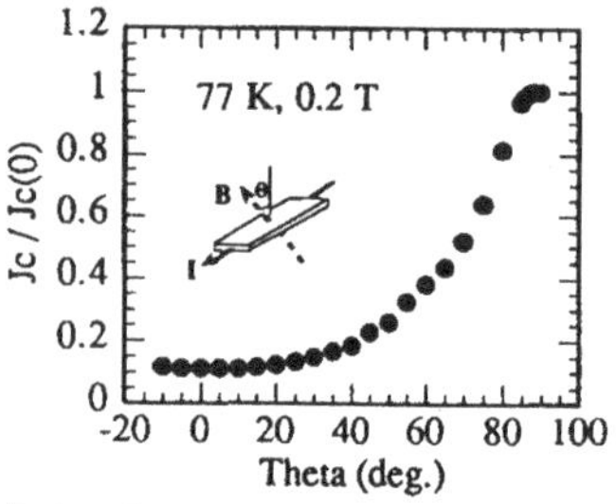

Fig. 5 Angular dependence of the relative J_C in magnetic field (0.2 T) at 77 K for multi-filament wire.

Observation of the microstructure of each filament indicated that microstructure at almost all parts of the wire was similer except at both edge parts of the wire. At the filaments in both edge parts, the orientation of superconducting grains was extremely bad. These results suggested that subdivided oxide cores provided not only an improvement in bending strain characteristics but also in the homogeneity of microstructure inside the wire. This means that increasing the intermediate compression force for multi-filament wire must be important in acheving better superconducting properties.

4. Conclusion

We investigated a relationship between superconducting properties and intermediate compression techniques. In a mono-filament wire, higher rolling force during the intermediate rolling process resulted in a smaller misalignment angle of superconducting grains and better superconducting properties. However, the misalighment angle was influenced by the rolling only in the area near the Ag-sheath. Also the misalignment angle was determined by the rolling process before heat treatment. In a multi-filament wire, high texture was realized in almost all the filaments, even if the wire was prepared by a conventional low rolling technique.

References

1. Y.Yamada, M.Sato, S.Murase, T.Koizumi and Y.Kamisada, *Proc. 5th Int. Symp. on Superconductivity* (1992) p717
2. T.Hikata, K.Muranakâ, S.Kobayashi, J.Fujikami, M.Ueyama, T.Kato, H.Mukai, K.Ohkura, N.Shibuta and K.Sato, *Proc. 1994 Int. Workshop on Superconductivity* (1994) p69
3. E.M.C.M. Reuvekamp, H. Rogalla, T.M. Dortmann and H.W. Zandbergen, *MRD Application Note by PHILIPS*
4. B.Hensel, J.C.Grivel, A.Jeremie, A.Perin, A.Pollini and R.Flukiger, *Physica C* **205** (1993) p329
5. Y.Yamada,M. Satou, T.Masegi, S.Nomura, S.Murase, T.Koizumi, and Y.Kamisada, *Proc. 6th Int. Symp. on Superconductivity* (1993) p609

TRANSPORT CRITICAL CURRENTS NORMAL TO TAPE PLANE IN Ag-SHEATHED $Bi_{1.8}Pb_{0.4}Sr_2Ca_{2.2}Cu_3O_x$ TAPES

J. Chikaba

Faculty of Engineering Kinki University ,Kyushu ,
Iizuka 820, Japan

ABSTRACT

Transport critical current densities Jcn normal to a tape plane in silver(Ag)-sheathed $Bi_{1.8}Pb_{0.4}Sr_2Ca_{2.2}Cu_3O_x$ superconducting tape have been studied at 77 K. The Jcn was measured directly by the four-terminal method using large copper blocks as a heat sink. Some contents of Ag is effective on the coupling strength between platelet- shaped grains. The peak value of ordinary transport critical current density Jc parallel to the tape plane was observed with increasing Ag addition. The anisotropy ratio Jc/Jcn is also discussed.

1. Introduction

It is expected at present that the Ag-sheathed Bi-system tape is the most promising superconductor at liquid nitrogen temperature. However, to obtain more stable and higher critical current density Jc, the following three problems (1) densification, (2) c- axis oriented microstructure and (3)large anisotropy should be mainly solved. Concerning the densification first, it was recently reported that the degree of cold working and heat treatments are effective on the densification[1]. Further, highly oriented microstructure can be observed near the boundary between Ag sheath and cores[2,3]. Lastly, highly anisotropy[4]is also a remarkable characteristic of the high Tc superconductor which has very complicated pinning mechanisms.

In this paper, in order to consider the above problems, a proper quantity of Ag was added to the Bi-system and the relation between Jcn and the degree of coupling between platelet grains.

2. Experiment

2.1 Sample Preparation

Ag-sheathed $Bi_{1.8}Pb_{0.4}Sr_2Ca_{2.2}Cu_3O_x$+Ag tapes were fabricated by the conventional powder-in-tube method. Ag_2O powder was added in the first step calcination. The Ag content of the tape are as follows: 9 wt% ~41wt%. After the calcination, well-pulverized powder was packed into the Ag tube(0.d=10mm, i.d=7mm, and l=50mm), which was swaged into a round wire (diam. 2mm). According to detailed reports on the Ag sheathed Bi(Pb)-2223 phase core [5,6], sintering condition was used as the temperature 840°C,50hrs for one step. Total sintering time was 250hrs . Uniaxial pressings were carried out 5 times. The final core dimensions were ~200μm thick,~ 6mm wide,and 40mm long, respectively.

2.2 *Measurement*

Usual transport critical current densities Jc parallel to a longitudinal direction of the tape and transport critical current densities Jcn normal to the tape plane were measured by the four-probe method for various contents of Ag tapes at 77 K. Electric field-criteria of the Jc and Jcn are 0.1μV/cm and 10μV/cm, respectively. Since large transport currents(300A<)was applied to the tape for the Jcn measurement, a small piece of the tape was sandwiched in between two copper blocks(30mm diam. 15mm thick). These copper block functioned as a good heat sink.The use of the Ag sheath as voltage terminals largely improved the problem related to the contact-surface resistivity in the previous report.[7]

3. Results and Discussion

3.1 *Effect of Ag Addition on Jc*

Since well oriented platelet-shaped grains were observed in the vicinity of the interface between the Ag sheath and core, some effects of Ag on the microstructure can be assumed. However, it has not yet been clarified whether the Ag is effective on the intergrains or intragrains.

In this work, relative large amount of Ag was added to the Bi(Pb)-system. Figure 1 shows the relation between the content of Ag and the critical current density Jc parallel to the longitudinal direction of the tape at 77K.

In the range of a small amount of Ag content, the value of Jc tends to decrease with increasing Ag content(<10wt%).

However, the value of Jc shows a maximum value at around 29wt% Ag. In the range of more than 30wt% Ag, the value of Jc was improved no more. Figure 2 shows the magnetic field dependence of Jc and Jcn in the 29wt% Ag tape. The magnetic field B were applied to the tape parallel and perpendicular to the plane. Remarkable differences between Jc and Jcn

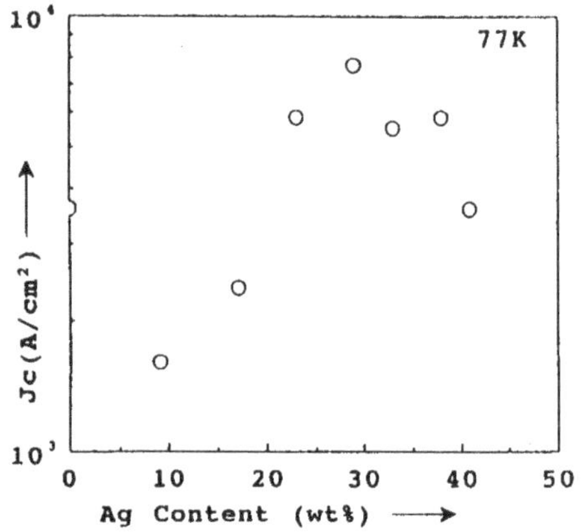

Figure 1. Ag content dependence of Jc at 77K.

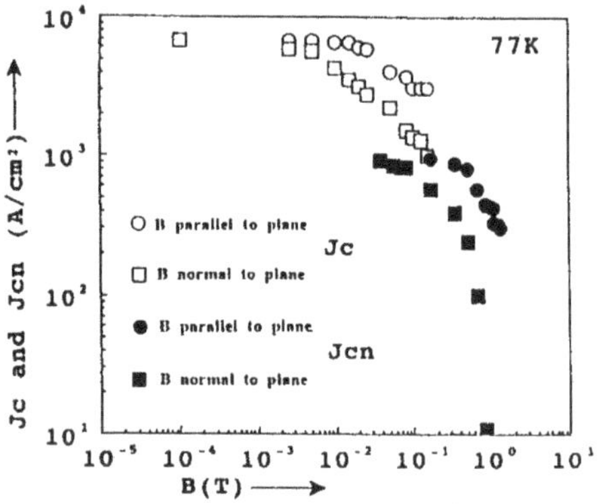

Figure 2. Magnetic field dependence of Jc and Jcn.

characteristics could be obtained as shown in Figure 2. The value of Jc attained to 6800A/cm², corresponding to 92A, in zero applied magnetic field. Although the value of Jcn in zero applied magnetic field could not measured on the limit of the current source, the roughly estimated value of Jcn is 1000~2000A/cm². Then, the anisotropy ratio Jc/Jcn is 3~7. This value is the same order of the magnitude as the result by Cho.[11]

3.2 Relation between Usual Transport Critical Current Jc and Jcn Normal to Tape Plane

The Jc of the Bi-system tapes generally attains to a high value in a high field at low temperature(T<20K), whereas it remarkably decreases with increasing field at high temperature(20K<T). Discussions on these experimental results so far may be classified into two models, namely, strong link model and brick-wall one.[8,9] Although it is difficult to specify the superconducting path , it was assumed that strongly coupled grains which grow up near the Ag sheath will give rise to the high Jc.[9] Then, Busch et al. considered the superconducting current path in Bi-2212 single crystals by using " transformer" configuration.[10] As a result, the resistivity ratio $\rho_{ab}/\rho_c \sim 10^5$ was obtained,where ρ_{ab} and ρ_c are resistivities parallel and perpendicular to the ab-plane. It can be assumed that current distribution must be nonuniform in such a single crystal. On the other hand, resistivities and anisotropy of transport critical current Ic in Ag-sheathed Bi-2223 tapes were measured by Cho et al..[11] Consequently, resistivity anisotropy was the order of 10. Hence, uniform superconducting current distribution can be assumed.

The direct measurement of the transport current normal to the tape plane was performed in this report. Figure 3 shows the relation between Jcn and Jc. The value of Jc is approximately proportional to Jcn in log-log scale.The ratio Jc/Jcn is 3~7 in zero applied magnetic field with 29wt% Ag as described above. However, the ratio Jc/Jcn seems to becomes large in low Jc tapes. The result shows that the grain coupling normal to the tape plane is extremely weak in such a low Jc tape.

SEM images of the plane of cleavage in Bi-2223 tape without Ag (a) and 29wt% Ag (b) are shown in Figure 4. In 0wt%Ag tape, platelet-shaped grains are not so strongly textured and also its average thickness is greater than 1μm . As for 29 wt% Ag tape, it consists of well aligned and strongly coupled thin grains (D<0.1μm), although it appears melt-like at first sight.

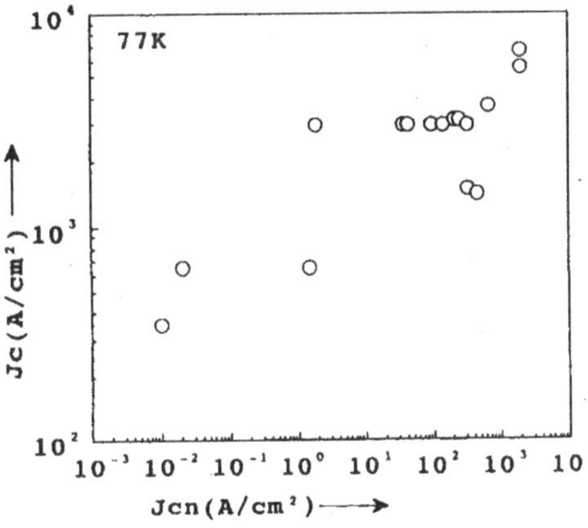

Figure 3. Relation between Jc and Jcn at 77K.

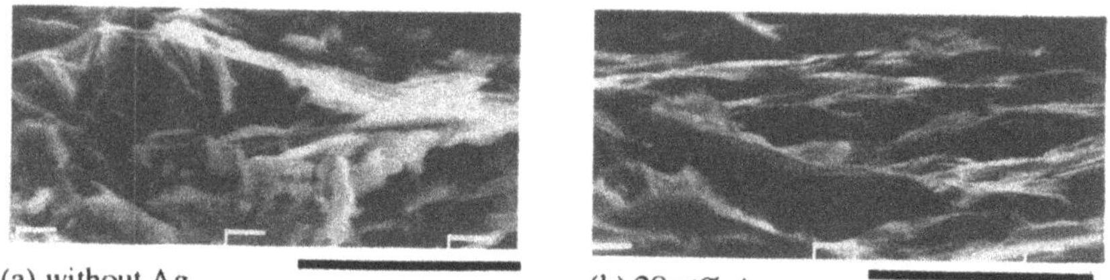

(a) without Ag 10 μm (b) 29wt% Ag 10 μm

Figure 4. SEM photographs of the plane of cleavage in the longitudinal direction of tapes.

4. Conclusion

It has been clarified that proper quantity of Ag has an effect on the strength of the grain coupling and densification. Concerning the influence of Ag on intragrain structure, it is necessary to perform another measurements such as magnetization with changing temperature for further quantitative discussion.

The author is grateful to M. Ozaki, T. Koyama, K. Shiozaki, E. Nakagawa, Y. Fujishiro, T. Tanabe, for assistance in measurements and also to M. Horimizu, K. Matsubara, T. Watanabe for photographing of SEM.

References

1. M. Satou, Y. Yamada, S. Murase, T. Kitamura and Y. Kamisada, Appl. Phys.Lett. **64** (5) (1994) 640.
2. Yi Feng, Y.E. High, D.C. Larbalestier, Y.S. Sung, and E.E.Hellstrom, Appl.Phys.Lett. **62**(13)(1993) 1553.
3. J.S. Luo, N. Merchant, V.A. Marouni, G.N. Riley, Jr. , and W.L. Carter, Appl. Phys. Lett. **63**(5) (1993) 690.
4. P. Schmitt, P. Kummeth, L. Schultz and G. Saemann-Ischenko, Phys. Rev. Lett.**67**(2) (1991) 267.
5. J.S. Luo, N. Merchant, V.A. Maroni, D.M. Gruen, B.S. Tani, W.L. Carter, G.N. Riley,Jr. , and K.H. Sandhage, J. Appl. Phys. **72**(6) (1992) 2385.
6. Ming Xu and D.K.Finnemore, J. Appl. Phys. **76**(2) (1994) 1111.
7. J. Chikaba, Adv.Cryo. Engi. **40** (1994) 185.
8. L.N. Bulaevskii, J.R. Clem, L.I. Glazman, and A.P. Malozemoff, Phys. Rev. **B45**(5) (1992) 2545.
9. L.N. Bulaevskii, L.L. Daemen, M.P. Maley, and J.Y. Coulter, Phys. Rev. **B48**(18) (1993) 13798.
10. R. Busch, G. Ries, H. Werthner, G. Kreiselmeyer, and G. Saemann-Ischenko, Phys. Rev. Lett. **69**(3) (1992) 522.
11. J.H. Cho, M.P. Maley, J.O. Willis, J.Y. Coulter, and L.N. Bulaevskii, Appl. Phys. Lett. **64**(22) (1994) 3030.

CRITICAL CURRENT DENSITY OF Bi2223 TAPES MADE BY USING AgCu ALLOY SHEATH DOPED WITH Ti, Zr, Hf OR Au

Y.Tanaka, F.Matsumoto and H.Maeda
Tsukuba Magnet Laboratories, National Research Institute for Metals,
1-2-1, Sengen, Tsukuba 305, Japan

and

M.Ishizuka
Research & Development Center, Sumitomo Heavy Industries,Ltd.,
63-30, yuhigaoka, Hiratsuka 254, Japan

Abstract

Bi2223 superconducting tapes have been prepared by the powder-in-tube technique using Ag-10at%Cu-x at%M (x = 0-1.0, M=Ti, Zr, Hf or Au) alloy sheaths. The alloy-sheathed tape samples showed considerably high critical current density,J_c values. J_c's at 4.2K, 14 T as high as 5 - 7x 10^4A/cm^2 were obtained for the doping of (0.03-0.1)at%Ti, 0.1at%Zr, 0.1at%Hf and 0.3at% Au. Micrographs on Ti-doped samples revealed a modified Bi2223 grain structure at the sheath/core interface and also a more dense and aligned microstructure. X-ray diffraction analysis, vibrating sample magnetometer measurement and ac susceptibility studies seem to suggest an enhancement of the Bi2223 phase formation with high T_c and improved texture, resulting in producing high J_c property.

1. Introduction

The silver-sheathed powder-in-tube technique has been the most promising fabrication process for Bi2223 conductors for high field superconducting magnets. A high critical current density, J_c, exceeding 1x10^5A/cm^2 at 1 T, 77k, on Bi2223 silver-sheathed tape has demonstrated[1] that applications of oxide superconductors at liquid nitrogen temperature may be feasible as well as at liquid helium temperature. Recently, a test-coil wound with Bi2223 monocore tape has generated magnetic field upto 2.6 T at 4.2K[2] and a coil wound with multi-filamentary Bi2223 conductor has also generated a magnetic field up to 1 T at 20 K[3]. However, J_c's in the coil windings have usually been reported to be very low in spite of the high J_c's in the short samples. We have found several factors that degrade the J_c value in both the silver-sheathed Bi2212 and Bi2223 tapes and effects of alloying the silver with copper on increasing the mechanical strength as well as improving the J_c reproducibility[4].

In the present paper an improved J_c property has been reported on the Bi2223 tapes by means of doping elements into the AgCu alloy sheath.

2. Experimental

The Ag- and AgCu alloy-sheathed tapes were prepared by the powder-in-tube technique. The Ag tube and Ag-10at%Cu- xat%M alloy tubes (x = 0-1.0, M = Ti, Zr, Hf or Au) with about 50 mm long, 8- and 5.7-mm outer and inner diameter, were filled with $Bi_{1.8}Pb_{0.4}Sr_{2.0}Ca_{2.2}Cu_{3.0}O_y$ (STD) calcined powder and $Bi_{2.0}Pb_{0.4}Sr_{2.0}Ca_{2.2}Cu_{2.4}O_z$(T2) calcined powder, respectively; hereafter these samples are referred to as Ag/STD, 10Cu/T2 (x = 0), and xM/T2. The AgCu alloys containing additional third element were prepared by melting weighed Ag, Cu metals with Cu-Ti, Cu-Zr or Cu-Hf mother-alloys or Au metal, respectively, in an argon gas atomosphere at about 1200°C and then casting into an iron-mould. These alloys were cold worked by grooved-roll and swaging machine to make bars of 8 mm in diameter, and the tubes were subsequently made by drilling. The powder-filled composites were initially swaged to 6 mm in diameter, then drawn to 1.04 mm in diameter, and finally flat-rolled to make tapes about 0.2 mm thick and 1.5-2 mm wide. The reduction ratios used in drawing or rolling pass were 5-20% at working speeds of 0.5-5 m/min. Short samples about 50 mm long, cut from these tapes were heat treated at 827°C initially for 100 h followed by 1 or 2 times repetition of intermediate pressing of 200 kgf/mm^2 or rolling and heating for 50 h, total heating time being 200 h. The powders and tape samples were characterized by a TG-DTA, SEM, ICP, EDX, and XRD. The superconducting properties were evaluated by VSM, ac Susceptibility, and I_c or J_c measurements at 4.2K up to 14 T by 4-probe resistive method. I_c was referred to a criterion of 1 μV/cm. The magnetic field was applied parallel to wide-surface of the tape and perpendicular to current direction. The critical current density (transport), J_c, was calculated to core cross-sectional area.

3. Results and Discussion

The I_c vs applied magnetic field curves are shown in Fig.1 for the tapes doped with Ti, Zr, and Hf into the Ag-10at%Cu alloy-sheath. The doped samples show considerably high I_c values compared with those of Ag/STD or 10Cu/T2 tapes. J_c's for the doped samples including Au are shown in Fig.2 as a function of nominal doped amount . It is noteworthy that J_c's as high as 5 - 7 x 10^4 A/cm^2, for Ti of 0.03-0.1, Zr and Hf of 0.1 and Au of 0.3, are higher than those obtained for Ag/STD or 10Cu/T2 samples by 2 or 3 times, respectively.

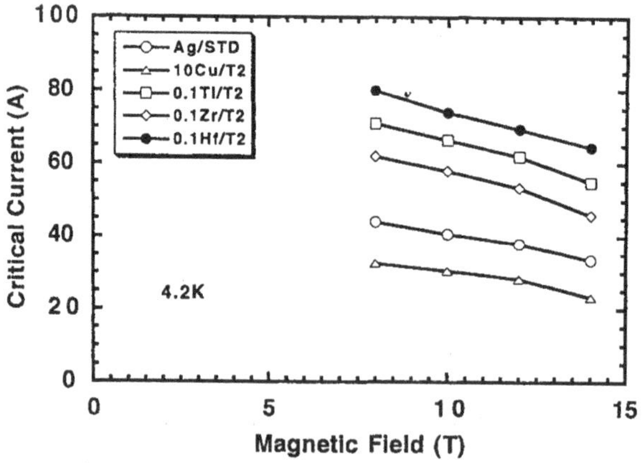

Fig.1. I_c vs magnetic field curves for the Ti, Zr, and Hf doped tapes compared with Ag/STD and 10Cu/T2 samples

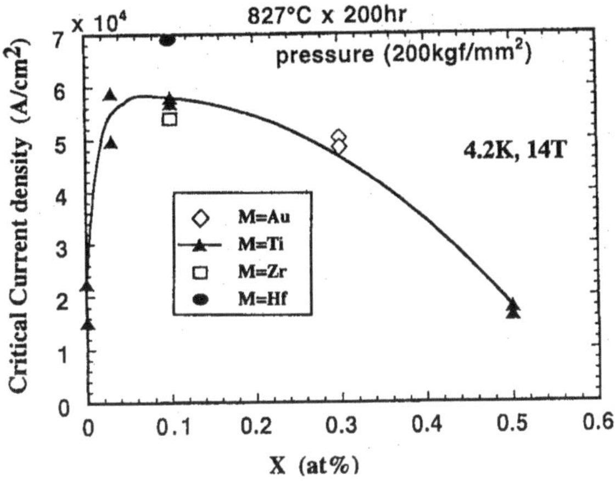

Fig.2. J_c vs nominal amount of the doped element into the AgCu alloy sheaths. Small quantity of doped element is effective in increasing the J_c

Hysteresis-curves on these samples obtained by the VSM are shown in Fig. 3. It is well known that magnetization difference ΔM between up and down hysteresis loop is proportional to J_c if one assume the Bean Model. In the present case large areas for both 0.03Ti/T2 and 0.1Ti/T2 tapes are responsible for the high transport J_c's as shown in Fig.2, except 0.03Ti/T2 (rolled) which is smaller than that of pressed one.

Transition curves, χ'', from ac susceptibility measurement revealed that steeper and narrower peak with higher transition temperature, T_C, seems to be characteristic to the

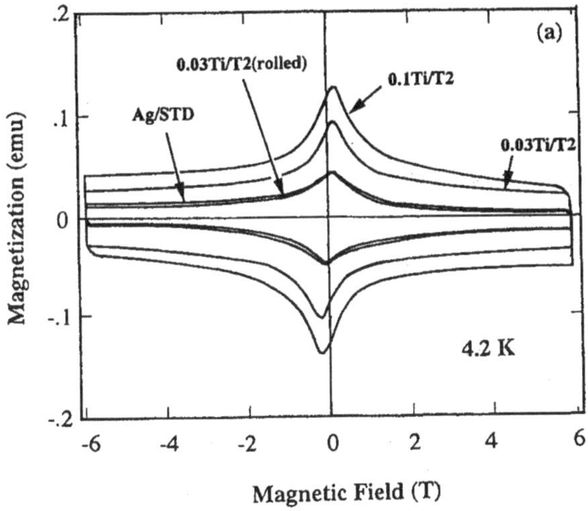

Fig.3. Hysteresis-roops obtained by VSM measurement at 4.2 K in the magnetic field -6T to 6T. Larger hysteresis roops are responsible for the higher transport J_c's and smaller curves for the lower transport J_c samples

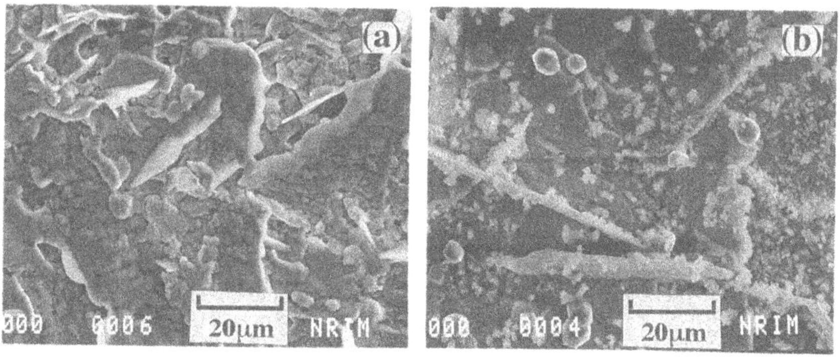

Fig.4. SEM micrographs of the Ag/STD (a) and 10Cu/T2 (b) tapes. Micrographs were taken on the core surface exposed by removing the half side of the tape by chemical process. Many fragment-like Bi2223 were observed on the Bi2223 grains in (b)

higher transport J_c. XRD profiles for Ti-doped samples taken on the core surface show a highly oriented microstructure with fewer low T_c phase or precipitate as Ca_2CuO_3. It is reasonable to think that the doping of these elements with a proper amount, although it is not precisely known yet, may at least make easier the formation of stoichiometric Bi2223 phase with highly oriented microstructure. Furthermore, doping Ti changes the microstructure at sheath/core interface and core inside i.e. a fragment-like Bi2223 particles, as shown in Fig.4(b), formed at the interface were disappeared and were formed larger Bi2223 grains than those found in Fig.4(a) with the core density improved. This modified grain structure at the sheath/core interface may produce good electrical contact as well as the dense and highly oriented microstructure. We tried ICP analysis to know distribution of Ti. Tape samples were cut opened at the center. ICP measurement was performed on three parts;(1) powder dug out from the tape, (2) powder adhered to sheath inside and (3) AgCu sheath. It was found that (a) the amount of Ti in the Ag-10at%Cu sheath was varied from sample to samle and was reduced to 1/2-1/10 from the nominal doped values and (b) the amount of Ti detected in the core was 1/10-1/100 to the amount actually obtained in the AgCu sheath. It seems promising to increase J_c by optimizing the doping quantity precisely

Acknowledgments

The authors would like to thank Dr. M. Fukutomi of National Research Institute for Metals for his competent performance in the ICP measurement.

References

1. K. Sato, T. Hikata, H. Mukai, M. Ueyama, N. Shibuta, T. Kato, T. Masuda, M. Nagata, K. Iwata, and T. Mitsui, IEEE Trans. Magn. vol, **27**, p1231, 1991.
2. P.Halder, J.G. Hoehn,Jr, I.R.motowidlo, U.Balachandran, Y.Iwasa, and M. Yunus, Proc. 6th Int. Symp. on Superconductivity(ISS'94), Oct 26-29, 1993, Hiroshima, Springer-Verlag, p.605.
3. K.Ohkura, H.Mukai, T.Hikata, M.Ueyama, T.Kato, J.Fujikami, K.Muranaka, and K.Sato, Proc. 6th Int. Symp. on Superconductivity(ISS'94), Oct 26-29, 1993, Hiroshima, Springer-Verlag, p.735.
4 Y.Tanaka, T.Yanagiya, F.Matsumoto, and H.Maeda, Adv. in Cryo.Eng.,Vol.40, p.153, 1994..

Temperature dependence of quench propagation velocity in silver sheathed Bi-based superconducting oxide tape

S.B.Kim, K.Aoki and A.Ishiyama
Department of Electrical Engineering, Waseda University
3-4-1 Ohkubo Shinjuku-ku, Tokyo 169,Japan

H.Okada, T.Yazawa and M.Satou
Toshiba Research and Development Center
4-1 Ukishima-cho Kawasaki-ku, Kawasaki 210, Japan

ABSTRACT

We report the effect of temperature dependence of quench propagation velocity of silver sheathed Bismuth-based(2223phase) superconducting oxide tape. We adopted operating temperature and transport current as parameter of measurement. We examined the influences of heater which led to trigger an initial quench of sample conductor, since it was predicted that in the case of oxide superconductor, the voltage occurs due to the difference of temperature caused by a steep slope of temperature change between voltage taps(Seebeck effect).

1. Introduction

Owing to the discovery of oxide superconductor with critical temperature over 100K, it is being made experimentally somewhere under the possibility that just by cooling liquid nitrogen is enough to operate the superconducting magnets. The quench propagation velocity is a useful parameter that magnet designers use as quantity of the natural ability of superconductor to protect itself from danger by quench. In general, quench propagation velocity of Low-Tc superconductors is very fast. A high quench propagation velocity implies that the quench volume grows rapidly which enables the stored electromagnetic energy to be dissipated over a large volume and thus reduces the risk of quench-induced damage. The issues of thermal stability and quench process of Low-Tc superconductors have been studied and used for the application of superconducting magnet[1][2]. However the quench propagation property of oxide superconductor, especially experimental data about thermal behavior have not been reported yet[3]. Accordingly, this experiment is carried out with Bi-based silver sheathed tape superconductor to investigate thermal and electromagnetic property and temperature dependence of quench propagation velocity while High-Tc superconductor is non-super transposited. The Bi-based silver sheathed tape is suitable for manufacturing of magnet, because its critical current density is high and can be made into long wire in oxide superconductors. We used operating temperature and transport current as parameter of experiments from assuming a case of magnet which was utilized with cryocooler.

2. Experiment

The sample conductor used in this study is $Bi_{1.7}Pb_{0.3}Sr_2Ca_2Cu_3O_x$(2223 phase)silver sheathed superconducting oxide tape made by OPIT-method. Overall length of sample conductor is 20cm. Table1 presents pertinent parameters of sample used in this study. It is measured that the critical temperature of sample is 105K. The schematic of experimental setup is shown in Fig.1. Operating temperature(or initial temperature) of sample is varried by GM cryocooler. The sample and location of lead wires are shown in Fig.2. Copper blocks are soldered at each end of sample, and two current leads with copper blocks at their ends. The capacity of the copper blocks are large enough to protect occurring end effect. Seven voltage taps are attached on sample conductor at every 10[mm] to measure quench propagation velocity, and thermocouples are placed between voltage taps. In this study, a strain gauge is used as an heater to trigger an initial quench because the sample conductor is type of tape. The quench propagation velocity V_{q23} between voltage tap2 and 3 can be calculated from $V_{q23}=L_{23}/\Delta t_{23}$ where L_{23} is the distance between voltage tap2 and 3, and Δt_{23} is the time lag when V_{23} and V_{34} are 50[μV/cm] respectively(where V_{23} is voltage between voltage tap2 and 3, and V_{34} is voltage between voltage tap3 and 4:Fig.4,6,8,10). Vq is the average value of V_{q23},V_{q34} and V_{q45}. We used value of 50[μV/cm] for the influences of thermoelectromotive force which can be ignored(but we used the value of 1[μV/cm] for measuring of critical current). The value of 50[μV/cm] is non-superconducting state but not fully normal state either. That is a characteristic of oxide superconductors. In Low-Tc superconductors, the transition from the fully superconducting state to the fully normal state is abrupt, but not in High-Tc superconductors, because of its large current sharing region.This is a remarkable difference between High-Tc and Low-Tc superconductors. If we made an observation up to the fully normal state of sample, the tape could have been burned out because the local temperature rose extremly, and actually, the tape was burned out, during some of the first experiments. As parameter of measurement, we varried initial temperature of sample conductors from 10K to 40K and measured transport current, a load factor(It/Ic:where It is transport current and Ic is critical current) from 0.4 to 0.9.

Table.1 Parameters of sample conductor

PARAMETER		SAMPLE
Tape thickness[mm]		0.15
Tape width[mm]		3.5
Tape length[mm]		200
Critical current density[A/cm²]	10[K]	41906
	15[K]	38763
	20[K]	30382
	30[K]	27763
	40[K]	22001

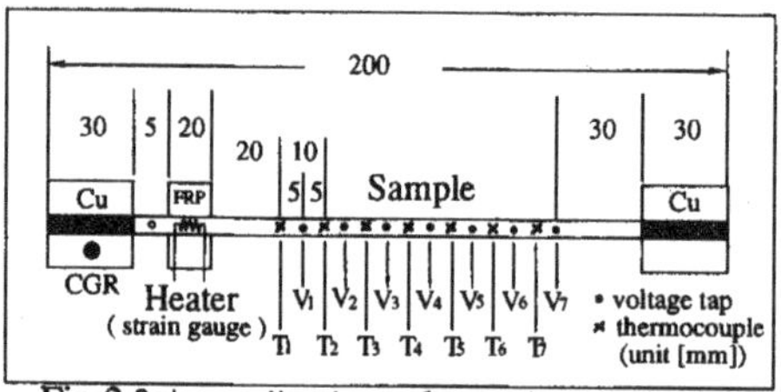

Fig.2 Schematic view of sample conductor and location of leads

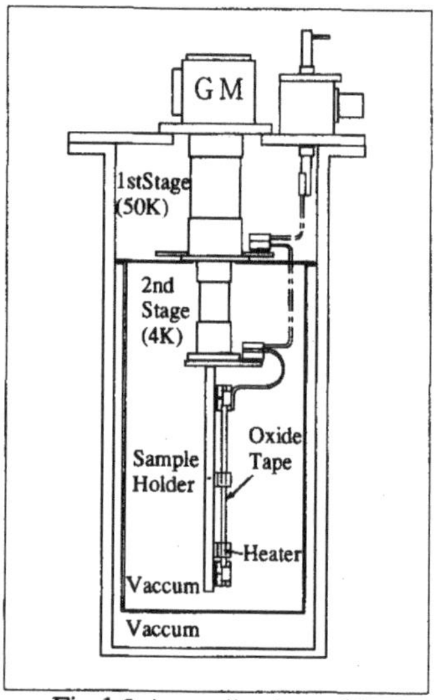

Fig.1 Schematic view of the experimental setup

3. Results

Fig.4 to Fig.11 indicate voltage and temperature traces at each operating temperature in the case of load factor 0.6. The quench propagation velocity was calculated from the voltage traces. The rising rate of temperature is varied by specific heat due to the temperature. The cross section of sample conductor and location of heater and thermocouple are shown in Fig.3. Thermocouples

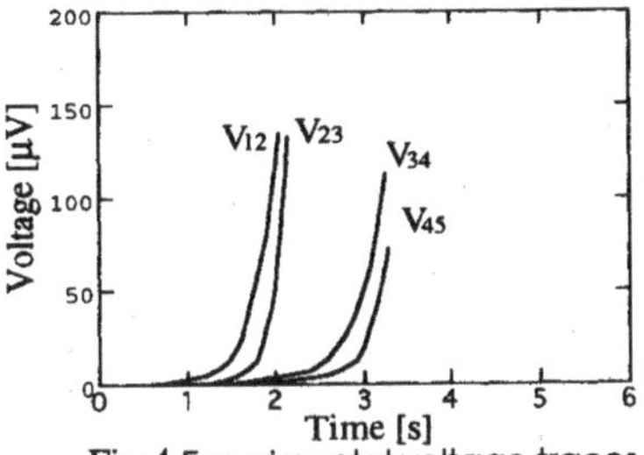

Fig.4 Experimental voltage traces
(Top:10[K], It:50[A])

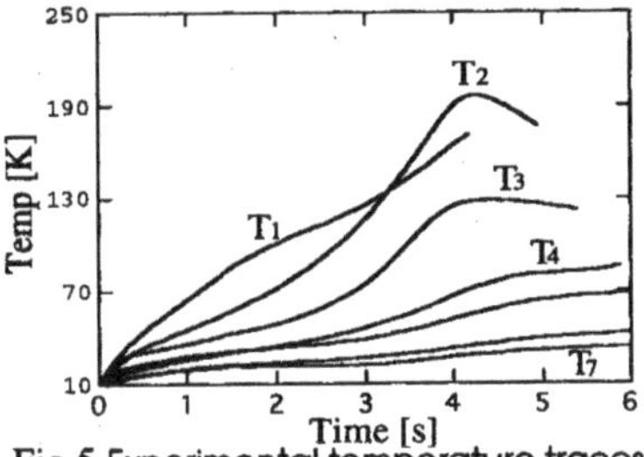

Fig.5 Experimental temperature traces
(Top:10[K], It:50[A])

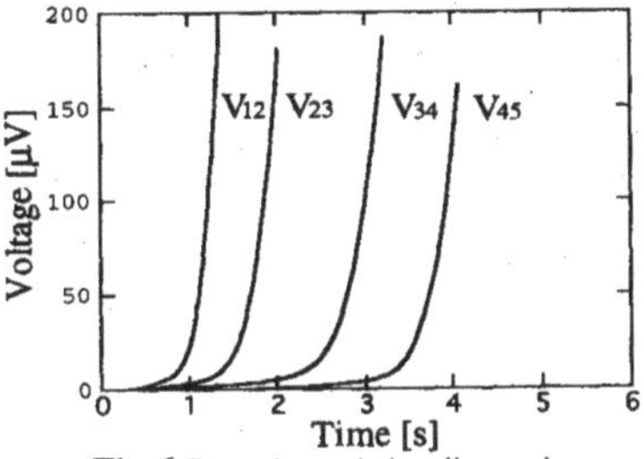

Fig.6 Experimental voltage traces
(Top:15[K], It:45[A])

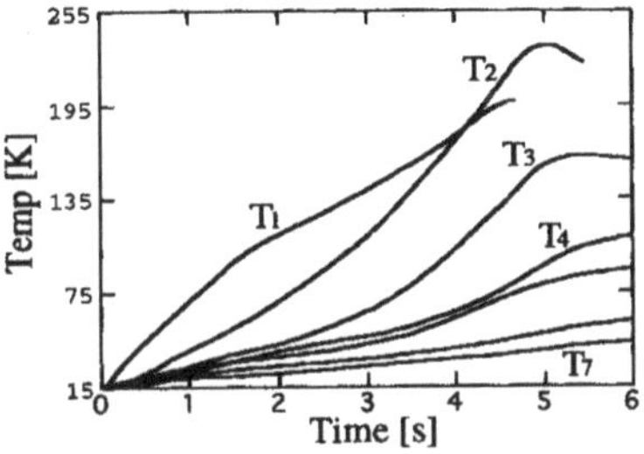

Fig.7 Experimental temperature traces
(Top:15[K], It:45[A])

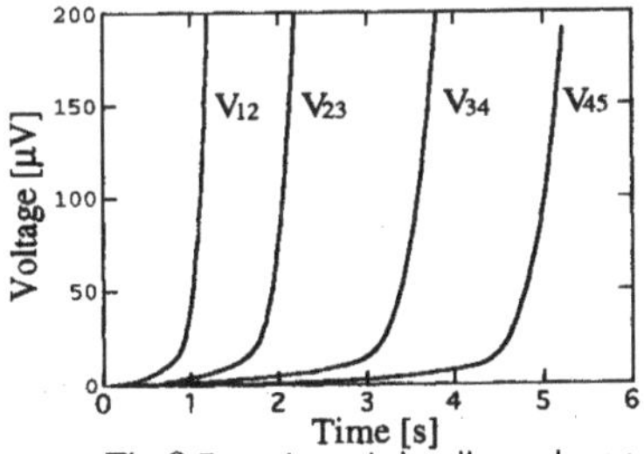

Fig.8 Experimental voltage traces
(Top:20[K], It:35[A])

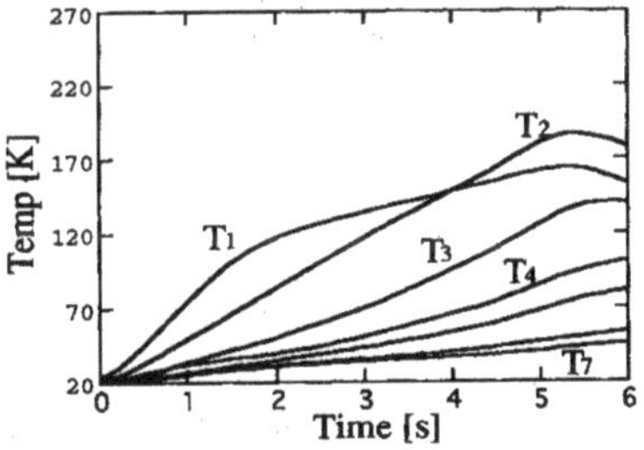

Fig.9 Experimental temperature traces
(Top:20[K], It:35[A])

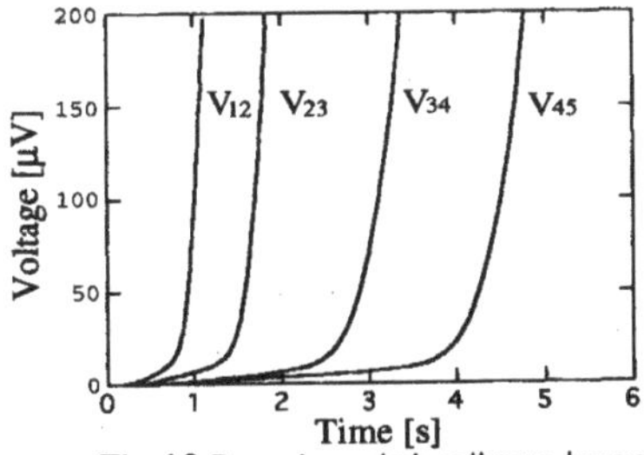

Fig.10 Experimental voltage traces
(Top:40[K], It:25[A])

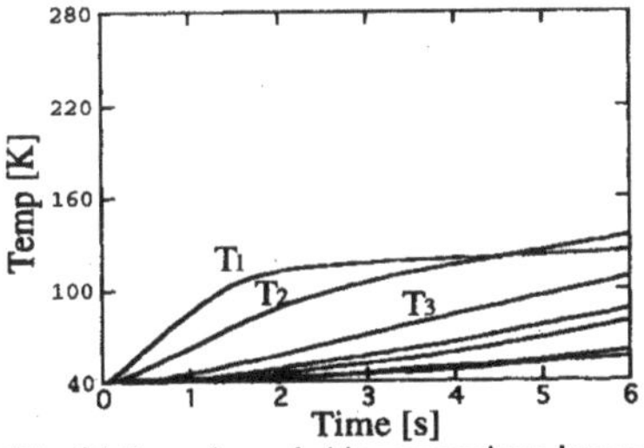

Fig.11 Experimental temperature traces
(Top:40[K], It:25[A])

are soldered on Ag sheath of opposite-side of heater. Accordingly, the heat, poured into the sample by heater, goes through the silver sheath whose thermal conductivity is higher than oxide , and propagates longitudinally. As thermal conductivity of silver, from 10K to 40K, becomes lower as temperature rises, the longitudinal thermal propagation velocity becomes slower. It is found, however, that at operating temperature of 40K, a rising rate of temperature is the lowest. It can be supposed that the specific heat of oxide superconductors becomes large according to the rise in temperature[4]. Therefore, the rising rate of temperature depends upon the specific heat of oxide superconductors. Fig.12 shows the quench propagation velocities at each operating temperature. It represents that x-axis is the load factor and y-axis is the quench propagation velocity,V_q. As the results of this experiment, we found that the quench propagation velocity in each operating temperature becomes a little faster as the load factor becomes higher. However, since the quench propagation velocity is from 0.5 to 1.5[cm/s] of magnitude, dependence of operating temperature is not large enough in the range of this measurment (It/Ic:0.4~0.9, T_{op}:10~40K). It is found that the quench propagation phenomena are divided into two groups, 10,15K and 20,30,40K. At 20K, the critical current of sample conductors changs abruptly.

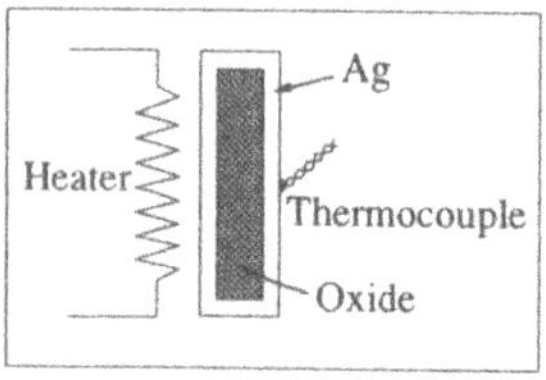

Fig.3 Schematic of cross section of sample conductor

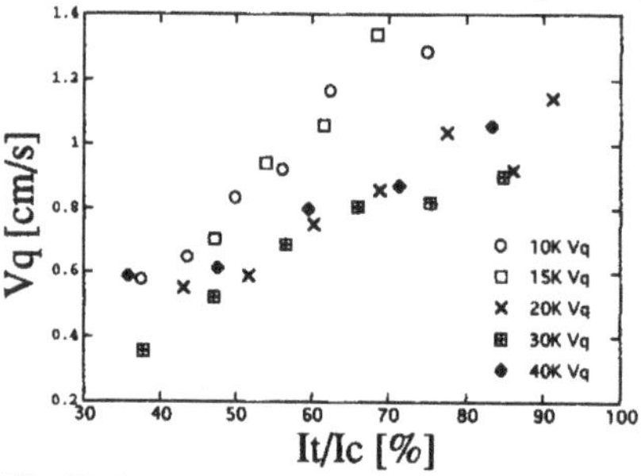

Fig.12 Quench propagation velocies in each temperature(Top:10~40(K))

4. Conclusions

- The voltage generation with quench propagation is very slow in comparison to Low-Tc superconductors.
- Since the quench propagation velocity is 0.5~1.5[cm/s], the dependence of operating temperature is not large enough in this study.
- As the quench propagation velocity of oxide superconductor is not temperature dependent, it is effective to operate at high temperature at which the stability margin is large.
- The longitudinal quench propagation velocity is two or three orders of magnitude lower in sample conductor used in this study than in Low-Tc superconductors.
- It is necessary that three dimensional consideration to be taken into account to analyze the quench propagation velocity of oxide superconductor tape, since the thermal conductivity of silver sheath is high, and the temperature of cross section of conductor is not uniform.

Reference

[1] A.Ishiyama and Y. Iwasa, *Quench propagation velocities in an epoxyimpregnated Nb3Sn superconducting winding mdel,* IEEE *Trans. Magn.* MAG-24(1988), P.1194

[2] A.Ishiyama, H.Matsumura, W.Takita and Y.Iwasa, *Quench propagation analysis in adiabatic superconducting winding,* IEEE *Trans. Magn.* MAG-27(1991), P.2092

[3] R.H.Bellis and Y.Iwasa,*Quench propagation in high Tc superconductor, Cryogenics*3 4(1994) No.2, P.129

[4] Y.Iwasa, *Design operational issues for 77-K superconducting magnets*, IEEE *Trans. Magn.* MAG-24(1988) No.2, P.1211

Effects of Ag Addition in Bi-2223 Bulk Using Current Leads

T.Honjo, S.Miyake, T.Hasegawa and Y.Kamisada
SHOWA ELECTRIC WIRE & CABLE CO. LTD., Kawasaki 210, Japan

ABSTRACT

Effects of Ag addition on microstructure and superconducting properties of Bi-2223 bulk fabricated by the cold isostatic pressing technique were investigated. The addition of 5 wt% Ag resulted in an increase in Jc value from 500 to 800 A/cm^2 and in a decrease in Rc value from 10^{-6} to 10^{-7} Ωcm^2 at 77K, 0T. The integral of heat conduction between 77K and 4.2K in this specimen was estimated to 80 W/m on an equality with that of a specimen without Ag. These results indicated that Ag addition improveed both increase in size of Bi-2223 grains and density of sintered bulk, but Ag did not affect the thermal conductivity since Ag particles dispersed homogeneously in a sintered Bi-2223 bulk. Furthermore, we obtained the maximum Jc value of 2500 A/cm^2 and Ic value of 1150A at 77K, 0T in a 5wt.% Ag added pipe-shaped specimen.

1. INTRODUCTION

Many efforts have been carried out to study a fabrication process of high-Tc superconductors(HTSC) and to improve the properties. HTSC in their sintered form is brittle and has remarkably low transport critical current density(Jc). In order for these superconductors to be used in a practical application, such as a current lead, it is important that Jc, contact resistance(Rc) and mechanical properties have to be improved. Especially, Jc of a $Bi_2Sr_2Ca_2Cu_3O_x$ (Bi-2223) bulk fabricated by sintering process is not only lower than those of $YBa_2Cu_3O_{7-x}$ (Y-123) and $Bi_2Sr_2CaCu_2$(Bi-2212) fabricated by melting processes but also decreases significantly in low magnetic fields at 77K. There are many reports on Jc enhancement by Ag-addition [1-3] in sintered Y-123 specimens and by intermediate press [4] in Bi-2223 sintered specimens. Bi-2223 bulk specimen with highly oriented microstructure was successfully prepared by the intermediate pressing technique and a significant increase in Jc were realized by this pressing technique. Additionally, Ag addition performed improvement of mechanical characteristics without lowering other superconducting properties. These results were considered to be mainly due to improvement of weak links between superconducting grains. However, as for Bi-2223 sintered bulk, there are a few reports [5,6] on effects of Ag addition.

In this paper, we investigated effects of Ag addition and intermediate cold isostatic press(CIP) on the superconductingand thermal properties of Bi-2223 bulk for a current lead.

2. EXPERIMENT

The superconducting powder with a cation ratio of $(Bi,Pb)_2Sr_2Ca_2Cu_3$ was prepared by the co-precipitation technique, and calcined at 1073K for 12h in air. The powder was mixed with Ag_2O powder in a ratio of 0, 5 or 10wt.%, and pressed into a rod or a pipe by the CIP technique under a pressure of 400 MPa. This composite was sintered at 1103K~1133K in air for 100h and cooled to room temperature. Afterwards, two pieces of Ag foil for electrodes were wound at both ends of the sample, and then the CIP and sintering process were repeated once more. The sintering temperature was determined by differential thermal analyzer(DTA) analysis of the calcined powder. Sintering was carried out at 1123K and 1108K in no-Ag-added and 5, 10wt.% Ag-added specimens, respectively. The specimens had approximate sizes of 50mm in length, 5 and 10mm in diameter for a rod (1) and (2), 11mm in outer diameter and 8mm in inner diameter for a pipe specimen. Jc and Rc measurements were performed by D.C. four and three-probe methods, respectively. The phases were determined technique(XRD) and their microstructual analysis was carried out using scanning electron microscope(SEM). Thermal conductivity was measured on rod specimens (diameter-5.5 mm and length- 50mm) using steady state heat flow method. The experiment was performed between 4.2K and 77K in the equipment in which liquid He was poured. One end of the specimen was attached to liquid He and the another was attached to a copper block containing a small heater. The specimen was protected heat radiation with the vacuum insulating and super-insulation. The difference temperature was measured by a C.G.R. thermocouple. The integral of thermal conduction was calculated by the following relationship:

$$Q=\frac{A}{L}\int_{TL}^{TH}\lambda(T)\,dT$$

where Q is the heat flux generated by the heater at the warm end, l(T) is the thermal conductivity, TH and TL are the temperature at the warm and cold end, respectively. A is the cross-section area and L is the length of the specimen.

3. RESULTS AND DISCUSSION

3.1 Effects of Ag addition

Figure 1 (A) and (B) show SEM photographs of a no and 5wt.% Ag-added specimen, respectively. The gray plate-shaped grains (appearing needle-like in the cross section) were Bi-2223 phase and bright areas were Ag particles. The Ag particles distributed homogeneously in discrete globules. Density of a sintered specimen was observed to increase by Ag addition. A density of ≈80% (based on the theoretical composite density) was obtained in the specimen with 5wt.% Ag content. Additionally, Bi-2223 grains in a Ag-added specimen had a average thickness of about 2μm and large numbers of Bi-2223 grains stuck together to form large aggregation. On the other hand, the grains in the no-Ag-added specimen were mostly isolated from each other and had an average thickness of 1μm. Bi-2223 grains of a Ag-added specimen were about twice as large as thoseof a Bi-2223 specimen without Ag. This increase in grain size was assumed to be due to the formation of much liquid phase by Ag addition. SEM observation indicated clearly that Ag addition changed the microstructure of these specimens.
Figure 2 shows effects of Ag content on Jc and Rc at 77K in zero magnetic field. Jc values of 500, 820 and 850 A/cm^2 were obtained for specimens with 0, 5 and 10 wt.% Ag content, respectively. Rc values of 1 0^{-6}Ωcm^2 in no-Ag-added specimen lowered to 10^{-7}Ωcm^2 by adding Ag. These effects of Ag addition might be due to better contact between superconducting grains, resulting from the increase in density.

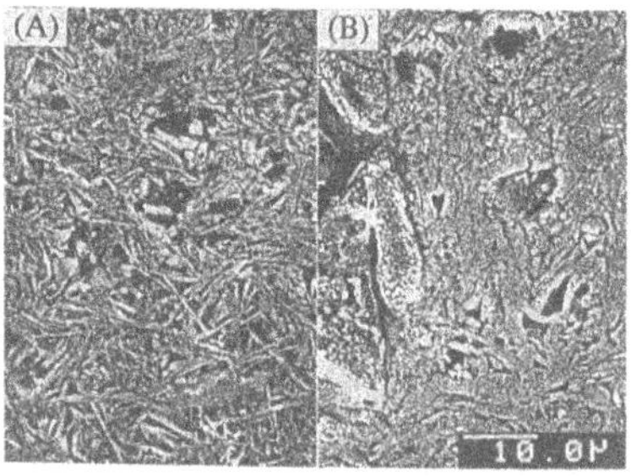

Fig. 1 SEM photographs of cross-section in no-Ag-added (A) and 5wt.% Ag-added (B) specimens.

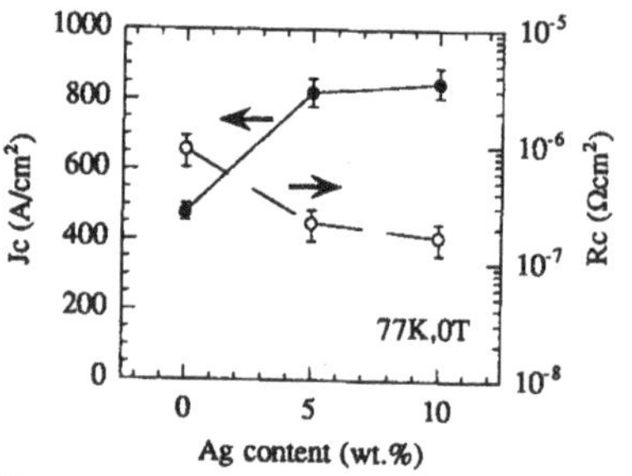

Fig. 2 The Jc and Rc dependence of Ag content in Bi-2223 sintered bulks at 77K,0T.

Table 1 shows the integral of thermal conduction in no and 5 wt.% of Ag-added specimens between 4.2K and 77, 40, 20K, respectively. Integral of thermal conduction did not change appreciably by Ag addition in a Bi-2223 bulk; 80 w/m was obtained in a specimen with 5 wt.% Ag content. This result suggested that the Ag particles were dispersed homogeneously in a specimen. As a comparison, the integral values of thermal conduction for different bulk superconductors [7] and other materials [8] are shown in this Table. The integrated thermal conduction of copper with RRR=50 was about three orders larger than that of Bi-2223 bulk superconductors.

Table. 1 Integral of heat conduction for the various materials.

TH (K)	$\int_{4.2K}^{TH} \lambda(T)\,dT$ (W/m)					
	Bi-2223 Ag5%	Bi-2223 Ag0%	Bi-2212*	Y-123*	Copper* RRR=50	Steel*
77	80	73	103	234	68600	314
40	20	18			40600	82
20	6	5			14000	16

3.2 Enhancement of critical current

Table 2 shows the times of intermediate press dependence on Jc (77K, 0T) and density for rod (1) and pipe specimens. Jc values of specimens which were pressed twice as intermediate CIP were much higher than those of specimens with once intermediate CIP. The maximum Jc of rod (1), (2) and pipe specimens were 1300, 600 and 2500 A/cm^2, respectively. Then the Ic value for the pipe specimen having a highest Jc was 1150A at 77K, 0T. The density of their specimens increased with increasing the times of intermediate press and the effect was remarkably in the pipe-shaped specimen.

Figure 4 shows the SEM photographs of the longitudinal cross-sections of rod (1) and pipe specimens with once (A), (B) and twice (C), (D) intermediate press. In rod specimens, the grains of the Bi-2223 did not oriented in both (A) and (B) but grain size of non-superconducting phase in the twice pressed specimen was smaller than that of the once pressed specimen. SEM photogragh with low magnification showed that a lot of pores and grains of the non-superconducting phase were observed in center parts of these specimens. On the other hand, in pipe-shaped spacemans, enhancement in the grain alignment of Bi-2223 crystals was realized by increasing the times of intermediate press. Highly textured grains were formed in the twice pressed specimen, while irregularly oriented plate-like cystals were observed in the once a pressed specimens. They clearly showed that the morphology was drastically changed between twice pressed and once pressed specimens.

Table. 2 Times of intermediate press dependence of the transport Jc(77K,0T) and the density for the 5wt.% Ag-added Bi-2223 bulks.

Sample	Imtermediate CIP (No.)	Ic (A)	Jc (A/cm^2)	Relative dencity (%)
Rod (A)	1	180	830	80
	2	265	1300	85
Pipe	1	430	960	75
	2	1150	2500	85

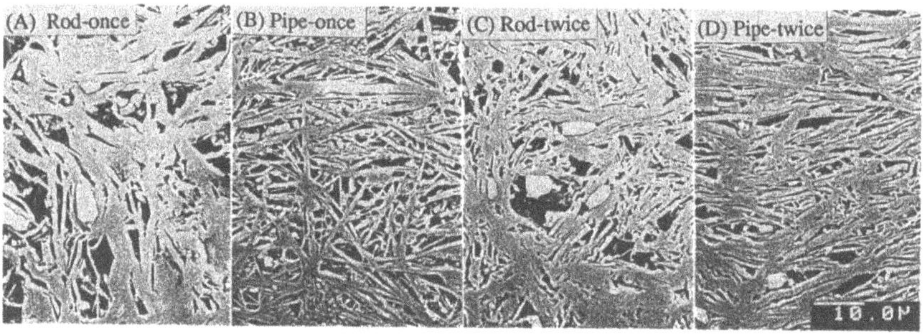

Fig. 4 SEM photographs taken on the longitudinal cross-sections of the rod and pipe specimens with once (A), (B) and twice (C), (D) intermediate press.

4. CONCLUSION

We observed that Jc and Rc properties were enhanced by a Ag-added Bi-2223 sintered bulk. Ag promotes the growth of Bi-2223 grains and increasing the density of sintered bulk. Addition of 5wt% Ag resulted in an increase in Jc value from 500 to 800A/cm^2 and in a decrease in Rc value from 10^{-6} to $10^{-7}\Omega cm^2$ at 77K, 0T. The integral of heat conduction in the spaceman was 80 W/m on an equality with that of a no-Ag-added specimen. These results indicated that thermal conductivity was not affected by Ag addition in Bi-2223 bulk, because Ag particles were dispersed discontinuously in porous bulk. Jc value of 2500 A/cm^2 and Ic value of 1150A at 77K, 0T were obtained in a pipe-type bulk of Bi-2223 superconductor prepared by a combined process of Ag addition, intermediate CIP and sintering.

ACKNOWLEDGEMENT

The authors are grateful to Takahashi, Yamamoto and Onoda of Toshiba Co.,LTD. and Kobayashi of Showa Electric & Wire and Cable Co.,LTD. for their cooperation in the performance of the experiments.

REFERENCES

[1] M.K.Malik, V.D.Nair, A.R.Biswas, R.V.Raghavan, P.Chaddah, P.K.Mishra, G.RaviKumar and B.A.Dasannacharya: Appl. Phys.Lett. 52 (1988) 1525
[2] J.P.Shigh, H.J.Leu, R.B.Poeppel, E.Van Voorhees, G.T.Goudey, K.Winsley and Donglu Shi: J. Appl. Phys. 66 (1989) 3154
[3] A.K.Pradhan, B.K.Roui, V.V.Rao and V.R. Kalvey: Cryogenics 33 (1993) 910
[4] T.Asano, Y.Tanaka, M.Fukutomi, K. Jikihara, J.Machida and H.maeda: Jpn. J. Appl. Phys. 27 (1988) 1652
[5] S.Jin,R.C.Sherwood,T.H.Tiefel,G.W.Kammilott, R.A.A.Fastnacht, M.E.Davis and S.M.Zahurak: Appl.Phys.Lett. 52 (1988) 1628
[6] Zhifeng Ren, Jui H.Wang: J. Mat. Sci. Lett.10 (1991) 1139
[7] P.F.Herrmann, C.Albrecht, J.Bock, C.Cottevieille et al.: IEEE Trans. Appl. Supercond. (1933) 923
[8] "Cryogenic hundbook": Uchida Rokakuho Press

DEPENDENCE ON TEMPERATURE AND MAGNETIC FIELD OF TRANSPORT CRITICAL CURRENT OF Bi-2223 BULK

T.HASEBE, T.TSUBOI, K.JIKIHARA, S.YASUHARA, J.SAKURABA and M.ISHIHARA
Research & Development Center, Sumitomo Heavy Industries Ltd.,
63-30 Yuhigaoka Hiratsuka, Kanagawa 254, Japan

and

Y.YAMADA
Faculty of Engineering, Tokai University,
1117 Kitakaname Hiratsuka, Kanagawa 259-12, Japan

ABSTRACT

The transport critical current of Bi-2223 bulk superconductors has been measured by a measurement unit utilizing Bi-based superconducting current leads and a cryocooler at a temperature between 20 K and 90 K in which an external magnetic field of up to 3 T can be applied by a cryocooler-cooled Nb_3Sn superconducting magnet. A specimen, fabricated by sintering and intermediate pressing process and almost entirely composed of Bi-2223 single phase, was measured and found to have a Jc value in excess of 1,000 A/cm^2 at 77 K without application of an external magnetic field in preliminary experimentation. It was found that critical current density increases drastically at temperatures below 50 K and that the flux pinning force density follows scaling relationship with the strength of the magnetic field.

1. Introduction

The investigation of temperature dependence of critical current density (Jc) of high critical temperature (Tc) superconductors within a magnetic field is one of the most important subjects both for research into the nature of high Tc superconductors and for their application. In this work, we investigated the dependence on temperature and magnetic field of critical current (Ic) of Bi-based cylindrical bulk superconductor for use as a current lead in a cryocooler cooled superconducting magnet [1]. In addition, we investigated the scaling law for Jc [2], which is a useful tool in the design of superconducting current lead for cryocooler cooled superconducting magnets.

2. Experimental

2.1. Sample production

The bulk sample of $(Bi,Pb)_2Sr_2Ca_2Cu_3O_{10}$ was prepared using a conventional powder sintering and intermediate pressing process [3], as follows.

Powders of Bi_2O_3, PbO, $SrCO_3$, $CaCO_3$ and CuO were mixed together in the ratio for Bi-2223 [i.e. $(Bi,Pb)_2Sr_2Ca_2Cu_3O_{10}$] , and calcined in air. A bulk tube of 13 mm external diameter, 10 mm internal diameter and 40 mm length was shaped from the calcined powder using CIP (Cold Isostatic Pressing) at 400 MPa and sintered for 100 hours at 810 - 840°C in an atmosphere of decreased oxygen partial pressure (Ar : O_2 = 12:1). After the first sintering, silver terminal of 15 mm width were formed on each ends of the bulk tube by plasma spray. Following application of the end terminals, intermediate CIP and post-sintering were each carried out twice following the parameters of the initial CIP and sintering. Two voltage taps were attached at 5 mm separation to the surface of the bulk tube using silver paste. Ic of the sample was initially measured in liquid nitrogen (77 K) using four probe method.

2.2. Measurement procedure

Figure 1 shows a schematic illustration of and measurement unit [4] and the superconducting magnet used for these experiments. As can be seen, liquid cryogen, such as liquid helium or liquid nitrogen are not used. Cooling of the measurement unit is achieved using a Gifford McMahon (GM) type cryocooler and Bi-2223 bulk current leads. The specimen, which is tested by four probe method, is cooled by conduction from the second cooling stage of the GM cryocooler, which has a refrigeration capacity of 4 W at 20 K. The specimen is contained inside a vacuum vessel. A pair of Bi-based superconducting current leads connect terminals on the first cooling stage to terminals on the second cooling stage. The warm end of the Bi-based leads is cooled to about 80 K, below its Tc, by the first cooling stage which has a refrigeration capacity of 27 W at 77 K. Ic of the bulk leads, which are fabricated by powder sintering and intermediate pressing, is grater than 1,000 A at 77.3 K in the absence of an external magnetic field [3]. As the thermal conductivity of the bulk lead is very low, the heat leak to the second cooling stage can be reduced by one order of the magnitude or more compared to conventional copper leads. The external magnetic field is applied by a Nb_3Sn wire wound superconducting magnet, also cooled by a GM cryocooler as shown in Fig.1 [1]. This cryocooler has a refrigeration capacity of 1.2 W at 10 K at the second cooling stage and 50 W at 67 K at the first cooling stage. In this Nb_3Sn superconducting magnet Bi-based bulk lead takes an important part of the magnet system.

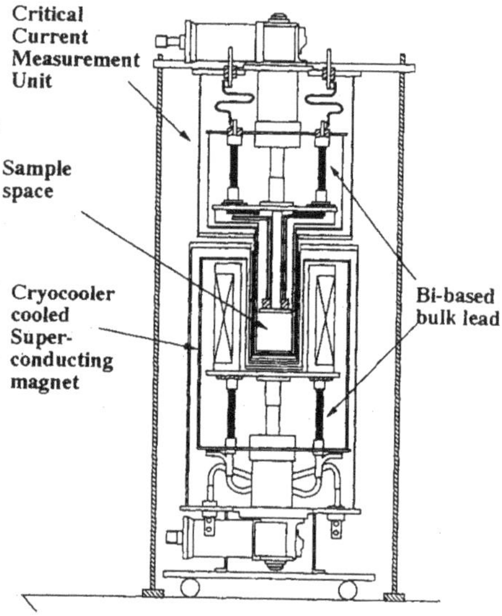

Fig.1 The testing apparatus equipped with the measurement unit and a superconducting magnet.

The temperature of the sample is measured using a CGR (Carbon-Glass Register) with an excitation current of 100 μA. The current applied to the specimen was increased at a sweep rate of 10 - 100 A/sec, and was decreased when the voltage across the specimen exceeded 50 μV, Ic thus being measured using an electrical field criterion of 2 μV/cm.

3. Results and discussions

3.1. Temperature and magnetic field dependence

Ic and Jc values for the sample of 550 A and 1,020 A/cm[2] respectively were obtained at 77 K in the absence of an applied external magnetic field in the initial experiment in liquid nitrogen. Measurement results for temperature ranges of 40 K to 63 K and 70 K to 90 K are shown in Fig.2-a and Fig.2-b, respectively. Characteristic field Bs, defined elsewhere [2] as the magnetic field at which the extrapolated value of Jc approaches zero (in this work 0.5 A/cm[2]) is shown in Fig.3. The reduced flux-pinning force density f_p (=Fp/Fp_{max}) is plotted against the reduced field b (=B/Bs) in Fig.4. Comparing the results with the scaling law written in the form, $f_p = b^m(1-b)^2$, we can show good correlation for m=0.3 as shown by solid-line in Fig.4.

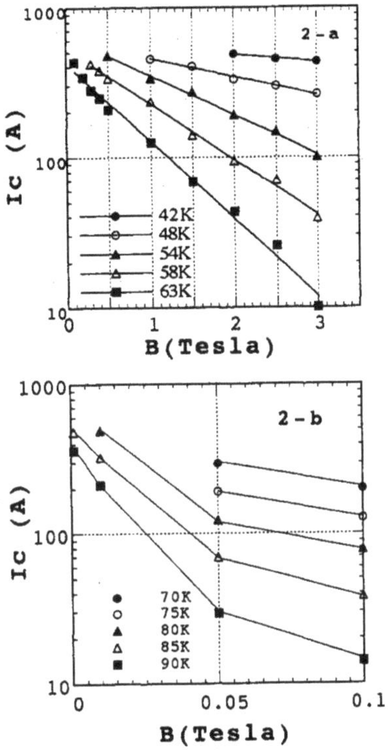

Fig.2 Measured Ic within the temperature range of 42 K to 90 K.

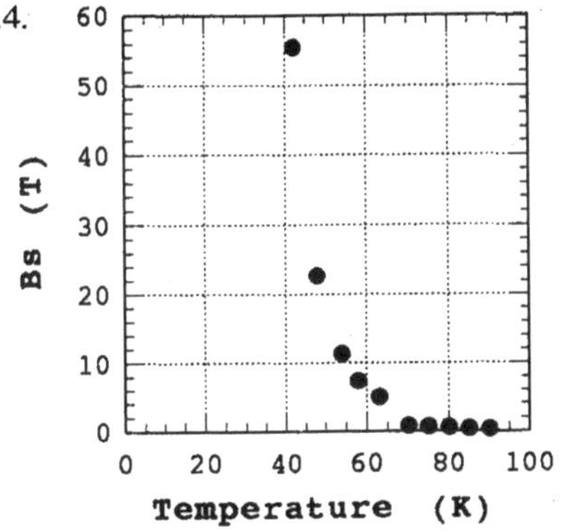

Fig.3 Characteristic field Bs. The magnetic field at which extrapolated values of Jc approach zero.

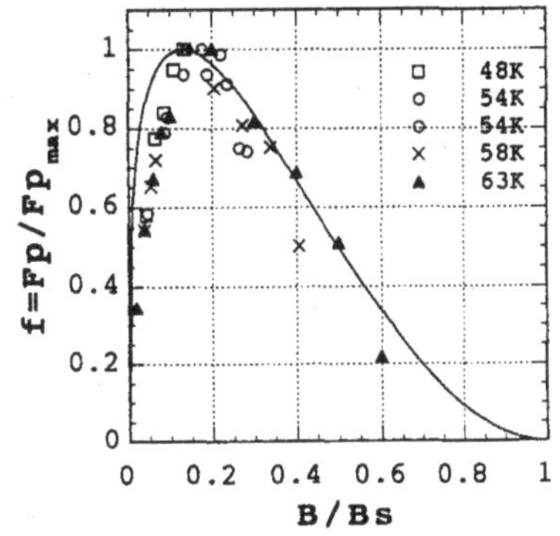

Fig.4 The reduced flux -pinning force f_p (=Fp/Fp_{max}) against the reduced field b (=B/Bs).

3.2. *Voltage-Current characteristics*

The voltage-current characteristics varied with variation of the temperature at 3 T as shown in Fig.5 (Logarithmic scale). The characteristics are consistent with those predicted by power law. The values of "n" derived from the slopes of the I-V curves are plotted against the applied magnetic field in Fig.6. It is clearly seen that the "n" remarkably increase as temperature lowered below 50 K.

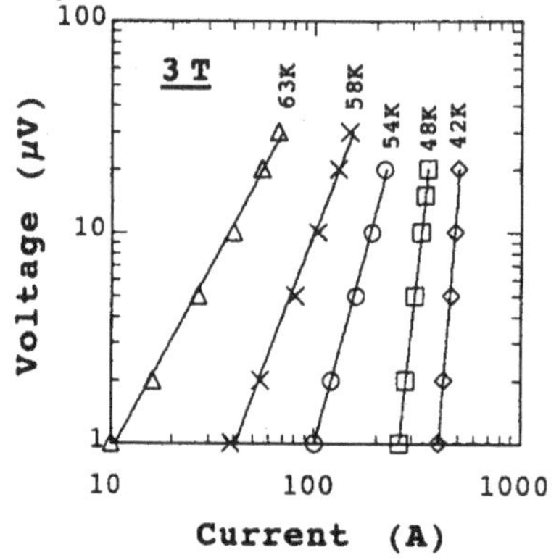

Fig. 5 Variation of voltage-current characteristics with respect to temperature (B = 3 T).

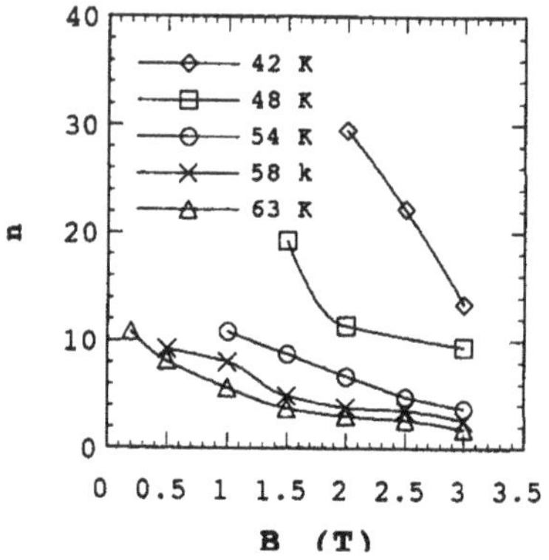

Fig.6 Value "n", obtained from Fig.5 plotted against magnetic field.

4. Conclusions

The dependence on temperature and magnetic field of Jc was determined for a Bi-2223 bulk specimen by the transport four probe method using measurement unit cooled by a GM cryocooler instead of liquid cryogen at temperature between 20 k and 90 K. The "n" value of the voltage-current characteristics and the temperature dependence on characteristic field Bs were examined. It was also determined that Ic properties of the specimen deteriorate drastically at temperatures above 50 K. The scaling law for fp was shown to have good correlation with the experimental results, so it has been demonstrated that the scaling law for Bi-2223 bulk superconductor is useful for the design of superconducting equipment.

References

1. K.Watanabe, Y.Yamada, J.Sakuraba, F.Hata, C.K.Chong, T.Hasebe and M.Ishihara, Jpn. J.Appl.Phys. 32(1993) pp.L488 Part 2, No. 4A, April 1993.
2. P.Kottman, H.Jones, V.Plechacek and M.Polak, Cryogenics 1992 Vol 32, No.11, pp.1005.
3. Y.Yamada, T.Yanagiya, T.Hasebe, K.Jikihara, M.Ishizuka, S.Yasuhara and M. Ishihara,IEEE Trans. Appl. Superconductivity Vol.3 (1993), No.1, pp.923.
4. T. Hasebe, T. Tsuboi, K.Jikihara, S. Yasuhara J. Sakuraba, and M. Ishihara, Appl. Superconductivity Conf. 1994 Boston LNC-15.

STRUCTURES AND SUPERCONDUCTING PROPERTIES OF ORIENTED Bi-2212 OXIDE LAYER SYNTHESIZED BY A DIFFUSION PROCESS

K. Tachikawa, Y. Yamada, M. Satoh and Y. Hishinuma

Faculty of Engineering, Tokai University, 1117 Kitakaname,
Hiratsuka, Kanagawa, 259-12, Japan

Abstract

A homogeneous Bi-2212 phase has been easily synthesized by the diffusion reaction. The structure of the diffusion layer shows dense plate-like grains aligned parallel to the diffusion direction with a high (h00) preferred orientation. The Ag added to the coating layer enhances the diffusion reaction and the preferred orientation. The transport critical current is strongly dependent on the reaction time which is the key parameter to determine the degree of orientation in the diffusion layer. The maximum transport critical current and critical current density of 2212 diffusion layer with about 150 μm in thickness are higher than 100 A and 14000 A/cm^2, respectively, at 4.2 K under self magnetic field.

1. Introduction

The high-Tc superconducting oxides synthesized by a diffusion process can be expected to show more homogeneous structure and higher density than that of oxides produced by a conventional sintering process [1)-4)]. In the Bi-Sr-Ca-Cu-O (BSCCO) system, we have successfully synthesized a thick and uniform superconducting layer of $Bi_2Sr_2Ca_1Cu_2O_{8+X}$ (2212) by diffusion reaction between the substrate of high melting point and the coating layer of low melting point. The addition of Ag to the coating layer has been found to enhance the diffusion reaction.

In the present study, the composition of substrate and coating layer are kept as Sr-Ca-Cu oxide (0212) and Bi-Cu oxide (2001) respectively because this combination of the diffusion couple is the most favorable to produce an uniform 2212 phase in the BSCCO system [3)]. In this paper, the effect of Ag addition on the diffusion reaction, and that of reaction time on the transport critical current and the structure of 2212 diffusion layer will be reported.

2. Experimental Procedures

Fig.1 illustrates the schematic diffusion process procedure. In the preparation of substrates, three starting powders ($SrCO_3$, $CaCO_3$, CuO) were mixed to obtain a Sr : Ca : Cu compositional ratio of 2 : 1 : 2 and calcined. The calcined powder was then pressed into rectangular bars 5 mm wide, 22 mm long and 1 mm in thickness, and sintered at 1000 ℃ in air. In the preparation of coating layer, Bi_2O_3 and CuO powders were mixed to have a composition of Bi : Cu = 2 : 1 and calcined with or without Ag_2O addition. The slurry of calcined powders

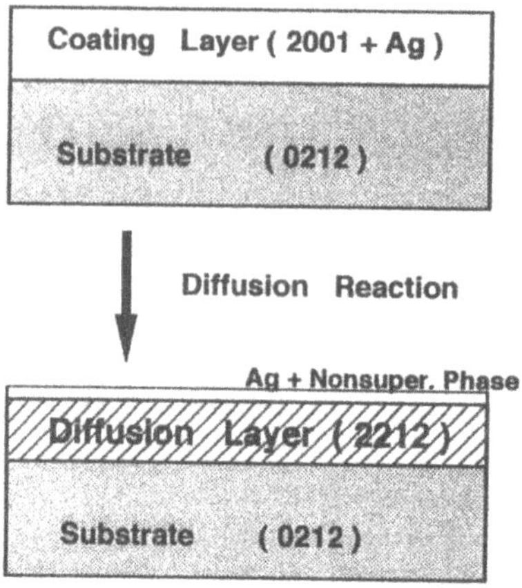

Fig. 1 Diffusion reaction process.

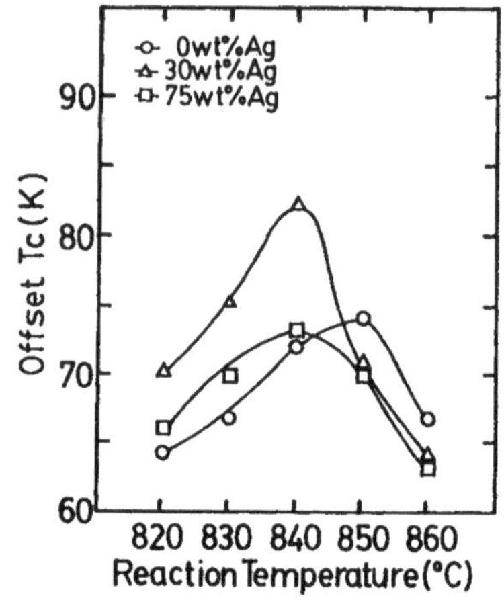

Fig. 2 Offset Tc versus reaction temperature of diffusion layer with or without Ag addition.

suspended in ethyl alcohol was coated on the surface of the substrate bars. The coating layer was about 150 μm in thickness. A two step heat treatment was performed to produce diffusion composites. The first heat treatment at lower temperature of 700 ℃ improved the bonding between the coating layer and the substrate. The second heat treatment at a higher temperature of 820 - 860 ℃ produced 2212 superconducting phase, which was followed by a slow cooling at a rate of 100 ℃/h to room temperature. The added Ag is pushed out from the diffusion layer and accumulates with a little nonsuperconducting phase on the surface of the specimen as shown in Fig.1.

3. Result and Discussions

Fig. 2 shows the relation between the offset Tc of the diffusion layer with or without Ag addition and reaction temperature. The diffusion layer with 30wt%Ag addition to the coating layer indicates an offset Tc of 82K at a reacting temperature of 840 ℃, while the diffusion layers with 75wt%Ag or without Ag addition show a lower offset Tc of 75 K. Thus, the appropriate Ag addition to the coating layer realizes a higher offset Tc and a sharp transition maybe due to the enhancement of the formation of 2212 phase.

The magnetic field dependence of transport Ic at 4.2 K for the 2212 diffusion layer reacted at 840℃ for 5 - 20 h is shown in Fig. 3. The transport Ic is strongly dependent on the reaction time. Under zero field, the transport Ic of the diffusion layer reacted for 10 h exceeds 100 A, while the Ic of those reacted for 5 and 15 h are about 60 A. The Ic of the specimen reacted for 20 h drastically decreases to

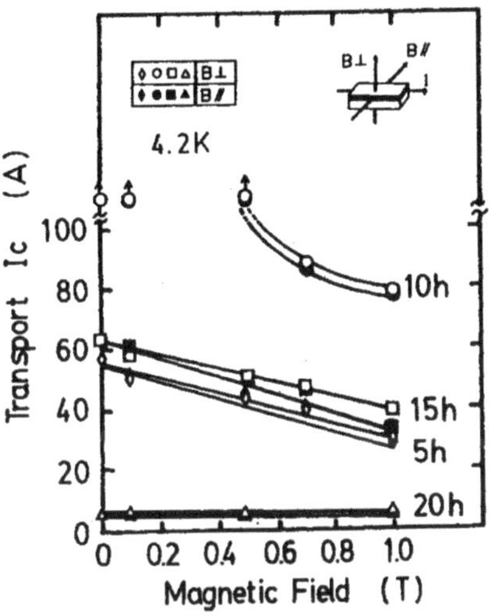

Fig.3 Magnetic field dependence of Ic at 4.2K for the diffusion layers reacted at 840℃ for 5-20h.

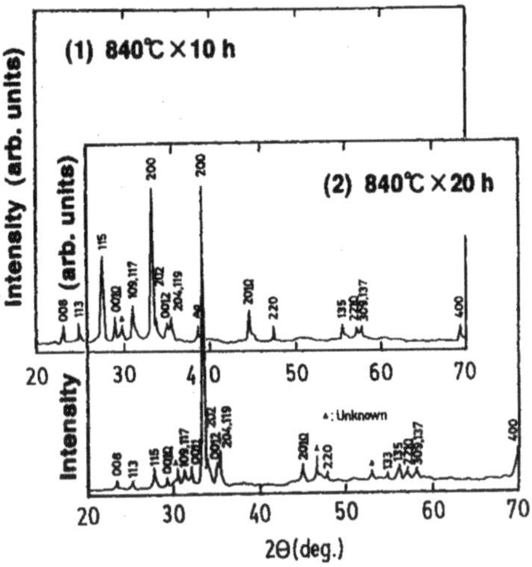

Fig.4 XRD patterns taken on the polished surface of the diffusion layer after quoted reaction time.

about 5 A. Moreover, the Ic of the diffusion layer reacted for 20 h scarcely depends on applied magnetic field up to 1 T, while the Ic of the specimen reacted for 10 h decreases with increasing magnetic field as shown in Fig. 3. The maximum transport Jc of the diffusion layer at 4.2 K under self magnetic field is more than 14000 A/cm^2 referring to the transport current and the thickness of the diffusion layer. The Ic of the specimen reacted for 5 - 20 h do not exhibit anisotropy with respect to the direction of applied field.

Fig. 4 presents X-ray diffraction (XRD) patterns taken on the polished surface of the 2212 diffusion layer reacted for 10 or 20 h. Both patterns are composed of diffraction peaks of the 2212 phase. The diffusion layer reacted for 20 h indicates unusually strong (200) peak in comparison with that for 10 h. The (200) peak intensity normalized to the (115) reflection is 16.4 in the diffusion layer reacted for 20 h, and is about 25 times higher than that in the random powder [5]. Thus the diffusion layer especially reacted for 20 h has (h00) highly-oriented texture. The results indicates that a axis is the most favorable direction which corresponds to the direction of preferential crystal growth in Bi-2212 phase. The (h00) oriented structure produced by the diffusion reaction may be useful for the electronic devices and the connection of high-Tc tape.

SEM micrographs of the fractured cross section of the diffusion composite reacted for 5, 10 and 20 h are presented in Fig. 5. The diffusion layer is composed of platelike crystal grains, aligned nearly parallel to the diffusion direction. According to the EPMA analysis, the thickness of diffusion layer is

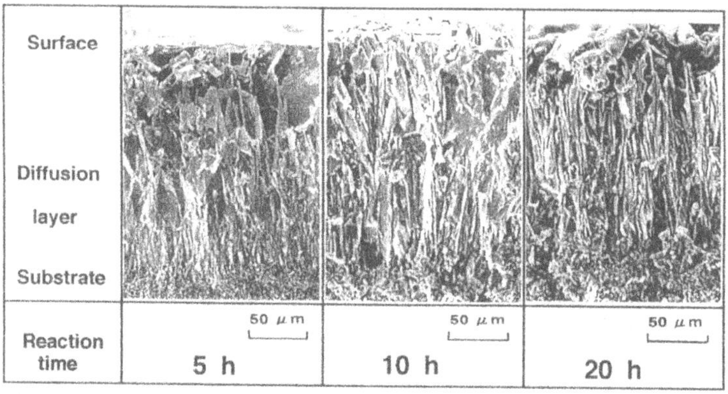

Fig. 5 SEM micrographs of the fractured cross section of the diffusion composites reacted for different time at 840 ℃.

about 150 μm in the specimens reacted for 10 h and 20 h, while is about 50 μm in that reacted for 5 h where unreacted layer is observed near the surface. Although the grain structure of the diffusion layer reacted for 10 h at 840 ℃ already shows a platelike morphorogy, it is much less oriented than that reacted for 20 h according to the XRD pattern as described before.

4. Conclusions

1. In the BSCCO system, 2212 superconducting phase has been easily synthesized by the diffusion reaction in a short time.
2. The Ag addition to the coating layer enhances the diffusion reaction and plays a role to improve the grain alignment.
3. The maximum transport Ic and Jc of the 2212 diffusion layer at 4.2 K under self magnetic field exceed 100 A and 14000 A/cm^2, respectively, for the layer thickness of 150 μm.
4. The transport Ic at 4.2 K under self- or applied- magnetic field is strongly dependent on diffusion reaction time since the structure of diffusion layer, in other words, the degree of orientation of the crystal grains highly depends on reaction time.

References

1) K.Tachikawa, N.Sadakata, M.Sugimoto and O.Kohno, *Jpn.J.Appl.Phys.* **27** (1988) L1501.
2) K.Tachikawa, T.Watanabe and T. Inoue, *Supercond. Sci. Technol.*. **3** (1990) 180.
3) K.Tachikawa, T.Watanabe, T.Inoue and K.Shirasu, *Jpn.J. Appl. Phys.* **30** (1991) 639.
4) K.Tachikawa,T.Inoue,K.Zama and Y.Hikichi, *Supercond.Sci. Technol.*. **5** (1992) 386.
5) M. Onoda, A. Yamamoto, E. Takayama-Muromachi and S. Takekawa, *Jpn. J. Appl. Phys.* **27** (1988) L833.

EFFECT OF F ADDITION ON THE SUPERCONDUCTING PROPERTIES AND STRUCTURES OF Tl-BASE OXIDE SYNTHESIZED BY A DIFFUSION PROCESS

K. Tachikawa, Y. Yamada, A. Kikuchi and T. Kinoshita
Faculty of Engineering, Tokai University, 1117 Kitakaname,
Hiratsuka, Kanagawa, 259-12, Japan

Abstract

The Tl-2223 phase has been easily synthesized by the diffusion process between a Ba-Ca-Cu oxide substrate and a Tl-base oxide coating layer within a short reaction time. The addition of fluorine(F) using TlF, BaF_2, CaF_2 or CuF_2 powder appreciably increases transport critical current (Ic) and reduces its degradation under magnetic field at 77K. The critical current density of the diffusion layer reaches about 10,000 A/cm^2 at 77K. Meanwhile, the annealing in O_2 atmosphere after diffusion reaction also increases the transport Ic at 77 K. No anisotropy in Ic and grain structure are observed in the diffusion layer. The XRD study reveals that the F addition promotes the phase transformation from Tl-2223 to Tl-1223 in the diffusion layer. The F addition from different fluorides yields different microstructures in the diffusion layer.

1. Introduction

The high-Tc superconducting oxides are easily synthesized by a diffusion process in Y-Ba-Cu-O (YBCO)[1)], Bi-Sr-Ca-Cu-O (BSCCO)[2)] and Tl-Ba-Ca-Cu-O (TBCCO)[3)] system, respectively. The diffusion composites reacted between the substrate of high melting point and the coating layer of low melting point show more homogeneous structure and higher density than that of composites produced by a conventional sintering process. In the TBCCO system, we have successfully synthesized $Tl_2Ba_2Ca_2Cu_3O_{10}$ (2223) diffusion layer within a short reaction time. In this paper, the effect of F addition on the transport Ic at 77K and the structure of the diffusion layer will be reported.

2. Experimental Procedure

In the preparation of diffusion substrates, three starting powders ($BaCO_3$, $CaCO_3$, CuO) were mixed to obtain a Ba : Ca : Cu ratio of 1 : 2 : 2 and calcined by a conventional process. The calcined powder was then pressed into small rectangular bars 5 mm wide, 22 mm long and 1 mm thick, and sintered at 900℃ for 12 h. In the present study, the composition of high melting point substrate are kept as Tl-free Ba-Ca-Cu oxide (0122). Meanwhile, another powders containing low melting point Tl oxide were mixed to obtain a appropriate composition ratio and suspended in ethyl alcohol to make a slurry. The slurry was coated on the surface of the bar-shaped substrate to form a diffusion composite. The thickness of the coating layer was about 100μm. The F addition to the coating layer was performed using TlF, BaF_2, CaF_2 or CuF_2 powder, while Ag was added using Ag_2O powder. The diffusion composites were reacted between 780 ℃ and 860 ℃, followed by slow cooling to room temperature. All heat treatment were performed in an open air. Then, some of the specimen were annealed at 550 - 600℃ in O_2 atmosphere after diffusion reaction. The phases indicated as 0122, 2100, 2212 etc. reflect the Tl : Ba : Ca : Cu atomic ratios.

3. Result and Discussions

The transport Ic at 77 K versus reaction time for the diffusion layer reacted with quoted diffusion couple and temperature is shown in Fig. 1. The Ag addition to the coating layer decreases the optimum reaction time as well as the reaction temperature to achieve large transport Ic. The F addition also decreases the optimum reaction time and temperature, and significantly enhances the transport Ic of the diffusion layer. The Ic of the diffusion layer, however, decreases after the prolonged reaction time. Here, 2(F)100 indicates the composition of the coating layer in which F is added using TlF powder.

Fig. 2 illustrates the magnetic field dependence of the transport Ic at 77 K for the 4102/0122, 4102+20%Ag/0122 and 21(F)00/0122 composites reacted at 850 ℃, 825 ℃ and 830 ℃, respectively. In Fig. 2, 21(F)00 indicates that F is added to the coating layer using BaF_2, and the diffusion composite post annealed in O_2 atmosphere after diffusion reaction is designated as 21(F)00 / 0122 (O_2). The magnetic field was applied parallel to the surface of the specimen and perpendicular to the transport current. The diffusion layers without F addition show a double-step degradation in Ic under magnetic field up to 1 T at 77 K. The first drop in Ic below 0.1 T followed by a plateau, and the second drop over 0.4 T are reported to be associated with weak link and flux flow behavior, respectively [4]. The Ic of the F free diffusion layers degrades by about two orders of magnitude under a magnetic field of 1T. A pronounced degradation of Jc under low magnetic field at 77 K is also reported for the Tl-1223/Ag tape [5]. On the other hand, the diffusion layers with F addition do not exhibit the second drop in Ic under magnetic field, even up to 1.5T. Thus, the degradation in Ic of the diffusion layer with F addition at 1T is less than one order of magnitude compared to the F free diffusion layers. The Ag addition is useless for the improvement of Ic under magnetic field at 77 K.

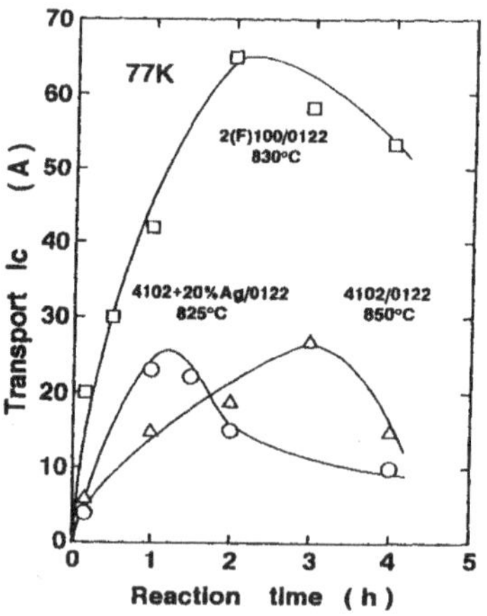

Fig.1 Transport Ic at 77 K versus reaction time for the diffusion layers with F or Ag addition, and without additional element at quoted temperature.

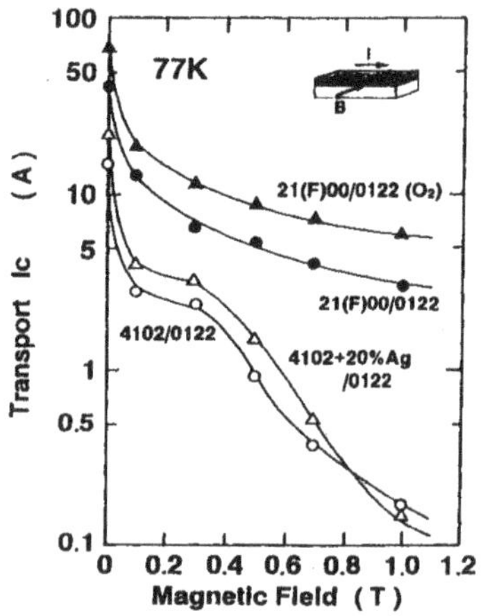

Fig.2 Magnetic field dependence of transport Ic at 77 K for the diffusion layers reacted in quoted diffusion composites.

The transport Ic of the diffusion layer post annealed in O_2 atmosphere reaches 78A which corresponds to the Jc of about 10,000 A/cm^2, referring the cross sectional area of the diffusion layer of 150μm in thickness. The post annealing in O_2 atmosphere seems to improve the density of the diffusion layer. The TBCCO diffusion layer with F addition shows much higher transport Jc at 77K than that of the YBCO and BSCCO diffusion ones. It may be remarkable, from a practical viewpoint, to achieve a relatively high Jc without any mechanical procedure, e.g. rolling or pressing.

Fig. 3 presents XRD patterns for the powder of diffusion layer reacted with additional TlF at 830 ℃ for 10 - 60 min. The 2212 phase is predominant in the diffusion layer reacted for 10 min. The 2212 phase transforms to 2223 phase and subsequently to 1223 phase with increasing reaction time. The 1223 phase becomes predominant in the diffusion layer reacted for 1 h. The improvement in transport Ic at 77K and the magnetic field dependence of Ic shown in Fig.2 seem to result from these phase transformations. The Tl-1223 phase is indicated to exhibit strong pinning under magnetic field at 77 K, although the weak link behavior still remains [6]. The F addition to the TBCCO diffusion layer promotes the phase transformation to 1223 phase without any element substitution.

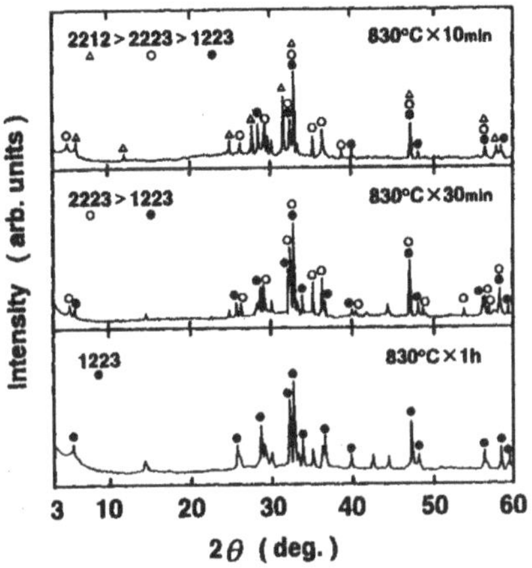

Fig.3 XRD patterns of diffusion layers reacted with TlF addition at 830 ℃ for 10 - 60 min.

Fig.4 shows AC susceptibility and resistance versus temperature for the diffusion layer with F addition using different F sources, i.e. TlF, BaF_2, CaF_2 and CuF_2. The F addition using any fluoride enhances the transformation from 2223 phase to 1223 phase referring the results obtained in the transition curves of AC susceptibility. The formation of 1223 phase causes a step around 115 K in the AC susceptibility versus temperature curve. However, the addition of F from different sources produces different microstructures in the diffusion layer.

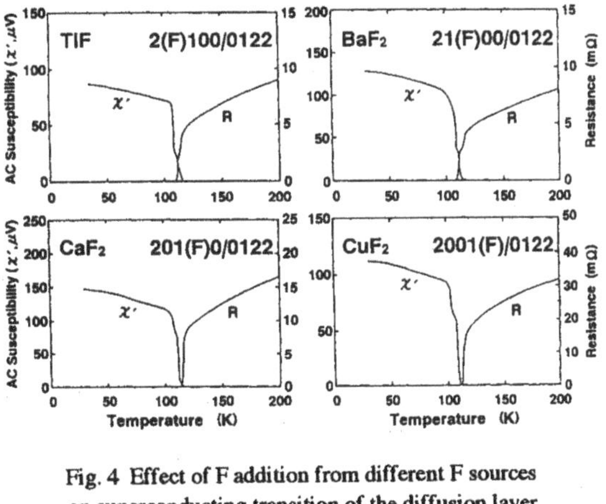

Fig. 4 Effect of F addition from different F sources on superconducting transition of the diffusion layer.

Fig. 5 shows the magnetic field dependence of transport Ic at 77 K for the diffusion layers with F addition using TlF, BaF_2, CaF_2 and CuF_2. The addition of F increases the transport Ic of the diffusion layer at 77K under 0 T compared to that of F free diffusion layer. The F addition from any sources eliminates the second drop in Ic beginning at around 0.4 T. Among the F sources, the F addition using TlF or BaF_2 yields better Ic performance at 77 K than using CaF_2 or CuF_2. The TBCCO diffusion composites with or without F addition do not exhibit the anisotropy in Ic with respect to the direction of applied field.

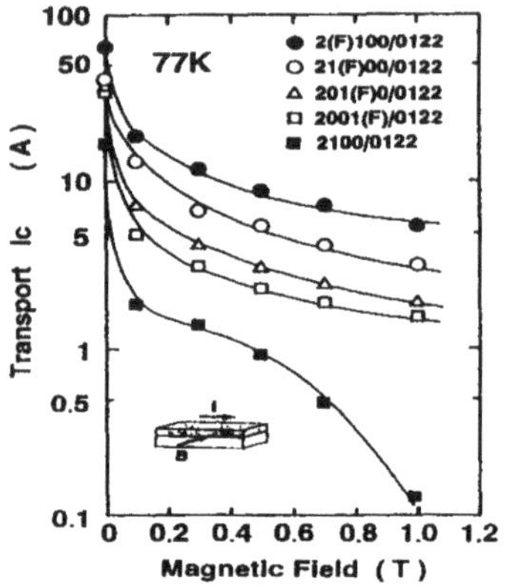

Fig. 5 Magnetic field dependence of the transport Ic at 77K for the diffusion layers with F addition from different F sources.

4. Conclusions

1. In the TBCCO system, the high-Tc 2223 phase has been easily synthesized by the diffusion reaction in a short time.
2. The F addition to the coating layer increases the transport Ic of the diffusion layer at 77K, and improves its degradation under magnetic field.
3. The F addition also promotes the transformation from 2223 phase to 1223 phase in the diffusion layer. The improvement of transport Ic and magnetic field dependence of Ic at 77 K results from the phase transformation.
4. The post annealing in O_2 atmosphere after diffusion reaction significantly increases the Ic of the diffusion layer at 77 K. The maximum transport Jc of about 10,000 A/cm2 has been achieved at 77 K in the diffusion layer with F addition and annealing in O_2.
5. The F addition using TlF or BaF_2 yields better performance of Ic of the diffusion layer at 77 K than using CaF_2 or CuF_2.

References

1) K.Tachikawa, N.Sadakata, M.Sugimoto and O.Kohno, *Jpn. J. Appl. Phys.* **27** (1988) L1501.
2) K.Tachikawa, T.Watanabe, T.Inoue and K.Shirasu, *Jpn. J. Appl. Phys.* **30** (1991) 639.
3) K.Tachikawa, K.Zama and A.Kikuchi, *Jpn. J. Appl. Phys.* **32** (1993) L654.
4) J.W.Ekin, T.M.Larson, A.M.Hermann, Z.Z.Sheng, K.Togano and H.Kumakura, *Physica C* **160** : 489 (1989).
5) M.Seido, J.Sato, T.Sasaoka, A.Nomoto,T.Kamo and S.Matsuda, *Ext. Abstr. of ISTEC/MRS nt. Workshop on Superconductivity*, Hawaii, (1992) p.236.
6) K.Aihara, T.Doi, A.Soeta, S.Takeuchi T.Iwasa, M.Seido, T.Kamo and S.Matsuda. *Cryogenics*, **32** (1992) 9356.

THERMAL EXPANSION OF $YBa_2Cu_3O_x$

H. Ogasawara, M. Matsukawa , K. Noto and H. Kimura
Faculty of Engineering, Iwate University, 4-3-5 Ueda, Morioka 020, Japan

ABSTRACT

The thermal expansion of sintered samples of $YBa_2Cu_3O_x$ has been measured between 77K and 293K using a differential transformer method. The relative length change $\Delta\ell/\ell$ are in good agreement with the averaged value of $\Delta\ell/\ell$ along the *a*, *b* and *c* axes which were determined from the X-ray diffraction. The Grüneisen constant γ obtained from the volume coefficient of the thermal expansion β is compared with that estimated from the thermal conductivity measurement.

1. Introduction

Thermal expansion coefficient gives significant information about the lattice dynamics. The Grüneisen parameter estimated from the thermal expansion measurement is related to the strength of the lattice-anharmonicity. Furthermore, the lattice-anharmonicity produces the phonon-phonon scattering (the Umklapp process) in the phonon thermal transport. Thus, it is found that the expansion coefficient is connected with thermal conductivity through the lattice-anharmonicity or the Grüneisen parameter.

In this paper, the thermal expansion of sintered samples of $YBa_2Cu_3O_x$ is studied between 77K and 293K using a differential transformer method. The Grüneisen parameter estimated from the thermal expansion measurement is compared with that calculated from the thermal conductivity measurement.

2. Experiment

The measured samples were prepared by the powder sintering method. The sample size is typically $3.5\times3.5\times6.6$ mm^3 with the density of ~80 %. The samples #1 and #2 were annealed at 600 ℃ in the oxygen atmosphere for 5 h and 20 h, respectively. The electrical transport measurement supported $T_c = 90.0$ K and $T_c = 92.7$ K for samples #1 and #2, respectively.

The thermal expansion of sintered samples of $YBa_2Cu_3O_x$ is measured between 77K and 293K using a differential transformer method.[1] The schematic diagram of the dilatometer is shown in Fig.1. The differential transformer was Shinko Denshi Co. type A22-3H, the lock-in amplifier , NF Circuit Block Co. model LI-572B. The sensitivity of this dilatometer is $\sim10^{-8}$. The thermal conductivity measurement is performed on the same sample from 10 K to 200 K with a steady-state heat-flow method.

3. Results and Discussion

The temperature dependence of relative length change $\Delta\ell/\ell$ for both samples #1 and #2 is shown between 77K and 293K in Fig. 2. In comparison, the thermal expansion data of the *a*, *b* and *c* axes from the X-ray measurement of You et al. [2] are cited. The present data of the polycrystalline sample from 180 K to 298 K are in good agreement with the averaged value of $\Delta\ell/\ell$ along the three principal crystalline axes. However, below 180 K the deviation from the reference data is gradually enhanced. This reason is not made clear at present. The corresponding linear thermal expansion coefficient $\alpha = (d\ell/\ell)/dT$ estimated from the present data are $0.99 \times 10^{-5}K^{-1}$ and $1.01 \times 10^{-5}K^{-1}$ for samples #1 and #2, respectively, at 200K.

The Grüneisen parameter γ is defined by the thermodynamic equation:

$$\gamma = \frac{V\beta B_T}{C_v} \tag{1}$$

where β ($=3\alpha$) is the volume thermal expansion coefficient, B_T the isothermal bulk modulus, C_V the heat capacity at constant volume, and V the volume of the specimen. Substituting the bulk modulus B = 563kbar (Salomons et al.[3]), and C_V = 215JK^{-1}mol^{-1} (Wette et al.[4]) into the eq. (1),then, we obtain γ = 0.82 and 0.84 for samples #1 and #2, respectively.

Next, if Umklapp processes are only operative in the phonon scattering processes, the expression of the phonon thermal conductivity κ is obtained by Leibfried and Schlomann [5] as follows,

$$\kappa = \frac{2.4 \cdot 4^{1/3}}{\gamma^2}\left(\frac{k}{h}\right)^3 Ma\Theta^2 \frac{\Theta}{T} \tag{2}$$

where γ is the Grüneisen parameter, *a* the lattice parameter, M the mass per atom and Θ the Debye temperature.

The temperature dependence of the thermal conductivity for sample #2 is shown from 10K to 200 K in Fig.3. A remarkable enhancement is observed and is associated with the superconducting transition. The electronic thermal conductivity is estimated to be less than ~10 % of the measured value using the Wiedemann-Franz law from the electrical transport measurement. Thus, the thermal carrier is mainly phonon and the contribution of electrons to thermal carrier is negligible.

The phonon thermal conductivity is discussed applying a relaxation-time approximation,

$$\kappa = AT^3 \int_0^{\Theta/T} \frac{x^4 e^x}{(e^x - 1)^2}\tau dx \tag{3}$$

where A, Θ and τ are a proportional constant, the Debye temperature and the relaxation time of phonon. In this calculation, phonon scatterers are considered to be boundaries, point-defects, sheetlike faults, excitations in the superconducting state and other phonons (Umklapp processes). The detailed expressions in this analysis are referred to in ref.6. Of course, the expression in eq.(3) is reduced to eq.(2) if phonon scatterers are negligible except for the Umklapp processes. The fitted curve shown in Fig.3 reproduces the experimental data. The Grüneisen parameter estimated from the present calculation become 1.97 for sample #2 (Θ=350 K).

4. Summary

(1) The thermal expansion of sintered samples of $YBa_2Cu_3O_x$ is studied between 77K and 293K using a differential transformer method. The sensitivity of the dilatometer in this measurement is ~10^{-8}. The relative length change $\Delta \ell / \ell$ for both samples #1 and #2 are in good agreement with the averaged value of $\Delta \ell / \ell$ along the three principal crystalline axes which were detemined from the X-ray diffraction. The Grüneisen parameters estimated from the thermal expansion measurement are 0.82 and 0.84 for samples #1 and #2, respectively.

(2) The thermal conductivity measurement is performed on the same sample of $YBa_2Cu_3O_x$ from 10 K to 200 K with a steady-state heat-flow method. The fitted curve by the relaxation time approximation of phonon thermal transport reproduces the experimental data. The Grüneisen parameter estimated from the present calculation including the Umklapp processes becomes 1.97 for sample #2.

(3) It is found that the Grüneisen parameter estimated from the thermal expansion measurement is comparable with that calculated from the thermal conductivity measurement.

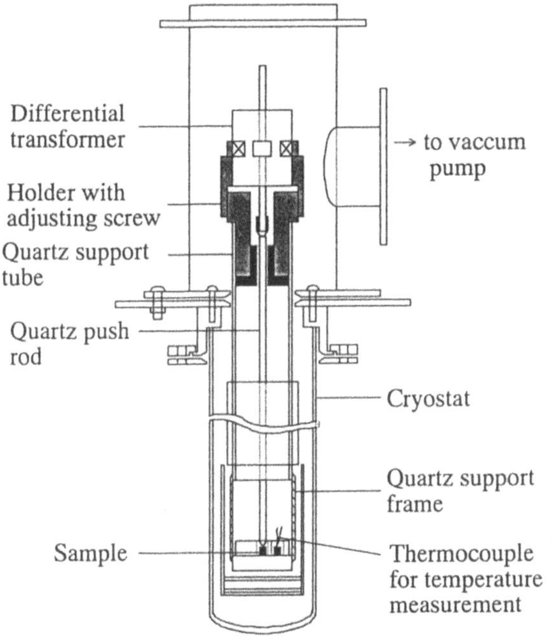

Fig.1 The schematic diagram of the dilatometer is shown.

References

1. M.Abe,K.Uchino,and S.Nomura, *Oyo Buturi* **44** (1975) 955. [in Japanese]
2. H.You,J.D.Axe,X.B.Kan,S.Hashimoto,S.C.Moss,J.Z.Liu,G.W.Crabtree and D.J.Lam, *Phys.Rev.* **B38** (1988) 9213.
3. E.Salomons,H.Hemmes,J,J,Scholtz,N.Koeman,R.Brouwer,A.Driessen, D.G.de Groot and R.Griessen, *Physica* **145B** (1987) 253.
4. F.W. de Wette and A.D.Kulkarni, *Phys.Rev.* **B42** (1990) 6707.
5. P.G.Klemens, *Solid State Physics* **Vol. 7** (1958) p46.
6. L.Tewordt and T.Wolkhausen, *Solid State Commun.* **70** (1989) 839.

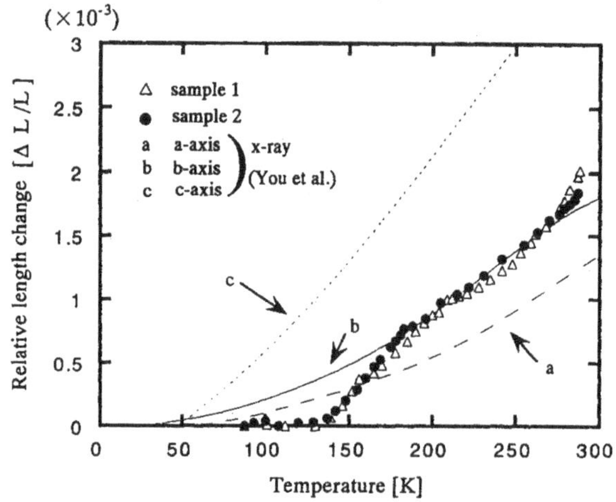

Fig. 2 The temperature dependence of relative length change $\Delta\ell / \ell$ for both samples #1 and #2 is shown between 77K and 293K.

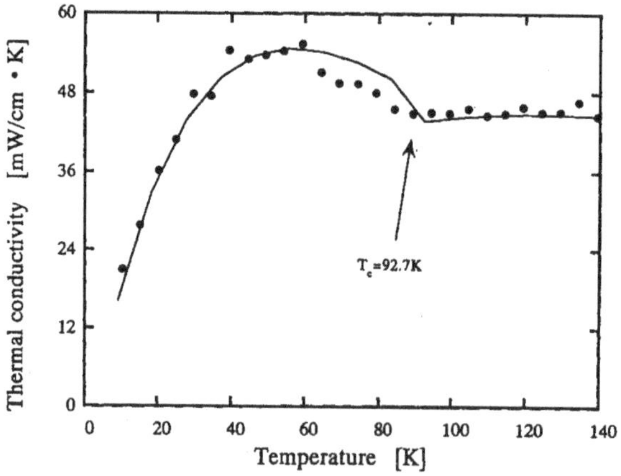

Fig. 3 The temperature dependence of the thermal conductivity for sample #2 is shown from 10K to 140 K The solid line denotes a calculation curve.

EFFECT OF SILVER RICH LAYER AROUND GRAIN ON J_c CHANGE UNDER WEAK MAGNETIC FIELD OF Bi-OXIDE

K. Ishizuka*, H. Yamamoto*, S. Fujino*, T. Ozaki*, K. Oguri** and Y. Nishi**
*Student of Materials Science, Tokai University, 1117, Kitakaname, Hiratuka, Kanagawa 259-12, Japan.
**Department of Materials Science, Tokai University,1117,Kitakaname, Hiratuka, Kanagawa 259-12, Japan.

ABSTRACT

The temperature dependence of J_c just below offset temperature of superconducting transition is observed under different magnetic fields of Bi-Sr-Ca-Cu-O which grains are coated with silver. Using a scaling-law $J_c = J_c^0 \cdot [1-(T/T_c)^2]^m$, we determines the exponent m. The exponent m is above 4 .

1. INTRODUCTION

In order to achieve high J_c values in HTS, many works have been reported[1-6]. The silver addition often increases the J_c value for HTS[7,8]. It may modify the weak link at grain boundary. Thus, we investigate the effect of silver rich layer around the grains on J_c of Bi-Pb-Sr-Ca-Cu-O.

2. EXPERIMENTAL PROCEDURE

Samples with nominal composition were prepared from high purity powders of CuO (99.9 mass %), Bi_2O_3 (99.99 mass %), Pb0 (99.9 mass %), $SrCO_3$ (99.99 mass %) and $CaCO_3$ (99.99 mass %). The powders were mixed and sintered in air at 1023 K for 6 h and then air-cooled. After crushing, the sintered powders were resintered for 18 h at 1073 K in air and then air-cooled. After crushing, the sintered powders were resintered for 100 h at 1123 K in air and then furnace cooled. Using silver mirror reaction[9], the resintered powders were coated with silver after crushing. A palletized tablets, 1 mm thick and 13 mm diameter, were sintered for 100 h at 1108 K in air and then air-cooled. The J_c was measured using a standard 4-probe technique and a Keithley 182 nanovoltmeter[10]. The temperature was measured by Au*Fe-normal silver thermocouple attached to the specimen in a cryostat at equilibrium temperature[11,12]. The structure of the sample was examined by means of x-ray diffraction.

3. RESULTS AND DISCUSSIONS

Figure 1 shows the temperature dependence of J_c just below offset temperature of superconducting transition for $Bi_{1.6}Pb_{0.4}Sr_2Ca_2Cu_3O_x$ coated with silver. These samples (■, ○, ▲ and □) show the temperature dependence of J_c under selected magnetic fields (0 G, 1.8 G, 11.8 G and 21.8 G).

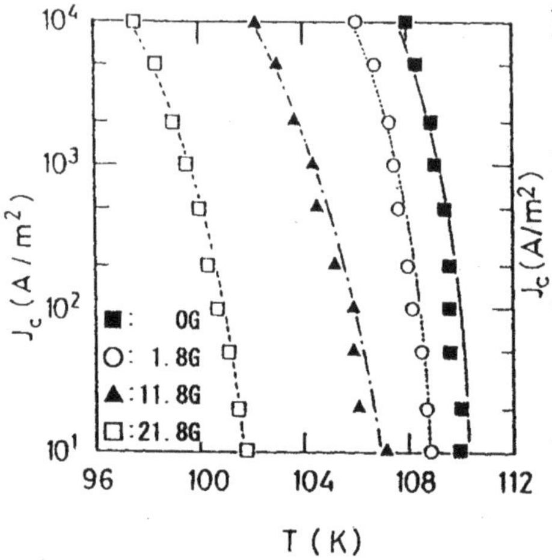

Fig 1. Temperuture dependence of J_c under selected magnetic field for $Bi_{1.6}Pb_{0.4}Sr_2Ca_2Cu_3O_x$ coated with silver.

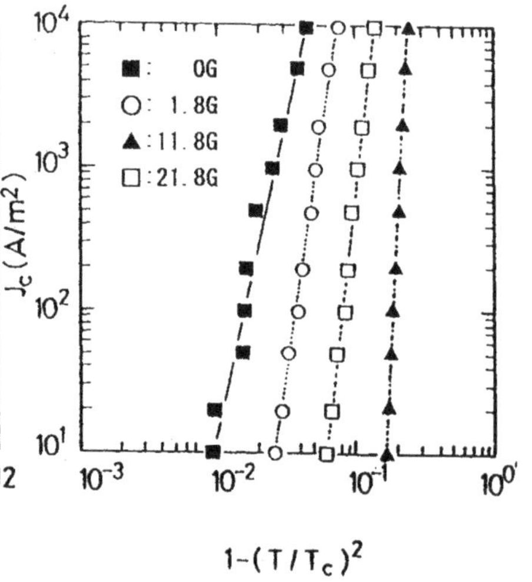

Fig 2. Change is J_c against $1-(1-T/T_c)^2$ under selected magnetic fields for $Bi_{1.6}Pb_{0.4}Sr_2Ca_2Cu_3O_x$ coated with silver.

The J_c is empirically known to be approximately expressed as[13]

$$J_c = J_c^0 \cdot [\,1-(T/T_c)^2\,]^m \qquad (1)$$

Equation (1) is determined from the pinning mechanism and then its value is useful to conjecture what kind of pinning centers contribute to the J_c value. Figure 2 shows the relationship between J_c and $1-(T/T_c)^2$ under selected magnetic fields in logarithmic scales. The J_c values under selected magnetic fields were approximately proportional to $1-(T/T_c)^2$. The exponent m value gives the temperature dependence in the conventional scaling law of the flux pinning force. In the case of planar pinning centers, such as twin and grain boundary, it is known that m takes a value of 1.5, while it is predicted to be 4 for point like defects[13].

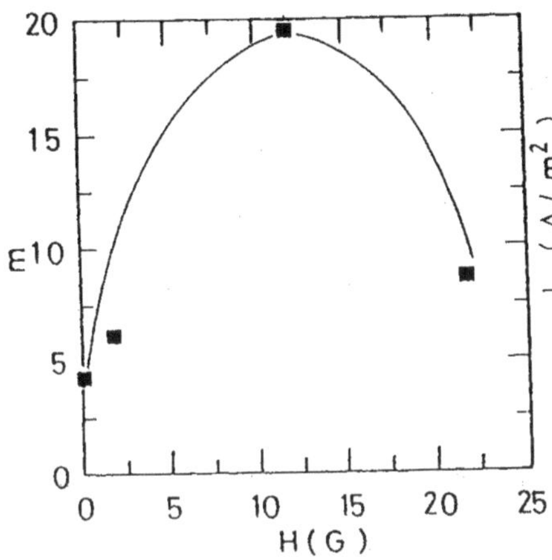

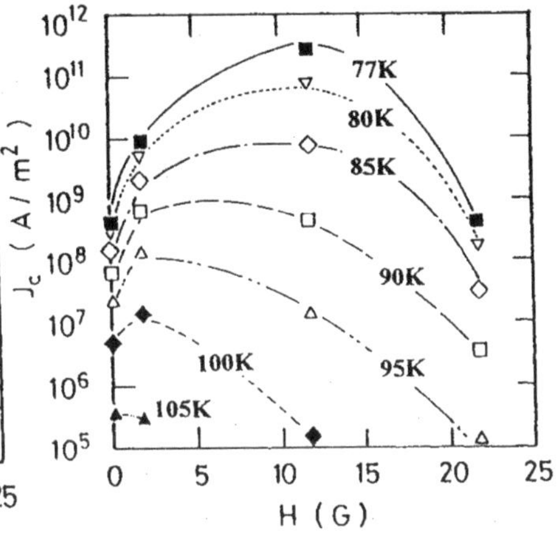

Fig 3 . Applied magnetic field H (G) and exponent m for $Bi_{1.6}Pb_{0.4}Sr_2Ca_2Cu_3O_x$ coated with silver.

Fig 4.Magnetic field dependence of calculated J_c value for $Bi_{1.6}Pb_{0.4}Sr_2Ca_2Cu_3O_x$ coated with silver.

Figure 3 shows the magnetic field dependence of the m value. However, the m value obtained in the experiment is much larger than the theoretical values. The J_c value above 77 K is estimated by using the equation (1). Figure 4 shows the magnetic field dependence of calculated J_c value. The weak magnetic field increases the J_c value. The maximum J_c value is found between 5 and 15 G. We suggest a possible explanation. The sample is mixture of the high H_c and low H_c phases. Namely, the low H_c phase exists in the sample. The low H_c phase converts to the normal state above 5 G. If fine particles of the low H_c phase exist , they act as additional pinning sites. Since the excess additional flux pinning sites enhance the J_c value above 5 G, the J_c value increases between 5 and 15 G[14,15]. The J_c value decreases above 15 G, because the normal state rapidly increases above 15 G. We often find the abnormal J_c increasing at high magnetic field of the HTS which shows the low J_c value[16-20]. It confirms our explanation. If we get new effective pinning site, the high J_c HTS will be developed.

4. CONCLUSION

In summary, the magnetic field dependence of J_c value is observed of $Bi_{1.6}Pb_{0.4}Sr_2Ca_2Cu_3O_x$ coated with silver. Using a scaling law $J_c = J_c^0 \cdot [\,1-(T/T_c)^2\,]^m$,the exponent m is estimated. The m values are above 4. Furthermore, the J_c value above 77 K is calculated for the sample. The J_c value increases against weak magnetic field. The maximum J_c value is found between 5 and 15 G.

5. REFERENCES

1. M. K. Wu, J. R. Ashburn, C. J.Torng, P. H. Hor, R. L. Meng, L. Gao, Z. J. Hung, Y. Q. Wang and C. W. Chu, *Phys. Rev. Lett*. 58, (1987) 90.
2. H. K. Maeda, Y. Tanaka, M. Fukumoto and T. Asano, *Jap. J. Phys*. 27, (1988) 209.
3. Z. Z. Sheng, A. M. Herman, A. E. Ali, C. Almasan, J. Estrada, T. Tadda and R. J. Matson, *Phys. Rev. Lett*. 60, (1988) 937.
4. S. Jin, T. H. Tiefel, R. C. Sherwood, R. B. van Dover, M. E. Davis, G. W. Kammlott, and R. A. Fastnacht. *Phys. Rev. B* 37, (1988) 7875.
5. S. Jin, T. H. Tiefel, R. C. Sherwood, R. B. van Dover, M. E. Davis, G. W. Kammlott, and R. A. Fastnacht, and H. D. Keith.. *App.Phys. Lett*. 52 (1988) 2074.
6. K. Yamamoto, M. Murakami, H. Fujimoto. N. Koshizuka and S. Tanaka. *J. Advan. Sci*. 4 (1992) 54.
7. T. Asaka, Y. Shiomi and K. Tachikawa, *J. Japan Inst. Metals*, 56, 6 , (1992) 722.
8. K. Tachikawa, T. Inoue and Y. Hikichi, *Supercond. Sci. Technol*. 5. (1991) 386.
9. B.Tollens, *Chem.Ber*, 15. (1882) 1635.
10. Y. Nishi, S. Moriya and S. Tokunaga, *Phys. Lett. A*, 126, (1987) 55.
11. Y. Nishi, K. Oguri,H. Ohinata, K. Tanioka, Y. Kita and N. Ninomiya, *Phys. Rev. B*, 41, (1990) 6520.
12. Y. Nishi, Y. Kita and K. Tanaka, *J. Mater. Sci*. 25, (1990) 5105.
13. A. M. Cambell, and J. E. Evetts, *Adv Phys*, 21, (1971) 358.
14. T. Matushita, M. Kimura, K. Miyamoto and K. Sawno, *Jpn. J. Appl. Phys*, 30, (1985) 342.
15. H. Kumakura, K. miura, *J. Appl. Phys*. 67, (1990) 3443
16. H. Kupfer, I. Apfelstedt, R. Flukiger, C. Keller, R. Meier-Hirmer, B. Runtsch, A. Turowski, U. Wiech, and T. Wolf, *Cryogenics* 29, (1989) 268.
17. M. Murakami, *J, Adv, Sci*. 2, , (1990) 2 .
18. H. Kupfer, I. Apfelstedt, R. Flukiger, R. Meier-Himer, H. Wuhl and H. Scheurer, *Z. Phy. B*, 69, (1987) 167.
19. J. W. Ekin and T. M.Larson. *Phsica. C*. 160. (1989) 489.
20. A. Kahan. *Supercond. Sci. Technol*. 6. (1993) 476.

5. Thin Films, New Materials and Others

Preparation of Highly Oriented Hg-Based Superconducting Films by Laser Ablation

Shoji Miyashita, Hiroko Higuma and Fusaoki Uchikawa
Materials & Electronic Devices Lab., Mitsubishi Electric Corp.
Sagamihara, Kanagawa 229

ABSTRACT

Pb doped Hg-1212 films were prepared on the (100)$SrTiO_3$ substrate by laser ablation and post-annealing. The Tc(mag) of these films was about 120K. These films exhibited the (00n) oriented X-ray diffraction pattern of the 1212 phase. The post-annealing temperature affected the form of the 1212 phase strongly. Highly oriented superconducting film was prepared by annealing at 760℃. The Ba-Ca-Cu-O phase and Hg-Ba-Ca-Cu-O phase (not 1212 phase) were formed at 740℃ and Ba-Cu-O phase was formed at 780℃ remarkably.

1. Introduction

Recently, the Hg-Ba-Ca-Cu-O superconductors have received much attention of many researchers because of high Tc over 120K, as high as that of the Tl-2223 superconductor[1,2]. So much efforts were made to make clear the physical properties and the structure for the Hg-Ba-Ca-Cu-O superconductor. But there were only a few reports on the Hg-based superconducting films[3,4,5].

We succeeded in preparing Hg-1212 superconducting films with Tc(ρ =0) of 97K[6] and to lower the post-annealing temperature from 850℃ to 760℃ by doping Pb to the film[7].

Our present research is to make clear the relation between the structure of the film and the annealing temperature for the Pb doped Hg-1212 superconducting film.

2. Experimental

Oxide powders of HgO, PbO, BaO, CaO and CuO were used as starting materials. The nominal cation ratio of oxide powders was Hg/Pb/Ba/Ca/Cu= 0.9/0.1/2/1/2 for a target. After mixing these powders in an agate mortar, the mixed powder was pressed into a disk as a target. We performed mixing and pressing operations in a glove box (the dew point was below -70℃) in Ar atmosphere.

KrF excimer laser (λ =248nm) was used as a source. The frequency of the laser pulse was 5Hz and the energy of irradiation was ~400mJ. A polished (100) $SrTiO_3$ single crystal was used as a substrate. The substrate was heated at 250℃ during irradiation. The distance between the target and the substrate was 40mm. The irradiation time was 45 minutes. Oxygen atmosphere in the irradiation chamber was kept to a pressure of 0.05Torr. The thickness of the as-deposited films was ~1 μm.

Then the film was immediately sealed with a Hg-Pb-Ba-Ca-Cu-O pellet in an evacuated quartz tube. The composition of the pellet was the same as that of the target. A quartz tube containing the as-deposited film and the pellet was annealed at 720℃~800℃ for 20minutes.

After annealing the materials, superconducting properties were measured and scanning electron microscopy (SEM) observation, electron-probe microanalysis (EPMA) and X-ray powder diffraction (XRD) were performed.

3. Results

3.1. Superconducting properties of the film

Figure 1 shows the typical temperature dependence of the magnetic susceptibility for the film annealed at 760℃ under the applied field of 10^{-3} T. The direction of magnetic field was perpendicular to the surface of the film with zero field cooling (ZFC) and field cooling(FC). The magnetic susceptibility of the film showed the transition point of 120K. There were variations of Tc(ρ =0) of 70~120K and it was difficult to prepare the film that had a smooth surface and low resistivity so far.

Figure 2 shows the temperature dependence of the M-H loops at 4.2K, 20K and 77K for the film annealed at 760℃. The field dependence of M at 4.2K was weak. But that at 77K was strong and ΔM was almost zero at 0.5T.

3.2. Structure of the annealed film

Figure 3 shows the XRD patterns of films annealed at 720℃~800℃ for 20 minutes. In the XRD pattern of the film annealed at 760℃, strong peaks of (00n) Hg-1212 and $SrTiO_3$ were observed. It was found that the c-axis of the 1212 phase was highly oriented perpendicular to (100)$SrTiO_3$. In the XRD pattern for the film annealed at 740℃, diffraction peaks of (00n)1212 were weak and that for the film annealed at 720℃ showed that the 1212 phase was not formed. The XRD pattern for the film annealed above 780℃ showed that unknown diffraction peaks at 2Θ=~26 and ~55 degrees became stronger in comparison with peaks of the 1212 phase. From these results, the optimum annealing temperature is 760℃.

Figure 4 shows SEM images of the film surface annealed at 740, 760 and 780℃. In the film annealed at 760℃, many platelike grains were laid on the substrate and a few platelike grains were formed on the other direction. The composition of this platelike grains was Hg/Pb/Ba/Ca/Cu = 0.65/0.04/2/1/2.1 and Pb was distributed uniformly in the grains by EPMA analysis.

In the film annealed at 740℃, the matrix was Ba-Ca-Cu-O phase and Hg-Ba-Ca-Cu-O phase was formed in the matrix. This Hg-Ba-Ca-Cu-O phase was not 1212 phase. In the film annealed at 780℃, Ba-Ca-Cu-O phase was formed mainly and Ba-Ca-O particles were formed.

By these analyses, it was clear that the annealing temperature had a strong effect on the condition of forming the 1212 superconducting films.

4. Conclusion

Pb doped Hg-1212 films were synthesized on the (100)$SrTiO_3$ substrate by laser ablation and post-annealing. The Tc(mag) of the film was about 120K. These films exhibited the (00n) oriented X-ray diffraction pattern of the 1212 phase. The post-annealing temperature influenced the formation of the 1212 phase strongly. Highly oriented superconducting film was prepared by annealing at 760℃. The Ba-Ca-Cu-O phase and Hg-Ba-Ca-Cu-O phase (not 1212 phase) were formed at 740℃ and Ba-Ca-Cu-O phase was formed at 780℃ remarkably.

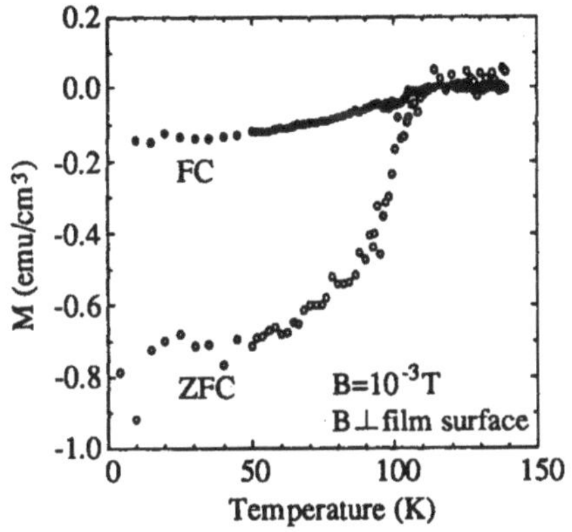

Fig. 1. Temperature dependence of magnetic susceptibility for the film annealed at 760℃ for 20 minutes.

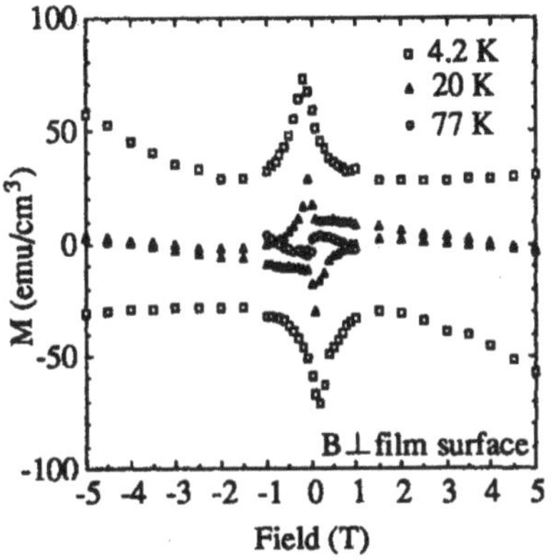

Fig. 2. Temperature dependenceof M-H loops at 4.2K, 20K and 77K for the film annealed at 760℃ for 20 minutes.

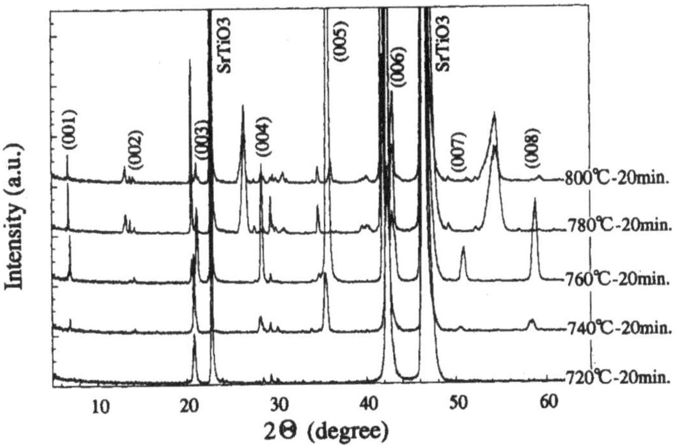

Fig. 3. X-ray diffraction patterns for annealed films.

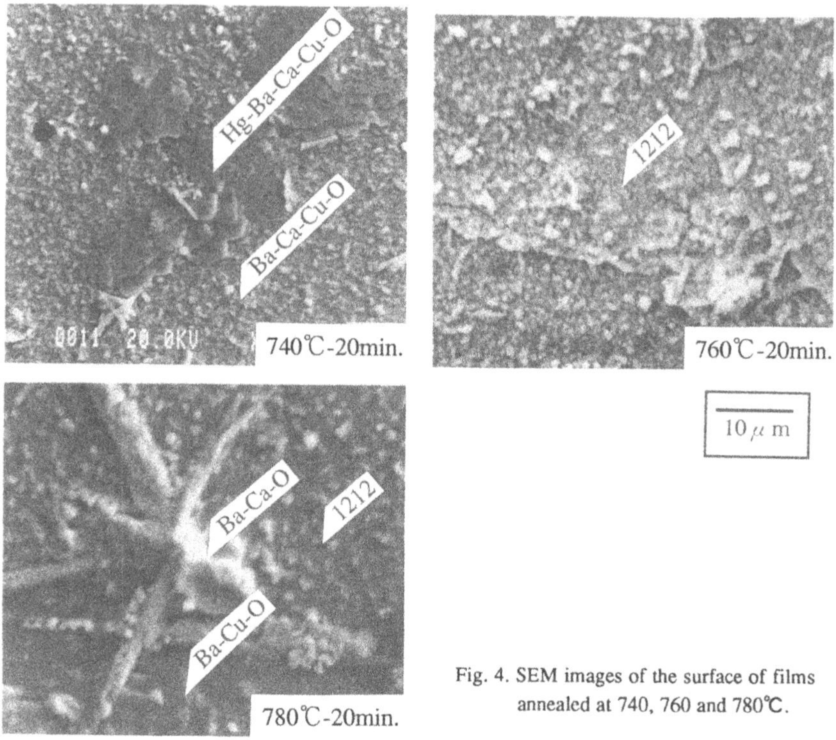

Fig. 4. SEM images of the surface of films annealed at 740, 760 and 780℃.

5. References

1. A. Schilling, M. Cantoni, J. D. Guo and H. R. Ott, *Nature* **363**(1993)56.
2. L. Gao, Z. J. Huang, R. L. Meng, J. G. Lin, F. Chen, L. Beauvais, Y. Y. Xue and C. W. Chu, *Physica C* **213**(1993)261.
3. H. Adachi, K. Mizuno, T. Satoh and K. Setsune, *Jpn. J. Appl. Phys.* **32**(1993)L1798.
4. Y. Q. Wang, R. L. Meng, Y. Y. Sun, K. Ross, Z. J. Huang and C. W. Chu, *Appl. Phys. Lett.* **63**(1993)3084.
5. C. C. Tsuei, A. Gupta, G. Trafas and D. Mitzi, *SCIENCE* **263**(1994)1259.
6. S. Miyashita, H. Higuma and F. Uchikawa, *Jpn. J. Appl. Phys.* **33** (1994)L931.
7. H. Higuma, S. Miyashita and F. Uchikawa, *Appl. Phys. Lett.* **65** (1994)743.

LASER ABLATION FOR THE FABRICATION OF MICROWAVE SUPERCONDUCTING COUPLED LINE BANDPASS FILTERS

Kwang-Yong Kang , Sang Yeol Lee and Seok Kil Han
Research Department, ETRI,
Yusong P.O.Box 106, Taejon, 305-600, KOREA

Dal Ahn
Dept. of Electronics Engineering, Soonchunhyang University,
P.O.Box 96, Onyang, 336-600, KOREA

and
Byung Ryong Moon
R&D Policy and Coordination office Ministry of Science and Technology
Gwachon, 472-760, KOREA

ABSTRACT

$YBa_2Cu_3O_{7-\delta}$ epitaxial thin films on (001) MgO substrate were grown *in situ* using laser ablation method and characterized by R-T curve, TEM, RBS, and AFM analysis. We design and simulated the optimal size and transmission responses using the microwave circuit analysis theory and commercially available design tools. The 2-pole parallel coupled-line bandpass filters with 7% bandwidth and 4-pole parallel coupled-line bandpass filters with 5% bandwidth have been fabricated YBCO films on MgO substrate (20 x 10 x 0.5 mm^3) respectively. The optimum circuits of these filters on the laser deposited YBCO epitaxial films were patterned using photolithographic process and etching process.

The performance of high-T_c superconducting parallel coupled-line bandpass filters were observed depending on temperature and frequency. The 2-pole and 4-pole parallel coupled-line bandpass filters showed microwave frequency responses of insertion losses<~1.0 dB with resonable flat passband(< 0.05 dB) in a range of 1~18 GHz below 77 K and revealed good agreement with the simulation results.

1. INTRODUCTION

A great deal of technological enhancement in the deposition techniques and physical properties of high temperature superconducting(HTS) epitaxial films open up exciting potentials for future RF and microwave system. Especially, HTS epitaxial films with use of an adequate substrate show the significantly lower surface resistance (R_s, around ~100 $\mu\Omega$ at 77 K an about 10GHz for YBCO epitaxial films) and intrinsically low microwave dissipation, which provide an ideal medium for miniaturization of device and achievement of high performance. Passive and active HTS devices such as microwave filters, mixers, antennas, and oscillators using the patterned YBCO HTS films of single layer and multilayer structures[1-3].

Most commonly used bandpass filters(BPF) are very important components in microwave circuits and system. The function of the BPF is to select a designated frequency band to let the signal pass through. Up to this time, three configuration of HTS-microstrip BPF were principally studied with diverse combinations of resonator(with matching network) and ground plane.[4] There have been some reports for multipole parallel coupled-line bandpass filters with stringent narrow bandwidths and of stable responses fabricated on high-quality YBCO films. However, cavity and dielectric resonator filters have drawbacks such as relatively large size and high cost[5].

In this paper, we will report on the design, fabriction, and performance of lowloss multipole(2-pole and 4-pole) parallel coupled-line HTS-BPF with metal films of Ti/Ag.

2. EXPERIMENT

The $YBa_2Cu_3O_{7-\delta}$(YBCO) superconducting thin films were grown on 0.5mm thick (100) MgO substrates by pulsed laser deposition processes. The incident excimer laser pulses of 308 nm wavelength were focused onto a rotating YBCO target. The laser pulses with a energy density of 1 J/cm^2 and a repetition rate of 5Hz were delivered to the YBCO target surface in an ambient of pure oxygen(5N). To grow the c-axis oriented films, the (100)MgO substrate was mounted on a substrate heater at a distance of 4 cm from the target surface. Before deposition processes, the target surface was smoothened by scraping or repolishing to enhance the reproducibility of the plume and the film quality. The YBCO on MgO films were deposited at 770°C in 200 mTorr of oxygen and the sample block temperature was lowered to 500°C in 500 Torr of oxygen after the film deposition. The characterization of ~5000Å YBCO/MgO thin films carried out by AFM, TEM, RBS, and T_c-measurements.

The 2-pole and 4-pole parallel coupled-line bandpass filters were designed for 0.02 dB ripple, a center frequency of ~11GHz and 6% bandwidth using microwave circuit analysis theory, namely transmission line and J-converter[5], and microwave design tools(Touchstone). Then they were arranged to fit on 20cm x 10cm MgO substrates. The 2-pole and 4-pole parallel coupled-line BPF-patterns are shown in Fig. 1. The BPF-patterns were prepared using photolithographic processes with e-beam mask and EDTA wet-etching processes. After that, Ti/Ag films were evaporated onto

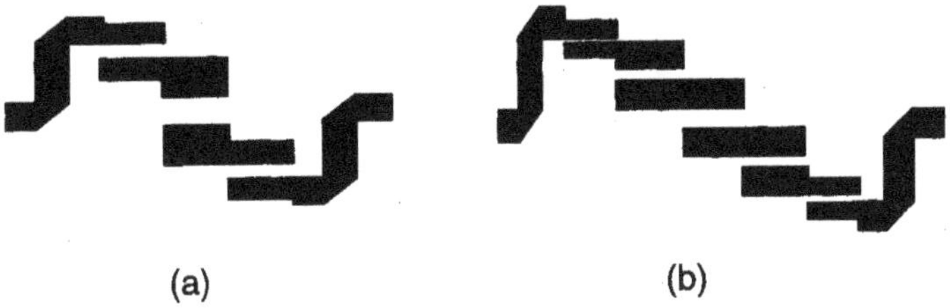

Fig. 1. Layout of 2-pole (a) and 4-pole (b)parallel coupled-line bandpass filters

the backside of substrate for good contact and matching with the text-fixture. The measurements of a packaged BPF were carried out over a frequency range of 1~18 GHz with an HP8510B vector network analyzer in a closed-cycle cryocooler(RMC-LTS-II) at temperature from 300K to 30K. The calibrations were performed to correct the insertion loss and to remove the unwanted mismatching due to both cable and connectors prior to frequency response measurement with an input power of 10 dBm.

3. RESULTS AND DISCUSSION

The crystallinity and surface morphology of YBCO on epitaxial MgO films was analyzed by TEM image, RBS analysis, and AFM image. Figure 2(a) shows the cross-sectional TEM image of c-axis oriented YBCO/MgO films, where the insertion is a SAD of laser ablated film. Figure 2(b) shows the temperature dependence of the resisrance of a 5000 Å thick YBCO film. Since the ratio of R(77K)/R(300K) is less than 0.1, YBCO/MgO films have strongly metallic behavior.

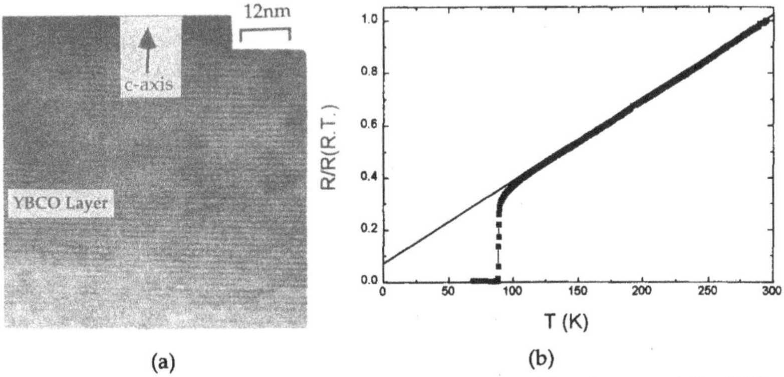

Fig. 2 .(a) Cross-sectional TEM image of YBCO/MgO film. (b) Resistivity *R(T)* of YBCO/MgO film. R(77K)/R(300K) is about 0.1.

The simulated frequency responses due to the geometry of the parallel coupled-line bandpass filters are presented in figure 3. Figure 4 shows the performance of the packaged 4-pole bandpass filter with Ti/Ag ground plane at 77K. The insertion loss around center frequency of 11 GHz, where the return loss with 4 minimum(one is buried) is better than 20dB, is less 0.5dB(maximal IL=0.02dB) The BPF shows symmetric passband shape with a steep skirt and a considerablely flat top.

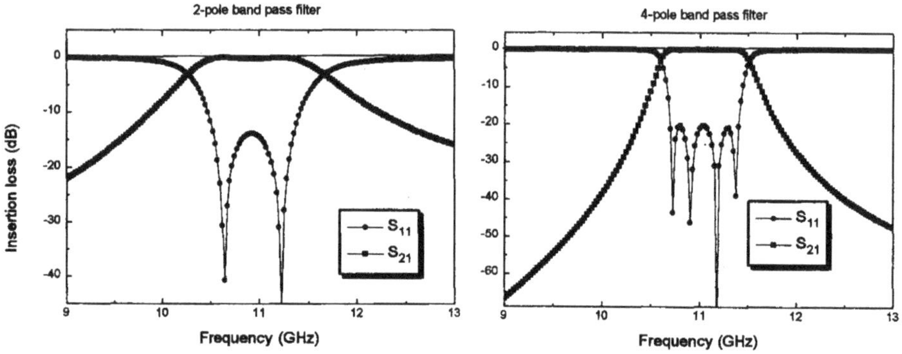

Fig. 3. Simulated frequency responses(S_{21} and S_{11}) of the parallel coupled-line bandpass filters. (a) 2-pole BPF. (b) 4-pole BPF.

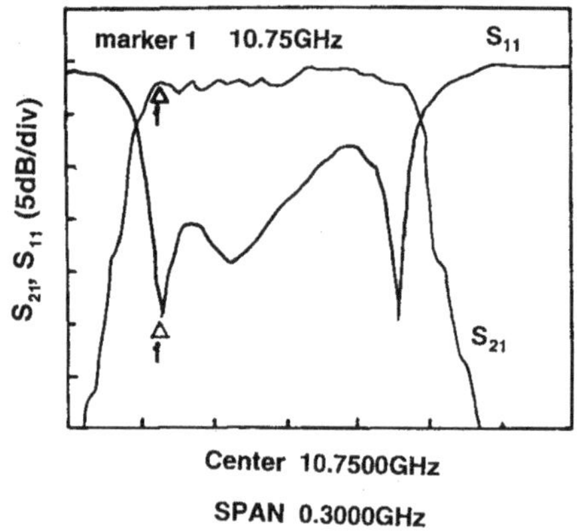

Fig. 4 . Measured frequency responses(S_{21} and S_{11}) of the 4-pole parallel coupled-line BPF.

As the temperature is increased from 30 K to 87 K of T_C, the insertion loss increases from 0.4 dB to 4 dB because of increasing microwave surface resistance. Also the center frequency shifts downwards as temperature increased. This phenomena was interpreted due to an increase in kinetic inductance and this tendency was presented in Figure 5.

From this study, we can prepare high-quality HTS-epitaxial films with suitable substrates using laser deposition system and fabricate high performance HTS-filters with normal metal ground plane.

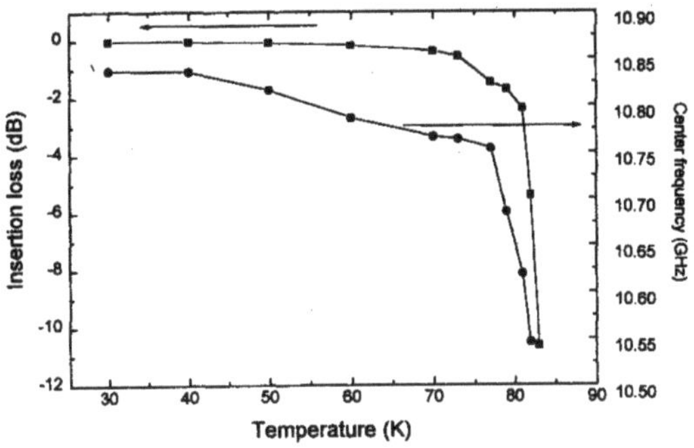

Fig. 5. Temperature dependence of insertion loss and center frequency shifts.

4. CONCLUSION

We have fabricated YBCO/MgO epitaxial thin films with a highly metallic behavior using pulsed laser deposition processes. The 2-pole and 4-pole parallel coupled-line bandpass filters with center frequency of 11GHz and 6% bandwidth were fabricated by photolithograpic and wet-etching processes. Especially, the frequency response measurements of packaged 4-pole BPF were carried out using HP-8510C vector network analyzer and RMC-cryocooler show that the insertion loss around center frequency of 11GHz, where the return loss with 4 minimum(one is buried) is better than 20 dB, is less 0.5 dB(maximal IL=0.02 dB). The BPF shows symmetric pass band shape with a steep skirt and a considerablely flat top. The potential existed to produce HTS-filters with much sharper filter skirts than bulky dielectric filters so that communication bands can be more effectively utilized.

5. REFERENCES

1. H. J. Chaloupka, M. A. Hein, and G. Muller, *"High-Tc Microwave Superconductors and Applications"* , SPIE Vol. 2156, (1994) 36.
2. N. Newman and W. G. Lyons, J. Supercond. 6(3), (1993) 119.
3. J. B. Ketterson, E. D. Rippert, S. N. Song, S. Lomatch, S. R. Maglic, H. C. Wang, D. J. Morgan, and M. P. Ulmer,*"High-Tc Microwave Supercond ductors and Applications"* , SPIE Vol. 2157, (1994) 274.
4. Zhi-Yuan Chen, "High Temperature Superconducting Microwave Circuits", Artech House, Boston (1994).
5. T. Yoshitake, H. Tsuge, T. Inui, and S. Suzuki, Jpn. J. Appl. Phys., 33(8B), (1994) L1156.
6. G. L. Matthaei, L. Young, E. M. T. Jones,*"Microwave Filters, Impedance-Matching Network, and Coupling Structure"*, Artech House Books Com. (1980).

HIGH-T_C SUPERCONDUCTING THIN FILMS FOR MULTIPOLE LOW-PASS FILTERS

Kwang-Yong Kang , Sang Yeol Lee and Seok Kil Han
Research Department, ETRI,
Yusong P.O.Box 106, Taejon, 305-600, KOREA

Dal Ahn
Dept. of Electronics Engineering, Soonchunhyang University,
P.O.Box 96, Onyang, 336-600, KOREA

and
Byung Ryong Moon
R&D Policy and Coordination office Ministry of Science and Technology
Gwachon, 472-760, KOREA

ABSTRACT

High temperature superconducting (HTS) $YBa_2Cu_3O_{7-\delta}/MgO$ epitaxial thin films were grown in situ at 770 ^0C using pulsed laser deposition technique and characterized by XRD, TEM and RBS for crystallinity, SEM and AFM for surface morphology, R vs T_c measurement for electrical analysis. The microstrip 3 pole, 5-pole, 7-pole lowpass filters have been fabricated on MgO substrate(20 x 10 x 0.5 mmt^3), after designing filters and simulating microwave responses, respectively. The optimum circuit-patterns of these filters on superconducting thin films were generated using photolithographic processes and EDTA-etching process. The performance of multipole HTS-lowpass filters were measured as a function of temperature and frequency. The 7-pole HTS-LPF with cutoff frequency of 8 GHz showed the insertion loss of ~1 dB and almost flat top in passband range.

1. INTRODUCTION

Recently high-temperature superconducting (HTS) films are much interest for application in passive microwave devices based on the linear conductivity properties of high-Tc superconductors. Especially, since HTS-epitaxial films with use of an adequate substrate show the significantly lower surface resistance(R_s) due to development of techniques of thin-film fabrication process. The high quality HTS epitaxial films show a promising applications of microwave and milimeter wave system.

Also, planar passive HTS-microwave components can be made more miniature in size than bulky ones and show better electron transport properties at frequencies below 100 GHz due to very low loss and low dispersion of HTS thin films. The potential applications for microwave system include low-loss bandpass filter(BPF) with narrow bandwidth, lowpass filter(LPF) with good skirt steepness, low phase noise oscillators, switched channelized filter bank(multiplexer), high efficiency antenna, and delay lines have used single layer structures in patterned HTS epitaxial films [1-2].

Besides, The S-N transition(from superconducting state to normal state) caused by control current can be utilized to fabricate the electrically controlled microstrip microwave LPF, namely the narrow distance between the stubs of microstrip LPF can be used as switching element controlled by dc bias current[3]. Planar superconducting microwave LPF which consist of microstrip stub and straightlines, are compact and can support a very sharp skirt-steepness to enhance frequency responses.

In this paper, we will report on the superconducting thin film science, fabrication of HTS-microwave LPF, and experimental results of a microstrip LPF with 3-component package structure, which we used the laser deposited YBCO superconducting thin films on (100) MgO substrates.

2. EXPERIMENT

Superconducting $YBa_2Cu_3O_{7-\delta}$(YBCO) epitaxial thin films were grown on (100)MgO substrates by pulsed laser deposition processes. The base pressure of the deposition was maintained at 10^{-6} Torr and incident excimer laser pulses of 308 nm wavelength were focused onto a rotating YBCO target with a repetition rate of 5Hz to generate a fluence of the energy density of 1 J/cm^2 were delivered to the target in an ambient of pure oxygen. Before depositing each film, the target surface was smoothened by scraping or repolishing to enhance the reproducibility of the plume and the film quality. To grow the superconducting YBCO epitaxial thin films, the MgO substrate was mounted on a substrate heater at a distance of 4 cm from the target surface and deposited at 770°C in 200 mTorr of oxygen. The film thickness was controlled by varing the laser shot number. The characterization of YBCO/MgO epitaxial films were carried out by AFM, TEM for analysis of surface morphology and crystallinity, and resistivity-temperature measurements.

Microstrip lowpass filter were fabricated on 5000Å YBCO thin films using photolithographic process(AZ 5214E) with e-beam mask and wet-etching process (EDTA) after design and simulation process. The optimal circuits-patterns of 3-pole microstrip LPF, 5-pole microstrip LPF, and 7-pole microstrip LPF were shown in Fig.1. The HTS-microstrip multipole LPF was packaged in a test-fixture made of brass to hold the HTS-circuit and provide an enclosure for it, associated with K-connectors(K-100) for microwave measurement. Low temperature microwave characterization was done over a frequency range of 1~18 GHz with an HP8510C vector network analyzer. The cable was calibrated to correct the insertion loss prior to frequency response measurements.

Fig. 1. Optimal test-patterns of multipole(3-pole, 5-pole, 7-pole) lowpass filters.

3. RESULTS AND DISCUSSION

AFM image for YBCO/MgO films deposited at 770°C is shown in Fig. 2. Using conventional four-point probe method, the critical temperature of YBCO thin films on MgO substrate was consistently found to be about 87 K under the various deposition conditions and showed a highly metallic behavior. The results of analysis for HTS-epitaxial thin film show the c-axis YBCO/MgO films, which have the device-qualtiy for fabrication of microwave devices.

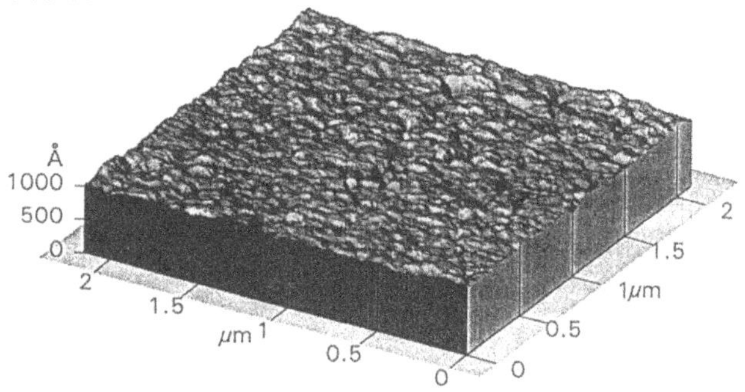

Fig. 2. AFM image of the YBCO/MgO film surface. AFM scanning area is 2μ x 2μ.

The designed and simulated results obtained from commercially available software (Touchstone) upon the multipole microstrip LPFs presented a good agreement with experimental results. Figure 3 showed the simulated transmission responses of designed microstrip multipole LPFs. As increasing poles, The skirt steepness of LPF is more sharper at cutoff frequency. A plot of frequency response of transmission (S_{21}) as a function of temperature for the HTS-LPF measured from the 300 K to 30K. Among the 3 kinds of microstrip LPFs, the measured frequency responses of 7-pole

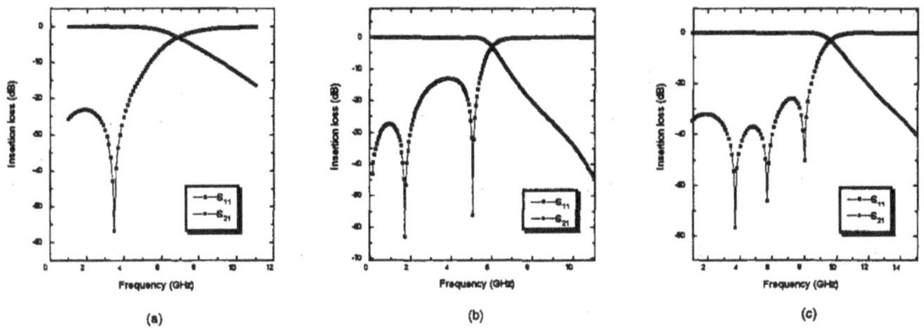

Fig. 3. Simulated microwave response of 3 kinds of microstrip lowpass filters.

LPF packaged in the test-fixture was shown in Fig. 4 with an input power of 10 dBm.

Microwave LPFs are important devices in the electrically controlled microwave components. As the number of poles increases, the skirt steepness of stopband range is decrease. In the range of passband, passband below the cutoff frequency of 8 GHz have very flat with very small ripple.

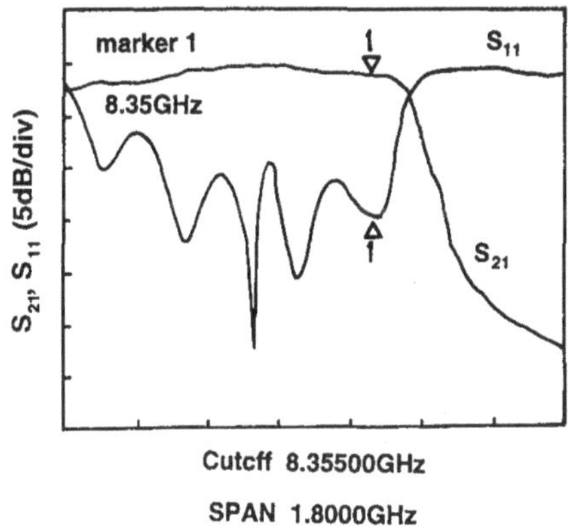

Fig. 4. Measured microwave responses(S_{21}, S_{11})of 7-pole microstrip lowpass filter.

4. CONCLUSION

YBCO epitaxial thin films with strongly metallic behavior were deposited on MgO substrates by pulsed laser deposition technique. Microstrip 7-ploe LPF with a cutoff frequency of 8 GHz was fabricated using above high-quality YBCO films with the Ti/Ag metal film ground plane of the backside of substrate. The frequency response measurement shows that

HTS-multipole LPF has a much lower passband insertion loss(S_{21}) of 0.1 dB below 77 K.

From the comparision of the experimental characteristics and simulated ones of the multipole lowpass filters, The frequency responses of 7-pole LPF is almost same as the simulated results of the designed 7-pole LPF. These results expected that the miniature and lowloss HTS microstrip LPF with 3-components text-jig and outstanding performance can be achieved for RF and microwave communications.

5. REFERENCES

1. K. Suzuki, K. Hayashi, A. T. Murphy, M. Fujimoto, K. Yamaguchi, S. Yoshikawa, T. Takennaka, and Y. Enomoto, "Advances in Superconductiivity VI", Proc. of ISS '93 (1994) 1129.
2. G. K. G. Hohenwarter, E. K. Track, R. E. Drake, and R. Patt, IEEE Trans on Appl. Supercond., vol.3(1), (1993) 2804.
3. R. Forse and S. Rohlfing, "High-Tc Microwave Superconductors and Applications', SPIE vol. 2156,(1994) 80.
4. S. A. Gal'chenko, D. I. Kaparkov, V. V. Kuznetsov, M. Leofgren, V. M. Pchelkin, A. W. Shchepak, and I. B. Vendik, Proc. of 23rd European Microwave Conf.,(1993) 848.

EXCIMER LASER APPLICATIONS FOR MICROSTRIP DUAL-MODE SUPERCONDUCTING RESONATORS WITH PERTURBED GEOMETRY

Sang Yeol Lee and KwangYong Kang
Research Department, ETRI,
Yusong P.O.Box 106, Taejon, 305-600, KOREA

and

Dal Ahn
Dept. of Electronics Engineering, Soonchunhyang University,
P.O.Box 96, Onyang, 336-600, KOREA

ABSTRACT

The comparison of microwave characteristics between the disk type and ring type dual-mode microstrip resonators (10 GHz) using $YBa_2Cu_3O_{7-x}$ (YBCO) superconducting thin films fabricated by pulsed excimer laser were carried out. We report on the microwave characteristics of microstrip dual-mode resonators and epitaxial YBCO thin films deposited on MgO substrate. Unlike the unperturbed case, perturbed dual-mode resonators showed different resonance properties. The fundamental resonant peak is splitting for the perturbed case.

1. INTRODUCTION

The discovery of high temperature superconductivity (HTS) in copper-oxide based ceramics has opened up great potentials for electronic device applications by using laser ablation technique [1]. One of the first applications of HTS materials include fabrication of passive microwave circuits using single layer high temperature superconducting (HTS) thin films [2]. The major advantages of HTS application for passive microwave circuit are the relatively simple fabrication process and the superior electron transport properties in thin films to those in the bulk. Also, passive microwave devices can be made more compact in the film than in the bulk and show better electron transport properties at frequencies below 100 GHz due to very low loss and low dispersion of HTS thin films [3]. Passive and active HTS devices such as microwave resonators [4], filters [5], antennas [6], and oscillators [7] have used single layer structures in patterned YBCO HTS films.

Very low loss and very narrow bandpass filters are required to efficiently utilize the limited frequency band. Superconducting thin films are attractive because superconducting microstrip lines show very low loss. Also, to meet the requirements of high quality microstrip filters with stringent narrow bandwidths and of stable responses, a proposed structure is dual-mode configuration. For high performance communication, satellite frequency multiplexers typically require use of dual mode cavity or dielectric resonator filters to realize quasi-elliptic responses which have pass bands less than 1% [8]. However, cavity and dielectric resonator filters have drawbacks such as relatively large size and high cost.

Superconducting thin film novel filters, which consist of dual mode resonators, are compact and can support a very sharp bandpass skirt to enhance frequency responses. Hence, HTS dual mode resonators have been intensively investigated in this experiment to obtain resonator-based high quality filter characteristics.

In this paper, we will report on the superconducting thin film preparation, frequency measurements, and experimental results of a dual mode disk resonator, which we have developed using YBCO superconducting thin films on (100) MgO substrates.

2. EXPERIMENT

YBCO superconducting thin films were grown on (100) MgO substrates by pulsed laser deposition. Excimer laser pulses of 308 nm wavelength were focused onto a rotating YBCO target as shown in Fig.1. The laser pulses had energy density of 1 J/cm^2 on the target surface and were delivered to the target at a repetition rate of 5 Hz in an ambient of

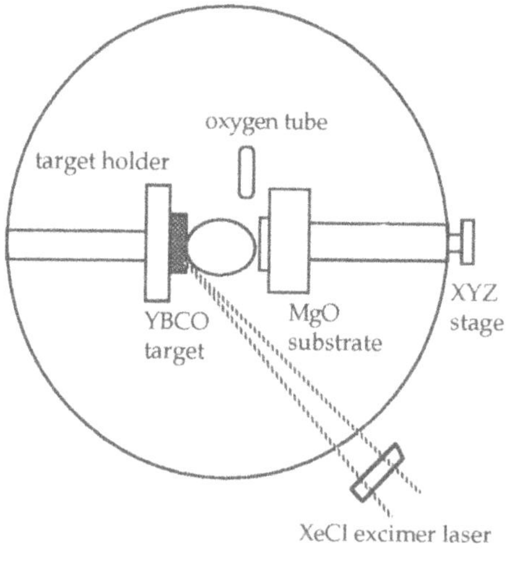

Fig. 1. Schematic diagram of pulsed laser deposition system.

pure oxygen. The MgO substrate was mounted on a substrate heater at a distance of 4 cm from the target surface. Before depositing each film, the target surface was smoothened by scraping or repolishing to enhance the reproducibility of the plume and the film quality. Superconducting thin films were deposited at 770°C in 200 mTorr of oxygen and then the sample block temperature was lowered to 500°C in 500 Torr of oxygen. Films thus prepared were found to be superconducting. The transition temperatures of the deposited superconducting films were measured about 87 K using conventional four point probe method. YBCO superconducting thin films on (100) MgO substrates were grown in various conditions to find an optimum deposition condition. The critical temperature of YBCO thin films was consistently found to be about 87 K under the above described deposition conditions. A detailed account of the laser ablation technique was published elsewhere [9].

Dual mode disk resonators (DMDRs) were fabricated on 5000 Å YBCO thin films using photolithographic pattern. The perturbation has been added to a dual mode resonator at a position of 135 degrees away from the axes of the coupling. The superconducting DMDR on MgO was packaged in a test jig, using K-connectors for microwave measurement. Testing was done over a frequency range of 1~26.5 GHz with an HP8510B vector network analyzer. The cable was calibrated to correct the insertion loss prior to frequency response measurement. A plot of frequency response of transmission (S_{21}) for the superconducting dual mode disk type resonators with a perturbation measured at 77 K was shown in Fig. 2 with an input power of 10 dBm.

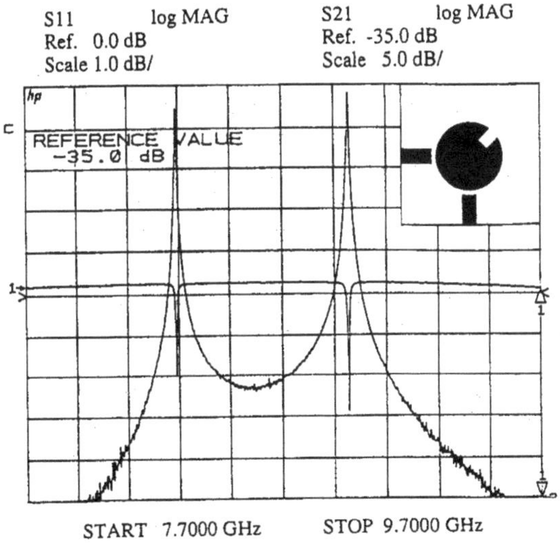

Fig. 2. Frequency response of superconducting perturbed DMDR measured at 77 K.

3. RESULTS AND DISCUSSION

The peak split of the fundamental resonance was clearly observed at 8.3 GHz and 8.97 GHz in a superconducting DMDR. This peak split is caused by introducing a perturbation into the DMDR structure. We could not observe any peak split in superconducting DMDR without a perturbation. The loaded quality factor (Q_L) was calculated to be 1,013 from 3 dB points.

To compare the characteristics of dual mode resonators of superconductor and those of normal metal, the disk-type dual mode copper resonators were made on printed circuit board. The measured frequency response shows almost same trend of superconducting DMDR except the low values of unloaded quality factors of 350.

4. CONCLUSION

We have fabricated a novel planar dual mode resonator with a perturbation using YBCO superconducting thin films. Epitaxial films of YBCO were deposited on MgO substrates by pulsed laser deposition technique. Perturbed superconducting DMDR revealed Q_{UL} value of 1,312 at 77 K. The frequency response measurement shows an obvious peak split at 8.29 GHz and 8.95 GHz of the fundamental resonance due to the introduction of perturbation into the superconducting DMDR. This type of superconducting DMDR can be used to implement dual mode resonator filters.

5. REFERENCES

1. J. G. Bednorz and K. A. Muller, Z.Phys.B 64, (1986) 189.
2. R.S. Withers, A.C. Anderson, and D.E. Oates, Solid State Tech., (1990) 83.
3. H. Asano, M. Sato, and T. Konaka, Appl.Phys.Lett. 58, (1991) 2981.
4. D. E. Oates and A. C. Acderson, IEEE Trans. Magn. MAG-27, (1991) 867.
5. D. Kalokitis, A. Fathy, V. Pendrick, E. Belohoubek, A. Findikoglu, A. Inam, X. X. Xi, T. Venkatesan, and J. B. Barner, Appl.Phys.Lett. 58, (1991) 537.
6. C. E. Gough, S. K. Khamas, T. S. M. Maulean, M. J. Mehler, N. McN Alford, and M. M. Hammer, IEEE Trans. Magn. MAG, vol. 25, 1313 (1989).
7. D. B. Laubacher, Z. Y. Shen, P. P. Pang, C. Wilker, and W. L. Holstein, Proc. Int. Symp. Superconductivity, Tokyo, (1991) 753.
8. J. A. Curtis and S. J. Fiedziuszko, 1991 IEEE Int. Microwave Symp. Digest 2, (1991) 443.
9. S. Y. Lee, K. Y. Kang, C. Y. Lee, and C. S. Yoon, Proc. SPIE, vol. 2156, (1994) 216..

STRAIN EFFECTS ON CRITICAL CURRENT DENSITIES OF SPUTTERED NbN FILMS

- The fabrication of NbN film on copper and the dependence of critical current density on the thickness and strain relation

D. Ishiko, M. Nunogaki, and T. Okada
ISIR, Osaka Univ., Mihogaoka, Ibaraki, Osaka 567, Japan

ABSTRACT

A tape specimen of NbN film coated on the copper substrate whose elastic region was small was fabricated and the characteristics of critical current density against strain was investigated experimentally. The NbN film was fabricated by means of the reactive sputtering process making use of a niobium target. The characteristics of strain was studied at about 4.2K in 8T. The value of $\varepsilon_{irrev.}$ of NbN films of thickness 0.5, 0.8, and 2 μm was 1.2, 0.8, and 0.3%, respectively. The thicknesses of NbN films gave the influence on the strain characteristics of NbN film. The disagreement of yielding points and each corresponding values of $\varepsilon_{irrev.}$ of the samples is different from the similar data obtained from NbN samples which were fabricated on other metal substrates[1~3].

1. Introduction

Many laborious studies about the strain effect of superconductor such as Nb_3Sn and NbTi have been carried out[4~6]. The strain effect is an essential parameter for large superconducting magnet design. Recently, the investigation for high Tc superconductor have been carried out[7], also. NbN has B1 (NaCl) type superconducting compound structure and very interesting characteristics of strain comparing with the practical superconducting wires. NbN films are produced by the reactive sputtering process using argon gas mixed with a very small quantity of nitrogen gas[8,9]. The upper critical field Bc_2 of NbN is over 20T, and the critical temperature, Tc about 16K[10]. NbN has a high radiation resistivity, too[11].

Some tape specimens of NbN films have been reported to show correlation between strain characteristics and yielding of substrates[1~3]. According to our experimental results, the strain characteristics of NbN films coated on copper varied with the thickness of films.

2. Experimental Procedure

Niobium Nitride films were fabricated by the reactive sputtering process using a niobium target. The sputtering apparatus used for fabrication of specimens were shown in Fig.1. The size of specimens including the copper substrate was $30x2x0.1mm^3$. Using argon plasma mixed with a very small quantity of nitrogen gas, a niobium target was sputtered with argon atoms by applying high voltage bias. NbN film was piled up to 0.5 μm thick on the copper substrate that was heated up to 500°C by niobium heater. Preparing conditions of specimen are shown in Table1.

We could use the specimen folder designed by Katagiri[6]. A specimen was soldered to the copper terminal. On applying uniaxial tensile strain, the critical current was measured under the constant field 8T. Jc was obtained from the critical current divided by the cross-section area of the film. The critical current was measured by the four-terminal resistivity method and its value was determined on occurrence of

Table 1 Preparing condition of specimen

Substrate		Cu
Substrate size		30x2x0.1 mm^3
Temperature		500 °C
Film thickness		0.5, 0.8, 2 μm
Gas pressure	Ar	3 mTorr
	N_2	0.1 mTorr
Target		Nb

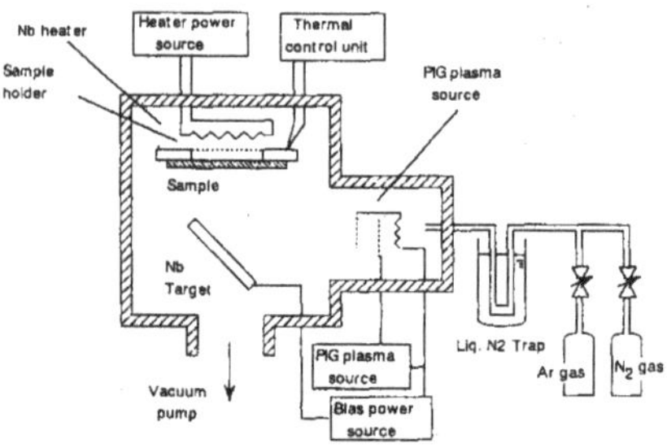

Fig.1 Schematic illustration of the reactive sputtering plasma processing apparatus

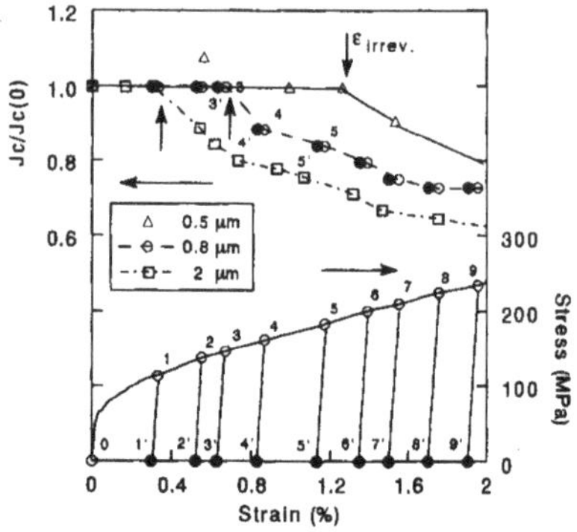

Fig.2 The characteristic curves of Jc versus strain and the stress-strain curve of NbN tape specimen coated on copper substrate. in 8T.

electric field 1μV/cm between the voltage taps. On measuring the dependence of Jc on tensile strain, load was removed halfway several times. The Jc reversibility on strain was examined. The value of reversible strain limit $\varepsilon_{irrev.}$ of the NbN films which suffered permanent damage was fixed as a criterion of degradation in the characteristics of Jc versus strain. Field was applied parallel to specimen surface and perpendicular to the direction of current.

3. Result and Discussion

The characteristic curves of Jc against strain of the specimens are shown in Fig.2. The thicknesses of NbN layer of three specimens used in the experiment were 0.5, 0.8, 2 μm and their substrates were copper of 0.1 mm thick. The numbers noted beside plots show the order of measurement of the tensile strain applied, and the number with prime indicated the order of measurement of loading after loading off once. Jc was still degraded even after loading off when the dtrain was over a reverible strain limit, ε_{irrev}. The $\varepsilon_{irrev.}$ value of NbN films whose thicknesses were 0.5, 0.8, and 2 μm were 1.2, 0.8, and 0.3%, respectively. The $\varepsilon_{irrev.}$ characteristics of specimen was dependent on the film thickness. The increase of film thickness made $\varepsilon_{irrev.}$ and the strain characteristics of Jc degraded. It is considered that long reactive time for increase the thick film promoted the growth of crystallized grain and made the film brittle[12].

According to the data in the past which used the other metal substrate[1~3], the occurrence of degradation of Jc caused by strain in NbN films was almost corresponding to the yielding of substrates, when strain gave rise to the plastic deformation to the substrate. In this work, however, the value of $\varepsilon_{irrev.}$ of the tape specimen composed with the copper substrate evidently different from yielding point.

The characteristics curve of Jc versus strain related to our tape specimens and the NbN specimens composed with the different substrate materials are shown in Fig.3. The value of $\varepsilon_{irrev.}$ of specimens composed with the copper substrate was larger than that of 0.3μm thickness of NbN film formed on 0.1mm thick SUS304 substrate by 0.4%, that of 0.3μm thickness of NbN film formed on 0.05mm Hastelloy B substrate by 0.8%, and that 0.6μm thickness of NbN film formed on 0.05mm niobium substrate by 0.3%.

Because of the difference of thermal contraction between film and substrate, NbN film is subjected to the compressive stress from substrate as the result of cooling from 500°C to LHeT. If the substrate is thicker than the film sufficiently, the prestrain of the NbN phase is expressed as follows,

$$\varepsilon = (\alpha_{NbN} - \alpha_{Sub}) \Delta T \qquad \text{- - - (1)}$$

where α is the thermal expansion coefficient, ΔT is the change of temperature. Calculated compressive prestrains of the NbN film with copper substrate, Hastelloy B and SUS304, and niobium substrate were 0.6%, 0.2%, 0.5%, and -0.26%[2], respectively.

The largest values of $\varepsilon_{irrev.}$ of NbN thin film with copper substrate was considered to be caused by the larger compressive prestrain than those of the others. It was observed that the characteristics of Jc versus strain was not corresponding to the yielding of the substrate directly.

After the tensile test up to 2% strain, the surface of 2 μm NbN film was observed by SEM as shown in Fig 4. The vertical direction of this photo is the tensile loading direction. The cracks in the surface were perpendicular to the

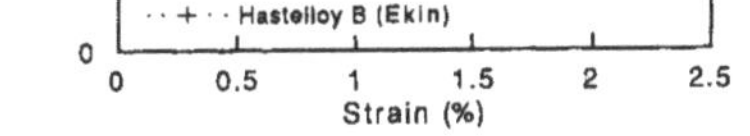

Fig.3 The comparison of the characteristics of critical current density versus strain of tape specimens with different substrate materials

Fig.4 The SEM photograph of the surface of 2 μm NbN film after the tensile test up to 2%. strain.

loading direction. The depths of these cracks were estimated not to divide the NbN layer perfectly, then, the superconducting current flowed still even after occurrence of cracks made by applied tensile strain. The cracks were seen to start from the spots formed with impurities which were distributed scatteringly over the film.

4. Summary

The thin layer of NbN was fabricated on the surface of copper substrate which was 0.1 mm in thickness and 2mm in width. Such specimens were investigated about the characteristics of Jc versus strain at 4.2 K in 8T. The obtained results are summarized as follows;

1. The disagreement of yielding point and $\varepsilon_{irrev.}$ was observed. It was different from the data reported until now which were obtained in experiment using the NbN films coated on the other kinds of metal substrates.
2. In comparison to the data of different metal substrates, it was considered that the reason of increase of $\varepsilon_{irrev.}$ value is because the compressive prestrain caused by the difference in thermal contraction on fabricating and testing was larger than that of the sample made with other substrates.
3. The values of $\varepsilon_{irrev.}$ of NbN films whose thicknesses were 0.5, 0.8, and 2 μm were 1.2, 0.8, and 0.3%, respectively. It may be that the long reactive time for obtaining a thick NbN layer made the crystalline grain big and the film brittle.
4. The crack in the NbN layer which was created in perpendicular to the loading direction was observed in the surface of specimens strained by SEM.

5. Acknowledgments

This investigation was carried out at Low Temperature Center in Osaka Univ. We thank Mr. Wakisaka and Mr. Makiyama who supplied liquid helium. We thank Dr. Nishijima and Professor Katagiri for lending us the experimental apparatus.

6. References

1. J. W. Ekin, J. R. Gavalar and J. Gragg, *Appl. Phys. Lett.* **41** (1982) 996.
2. J. R. Gavaler, J. Greggi, R. Wilmer and J. W. Ekin, *IEEE Trans. Magn.* **MAG-19** (1983) 418.
3. M. Suzuki, T. Kibosi, T. Sakurai and A. Nagata, *TEIONKOUGAKU* **25** (1990) 383
4. J. W. Ekin, *Adv. Cryog. Eng. Mater.* **30** (1984) 823.
5. J. W. Ekin, *Cryogenics* **20** (1980) 611.
6. K. Katagiri, M. Ohgami, T. Okada, T. Fukutsuka, K. Matsumoto, M. Hamada, K. Noto, K. Watanabe and A. Nagata, *Adv. Cryog. Eng. Mater.* **36** (1990) 85.
7. T.Kuroda,M.Yuyama,K.Itoh and H.Wada, *Adv.Cryog.Eng. Mater.* **38** (1992) 1045.
8. T.Mitszuoka,T.Yamasita,T.Nakazawa and Y.Onodera, *J.Appl.Phys.* **39** (1968) 4788.
9. K. S. Keskar, T. Yamasita and Y. Onodera, *Jpn. J. Appl. Phys.* **10** (1971) 370.
10. T.Yamasita,S.Kitahara,Y.Onodera,Y.Goto and T.Aso, *J.Appl.Phys.* **43** (1972) 4749.
11. H. Weber, P. Gregshammer, K. E. Gray and R. T. Kampwirth, *IEEE Trans. Magn.* **MAG-25** (1989) 2080.
12. M. Ashkin, J. R. Gavaler, J. Greggi and M. Decroux, *J. Appl. Phys.* **55** (1984) 1044

THE FABRICATION OF Nb_3Al THIN FILM SUPERCONDUCTOR BY RAPID THERMAL PROCESSING

J. Y. Shim, K. W. Kim, H. D. Han, H. K. Baik,
Department of Metallurgical Engineering, Yonsei University, Seoul, Korea

and

S. S. Oh, D. W. Ha, K. S. Ryu
Korea Electrotechnology Research Institute, Changwon, Korea

ABSTRACT

Interfacial reactions of Nb/Al/Nb thin films on Nb substrate have been studied through rapid thermal annealing. It is expected that RTP can be applied for superconductor fabrication process. Nb_3Al is formed simultaneously with Al_2O_3 at 850℃. We confirmed that the oxygen source for the formation of Al_2O_3 came from the surface and the Nb/Al interface. The Al-oxides greatly retarded the reaction between Nb and Al and, thus, degraded the superconducting property of Nb_3Al thin film. It is considered that oxygen impurities can affect phase formation sequence in Nb/Al system. $NbAl_3$ with the greatest negative value of effective heat of formation was predicted to be the first crystalline phase using the EHF model. When the atomic concentration ratio of Nb to Al was larger than 3, the expected phase sequence was $NbAl_3 \rightarrow Nb_2Al \rightarrow Nb_3Al$.

1. Introduction

The superconductor Nb_3Al has been considered as an alternative to Nb_3Sn, since J_c of Nb_3Al at high magnetic field is comparable to that of Nb_3Sn[1]. It is well known that Nb_2Al and $NbAl_3$ as well as Nb_3Al can be formed during the Nb_3Al superconductor fabrication process. During these phases, the grain size of Nb_3Al, varing with heat treatment conditions, is crucial for determining J_c. Thus, reactions at Nb/Al interface are important in the fabrication process of A15 superconductor.

Many researchers have studied the reactions in bulk and thin film diffusion couples of Nb and Al[2-5]. It is difficult to examine the reaction in bulk diffusion couple where many phases are simultaneously formed, consistent with the equilibrium phase diagram. On the contrary, only a single phase is formed in thin film couples because the thin film interface is in non-equilibrium state, therefore, Nb/Al multilayer thin film is suitable to study the phase formation sequence in the Nb/Al interfacial reaction.

Im *et al.* observed that the Nb_3Al phase is formed first following the formation of $NbAl_3$[3]. K. Barmak and his coworkers reported that the phase formation sequence varies with the atomic concentration ratios of Nb/Al multilayer thin films. When enough Nb was supplied[5], the phase formation sequence was $NbAl_3 \rightarrow Nb_2Al \rightarrow Nb_3Al$.

It is well known that the Nb_3Al phase is formed at temperature above 700℃ in multilayer thin films but the formation of Nb_3Al is retarded to a high temperature when oxygen in Nb substrate is introduced into the Nb/Al thin film interface[2,4]. However, the effect of oxygen on the phase formation sequence has not been reported.

In this study, we presented the results of the superconducting properties of Nb_3Al thin film and the interfacial reactions in sputtered Nb/Al multilayer thin films through the heat treatments. These experimental results were compared with the theoretical results from the effective heat of formation (EHF) model which predicted the first phase formed by solid state reaction in binary thin film[6,7].

2. Experimental Methods

The Nb/Al/Nb multilayer thin films were sequentially deposited on Nb substrates using r. f. magnetron sputtering system. The vacuum level was set below 3.0×10^{-6}Torr and was maintained at 5.0×10^{-3}Torr during deposition. The purities of Nb and Al targets with 4-inch diameter were 99.9% and 99.99%, respectively. Nb sheet with 0.025mm thickness was used as substrate and contained oxygen about 200 ppm. To remove native oxide on the Nb sheet surface, Nb substrate was dipped in HF solution and was sputter cleaned with -400V bias. The thicknesses of Nb/Al/Nb layers on the Nb substrate were 126nm/100nm/200nm and the Nb to Al atomic concentration ratio of the total thickness was set at 3 to 1.

The films were annealed in the temperature range 850-1100℃ in an evacuated quartz tube furnace($p<6.0\times10^{-6}Torr$). To investigate Nb/Al thin film reactions and enhance superconducting properties, rapid thermal annealing was carried out.

3. Results and Discussion

Fig. 1 shows the results of XRD pattern for the films annealed at different temperatures. The three major peaks came from the Nb sheet substrates, regardless of annealing conditions. For each film it could be seen that the Nb_3Al and Al-oxide peaks appeared in all annealing conditions and the intensities of Nb_3Al peaks did not increase with increasing annealing temperature. However, the (111), (211) and (222) Nb_3Al peaks at 850℃ disappeared in the higher temperatures, which was caused by the reaction between oxygen and aluminium in Nb_3Al. In addition, the Al-oxides that retarded the reaction between Nb and Al simultaneously formed with Nb_3Al in all temperatures.

Auger depth profiles of the as-deposited and annealed films are shown in Fig. 2 and 3. In Fig. 2, significant amounts of oxygen existed in the film and the interface between Al and Nb layers. Oxygen diffused into the Al layer and the reaction between oxygen and aluminium resulted in the formation of Al_2O_3. K. Barmak *et al.* reported that the formation of an Al_2O_3 layer from interstitial oxygen in the Nb substrate could retard the formation of Nb_3Al[6]. We concluded that the oxygen source for the formation of Al_2O_3 came from the surface and the interface between

Al and Nb layer and, thus, it affected the phase formation sequence in Nb/Al system.

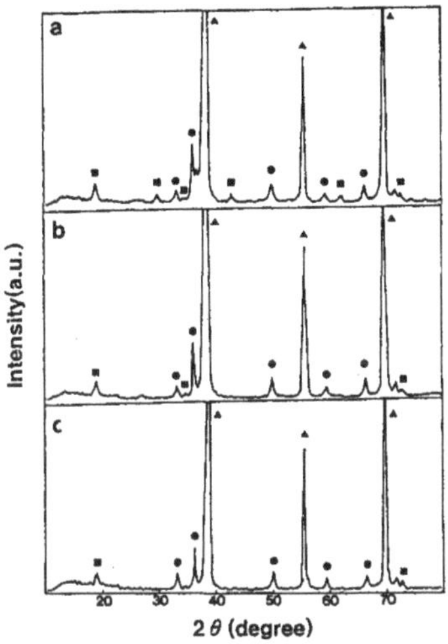

Fig. 1 XRD patterns of the films annealed at (a) 850℃, 1 min, (b) 900℃, 1 min, (c) 1000℃, 1 min. The phases are Nb(▲), Nb_3Al(■) and Al_2O_3(●)

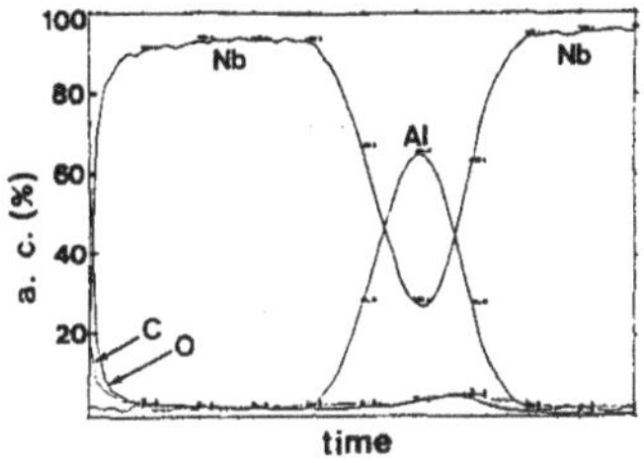

Fig. 2 Auger depth profile of the as-deposited Nb/Al/Nb film

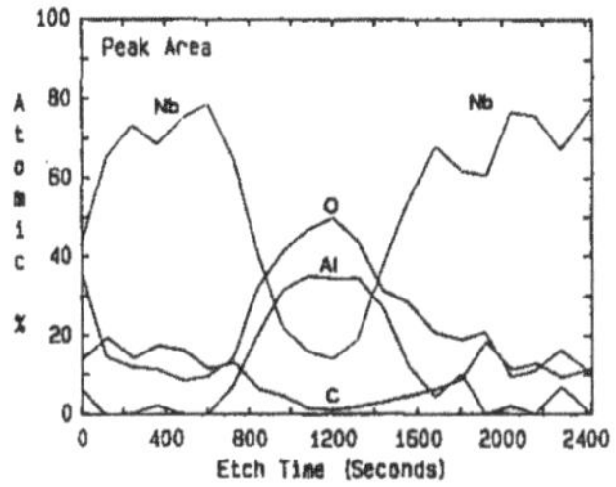

Fig. 3 Auger depth profile of the film annealed at 1000℃, 3 min

In the present study, the first phase and phase sequence, which have been identified through heat treatments and XRD analysis, were compared with predictions from EHF model. As seen in equation (1), the concept of EHF recognizes the heat of formation as a function of effective concentration[6,7].

$$\Delta H' = \Delta H\frac{C_e}{C_c} = \Delta H\cdot\frac{\textit{effective concentration limiting element}}{\textit{compound concentration limiting element}} \qquad (1)$$

The effective concentration of thin film interface in non-equilibrium state and element lacking in supply have been taken into consideration. Also, the actual number of atoms participating in solid state reaction has been applied to phase formation process. In order to predict first phase formed in Nb/Al system, the EHF of all intermetallic compounds in equilibrium phase diagram have been calculated. These are shown in Table 1 and Fig. 4.

The effective concentration in Nb/Al system is $Nb_{0.020}Al_{0.980}$, which is the concentration at lowest eutectic temperature. As shown in Table 1 and Fig. 4, the heat of formation of $NbAl_3$ at effective concentration was -2.64kJ/g · atom. Thus, $NbAl_3$ with the greatest negative value of effective heat of formation was predicted to be the first phase. Inconsistent with the first phase predicted by the EHF model,

Nb_3Al was formed first with Al_2O_3 due to the high annealing temperature.

On the other hand, the phase sequence according to the atomic concentration ratio of Nb to Al may be interpreted with the EHF diagram. The interface structure of Nb/Al diffusion couple became Nb/$NbAl_3$/Al with the formation of $NbAl_3$ as a first phase. When the atomic concentration ratio of Nb to Al is larger than 3, Al consumed to form $NbAl_3$ and, then, Nb_2Al formed at the Nb/$NbAl_3$ interface. Nb_3Al formed at the Nb/Nb_2Al interface after the complete transformation of $NbAl_3$ to Nb_2Al. Consequently, the phase sequence became $NbAl_3$→Nb_2Al→Nb_3Al when the atomic concentration ratio of Nb to Al was larger than 3.

Table 1. Effective heats of formation for Nb/Al system.

Liquidus minimum = $Nb_{0.020}Al_{0.980}$					
phase	congruency	limiting element	composition	ΔH (kJ/g · at.)	$\Delta H'$ (kJ/g · at.)
Nb_3Al	NC	Nb	$Nb_{0.750}Al_{0.250}$	-19	-0.51
Nb_2Al	NC	Nb	$Nb_{0.667}Al_{0.333}$	-25	-0.75
$NbAl_3$	C	Nb	$Nb_{0.250}Al_{0.750}$	-33	-2.64
Predicted first phase = $NbAl_3$					

NC:noncongruency, C:congruency

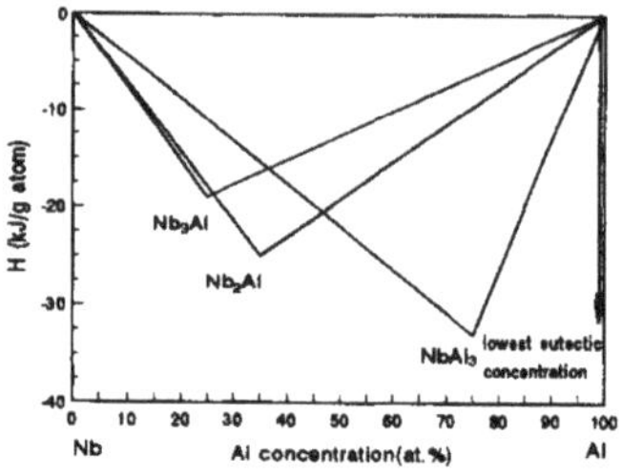

Fig. 4 Effective heats of formation diagram for Nb/Al system

4. Conclusions

We tried to develop the RTP for superconductor fabrication process. Interfacial reactions of Nb/Al have been investigated. Nb_3Al was formed simultaneously with Al_2O_3 at 850℃. The Al-oxides greatly retarded the reaction between Nb and Al and, thus, degraded the superconducting property of Nb_3Al thin film. We believe that oxygen impurities can affect phase formation sequence in Nb/Al system. When the atomic concentration ratio of Nb to Al was larger than 3, the phase sequence was predicted to be $NbAl_3$→Nb_2Al→Nb_3Al. The experimental result was different from the theoretical result which predicted Nb_3Al as the first phase.

References

1. H. Kumakura and K. Tachikawa, *Appl. Phys. Lett.* **49(1)** (1986) 46
2. J. R. Gavaler and J. Greggi, *J. Appl. Phys.* **61(2)** (1986) 659
3. Y. S. Im and J. W. Morris, Jr., *J. Appl. Phys.* **64(7)** (1988) 3487
4. K. Barmak and D. A. Rudman, *J. Appl. Phys.* **67(8)** (1990) 3780
5. K. Barmak and D. A. Rudman, *J. Appl. Phys.* **67(12)** (1990) 7313
6. R. Pretorius, *MRS Symp. Proc.* **25** (1984) 15
7. R. Pretorius and F. W. Saris, *J. Appl. Phys.* **70** (1991) 3636

MICROSTRUCTURE AND SUPERCONDUCTING PROPERTIES OF Y-T(=Ni,Pd,Pt)-B-C SYSTEM

K.TOGANO, H.FUJII, S.ARISAWA, H.KITAGUCHI,
K.HIRATA, H.KUMAKURA AND K.KADOWAKI
National Research Institute for Metals,
1-2-1, Sengen, Tsukuba 305, Japan

ABSTRACT

The investigations to achieve higher volume fraction of superconducting phase in the newly discovered Y-T(=Ni,Pd,Pt)-B-C system have been carried out. We found that the substitution of Pt for Pd in Y-Pd-B-C system stabilizes the YPd_2B_2C(1221) phase and a single phase of 1221 could be obtained by applying proper heat treatment, although Tc was depressed by the substitution. We are also applying deposition techniques to synthesize a single phase. Thin film of YNi_2B_2C with strong c-axis alignment was successfully synthesized by the magnetron sputtering and post annealing, which showed Tc~15K. Jc measurements in applied magnetic field suggested that the film has anisotropies in pinning force as well as upper critical field Hc2.

1. Introduction

The discovery of superconductivity in borocarbide compounds (RET_2B_2C(RE=rare earth, T=Ni,Pd,Pt) (1221)) has stimulated considerable interest as a new family of superconductors[1,2]. Particularly, 23K superconductivity observed for YPd_2B_2C has given us a new possible road to high Tc superconductivity[3]. However, the complex phase diagram of this type of materials makes it difficult for us to obtain a single phase of 1221 superconducting compound. This paper gives our investigation to achieve higher volume fraction of superconducting 1221 phase in the borocarbide systems.

2. Substitution of Pt for Pd in YPd_2B_2C

The superconducting YPd_2B_2C phase with Tc~23K can not be formed from the same nominal composition by conventional arc melting method[3,4]. However, one can expect that the 1221 phase would be stabilized by the substituion of Pt, since Pt has almost same atomic size with Pd and the 1221 phase is much more stable in the Pt system than in Pd system[5]. Therefore, we prepared samples with the compostions of $Y(Pd_{1-x}Pt_x)_2B_2C$ by the conventional arc melting and studied their microstructures and superconducting properties.

The x-ray diffraction and magnetization measurement showed that the x=0 sample of YPd_2B_2C does not contain superconducting 1221 phase as reported in the previous papers[3,4]. However, as the increase of x, the peaks from 1221 phase in the diffraction pattern beginned to appear at around x=0.2 and the peak intensity of 1221 phase increased with x as shown in Fig. 1(a). The signal of diamagnetism also increased with x, although the onset Tc became lower. The onset Tc for x=0.2 was 14K and dropped monotonically with x to ~7K at x=1. However, even for the YPt_2B_2C(x=1), the diffraction pattern indicated that the sample contained a fairly large amount of impurity phases together with the major phase of 1221.

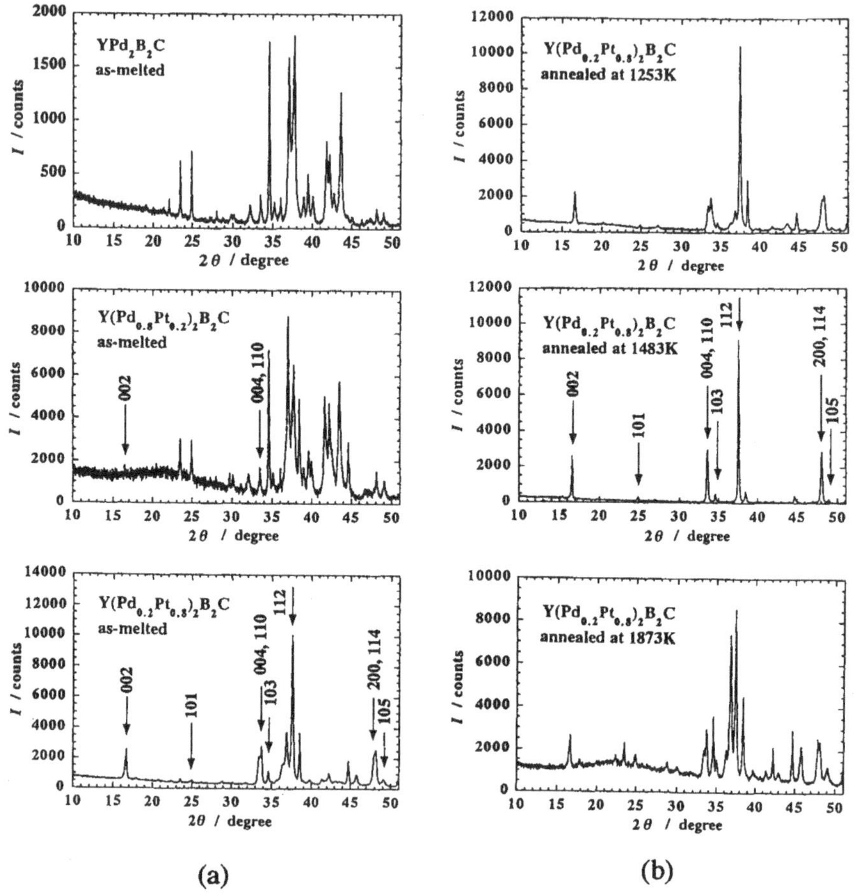

Fig. 1 (a) X-ray diffraction patterns for the x=0, 0.2 and 0.8 samples of $Y(Pd_{1-x}Pt_x)_2B_2C$ after the arc melting.
(b) x-ray diffraction patterns of the x=0.8 sample after annealed at 1253K, 1483K and 1873K.

The arc melted samples were then heat treated in vacuum at the temperatures between 800°C - 1800°C. We found that the annealing obviously increased the volume fraction of 1221 phase when the samples were annealed at the temperature between 1200-1300℃. Figure 1(b) shows the result of the x=0.8 sample. Both lower and higher temperature annealings than these optimal temperatures resulted in the decrease of 1221 volume fraction. The Tc transition became sharp by the annealing. These results indicated that the 1221 compounds is stable only at a certain temperature region at high temperature.

3. Deposition of YNi_2B_2C

Vapour deposition is one of the most effective method to form metastable phases or to retain the high temperature stable phase to room temperature. However, our attempts to form the films of YPd_2B_2C with Tc~23K by the vapour deposition technique have not yet been successful. In this paper, we introduce the successful formation of YNi_2B_2C thin film by the vapour deposition.

The deposition was carried out by a conventional magnetron sputtering apparatus.The target was powder materials prepared by milling the arc-melted ingot. The sputtered material was deposited on MgO(100) substrates maintained at room temperature under pressure of 0.5Pa of Ar. The substrates were arranged in on-axis geometry to the target. The deposition rate was approximately 0.08nm/s and the total thickness of the films was 300nm. The sample was then wrapped in a titanium foil and annealed in a vacuum furnace at several temperatures up to 1050°C for 20min. Details of the sample preparation was given in our previous paper[6].

The as deposited films were completely amorphous which is evidenced by a single broad peak. Below the annealing temperature of 900°C, only very weak reflections, which could not be identified as the 1221 phase, were observed in the diffraction pattern. However, the sample annealed at 1050°C showed obviously the transformation into crystalline state in the diffraction pattern which is shown in Fig. 2. The two sharp peaks can be indexed as (002) and (004) reflections of YNi_2B_2C structure, indicating that the film has a strong c-axis orientation.

The as deposited and those annealed at temperatures below 900°C showed no trace of superconductivity. However, the sample annealed at 1050°C showed a sharp superconducting transition with the onset of 15K, in both resistive and magnetization measurements, which was almost the same with the value reported for bulk YNi_2B_2C sample. Figure 3 shows the resistive vs. temperature curve. The critical current density Jc measured at 4.2K is ~$4 \times 10^5 A/cm^2$ in zero bias field. The Jc measurement in applied magnetic field suggested that the film has anisotropies in Jc as well as in upper critical field H_{c2}.

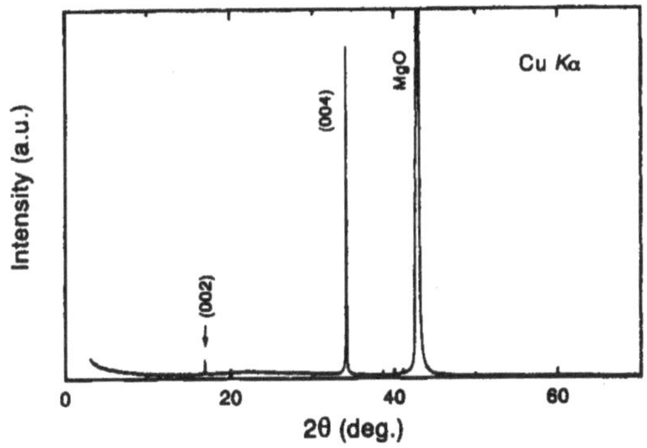

Fig. 2 X-ray diffraction pattern of the YNi_2B_2C thin film synthesized by magnetron sputtering and post-annealing.

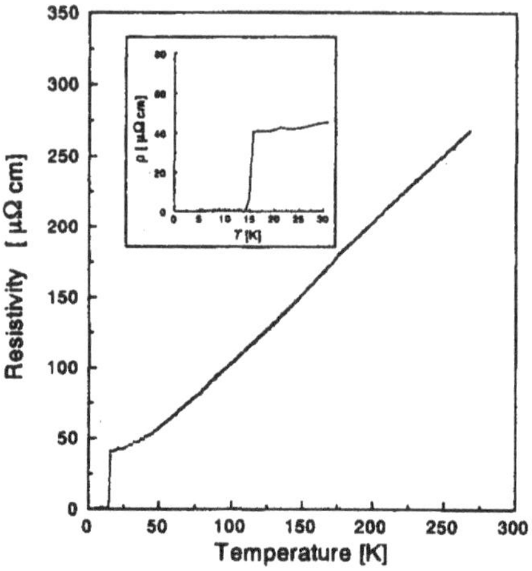

Fig. 3 Resistivity vs.temperature curve for the YNi_2B_2C thin film synthesized by magnetron sputtering and post-annealing.

4. Conclusion

We found that the substitution of Pt for Pd in Y-Pd-B-C system stabilizes the 1221 phase and a single phase of 1221 could be obtained by applying proper heat treatment, although Tc was lowered from 23K by the substitution. We are also applying deposition techniques to synthesize a single phase. Thin film of YNi_2B_2C with strong c-axis alignment was successfully synthesized by the magnetron sputtering and post annealing, which showed Tc~15K. The formations of YNi_2B_2C film without post annealing and YPd_2B_2C film with higher Tc are now in progress.

References

1. R.Nagarajan, C.Mazumdar, Z.Hossain, S.K.Dhar, K.V.Gopalalkrishnan, L.C.Gupta, C.Godart, B.D.Padalia and R.Vijayaraghavan, Phys. Rev. Lett. 72(1994), 274.
2. R.J.Cava, H.Takagi, H.W.Zandbergen, J.J.Krajewski, W.F.Peck,Jr., T.Sieglist, B.Batlogg, R.B.van Dover, R.J.Felder, K.Mizuhashi, J.O.Lee, H.Eisaki and S.Uchida, Nature, 367(1994), 252.
3. R.J.Cava, H.Takagi, B.Batlogg, H.W.Zandbergen, J.J.Krajewski, W.F.Peck,Jr., R.B.van Dover, R.J.Felder, T.Sieglist, K.Mizuhashi, J.O.Lee, H.Eisaki, S.A.Carter and S.Uchida, Nature, 367(1994), 146.
4. H.Fujii, S.Ikeda, T.Kimura, S.Arisawa, K.Hirata, H.Kumakura, K.Kadowaki and K.Togano, Jpn. J. Appl. Phys. 33(1994), L590.
5. R.J.Cava, B.Batlogg, T.Sieglist, J.J.Krajewski, W.F.Peck,Jr., S.Carter, R.J.Felder, H.Takagi and R.B.van Dover, Phys. Rev. B49, 12384.
6. S.Arisawa, T.Hatano, K.Hirata, T.Mochiku, H.Kitaguchi, H.Fujii, H.Kumakura, K.Kadowaki, K.Nakamura and K.Togano, Appl. Phys. Lett, 65(1994), 1299.

PHASE COHERENCE ENERGY VERSUS CHARGING−ENERGY IN HIGH Tc LAYERED SUPERCONDUCTORS

C.KAWABATA*, S.R.SHENOY** and A.R.BISHOP***
*Okayama University Computer Center, Okayama 700, Japan
**Condensed Matter Group, International Center for Theoretical Physics, Trieste 34100, Italy
***Theoretical Division, Los Alamos National Laboratory, Los Alamos, NM87545, U.S.A.

ABSTRACT

We consider the issue of competing phase coherence and charging−energy induced phase fluctuations in high T_c layered superconductors. Our study is aimed at suggesting the essential phenomenology of high T_c layered superconductors. We estimated the critical temperature T_c of layered superconductors as a function of the main interlayer spacing d_1, the carrier fraction x per layer, the interplane effective dielectric constant ε and the number of subsidiary intercalated extra layers N, by using a 3D Josephson−Junction array model and vortex loop scaling ideas.

1. Introduction

First, we review our studies so far,for background. About 5 years ago, Kawabata et al. presented the results of image processing analysis on the photographs of high T_c copper oxide TL−,Bi−,Y−based layered superconductors taken by a high−resolution transmission electron microscope[1]. There, it was shown and emphasized that the critical temperature T_c of the layered high Tc materials depends on the main spacing d_1 between CuO_2 layers along the c−axis for the small range of distances around 4Å > d_1 > 3Å below which superconductivity appears.

Currently, there is a growing interest in applying Kosterlitz−Thouless phase transition phenomenology to Josephson−Junction arrays and high T_c superconductors[2−5]. Williams and Shenoy extended the generalized 2D−XY vortex unbinding to the 3D−case for application to systems such as superfluid He^4 and 3D−XY or Josephson Junction array (JJA) models. We studied the vortex loops in the 3D−anisotropic XY model and investigated the relation between vortex loops and critical temperature Tc estimated from the susceptibility by Monte Carlo simulation[6].

In addition, we argued that the Josephson−Junction coupling energy via Cooper pair tunneling between CuO_2 layers enhances the critical temperature T_c for small main spacing d_1 in the range around 3Å, while the charge coupling energy for the capacitance between CuO_2 layers depresses the critical temperature T_c[6, 7].

2. Josephson Junction Array Model

Josephson−junction arrays with small junction capacitances C have large junction charging energies $E_c = q^2/2C$, that can destroy phase coherence even at temperature T = 0. We discuss here a simple 'back−of the−envelope' estimate for the T_c of a layered quantum–capacitive JJA with Angstrom–scale parameters. This estimate shows qualitative features similiar to trends observed in high T_c superconductors[8, 9].

The basic idea is that since number and phase are conjugate, a definite phase state must have sufficient coherence energy to overcome charging–energy induced phase fluctuations. The essential result is that, for high T_c – related parameters, the charging energy is of order$\sim$10000° K, so coherence energies must be at least of this order, to give a $T_c \sim 100°$ K. We will estimate T_c as a function of an interplane distance d_1, planar carrier concentration x, and interplane–atom effective dielectric constant ε. The above considerations should have some relevance, regardless of the detailed paring mechanism, since they arise from a general competition between phase coherence energies, and charging energies. To raise T_c, we must reduce charging energies.

We model the ceramic high T_c materials as a JJA, with weakly coupled superconducting grains on a 3D lattice. However, we have in mind a more microscopic picture of a high T_c single crystal, with some microscopic mechanism generating local coherent regions, that themselves are more weakly coupled within, and between the crystal lattice atomic layers. The JJA model has Angstrom–scale parameters.

We have shown, in our previous studies with the computer–analyzed electron–micrograph data, that d_1 is an important distance. (eg. Y atom region in 123) T_c falls sharply, and quasi linearly, as d_1 is increased, in families of high Tc materials. We assume that this interplanar distance d_1 is the region of weakest phase coherence, with strongest charging energies (quantum fluctuations) across it. The other 'charge neutralizing' blocks d_2 have more rapid charge transfers across them, and are treated as phase–locked. We therefore consider a layered 3D quantum capacitive JJA, of layer separation d_1, w determined by the carrier fraction per plane x, is the separation between inplane coherent regions. There is Josephson coupling $E_{J//}(w)$ between these regions in the plane. There is a coupling $E_{J\perp}(d_1)$ between coherent regions in different layers, i.e. between unit cells. The charging energy that dominates is $E_{c\perp}(d_1)$. It turns out, as mentioned below, that the effective competing coherence energy is the geometric mean, $E_J = \sqrt{E_{J\perp} E_{J//}}$.

First, we consider a classical, anisotropic 3D XY model, i.e. cosine coupled classical spins.

$$\beta H_{cos}(\{\theta_i\}) = -\sum_{\mu=x,y,z} \sum_i K_\mu \mathrm{Cos}(\Delta_\mu \theta_i) \qquad (1)$$

where Δ_μ is a discrete derivative in directions μ =x,y,z; {i} are superconducting grain (phase) sites on a cube lattice; $\{\theta_i\}$ are phase variables $-\pi < \theta_i < \pi$, and dimensionless layered couplings are $K_x = K_y \equiv K_{//} = \beta E_{J//}$, $K_z \equiv K_\perp = \beta E_{J\perp}$.

This can be described in terms of vortex loop excitations, in a generalization of Kosterlitz–Thouless scaling. (A loop blowout phenomenon[3] has been seen in simulations.)

The results of anisotropic loop scaling are that

(1) For strong anisotropy, quasi−2D Kosterlitz−Thouless like behavior, similiar to experiment, is found, outside a 3D critical region.

(2) The effective coherence energy, that sets the scale, is $E_J = \sqrt{E_{J\perp}E_{J//}}$.

(3) The critical coupling (or inverse critical temperature) is a function of $E_{J\perp}/E_{J//}$ anisotropy, that can be evaluated numerically and agrees with MC simulations.

Now consider quantum effects. We would have to work with an effective Hamiltonian in a path integral representation. The model would be in (3+1) dimensions, with the ' τ ' label being the extra dimension.

3. Estimates of Critical Temperatures T_c

Since the generalization of vortex loop scaling to three dimensions, with vortex tubes needs further work, we make a simple 'back of the envelope' estimate of T_c.

The unknown function $f(E_{J\perp}/E_{J//}, E_{c\perp}/E_J)$ must depend on the dimensionless ratio of the charging and effective coherence energies. For $E_{c\perp} = 0$, it reduces to the previous, classical case. We make a Taylor expansion of the critical coupling, and set order−unity Taylor coefficients equal to unity, for simplicity. Then T_c is as shown in the following;

$$k_B T_c = 2.2E_J[1 + E_{c\perp} / E_J(N+1)]^{-1} \qquad (2)$$

The charging energy $E_{c\perp}$ contains the parameters ε and d_1. The coherence energy E_J contains x, the carrier fraction in a plane. The Josephson coupling energies are in terms of the wave function overlaps: Between the planes, Cu_{dz^2} orbitals for $E_{J\perp}$; Within the planes $Cu_{dx^2-y^2}$ orbitals and oxygen p orbitals, for $E_{J//}$. (We approximate this by $Cu_{dx^2-y^2}$ alone)

If N extra CuO_2 layers are intercalated, each of separation d_1, then the coherence−breaking 'voltages' between unit cells are disributed, and reduced across any one. Effectively, $E_{c\perp} \rightarrow E_{c\perp}/(N+1)$ in the main result. Finally, we have T_c as a function of ε, x, N, and d_1, with T_J as a fitting parameter describing the coherence energy.

We used this simple Tc estimate for HTS material parameters such as the specific La−, Y−, TL−, Bi− and Hg−series. T_J is fitted to give the observed optimum T_c, and is of order ~10000 K.

We find then that T_c rises with decreasing d_1, increasing coherence T_J, and more layers N. It rises with increasing x, as the $I_{//}(w(x))$ overlap integral rises, and then falls, as the charging energy ($\sim x^2$) fluctuations destroy coherence.

Fig.1 shows a plot of scaled T_c versus scaled charging energy $E_{c\perp}$, for various HTS series. The star asterisk denotes material parameters as above for specific HTS materials.

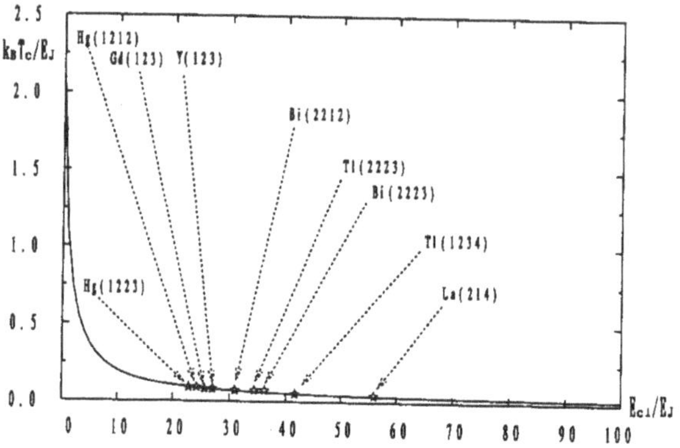

Figure 1

scaled k_BT_c/E_J
vs. scaled $E_{c\perp}/E_J$

4. Conclusion

One must reduce interplanar charging energies (small atoms = small d_1, or more polarizable, i.e. larger ε) in order to creep up along the curve: Lower $E_c \sim$ higher T_c.

We have suggested that a quantum capacitive Josephson–Junction array model with Angstrom scale parameters could capture the essential physics of layered high T_c superconductors, acting as an intermediate–level phenomenology. The ideas could also be explored in real arrays, and be useful in the design of high T_c superconducting electronic devices.

5. References

1. C.Kawabata and T.Nakanishi, Read at Japan Physical Society Annual Autumn Meeting at Kagoshima,Japan,in October 5th,1989 and Satellite Symposium 4 on High Temperature Superconductors of the 7th CIMTEC,Trieste,Italy,July 2–5,1990 and J.Phys.Soc.Jpn.59(1990)3835 and references therein.
2. G.Williams,Phys.Rev.Lett.59(1987)1926.
3. S.R.Shenoy,Phys.Rev.B40(1989)5056 and Phys.Rev.B42(1990)8692.
4. F.Falo,A.R.Bishop and P.S.Lomdahl,Phys.Rev.B41(1990)10983 and Phys.Rev. B46(1992)11149.
5. B.Chattopadhyay and S.R.Shenoy,Phys.Rev.Lett.72(1994)400.
6. C.Kawabata,M.Takeuchi,A.R.Bishop and S.R.Shenoy, IUMRS–ICAM–93, Symposium D(Tokyo,Japan,1993).
7. C.Kawabata,S.R.Shenoy and A.R.Bishop,Proc,6th Int.Symp.on Super. (ISS'93, Hiroshima,Japan,1993).
8. C.Kawabata, S.R.Shenoy and A.R.Bishop, Proc.of ETL Workshop on High Temperatures Superconductors Bulletin of Electrotechnical Laboratory Vol.58, No.5,(1994)p17 and references therein.
9. C.Kawabata, S.R.Shenoy and A.R.Bishop, 8th CIMTEC Forum on new Materials Topical Symposium IV (Superconductivity & Superconducting Materials Technologies) Florence,Italy,July 2–4,1994 and references therein.

SINGLE CRYSTAL X-RAY STRUCTURE ANALYSIS OF NEW Cu-BASED SUPERCONDUCTORS

J. AKIMOTO, Y. GOTOH and Y. OOSAWA
National Institute of Materials and Chemical Research
Higashi, Tsukuba, Ibaraki 305, Japan

and

K. TOKIWA, A. IYO and H. IHARA
Electrotechnical Laboratory,
Umezono, Tsukuba, Ibaraki 305, Japan

ABSTRACT

There is now a new series of high-temperature superconductors with the maximum T_c = 116 K that may be represented as a general formula of $Cu_{1-x}Ba_2Ca_{n-1}Cu_nO_{2n+4-y}$. Single crystals (maximum size: 0.1 × 0.1 × 0.01 mm) of Cu-1234 (n=4) and Cu-1245 (n=5) phases have been synthesized using a high-temperature and high-pressure technique. The X-ray precession photographs showed the basic structures may be chosen as tetragonal primitive lattices; a = 3.8526(18) Å, c = 17.973(5) Å for Cu-1234, and a = 3.8577(5) Å, c = 21.1546(14) Å for Cu-1245, respectively. The weak superlattice spots were observed in both compounds, which suggested a micro-twinning of the superlattice structures. The crystal structures were determined from single-crystal X-ray diffraction method using intensity data collected on the four-circle diffractometer to the conventional values R = 0.047 for Cu-1234 and R = 0.050 for Cu-1245, respectively. The existence of the deficient CuO_2 sheet between two BaO layers in these compounds has been confirmed.

1. Introduction

Recently, Ihara *et al.*[1] and other researchers[2-4] independently reported new superconducting oxide family having the structural formula of $Cu_{1-x}Ba_2Ca_{n-1}Cu_nO_{2n+4-y}$ and oxycarbonate family of $(Cu_{0.5}C_{0.5})Ba_2Ca_{n-1}Cu_nO_{2n+3}$. All of them were synthesized using a high-temperature and high-pressure technique, and have the maximum critical temperature T_c of 116 K in the case of n = 4. The advantage of these Cu-based superconductors is in their non-toxicity, simple composition and inexpensive source materials. The primitive crystal structures of $Cu_{1-x}Ba_2Ca_nCu_{n-1}O_{2n+4-y}$ are thought to be very similar to those of the Thallium-based superconductors $TlBa_2Ca_{n-1}Cu_nO_{2n+3}$[5], except for the further Cu-O sheet instead of the Tl-O sheet. However, the precise crystal structures of $Cu_{1-x}Ba_2Ca_{n-1}Cu_nO_{2n+4-y}$ have not been determined yet, because of the difficulty in the synthesis of the pure Ba-Ca-Cu-O system superconductors. For this point of view, we intended to synthesize single crystals of $Cu_{1-x}Ba_2Ca_{n-1}Cu_nO_{2n+4-y}$ family superconductors in order to determine the precise crystal structures. In this paper, we report the high-pressure single-crystal growth and the results of the crystal structure analyses of two new Cu-based superconductors; $Cu_{1-x}Ba_2Ca_3Cu_4O_{12-y}$ (Cu-1234; T_c=116 K[1]) and $Cu_{1-x}Ba_2Ca_4Cu_5O_{14-y}$ (Cu-1245; T_c=111 K[6]).

2. Experimental and Results

2.1. Single Crystal Description

The source material for high-pressure single-crystal synthesis was a mixture of a precursor material $Ba_2Ca_2Cu_5O_9$, peroxide CaO_2 and additional CuO. The precursor material was prepared by calcining a well-ground mixture of $BaCO_3$, $CaCO_3$ and CuO powders with the nominal composition at 1203 K for 20 hours in O_2. After regrinding and mixing with powdered CaO_2 and CuO in the correct proportions of Cu-1234 and Cu-1245 compositions, the pressed pellets were sealed in gold capsules of 4.4 mm diameter and 4.1 mm length. The sample capsules were heated at 1373-1423 K for 3-10 hours under a high pressure of 5 GPa, and rapidly cooled to room temperature before the pressure was released. Melting occurred partly and the growth of small platelets of $Cu_{1-x}Ba_2Ca_3Cu_4O_{12-y}$ (Cu-1234) and $Cu_{1-x}Ba_2Ca_4Cu_5O_{14-y}$ (Cu-1245) single crystals was prompted. They were mechanically separated and used for further characterization and structure determination. The maximum sizes of the single crystals are 0.03 × 0.03 × 0.01 mm for Cu-1234, and 0.12 × 0.10 × 0.01 mm for Cu-1245, respectively. The chemical compositions were measured by SEM-EDX. The average analytical values of the atomic ratios were Ba : Ca : Cu = 2 : 3.0 : 4.6 for Cu-1234, and Ba : Ca : Cu = 2 : 4.0 : 5.7, respectively, where the Ba ratio was fixed to 2.

The X-ray precession photographs showed the basic lattice may be chosen as a tetragonal primitive cell of a = 3.85 Å and c = 18.00 Å for Cu-1234, and a = 3.86 Å and c = 21.15 Å for Cu-1245, respectively. The possible space group are P4/mmm, P422, P4mm, P42m or P4m2. Some weak superlattice spots were observed in both compounds. They were explained by the 90° micro-twin structure, determined by electron diffraction[7,8], in which c-axes are parallel, but a-axes are perpendicular to each other. However, they were very weak in X-ray intensity, so we have chosen the primitive lattices for the subsequent structure determination. The primitive tetragonal lattice parameters, determined by least-squares refinement using Mo-$K\alpha_1$ radiation (λ = 0.70926 Å) and 25 well-centered reflections in the range $50 \le 2\theta \le 70°$ measured on an automated Rigaku AFC-5 four-circle diffractometer, are a = 3.8526(18) Å, c = 17.974(5) Å for Cu-1234, and a = 3.8577(5) Å, c = 21.1546(14) Å for Cu-1245, respectively.

2.2. Data Collection and Refinement

Integrated intensity data were collected on the four-circle diffractometer with graphite monochromatized Mo-$K\alpha$ radiation (λ = 0.71069 Å). The ω scan mode of the four-circle diffractometer was selected with a scan rate of 1°/min. The intensity data were corrected for Lorentz and polarization effects. Absorption correction was made by Gaussian method. The number of the measured reflections was 760 for Cu-1234, and 1593 for Cu-1245, respectively. All calculations were carried out using the *XTAL3.2* program[9]. Full-matrix least-squares refinements confirmed the deficient model of $Cu(1)O_2$ sheet between two BaO layers in both compounds. The occupancy parameters for Cu(1) site were refined to 0.60(1) for Cu-1234, and 0.68(1) for Cu-1245, respectively, which were consistent with the EDX results. The structure refinements were converged to $R = 0.047$ and $wR = 0.032$ [$w = 1/\sigma^2(Fo)$] for observed 278 reflections for Cu-1234, and $R = 0.050$ and $wR = 0.041$ for observed 604 reflections for Cu-1245, respectively. The atomic coordinates for Cu-1234 and Cu-1245 are given in Tables 1 and 2, respectively.

3. Discussion

The crystal structures of $Cu_{0.60}Ba_2Ca_3Cu_4O_{10.80}$ and $Cu_{0.68}Ba_2Ca_4Cu_5O_{12.06}$, can be regarded approximately as an intermediate between infinite-layered structure such as $(Ca,Sr)CuO_2$[10] and Tl-based 1234 and 1245 structure. From the view point of the infinite layered structure, these compounds are expressed by structural formulae $[(BaCu_{1-x}O_{3-y})(BaCuO_3)(CaCuO_2)_3]_n$ for Cu-1234, and $[(BaCu_{1-x}O_{3-y})(BaCuO_3)(CaCuO_2)_4]_n$ for Cu-1245, respectively.

Table 1. Atomic Coordinates for $Cu_{0.60}Ba_2Ca_3Cu_4O_{10.80}$ (Cu-1234)

	x/a	y/b	z/c	U	s.o.f.
Ba(1)	1/2	1/2	0.12210(8)	0.0126(3)	1
Ca(1)	1/2	1/2	0.3197(2)	0.0065(8)	1
Ca(2)	1/2	1/2	1/2	0.008(1)	1
Cu(1)	0	0	0	0.013(2)	0.60(1)
Cu(2)	0	0	0.2338(1)	0.0070(6)	1
Cu(3)	0	0	0.4113(1)	0.0061(5)	1
O(1)	0	1/2	0	0.12(2)	0.60
O(2a)	0	0	0.100(1)	0.015(6)	0.60
O(2b)	0	0	0.057(5)	0.03(3)	0.20
O(3)	0	1/2	0.2381(5)	0.010(2)	1
O(4)	0	1/2	0.4122(5)	0.013(2)	1

Table 2. Atomic Coordinates for $Cu_{0.68}Ba_2Ca_4Cu_5O_{12.06}$ (Cu-1245)

	x/a	y/b	z/c	U	s.o.f.
Ba(1)	1/2	1/2	0.10391(5)	0.0113(2)	1
Ca(1)	1/2	1/2	0.2717(1)	0.0047(4)	1
Ca(2)	1/2	1/2	0.4243(1)	0.0055(4)	1
Cu(1)	0	0	0	0.0053(7)	0.68(1)
Cu(2)	0	0	0.19861(7)	0.0036(3)	1
Cu(3)	0	0	0.34944(7)	0.0034(3)	1
Cu(4)	0	0	1/2	0.0033(4)	1
O(1)	0	1/2	0	0.1087	0.60(4)
O(2)	0	0	0.085(1)	0.009(5)	0.43(4)
O(3)	0	1/2	0.2026(3)	0.008(2)	1
O(4)	0	1/2	0.3512(3)	0.007(2)	1
O(5)	0	1/2	1/2	0.006(2)	1

Details of the individual Cu-O coordination and bond distances will be discussed in other papers[11,12]. The existence and the ordering of the deficient $Cu(1)O_2$ sheet in both compounds may be the origin of the superlattice structure observed by single-crystal X-ray precession method. In addition, the deficiency of $Cu(1)O_2$ sheet effects the position and occupation of the O(2) site of BaO layers that reside just above and below the deficient sheet in both compounds. At the present structure analyses of Cu-1234 and Cu-1245, no other ordered significant residual electron peaks such carbon sites in $(Cu_{0.5}C_{0.5})Ba_2Ca_3Cu_4O_{11\pm\delta}$[4] could be observed in the final difference Fourier maps. The total oxygen content in the half of the unit-cell by the present refinement is determined to be 10.80 for Cu-1234, and 12.06 for Cu-1245, respectively. From the charge compensation, the average copper valence state is estimated to be +2.52 for Cu-1234, and +2.13 for Cu-1245, respectively, the former value is slightly higher than the other superconducting cuprates. It should be noted that the T_c values of this superconductor family would be increased further by optimizing carrier concentration due to the control of oxygen concentration, or by increasing the number of CuO_2 layers, because the optimum average valence state of Cu is around +2.33 for high-T_c superconducting cuprates.

4. References

1. H. Ihara, K. Tokiwa, H. Ozawa, M. Hirabayashi, A. Negishi, H. Matuhata and Y. S. Song, *Jpn. J. Appl. Phys.* **33**, (1994) L503.
2. M.A. Alario-Franco, C. Chaillout, J.J. Capponi, J.-L. Tholence and B. Souletie, *Physica C* **222**, (1994) 52.
3. C.-Q. Jin, S. Adachi, X.-J. Wu, H. Yamauchi and S. Tanaka, *Physica C* **223**, (1994) 238.
4. T. Kawashima, Y. Matsui and E. Takayama-Muromachi, *Physica C* **224**, (1994) 69.
5. H. Ihara, R. Sugise, M. Hirabayashi, N.Terada, M. Jo, K. Hayashi, A. Negishi, M. Tokumoto, Y. Kimura and T. Shimomura, *Nature* **334**, (1988) 510.
6. H. Ihara, K. Tokiwa and A. Iyo, to be submitted.
7. X.-J. Wu, S. Adachi, C.-Q. Jin, H. Yamauchi and S. Tanaka, *Physica C* **223**, (1994) 243.
8. H. Matsuhata, H. Ihara, A. Iyo, K. Tokiwa and H. Ozawa, to be submitted.
9. S.R. Hall, H.D. Flack and J.M. Stewart, Eds. *Xtal3.2 Reference Manual.* Universities of Western Australia, Geneva and Maryland (1992).
10. T. Siegrist, S.M. Zahurak, D.W. Murphy and R.S. Roth, *Nature* **334**, (1988) 231.
11. J. Akimoto, Y. Oosawa, K. Tokiwa, M. Hirabayashi and H. Ihara, *Physica C,* (1994) in press.
12. J. Akimoto, Y. Gotoh, Y. Oosawa, K. Tokiwa, A.Iyo and H. Ihara, to be submitted.

HIGH PRESSURE STUDY OF $TbSr_2Cu_{2.7}Mo_{0.3}O_y$ AND RELATED OXIDE SUPERCONDUCTORS

S. YOMO and M. KAWAKAMI

Department of Electronic and Information Engineering, Hokkaido Tokai University
Minami-ku, Sapporo 005, Japan

and

H.H. Feng, J.W. Chu, P.H. Hor and C.W. Chu

Department of Physics and Texas Center for Superconductivity,
University of Houston, Texas 77204-5932, U.S.A.

and

N. Môri

Institute for Solid State Physics, Roppongi, Minato-ku, Tokyo 106, Japan

Terbium was stabilized in the Sr-based 123 compounds $R\,Sr_2Cu_{2.7}Mo_{0.3}O_y$ with R = Sm, Eu, Gd, Tb, Dy, Ho, Er, Tm and Yb. Terbium is known to be a mixed valent element, therefore high pressure study is interesting to investigate the stability of the valence state and the superconducting state. We have studied a quasi hydrostatic pressure effect on the T_c of Tb-123 system up to 12 GPa and found that the T_c (= 30 K at ambient pressure) is enhanced to 70 K by the pressure of 6 GPa and saturates above 6 GPa. Our preliminary measurements of the Hall coefficient (R_H) in this compound up to 1.2 GPa shows that the carrier is a hole and the R_H, which is about $6\times10^{-9}(m^3/C)$ at ambient temperature and pressure, decreases with increasing pressure with a rate of −12 %/GPa.

1. Introduction

The 123 structure for high temperature superconductivity (HTS) was successfully formed in the Sr-based $R\,Sr_2Cu_3O_y$ (RSCO) compounds by the partial replacement of Cu by Mo in the standard solid-state reaction technique [1]. Among them it is noteworthy that Terbium was stabilized in this Sr-based 123 compounds $R\,Sr_2Cu_{2.7}Mo_{0.3}O_y$ with R = Sm, Eu, Gd, Tb, Dy, Ho, Er, Tm and Yb. Terbium is known to be a mixed valent element, therefore high pressure study is interesting to investigate the relation between the stability of the valence state and the superconducting state. Hydrostatic pressure effect on the $R\,Sr_2Cu_{2.7}Mo_{0.3}O_y$ up to 1.6 GPa showed that the superconducting transition temperature (T_c) increases rapidly with increasing pressure (P) and there is a distinct jump in dT_c/dP vs. ionic radius curve below the ionic radius of Terbium [2]. We have studied a quasi-hydrostatic pressure effect on the T_c of $Tb\,Sr_2Cu_{2.7}Mo_{0.3}O_y$ system (Tb-123) up to 12 GPa and a hydrostatic pressure effect on the Hall coefficient (R_H) of the same compound up to 1.2 GPa, in order to clarify the possible effect of the mixed-valent element Tb.

2. Experiment

Polycrystalline samples of $RSr_2Cu_{2.7}Mo_{0.3}O_y$ with R = Tb, as well as the other rare earth elements Yb, Tm, Er, Ho, Dy, Gd, Eu and Sm, were prepared by a solid-state reaction as described previously [1,2]. The quasihydrostatic pressure measurements have been done in the Bridgman-anvil clamp apparatus with a face diameter of 3.5 mm and with steatite as a pressure medium, as described elsewhere[3]. The sample space was about 0.05 x 1.0 x 1.0 mm. The resistance measurements in the four-lead method was used to detect the onset of the superconducting transition. The Hall coefficient was measured in a piston-cylinder equipment with the Swenson-type, as described in ref.4. The sample was embedded in a Teflon cell with outer diameter of 8 mm and pressure medium of a mixture of Fluorinert #FC70 and #FC77. The sample was 4 x 2.6 x 0.3 4 mm in dimension and placed in a magnetic field of 4 T which was perpendicular to the sample face. We have used gold wires of 50 μm in diameter attached to the sample with gold paste to take the electrical contact.

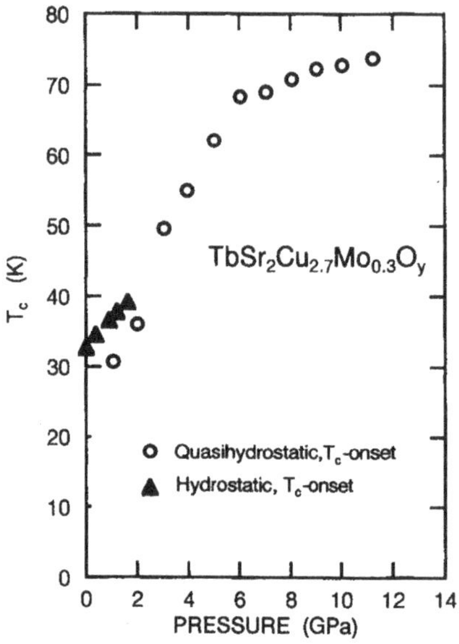

Fig. 1 Pressure dependence of T_c of $TbSr_2Cu_{2.7}Mo_{0.3}O_y$.

3. Results and Discussion

Figure 1 shows the pressure dependence of $RSr_2Cu_{2.7}Mo_{0.3}O_y$ under quasihydrostatic pressure up to 12 GPa and hydrostatic pressure up to 2 GPa. In the quasihydrostatic pressure, the transition width was broad partly due to a pressure distribution in a solid pressure medium, and therefore only the onset of the transition could be defined. This will be the main reason for the difference of T_c from the hydrostatic one. However, we notice nearly parallel pressure dependence in the 2 GPa region. The T_c increases rapidly with increasing pressure up to 6 GPa and tends to saturate above 6 GPa.

The initial slope of the T_c vs. pressure curve is dT_c/dP = 4 K/GPa and $d\ln T_c/dP$ = 1 2 %/GPa. As described previously[2], the $d\ln T_c/dP$ ranges between 12 %/GPa and 18

%/GPa in the $RSr_2Cu_{2.7}Mo_{0.3}O_y$. By comparing with $dlnT_c/dP$ vs. Tc curve for the 123 system, these high values belongs to the group in which the chain site Cu (Cu1) is substituted[5]. This strongly indicates that Mo introduced to stabilize the 123 structure in Sr-based $RSr_2Cu_{2.7}Mo_{0.3}O_y$ is located in the Cu1 site.

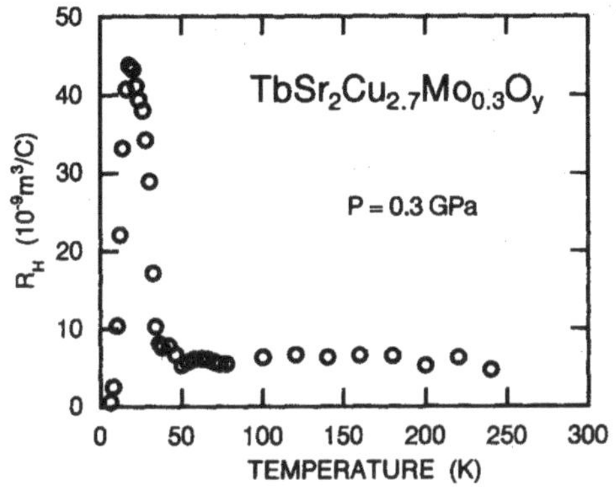

Fig. 2 Temperature dependence of the Hall coefficient under the pressure of 0.3 GPa.

The change of the nature of the pressure dependence in Fig.1 around 6 GPa might suggest a possibility of a valence change in Tb ions. If the valence state of the Tb ions changes from 3+ to 4+ or the intermediate state, the Cu ions need to be reduced in order to maintain the average charge balance. As described later, the Hall coefficient decreases with pressure, then the number of holes will increase if we adopt a simple picture. The reduction, if occurred in the plane site of Cu (Cu2), would suppress the increase of hole numbers, hence suppress the T_c enhancement.

However, if we ignore the details around 6 GPa, such behavior as in Fig. 1 has been observed in some of other systems[4,5], in which T_c is suppressed under high pressure and has a maximum as a function of pressure. This could be a manifestation of a universal feature that the T_c vs the number of holes relation has a bell shaped structure around an optimum hole concentration. This picture would be also failed if we could not take the $1/eR_H$ as a measure of the hole concentration, where the R_H is the Hall coefficient.

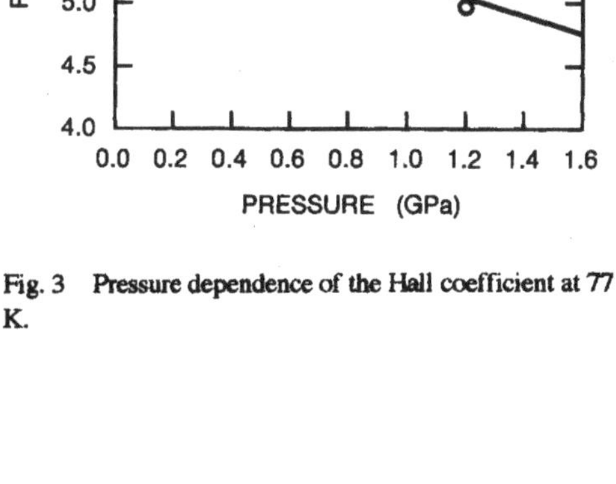

Fig. 3 Pressure dependence of the Hall coefficient at 77 K.

The temperature dependence of the Hall coefficient of $TbSr_2Cu_{2.7}Mo_{0.3}O_y$ was measured at a pressure of 0.3 GPa, as shown in Fig. 2. There is a weak temperature dependence above T_c and a peak below the T_c. The origin of the peak is not clear but a similar behavior was reported also in $YBa_2Cu_4O_8$ system[6].

The value of R_H is positive in the experimental region and about 6 x 1 0^{-9} m^3/C in the weak temperature dependent region. This corresponds to 0.18 holes/unit cell or 0.06 holes/Cu, if we simply take the $1/eR_H$ as the number of carriers. By comparison with the transition metal substituted and oxygen deficient Y123 systems[7], this value with the T_c of 31 K lies in the underdoped region and well on the line of those substituted by 3+ ions of Fe and Co in Cu1 sites and those with deficient oxygen in the chain, if we take the abscissa as holes/Cu in Fig.4 of ref. 7.

The Hall coefficient decreases with pressure, as shown in Fig. 3 at 77 K, in a rate of $d\ln R_H/dP = -12$ %/GPa which is large enough to explain the large pressure effect of T_c. However, this value of $d\ln R_H/dP$ is comparable to those in other systems such as $YBa_2Cu_3O_7$, $YBa_2Cu_4O_8$, $Bi_2Sr_2CaCu_2O_{8+y}$ and $Tl_2Ba_2CuO_{6+y}$, although $La_{2-x}Sr_xCuO_4$ system has no pressure effect on R_H. This indicates that the existence of possible valence fluctuating element Tb in $TbSr_2Cu_{2.7}Mo_{0.3}O_y$ does not affect appreciably the pressure effect on the superconductivity up to 2 GPa.

4. Conclusion

High pressure effect on the $TbSr_2Cu_{2.7}Mo_{0.3}O_y$ compound was studied and the result was compared with the other HTS compounds. The possible change in the valence state of Tb under pressure was not clear within the present experimental data. Further study of Hall coefficient at higher pressure region will be required.

5. Acknowledgments

One of the authors (S.Y.) was indebted to the Hokkaido Tokai University Foundation for Research and Education and the Grant-in-Aid for Scientific Research on Priority Areas, the Ministry of Education, Science and Culture. This work was also funded in part by the NSF(DMR 9122043), DARPA (MDA 972-90-J-1001) and the state of Texas through the Texas Center for Superconductivity at the University of Houston.

6. References

1. P.H.Hor, J.W.Chu, Y.Wang, H.H.Feng, Y.Y.Sun, K.Matsuishi, Q.Xiong and C.W.Chu, in *Proceedings of the 1992 TCSUH Workshop on HTS Materials, Bulk Processing, and Bulk Applications, ed. C.W.Chu, W.K.Chu, P.H.Hor and K.Salama* (World Scientific, Singapore,1992).
2. Q.Xiong, Y.Y.Xue, J.W.Chu, Y.Y.Sun, Y.Q.Wang, P.H.Hor and C.W.Chu, *Phys. Rev.* **B47** (1993) 11337.
3. S.Yomo, L.Gao, R.L.Meng, P.H.Hor, C.W.Chu and J.Susaki, *J.Mag. Mag. Mat.* **76&77** (1988) 257.
4. N.Môri, C.Murayama, H.Takahashi, H.Kaneko, K.Kawabata, Y.Iye, S.Uchida, H.Takagi, Y.Tokura, Y.Kubo, H.Sasakura and K.Yamaya, *Physica C* **185-189** (1991) 40.
5. S.Yomo, L.Gao, Y.Y.Xue, R.L.Meng, P.H.Hor and C.W.Chu, in *Research Report on Mechanism of Superconductivity* (Science Research on Priority Areas No.031, 1989) p.108.
6. C.Murayama, Y.Iye, T.Enomoto, A.Fukushima, N.Môri, Y.Yamada and T.Matsumoto, *Physica C* **185-189** (1991) 1293.
7. N.Okazaki, S.Kambe, A.Kishi, N.Kanazawa, A.Ohtomo, A.Fukuoka, T.Hasegawa, K.Kishio, K.Kitazawa and K.Fueki, in *Proceedings of the MRS meeting* (Tokyo, 1988).

IMPROVEMENT IN T_c BY Ba ADDITION IN HIGH T_c PHASE OF Bi OXIDE

Kaoru Usui*, Manabu Suzuki*, Kazuya Oguri** and
Yoshitake Nishi**
*Student of Materials Science, Tokai University, 1117, Kitakaname, Hiratsuka,
Kanagawa, 259-12, Japan
**Department of Materials Science, Tokai University, 1117, Kitakaname, Hiratsuka,
Kanagawa, 259-12, Japan

ABSTRACT

The effects of Ba addition on T_c are investigated for $Bi_{1.6}Pb_{0.4}Sr_2Ca_2Cu_3O_x$. Although the small amount ($0.01<z<0.2$) of Ba addition increases the T_c^{mid} values, the large amount ($z=0.4$ and 0.6) of Ba addition decreases the T_c^{mid} values for the $Bi_{1.6}Pb_{0.4}Ba_zSr_{2-z}Ca_2Cu_3O_x$. Based on results of the electrical resistivity and X-ray diffraction, the T_c changes are explained.

1. INTRODUCTION

The high T_c has been reported above 100 K of the Bi-oxides[1-3]. From an engineering point of view, it is important to investigate the effects of solutes on T_c. The element Ba is in the group of the alkaline earth metal group as Sr and Ca. The electronegativity difference between the oxygen and metal is one of the convenient indicators to evaluate the ionic bonding force. Based on the electronegativity value, the bonding force of Ba against oxygen ion may be larger than those of other metallic ions in Bi-oxide. On the other hand, the T_c value often changes against lattice constants[4,5]. Namely, since the Ba substitution for Sr may shorten the average distance between oxygen ion and cation, it is possible to increase the T_c value. Thus, the T_c change is investigated for the effect of Ba addition in high T_c phase of $Bi_{1.6}Pb_{0.4}Sr_2Ca_2Cu_3O_x$.

2. EXPERIMENTAL PROCEDURE

Samples with nominal composition were prepared by high purity powders of Bi_2O_3 (99.9 mass%), PbO (99.9 mass%), $SrCO_3$ (99.9 mass%), $CaCO_3$ (99.9 mass%), CuO (99.9 mass%) and $BaCO_3$ (99.9 mass%). Ba is try to be substituted for Sr of nominal composition. The powders were mixed and sintered for 6h at 1023 K in air and then air-cooled. After crushing, the sintered powders were resintered for 18 h at 1073 K in air and then air-cooled. After crushing, pelletized tablets, 1 mm thick and 13 mm in diameter, were sintered for 100 h at 1123 K in air and then furnace cooled. The electrical resistivity was measured using a standard 4-probe technique and a Keithley 182 nanovoltmeter. The temperature was measured by Au*Fe-chromel thermocouple attached to the specimen in a cryostat[3,5]. The structure of the sample was examined means of X-ray diffraction.

Figure 1 shows temperature dependence of the electrical resistivity (R) of $Bi_{1.6}Pb_{0.4}Sr_2Ca_2Cu_3O_x$. The deviated point is taken at $d(R/R^{300K})/dT = 0.004$, where $d(R/R^{300K})/dT$ and R^{300K} are slope reduced electrical resistivity (R/R^{300K}) against the temperature graph and the resistivity at 300 K, respectively. The point of electrical resistivity is R^{on} at onset temperature of the superconducting transition. T_c^{mid} is designated the temperature of 1/2 R^{on}.

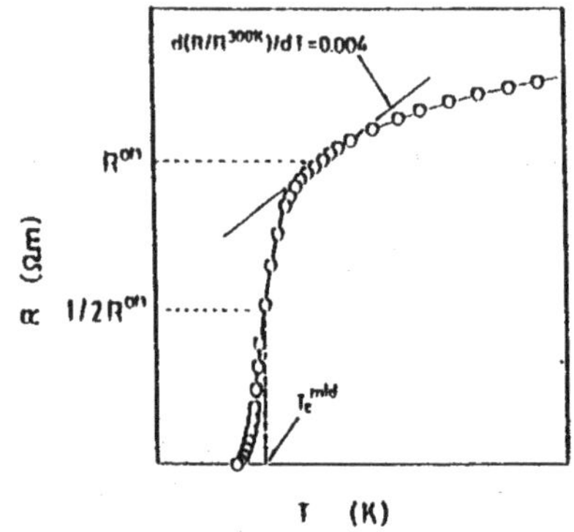

Fig. 1. Temperature dependence of the electrical resistivity (R) of $Bi_{1.6}Pb_{0.4}Sr_2Ca_2Cu_3O_x$.

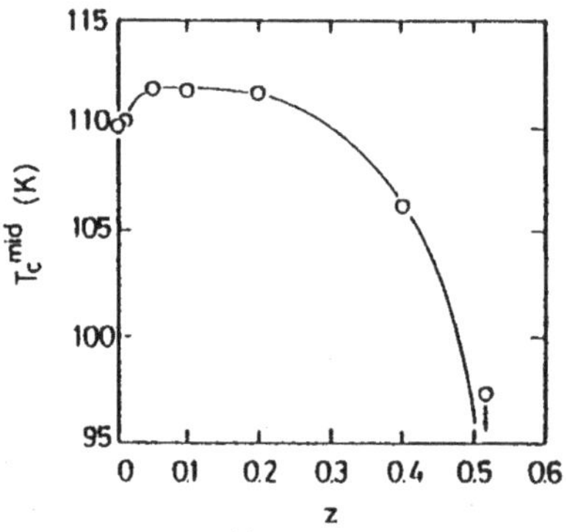

Fig.2. Changes in T_c^{mid} against Ba concentration (z).

3. RESULTS AND DISCUSSIONS

Figure 2 shows changes in T_c^{mid} at 10^3 A/m^2 against the Ba concentration (z) for $Bi_{1.6}Pb_{0.4}Sr_{2-z}Ba_zCa_2Cu_3O_x$ ($0.01<z<0.6$). The small amount ($0.01<z<0.2$) of Ba substitution increase the T_c^{mid} values. A larger amount of Ba substitution (z=0.6) decreases T_c^{mid}. In order to explain the T_c^{mid} changes, electronic discussions are attempted.

A carrier concentration (C*) is one of the dominant factors that control the T_c. C* can be expressed by the following equation[5,7,8].

$$C^* = N/\exp[(\pm eQ/k) - A] \quad (1)$$

Here, N, Q, k, e, and A are the density of state, Seebeck coefficient, Boltzmann constant, electron change and constant, respectively. Since the band gaps for the high-T_c oxides are smaller than those of semiconductors, A is assumed to be about zero[7,8]. Since the Q value of high-T_c oxides is approximately zero, the exponential term of the samples is about 1. Therefore, C* is dominated by the density of states (N). The density of states (N) can be estimated using the electrical resistivity (R). N is expressed by the following equation[9].

$$N = A'/R^{1/2} \quad (2)$$

Here, A' is a constant, which is dominated by the electron mass, charge, Plank constant, and mean free path. If we assume that the Ba substitution does not largely change the mean free path, the N value is inversely proportional to the $R^{1/2}$ value. Fig. 3 shows the changes in the electrical resistivity (R) at 200 K against Ba concentration (z) for

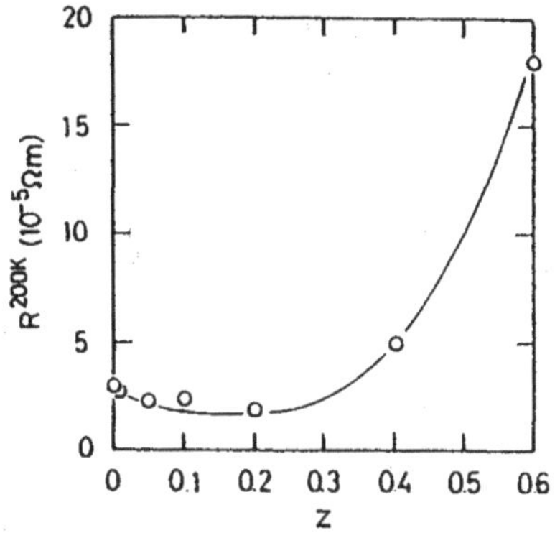

Fig.3. Changes in electrical resistivity (R) at 200 K against Ba concentration (z).

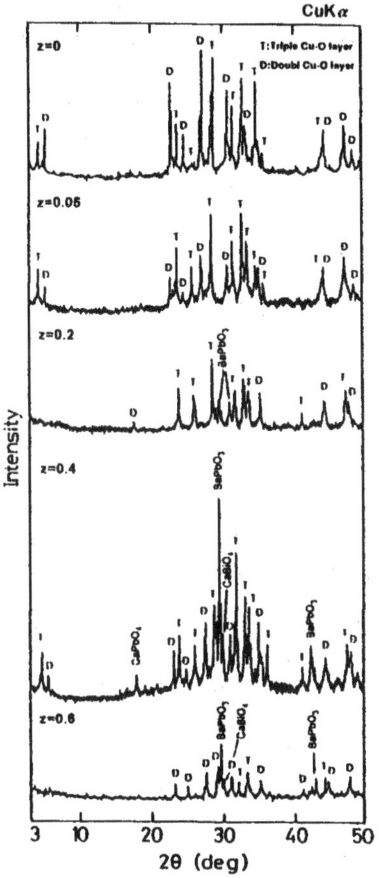

Fig.4. X-ray powder diffraction peaks of $Bi_{1.6}Pb_{0.4}Sr_2Ca_2Cu_3O_x$ samples (z=0, 0.05, 0.2, 0.4 and 0.6).

$Bi_{1.6}Pb_{0.4}Sr_{2-z}Ba_zCa_2Cu_3O_x$. The changes in the R values approximately agree with the changes in the T_c^{mid} values. Therefore, the lower the R values (see fig.3), the lager the N and C^* values become, resulting in a higher T_c (see fig.2).

Figure 4 shows X-ray powder diffraction peaks of the $Bi_{1.6}Pb_{0.4}Sr_{2-z}Ba_zCa_2Cu_3O_x$ samples (z=0, 0.05, 0.2, 0.4 and 0.6). Most peaks can be identified by a tetragonal structure. The low-T_c phase (double Cu-O layer) and the high-T_c phase (triple Cu-O layer) are found in the $Bi_{1.6}Pb_{0.4}Sr_{2-z}Ba_zCa_2Cu_3O_x$ sample (z=0, 0.05, 0.2, 0.4 and 0.6). The peaks of $BaPbO_3$ (z=0.2, 0.4, 0.6) are found[6], as shown in fig .4.

Since the excess Ba addition (z=0.4 and 0.6) generates a remarkable $BaPbO_3$ peaks, it decreases the T_c value.

Figure 5 shows the changes in FWHM (the value of the full width at half maximum intensity) for the high-T_c phase of the (002) plane against the Ba concentration (z) for $Bi_{1.6}Pb_{0.4}Sr_{2-z}Ba_zCa_2Cu_3O_x$ ($0.01<z<0.6$). The high-T_c sample approximately shows small FWHM values of the peaks. The change in the FWHM values approximately agrees with the change in the T_c^{mid} values. Namely, the high T_c values can be maintained by the high crystal perfection of the crystal lattice ordering, although the Ba solute is added.

Figure 6 shows the changes in the lattice constants (c) for the high-T_c phase against the Ba concentration (z). The ionic radii are 0.14 nm for Ba^{2+} and 0.13 nm for Sr^{2+}. Ba addition (z=0.01, 0.05) decreases the lattice constant (c). Since the change in the lattice constant of the (002) phase largely depends on the ionic radii of the elements, we conclude that the Ba atom probably substitutes for the Sr site.

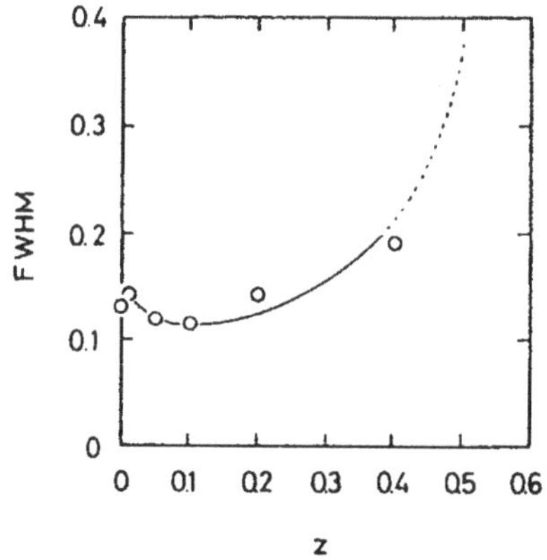

Fig.5. Changes in FWHM of high-T_c phase against Ba concentration (z).

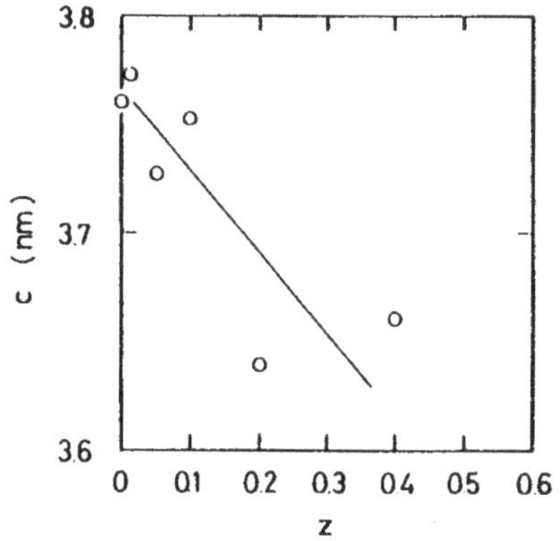

Fig.6. Changes in lattice constants (c) for the high-T_c phase against the Ba concentration (z).

4. CONCLUSION

In summary, the small amount of Ba addition in high T_c phase increases T_c^{mid} for $Bi_{1.6}Pb_{0.4}Sr_{2-z}Ba_zCa_2Cu_3O_x$ ($0.01<z<0.2$). The high crystal perfection (low FWHM of X-ray diffraction) of the triple Cu-O layer phase and the low electrical resistivity (the high density of states and high carrier concentration) are found in the high-T_c samples.

5. REFERENCES

1. U. Endo, S. Koyama and T. Kawai *Jpn. J. Appl. Phys.* **27** (1988) L1476
2. H. Maeda, Y. Tanaka, M. Fukutomi and T. Asano *Jpn. J. Appl. Phys.* **27** (1988) L209
3. Y. Nishi, Y. Kita and K. Tanioka *J. Mater. Sci.* **25** (1990) 5105
4. Y. Nishi, S. Noriya and S. Tokunaga *J. Mater. Sci. Lett* **7** (1988) 596
5. Y. Nishi, K. Oguri, H. Ohinata, K. Tanioka, Y. Kita and N. Ninomiya *Phys. Rev. B* **41** (1990) 6520
6. *Powder Diffraction File*, edited by Leonard G. Berry (The Joint Committee on Powder Diffraction Standards, Philadelphia, PA, 1991), Sets 12-664, p. 373.
7. N. M. Tallan and I. Bransky *J. Electrochem. Soc.* **118** (1971) 345
8. R. F. Brebrick, in *Defects in Solids*, Vol. 2 of *Treatise on Slid State Chemistry*, edited by N. B. Hannay (Plenum, New York, 1975), pp. 357-362.
9. N. F. Mott, *Philos. Mag.* **19** (1969) 835

CHANGE IN T_c BY Zr ADDITION IN HIGH-T_c PHASE OF Bi OXIDE SYSTEM

K.Fukushima*, K.Takahashi*, Y.Ohmori*, K.Oguri**, T.Kitamura* and Y.Nishi**

*Student of Department of Materials Science, Tokai University, 1117 Kitakaname, Hiratsuka, Kanagawa, 259-12, Japan.

**Department of Materials Science, Tokai University, 1117 Kitakaname, Hiratsuka, Kanagawa, 259-12, Japan.

ABSTRACT

Change in T_c by Zr addition in high-T_c phase are studied for the $Bi_{1.6}Pb_{0.4}Sr_2Ca_2Cu_3O_x$. Although the large amount (z=0.6) of Cu diminution decreases the T_c^{mid} value for the $Bi_{1.6}Pb_{0.4}Sr_2Ca_2Cu_{3-z}O_x$, the large amount (z=0.6) of Zr addition does not decrease the T_c^{mid} value for the $Bi_{1.6}Pb_{0.4}Sr_2Ca_2Cu_{3-z}Zr_zO_x$. Using the x-ray diffraction and EDX analysis, the effects of Zr addition on the T_c^{mid} values are discussed.

1. Introduction

Many HTSC materials have been reported[1-3]. From science and engineering points of view, it is important to know the effects of solutes on their supercunducting performance. We have recently found that the large amount of titanium (Ti) addition does not greatly decrease the T_c value[4,5]. Furthermore, it improves ductility[6]. Therefore it is possible to use the substrate and cover materials including Ti for the Bi system. Thus, we undertake to obtain information of the effects of zirconium (Zr) addition on the T_c values of Bi-Pb-Sr-Ca-Cu-O.

2. Experimental procedure

The samples with nominal composition $Bi_{1.6}Pb_{0.4}Sr_2Ca_2Cu_{3-z}O_X$ and $Bi_{1.6}Pb_{0.4}Sr_2Ca_2Cu_{3-z}Zr_zO_X$ were prepared by high purity powders Bi_2O_3 (99.99 mass %), PbO (99.99 mass %), $SrCO_3$ (99.9 mass %), $CaCO_3$ (99.99 mass %), CuO (99.9 mass %) and ZrO_2 (99.9 mass %). The Zr ion is try to be substituted for the Cu ion. The Cu poor sample ($Bi_{1.6}Pb_{0.4}Sr_2Ca_2Cu_{3-z}O_X$) is prepared at the same time to observe the effect of Zr substitution for Cu. The powders were mixed and sintered for 6 h at 1023 K in air and then air-cooled. The sintered powders were resintered for 18 h at 1073 K in air and then air-cooled. After crushing, a pelletized tablets, 1 mm thick and 13 mm in diameter, was sintered for 100 h at 1123 K in air and then furnace-cooled.

The electrical resistivity was measured using a standard 4-probe technique and

Keithley 182 nanovoltmeter[1,3]. The temperature was measured by Au*Fe-chromel thermocouple attached to the sample in a cryostat at equilibrium temperatures[1,3]. It was calibrated by standard Pt resistance thermometer. The structure of the sample was examined by means of x-ray diffraction of a Cu target.

3. Results and discussion

Figure 1 shows the temperature (T) dependence of the electrical resistivity (R) for the $Bi_{1.6}Pb_{0.4}Sr_2Ca_2Cu_{2.4}O_x$ (O) and $Bi_{1.6}Pb_{0.4}Sr_2Ca_2Cu_{2.4}Zr_{0.6}O_x$ (●). The lower the temperature, the lower the electrical resistivity becomes. Onset T_c^{on} of the transition is defined as the temperature corresponding to the deviation of the electrical resistivity. The deviated point is taken at $d(R/R^{300\,K})/dT=0.004$, where $d(R/R^{300\,K})/dT$ and $R^{300\,K}$ are slope of the reduced electrical resistivity $d(R/R^{300\,K})$ against the temperature graph and the resistivity at 300 K, respectively. The T_c^{mid} is designated the temperature of $1/2R^{on}$.

Figure 2 show the changes in T_c^{mid} values against the amount of the Cu diminution (z) for the $Bi_{1.6}Pb_{0.4}Sr_2Ca_2Cu_{3-z}O_x$ (O) and the amount of the Zr ion substituted for the Cu ion for the $Bi_{1.6}Pb_{0.4}Sr_2Ca_2Cu_{3-z}Zr_zO_x$ (●). Although the large amount (z=0.6) of Cu diminution decreases the T_c^{mid} value for the $Bi_{1.6}Pb_{0.4}Sr_2Ca_2Cu_{3-z}O_x$, the large amount (z=0.6) of Zr substituted for Cu does not decrease the T_c^{mid} value for the

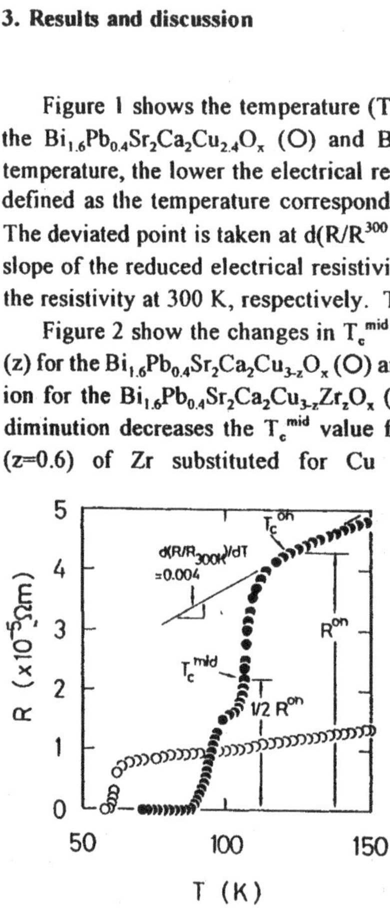

Fig. 1. Temperature (T) dependence of electrical resistivity (R) for $Bi_{1.6}Pb_{0.4}Sr_2Ca_2Cu_{2.4}O_x$ (O) and $Bi_{1.6}Pb_{0.4}Sr_2Ca_2Cu_{2.4}Zr_{0.6}O_x$ (●).

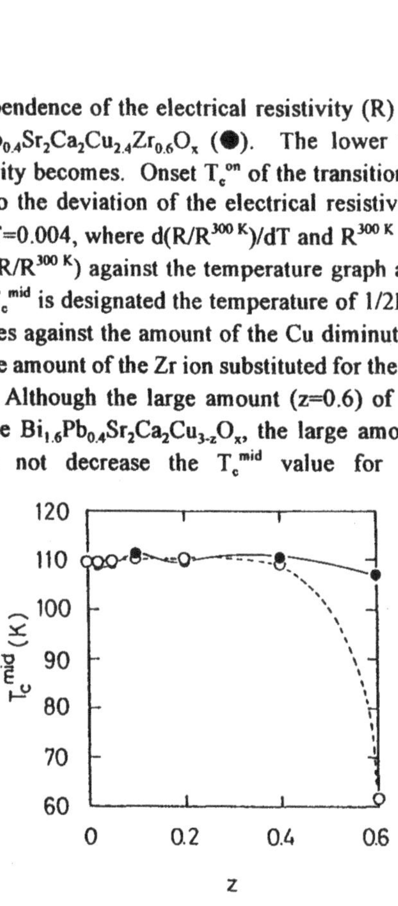

Fig. 2. Changes in T_c^{mid} values against amount of Cu diminution (z) for $Bi_{1.6}Pb_{0.4}Sr_2Ca_2Cu_{3-z}O_x$ (O) and Zr substitution for Cu for $Bi_{1.6}Pb_{0.4}Sr_2Ca_2Cu_{3-z}Zr_zO_x$ (●).

$Bi_{1.6}Pb_{0.4}Sr_2Ca_2Cu_{3-z}Zr_zO_x$. We found a critical amount of Cu diminution (z=0.4) and Zr substitution (z≧0.6) to maintain the high-T_c^{mid} value, respectively. It is interesting that adding such a large amount of Zr addition does not decrease T_c^{mid} for the high-T_c oxides. In order to explain the T_c^{mid} changes, morphological discussions are attempted.

Figure 3 shows the x-ray powder diffraction peaks of the $Bi_{1.6}Pb_{0.4}Sr_2Ca_2Cu_{2.4}O_X$ and $Bi_{1.6}Pb_{0.4}Sr_2Ca_2Cu_{2.4}Zr_{0.6}O_X$ samples. The powder method is useful to evaluate the volume change of high and low T_c phases because the preferred orientation does not affected the peaks height. Most peaks can be identified by a tetragonal structure. Although there is almost low-T_c phases (double Cu-O layer[7]) for the $Bi_{1.6}Pb_{0.4}Sr_2Ca_2Cu_{2.4}O_X$, the low-$T_c$ phase and the high-T_c phase (triple Cu-O layer[8]) are found for the $Bi_{1.6}Pb_{0.4}Sr_2Ca_2Cu_{2.4}Zr_{0.6}O_X$. If it assumed that the Zr ion is substituted for the Cu ion for the $Bi_{1.6}Pb_{0.4}Sr_2Ca_2Cu_{2.4}Zr_{0.6}O_X$, it can explain to exist the high-T_c phases. Therefore, the high-T_c^{mid} value is kept.

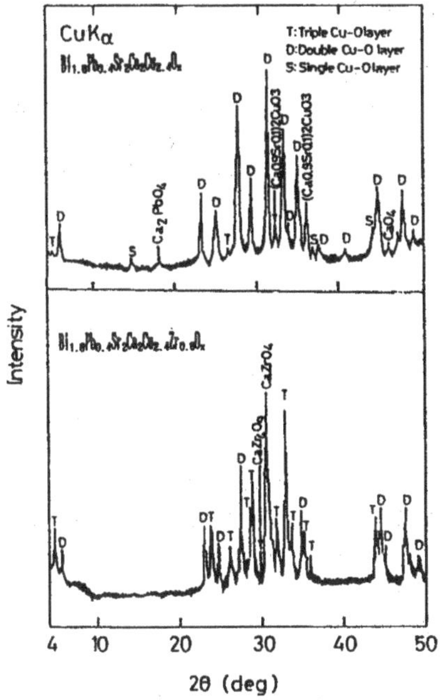

Fig. 3. X-ray powder diffraction peaks of $Bi_{1.6}Pb_{0.4}Sr_2Ca_2Cu_{2.4}O_X$ and $Bi_{1.6}Pb_{0.4}Sr_2Ca_2Cu_{2.4}Zr_{0.6}O_X$ samples.

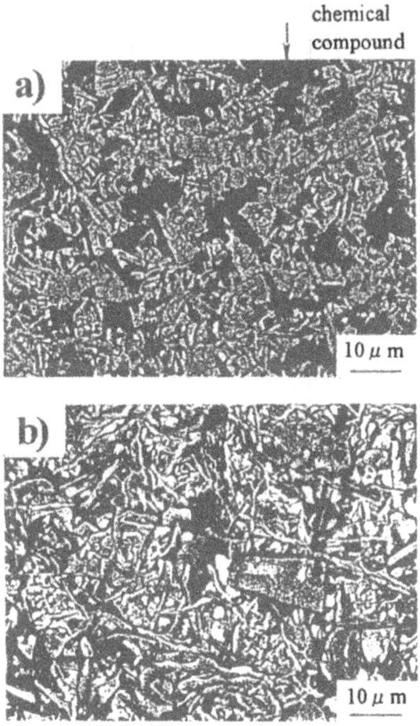

Fig. 4. Photograph of sample surface by means of EDX analysis for $Bi_{1.6}Pb_{0.4}Sr_2Ca_2Cu_{2.4}O_X$ (a) and $Bi_{1.6}Pb_{0.4}Sr_2Ca_2Cu_{2.4}Zr_{0.6}O_X$ (b).

Figure 4 shows the photograph of the sample surface by means of EDX analysis for the $Bi_{1.6}Pb_{0.4}Sr_2Ca_2Cu_{2.4}O_X$ (a) and $Bi_{1.6}Pb_{0.4}Sr_2Ca_2Cu_{2.4}Zr_{0.6}O_X$ (b). The brack area is chemical compound of $Sr_4Ca_5Cu_{10}O_x$. The large amount of chemical compound is found in the $Bi_{1.6}Pb_{0.4}Sr_2Ca_2Cu_{2.4}O_X$ sample. On the other hand, the small amount of chemical compound is found for the $Bi_{1.6}Pb_{0.4}Sr_2Ca_2Cu_{2.4}Zr_{0.6}O_X$. Writing the formula of the Bi-Sr-Ca-Cu-O series as 2-2-(n-1)-n, the removal of Cu obviously results in a shift in phase equilibrium of the remaining mixture towards lower n values (n=3→2), therefore the T_c^{mid} decrease for the $Bi_{1.6}Pb_{0.4}Sr_2Ca_2Cu_{2.4}O_X$. Based on the results of x-ray diffraction and EDX analysis, it is obvious that added Zr played an important role to maintain the structure of high-T_c (2223) phase.

4. Conclusion

In summary, the effects of Zr addition on T_c have been investigated for high-T_c $Bi_{1.6}Pb_{0.4}Sr_2Ca_2Cu_3O_x$. Although the large amount (z=0.6) of Cu diminution decreases the T_c^{mid} value for the $Bi_{1.6}Pb_{0.4}Sr_2Ca_2Cu_{3-z}O_x$, the large amount (z=0.6) of Zr addition does not decrease the T_c^{mid} value for the $Bi_{1.6}Pb_{0.4}Sr_2Ca_2Cu_{3-z}Zr_zO_x$. It is interesting that adding such a large amount of Zr addition does not decrease T_c^{mid} for the high-T_c oxides. Based on the results of x-ray diffraction and EDX analysis, it is obvious that added Zr played an important role to maintain the structure of high-T_c (2223) phase.

5. References

1 Y. Nishi, S. Moriya, T. Manabe, *J. Appl. Phys.* 65 (1989) 2389.
2 Y. Nishi, K. Oguri, K. Ohinata, K. Tanioka, Y. Kita, *Phys. Rev. B.* 41 (1990) 6520.
3 Y. Nishi, Y. Kita, K. Tanioka, *J. Mater. Sci.* 25 (1990) 5105.
4 Y. Nishi, K. Nozaki, T. Kurotaki, Y. Kita, K. Oguri, *J. Appl. Phys.* 71 (1992) 350.
5 Y. Nishi, K. Nozaki, T. Kurotaki, Y. Kita, K. Oguri, *Phys. Lett. A.* 163 (1992) 465.
6 K. Nozaki, K. Shiraishi, S. Uchida, M. Shimabara, Y. Nishi, *Pro. The 3rd IUMRS International Conference on Advanced Materials* (1993).
7 M. Onoda, A. Yamamoto, E. Takayama-Muromachi and S. Takekawa, *Japan. J. Appl. Phys.* 27 (1988) L833.
8 U. Endo, S. Koyama and T. Kawai, Japan. *J. Appl. Phys.* 27 (1988) L1476.

INFLUENCE OF LOW GAS PRESSURE ON T_c OF Bi OXIDE.

K. OGURI, K.FUKUSHIMA* and Y. NISHI
Department of Materials Science, Tokai University, 1117 Kitakaname, Hiratsuka, Kanagawa, 259-12, Japan.
*Student of Materials Science, Tokai University, 1117 Kitakaname, Hiratsuka, Kanagawa, 259-12, Japan.

ABSTRACT

Effect of pressure on T_c is studied for $Bi_{1.6}Pb_{0.35}Sb_{0.05}Sr_2Ca_2Cu_3O_x$. The pressure is performed by helium gas. The gas pressure increases the T_c value. The T_c increase is 0.3 K at 1.00 kgf/mm^2.

1. Introduction

From an engineering point of view, it is important to know the effect of stress on T_c. Many workers have reported effects of stress on T_c of high-T_c superconductors[1-3]. It was known that the stress decreases lattice constant and increases carrier concentration. However the large plastic deformation in the former works induces the microcrack and heterogeneous stress. We need the T_c value under the homogeneous stress by gas pressure.

Effects of small compressive stress on T_c have been reported for $YBa_2Cu_3O_{7-y}$[4,5], $Bi_{1.6}Pb_{0.4}Sr_2Ca_2Cu_3O_x$[6] and $Bi_{1.6}Pb_{0.35}Sb_{0.05}Sr_2Ca_2Cu_3O_x$[7]. However, the effect of small equiaxial stress on T_c has never been reported. Thus, we undertaken to obtain information of an influence of the gas pressure on T_c of the high-T_c Bi-Pb-Sb-Sr-Ca-Cu-O.

2. Experimental Procedure

Samples with nominal composition $Bi_{1.6}Pb_{0.35}Sb_{0.05}Sr_2Ca_2Cu_3O_x$ were prepared by high-purity powders of CuO (99.9 mass%), PbO (99.9 mass%), Sb_2O_3 (99.9 mass%), $CaCO_3$ (99.9 mass%), $SrCO_3$ (99.9 mass%) and Bi_2O_3 (99.99 mass%). The powders were mixed and sintered in air at 1023 K for 8 h and then air cooled. After crushing, the sintered powders were resintered in air at 1073 K for 18 h and then air cooled. After crushing, the sintered powders were resintered in air at 1123 K for 50 h and then air cooled. After crushing, a pellet-shaped tablet, 1 mm thick and 13 mm diameter, was sintered in air at 1123 K for 50 h and then furnace cooled.

The low pressure was provided by He gas (see Fig.1). The electrical resistivity was measured using a standard four-probe technique and Keithley 181 nanovoltmeter. The temperature was measured by Au*Fe-chromel thermocouple attached to the specimen in a cryostat at equilibrium temperature[8]. Figure 2 shows the pressure dependence of temperature change in probe (sample free). It does not change by the pressure (< 1.00 kgf/mm^2).

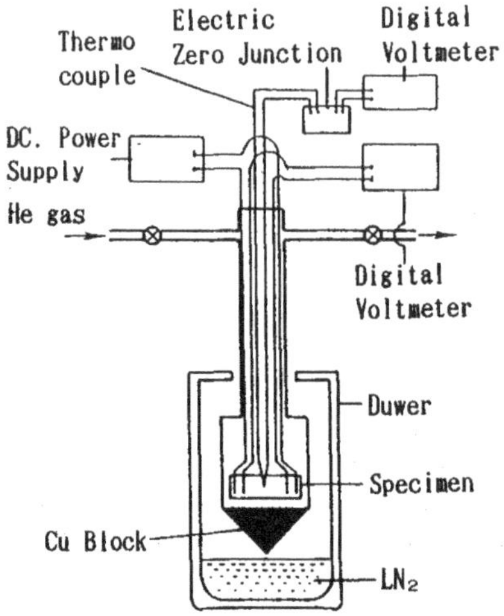

Fig.1 Schematic drawing of gas pressure apparatus.

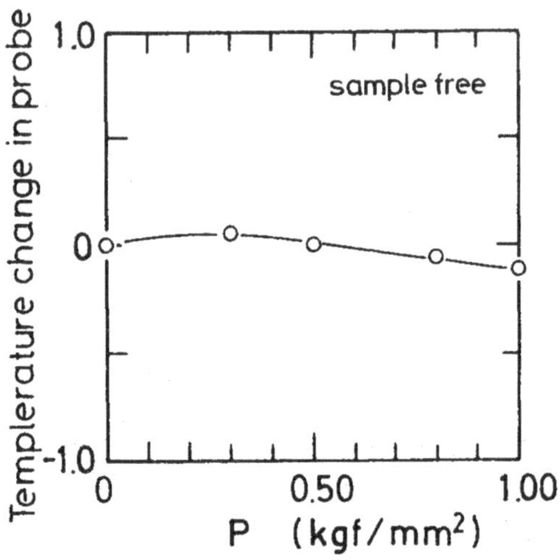

Fig. 2 Pressure dependence of temperature change in prove (sample free).

3. Results and Discussion

Figure 3 shows x-ray diffraction pattern of $Bi_{1.6}Pb_{0.35}Sb_{0.05}Sr_2Ca_2Cu_3O_x$ sample. The sample is the high-T_c phase.

Figure 4 shows the temperature dependence of electrical resistivity (R) of the unloaded and loaded (1.00 kgf/mm^2) $Bi_{1.6}Pb_{0.35}Sb_{0.05}Sr_2Ca_2Cu_3O_x$. The lower the temperature, the lower the electrical resistivity becomes. Offset T_c of the transition is defined as the temperatures corresponding to the achievement of zero resistivity (below 10^{-9} Ωm).

Figure 5 shows changes in T_c against pressure. The small pressure enhances T_c. The dT_c/dP value is 0.04 (K mm^2/kgf). The maximum T_c is about 105.2 K at 1.00 mA/mm^2 of the current density which T_c is 104.9 K before the loading. T_c does not increase above endurance limit (0.80 kgf/mm^2).

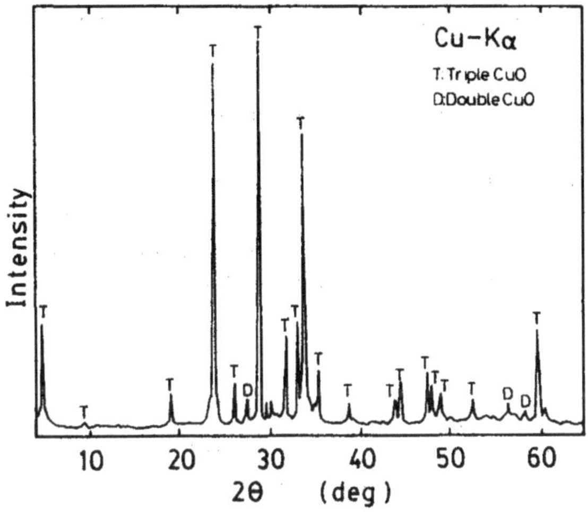

Fig. 3 X-ray diffraction pattern for $Bi_{1.6}Pb_{0.35}Sb_{0.05}Sr_2Ca_2Cu_3O_x$ sample.

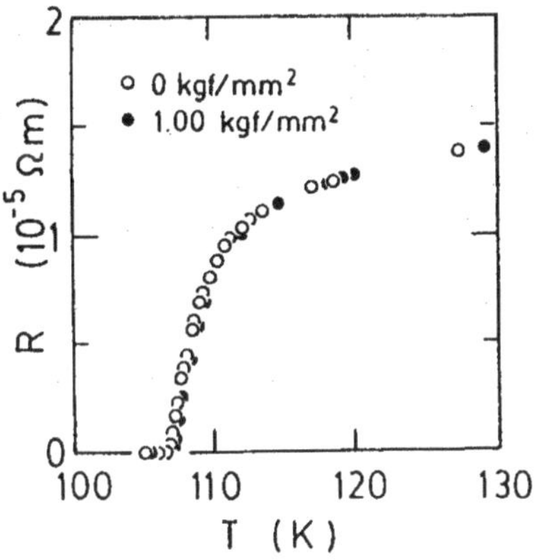

Fig. 4 Temperature dependence of electrical resistivity (R) of unloaded and loaded (1.00 kg/mm^2) for $Bi_{1.6}Pb_{0.35}Sb_{0.05}Sr_2Ca_2Cu_3O_x$.

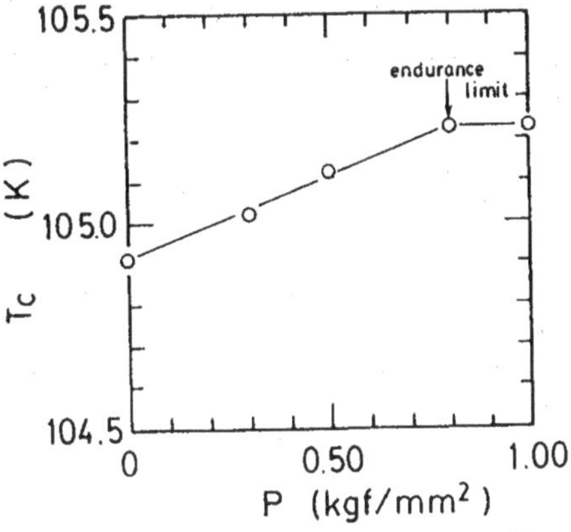

Fig. 5 Changes in T_c at 1.00 mA/mm^2 of current density against pressure.

T_c is expressed by the following equation as a function of compressive stress (P;kgf/mm^2).

$$T_c = 104.9 + 0.04 \times P, (0 < P < 0.80 \text{ kgf/mm}^2) \qquad (1)$$

There are not change that the lattice constant and the half value width of x-ray diffraction for unloaded and after loaded (< 1.00 kg/mm^2) sample. Namely, it shows that the pressure is loaded within elastic deformation. It is important to know the increasing on T_c of $Bi_{1.6}Pb_{0.35}Sb_{0.05}Sr_2Ca_2Cu_3O_x$ within elastic deformation.

4. Conclusion

In summary, the effect of gas pressure on T_c is studied for $Bi_{1.6}Pb_{0.35}Sb_{0.05}Sr_2Ca_2Cu_3O_x$. The low gas pressure was found to increase the T_c value. The T_c value is expressed by following equation.

$$T_c = 104.9 + 0.04 \times P, (0 < P < 0.80 \text{ kgf/mm}^2)$$

5. References

1. M.K.Wu, J.R.Ashburn and C.T.Torng, *Phys.Rev.Lett.*, 58 (1987) 908.
2. S.Yomo, C.Murayama, W.Utsumi, H.Takahashi, Y.Yagi N.Mori, T.Tamegai, A.Watanabe and Y.Iye, *Jpn.J.Appl.Phys.Suppl.*, 26 (1987) 1107.
3. C.C.Kim, E.F.Skelton and M.S.Osofsky, *Phys.Rev.B*, 48 (1993) 6431.
4. Y.Nishi, N.Ninomiya, and S.Tokunaga, *J.Mater.Sci.Lett.*, 7 (1988) 361.
5. Y.Nishi, N.Ninomiya, K.Oguri and S.Tokunaga, *J.Appl.Phys.*, 65 (1989) 3972.
6. Y.Nishi, K.Oguri, H.Ohinata, K.Tanioka, Y.Kita and N.Ninomiya, *Phys.Rev.B* 41 (1990) 6520.
7. K.Oguri, K.Kubota, F.Kato, T.Zakoji, H.Sasaki and Y.Nishi, *J.Mater.Sci.Lett.*, in press.
8. Y.Nishi, A.Igarashi, S.Tokunaga and S.Moriya, *J.Mater.Sci.Lett.*, 7 (1988) 71.

THERMAL ANALYSIS OF LIQUID-QUENCHED $Bi_{1.6}Pb_{0.4}Sr_2Ca_2Cu_{3-z}Ti_zO_x$ GLASSES (0.003 < z < 1.5)

T. Aida, Y. Kawaguchi and M. Shimabara
Graduate Student of Materials Science, Tokai University, Hiratsuka, Kanagawa 259-12, Japan.

Y. Nishi*
** Department of Materials Science, Tokai University, Hiratsuka, Kanagawa 259-12, Japan.*

ABSTRACT

It is necessary for crystal growth to know the T_g, T_x and T_p values. A remarkable effect of the Ti substitution for Cu does not lower T_g, T_x and T_p values below 1.0 of z. A large amount of Ti addition (substitution for Cu) increases T_g, T_x and T_p values above 1.0 of z. Based on results of the x-ray diffraction, changes in T_g, T_x and T_p values are discussed.

1. Introduction

Since the melting point of high-T_c Bi oxide is low, it is easy to be melted and liquid-quenched. The density of the liquid-quenched sample is apparently higher than that of the sintered sample, which may improve their superconducting performance [1,2]. A high T_c is found for liquid-quenched Bi oxide glass after heat treatments[1]. The crystallization occurs during the annealing [3]. It is important to know the thermal behaviors of the crystallization. Meanwhile, the Ti addition improves ductility of the Bi oxide [4], although it does not greatly decrease the T_c value [5]. Thus, the purpose of present work is to investigate the change in glass transition temperature (T_g) and crystallization temperatures (T_x and T_p) of liquid-quenched $Bi_{1.6}Pb_{0.4}Sr_2Ca_2Cu_{3-z}Ti_zO_x$ glasses (0.003 < z < 1.5).

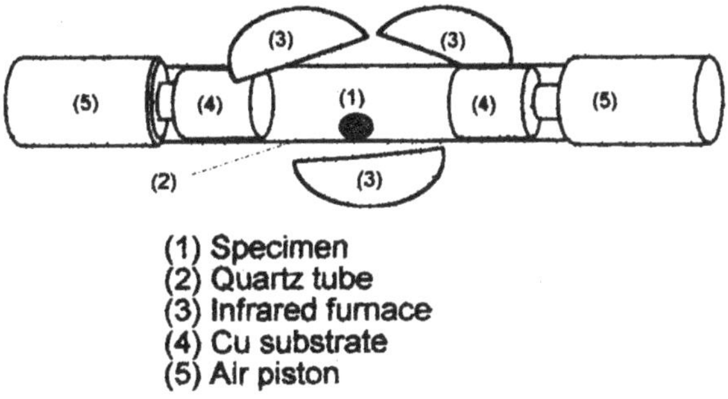

Fig. 1. Schematic diagram of a twin-type piston-anvil apparatus.

2. Experimental procedure

Samples with nominal composition were prepared

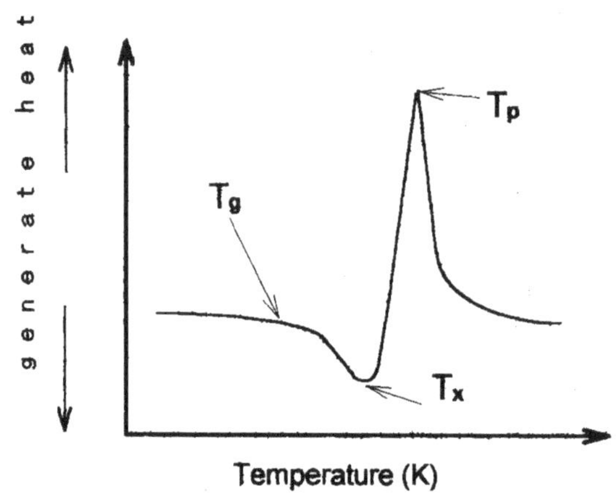

Fig. 2. DSC curve at 20 K/min heating rate.

from high-purity powders of CuO (99.9 %), TiO_2 (99.9 %), Bi_2O_3 (99.99 %), PbO (99.9 %), $SrCO_3$ (99.9 %) and $CaCO_3$ (99.9 %). The powders were mixed and sintered in air at 1023 K for 6 h and then air-cooled. After crushing, the sintered powders were resintered in air at 1073 K for 18 h and then air-cooled. After crushing, a pelletized tablet, 1 mm thick and 13 mm in diameter, was sintered in air at 1123 K for 12 h and then furnace cooled. Liquid-quenched glassy foils were prepared from chips cut form sintered tablets, using a twin-type piston-anvil apparatus [6-9]. Figure 1. shows the apparatus. It was constructed to quench the molten sample. The weight of the sintered tablet melted by an infrared furnace in one run was about 0.2 g. The molten sample was quenched rapidly. The quenching rate R (K/s) depended on the sample mass [10].

Figure 2 shows a differential scanning calorimeter (DSC-50). It was performed in Ar gas at 20 K/min of a heating rate. This apparatus was calibrated with In (99.99 %), Pb (99.9 %), Zn (99.9 %) and Sn (99.9 %). The structure of the sample was examined by means of x-ray diffraction (RU-200B, Rigaku Denki, Tokyo). The diffraction measurements were performed under a step scanning mode (Cu K_α target, 40 kV, 50 mA).

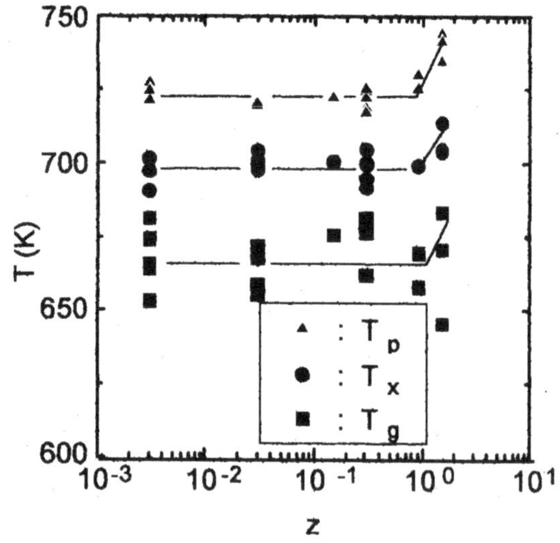

Fig. 3. Change in temperature (T_g, T_x and T_p) at 20 K/min plotted against titanium concentration (z) for $Bi_{1.6}Pb_{0.4}Sr_2Ca_2Cu_{3-z}Ti_zO_x$.

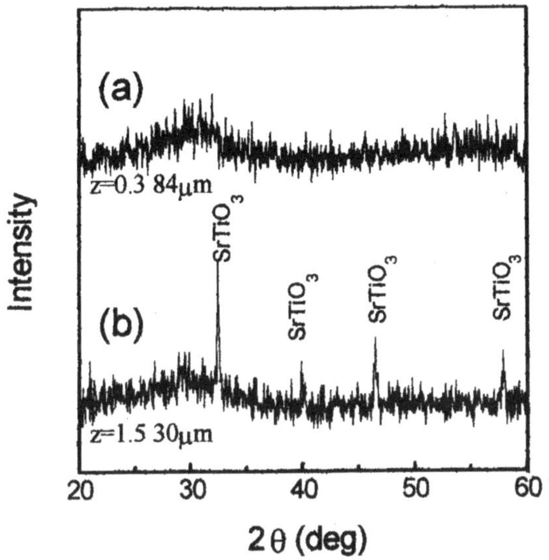

Fig. 4. X-ray powder diffraction patterns of (a) $Bi_{1.6}Pb_{0.4}Sr_2Ca_2Cu_{2.7}Ti_{0.3}O_x$ (84 μm) sample and (b) $Bi_{1.6}Pb_{0.4}Sr_2Ca_2Cu_{1.5}Ti_{1.5}O_x$(30 μm) sample.

3. Results and discussion

Figure 3 shows the changes in temperatures T_p (▲), T_x (●) and T_g (■) against z for the $Bi_{1.6}Pb_{0.4}Sr_2Ca_2Cu_{3-z}Ti_zO_x$, where z is the composition ratio of Ti to Cu and no change in the T_g, T_x and T_p values below z = 1.0.

Figure 4 shows x-ray powder diffraction patterns of $Bi_{1.6}Pb_{0.4}Sr_2Ca_2Cu_{2.7}Ti_{0.3}O_x$ and $Bi_{1.6}Pb_{0.4}Sr_2Ca_2Cu_{1.5}Ti_{1.5}O_x$ samples (z = 0.3 and z = 1.5). The glassy structure can be found for the $Bi_{1.6}Pb_{0.4}Sr_2Ca_2Cu_{2.7}Ti_{0.3}O_x$ sample of 84 μm in thickness. On the other hand, the $Bi_{1.6}Pb_{0.4}Sr_2Ca_2Cu_{1.5}Ti_{1.5}O_x$ powder sample (30 μm) has $SrTiO_3$ peaks, as shown in Fig. 4. If the $SrTiO_3$ [11] phase prevents the crystal nucleation, we can explain that the Ti addition (z = 1.5) increases the T_g, T_x and T_p values (see Fig. 3).

Figure 5 shows activation energy of Kissinger plots of crystallization temperature T_x and T_p [12, 13]. The composition dependence of E_x and E_p values approximately agree with those of T_x and T_p in the Fig. 3. They show that the apparent activation energy of crystallization doesn't depend on z (Titanium concentration) below 1.0 of z. The excess Ti addition above 1.0 of z rapidly decreases the apparent activation energies of heterogeneous crystallization because of the $SrTiO_3$ formation (See Fig. 4.)

4. Conclusion

In summary, we found that the Ti addition doesn't change T_g, T_x, T_p, E_x and E_p values of below z = 1.0 for the $Bi_{1.6}Pb_{0.4}Sr_2Ca_2Cu_{3-z}Ti_zO_x$ samples. However, a large amount of Ti addition (substitution for Cu; z = 1.5) increases T_g, T_x and T_p values and decreases E_x and E_p. The excess Ti addition above 1.0 of z generates the $SrTiO_3$ phase. Namely, it probably prevents the crystal nucleation at low temperature (see results of T_g and T_x in Fig. 3) and decreases the apparent activation energy of E_x and E_p (see Fig. 5).

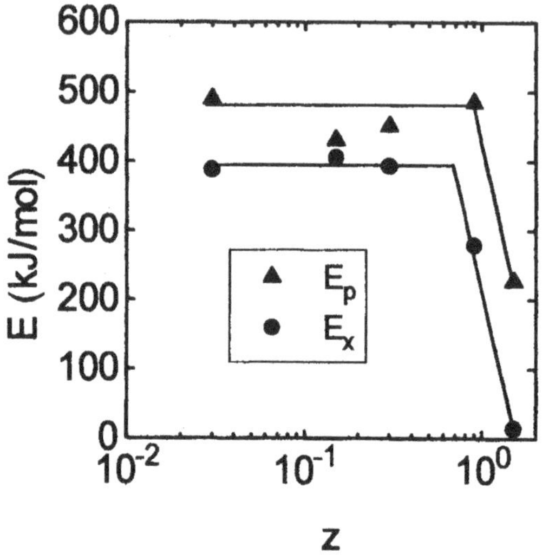

Fig. 5. Activation energy (E_x and E_p) of Kissinger plots of crystallization temperature T_x and T_p.

5. References

1. Y. Nishi, S, Moriya, and T. Manabe, *J. Appl. Phys.* **65**, 2389 (1989).
2. Y. Nishi, Y. Kita, and K. Tanioka, *J. Mater. Sci.* **25**, 5105 (1990).
3. Y. Nishi, N. Ninomiya, F. Kanai, S. Uchida, and S. Moriya, *J. Appl. Phys.* **66**, 2069 (1989).
4. K. Nozaki, K. Shiraishi, S. Uchida, M. Shimabara, and Y. Nishi, *in Proceeding of the third IUMRS International Conference on Advanced Materials* (1993), in press.
5. Y. Nishi, K. Nozaki, and T. Kurotaki, *J. Appl. Phys.* **71**, 350 (1992).
6. Y. Nishi, K. Suzuki, and T. Masumoto, *J. Jpn. Inst. Met.* **45** 818 (1982).
7. Y. Nishi, H. Harano, H. Ishizuki, M. Kawakami, and E. Yajima, *Mater. Sci. Eng.* **98**, 505 (1988).
8. Y. Nishi, T. Kai, and K. Kitago, *Wear* **126**, 191 (1988).
9. Y. Nishi, and H. Harano, *J. Appl. Phys.* **63**, 1141 (1988).
10. Y. Nishi, M. Tachi, and Yajima, *Scr. Metall.* **19**, 865 (1985).
11. L. G. Berry, ed., *Power diffraction files*, Sets 35-734 (Joint Committee on Powder Diffraction Standards, Philadelphia, 1991) p. 263.
12. Y. Nishi, T. Manabe, S. Watanabe, A. Igarashi, and K. Mikagi, *J. Appl. Phys.* **70**, 5684 (1991).
13. Y. Nishi, T. Manabe, H. Kameyama. S. Moriya, T. Katagiri, S. Uchida, and M. Kwakami, *J. Mater. Sci. Lett.* **9** 801 (1990).

TEMPERATURE DEPENDENT THERMOELECTRIC POWER OF HIGH-T_c BI OXIDE

Y.OHMORI*, M.SHIMABARA*

*Student of Materials Science, Tokai University, 1117 Kitakaname,Hiratuka, Kanagawa 259-12, Japan.

K.OGURI**, Y.NISHI**

**Department of Materials Science. Tokai University,1117 Kitakaname,Hiratsuka, Kanagawa 259-12, Japan.

ABSTRACT

The temperature dependent thermoelectric power is studied of high-Tc Bi oxide above 120 K of onset of superconducting transition.. The lower the temperature above the onset temperature (T_c^{on}) of superconducting transition, the higher the Seebeck coefficient becomes.

1. Introduction

Various properties, such as electric, magnetic and mechanical properties, have been reported in high-T_c superconductors[1-4]. Especially, it is important to know the relationship between the carrier and superconducting properties. Seebeck coefficeiceint is one of the important factor to indicate the carrier condition[5-7]. Thus, we investigate the temperature dependence of the Seebeck coefficeint of Bi oxide.

2. Experimental Procedure

Samples with nominal composition $Bi_{1.6}Pb_{0.35}Sb_{0.05}Sr_2Ca_2Cu_3O_x$ were prepared by high - purity powders of Bi_2O_3 (99.99%), PbO (99.99%), $SrCO_3$ (99.9%), $CaCO_3$ (99.9%),CuO (99.9%), and Sb_2O_3 (99.9%). The powders were mixed and sintered for 6 h at 1023 K in air and then air cooled. After crushing, the sintered powers were resintered for 18 h at 1073 K in air and then air cooled. After crushing, a pelletized tablet, 1 mm thick and 13 mm diameter, was sintered in air at 1123 K and then furnace cooled.

The electrical resistivity was measured in a cryostat at equilibrium temperatures[8]. The electrical resistivity was measured using a standard four-probe technique and a Keithley 181 nanovoltmeter. The temperature was measured by Au∗Fe-chromel thermocouple attached to the specimen[9]. The temperature of both sides to the sample was measured by means of Au∗Fe-chromel thermocouple. The thermal emf of the sample was measured by a Keithley 182 nanovoltmeter.

3. Results and Discussion

Figure 1 shows the temperature dependence of the electrical resistivity of high-T_c Bi system. The lower the temperature, the lower the electrical resistivity becomes. Onset T_c^{on} of the transition is defined as the temperature corresponding to the deviation of the electrical resistivity. The deviated point is taken at $d(R/R^{300K})/dT=0.004$, where $d(R/R^{300K})/dT$ and R^{300K} are the slope of the reduced electrical resistivity (R/R^{300K}) against the temperature graph and the resistivity at 300K. R^{on} is resistivity at the deviated point.

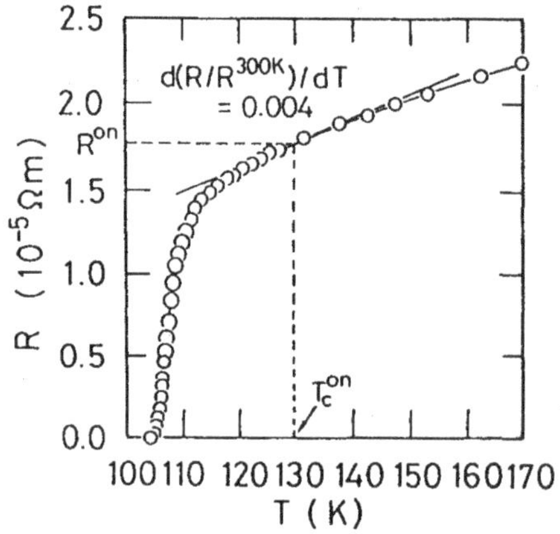

Fig. 1 Temperature dependence of electrical resistivity of high-Tc Bi system.

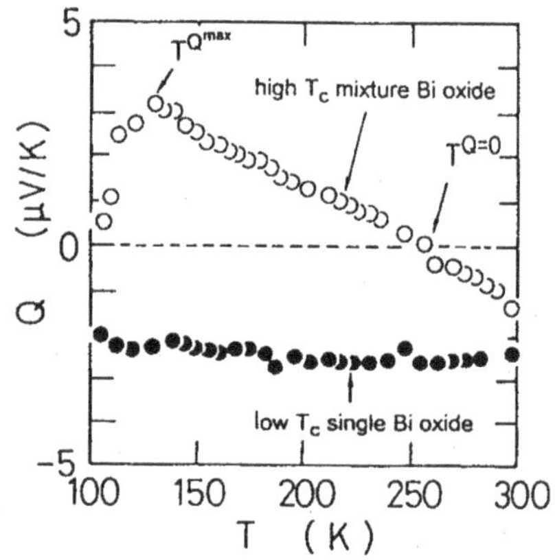

Fig. 2 Temperature dependence of Seebeak coefficient of sample.

Figure 2 shows the temperature dependence of the Seebeak coefficient (Q) of the sample. The Seebeck coefficient rapidly decreases at near the T_c^{on}. The sign of the Seebeak coefficient at room temperature is negative for the low T_c single Bi phase. The lower the temperature above the onset temperature (T_c^{on}) of superconductivety transition, the higher the Seebeck coefficient becomes. The sign of Q becomes positive below the temperature at TQ=0. TQ=0 is the temperature which changed the sign of the Seebeak cofficient.

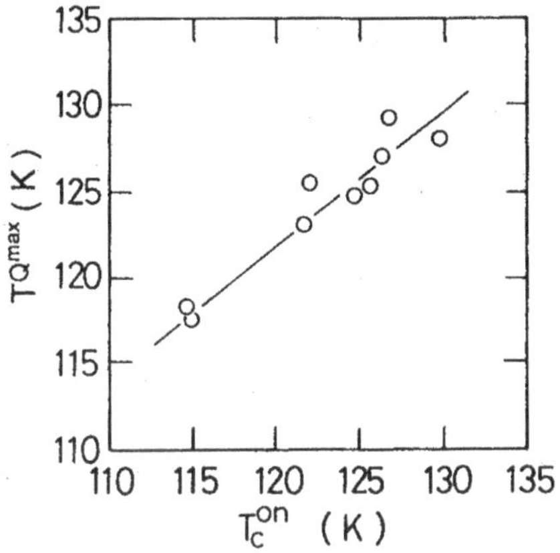

Fig. 3 Relationship between the T_c^{on} value and the temperature (TQ^{max}) at the maximum Seebeck coefficient value.

Figure 3 shows the relationship between the T_c^{on} value and the temperature (TQ^{max}) at the maximum Seebeck coefficient value. The T_c^{on} value depends on the TQ^{max} value. The higher the T_c^{on} value, the higher the TQ^{max} becomes. The TQ^{max} values are approximately equal to the T_c^{on} values.

It is well-known that high-Tc superconductors have something in common with the positive carrier. Figure 4 shows the temperature dependence of schematic model. The lower the temperature, the lower the Fermi energy (E_f) level of positive carrier becomes[10-12].

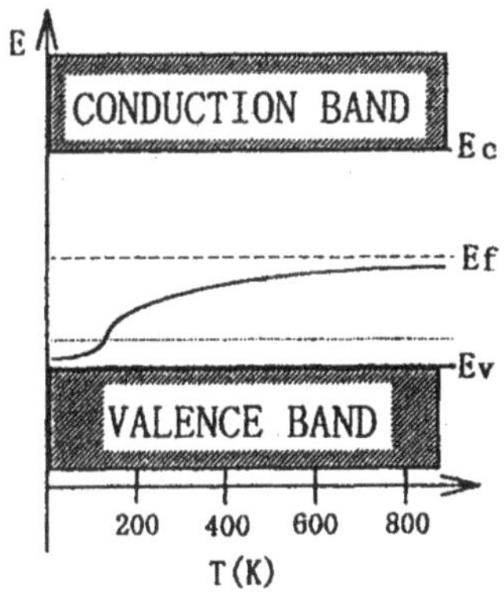

Fig. 4 Temperature dependence of schematic model for the positive carrier.

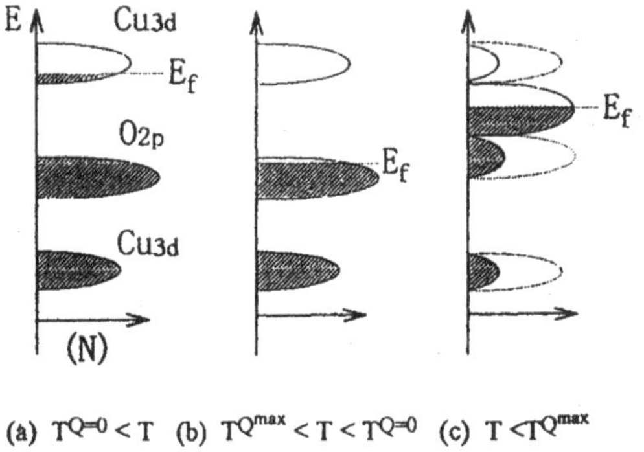

Fig. 5 Schematic model of the bands near the Fermi energy (E_f) for the high-T_c Bi oxide.

Figure 5 shows schematic model of the bands near the Fermi energy (E_f) for the high-T_c Bi oxide[13-17]. (a), (b) and (c) is schematic model of the bands above $TQ^{=0}$, between $TQ^{=0}$ and TQ^{max}, and below TQ^{max}, respectively. If the Fermi energy (E_f) sifts to the side of the low energy level, the change of thermoelectric power is explained. Thus, we conclude that it is one of the important factors that the thermoelectric power know the mechanism of superconductivety transition.

4. Conclusion

The relationship between themal power and T_C is studied for the high-T_C Bi oxide. The lower the temperature above the onset temperature (T_c^{on}) of superconducting transition, the higher the Seebeck coefficient becomes. The T_c^{on} values approximately agree with the TQ^{max} values.

5. References

1. H. Maeda, Y. Tanaka, M. Fukutomi, T. Asano, *Jpn. J. Appl. Phys.* **27**, (1988) 209.
2. M. Xu, A. Umezawa, W. G. Crabtree, *Phys. Rev. B.* **46**, (1992) 11928.
3. Y. Nishi, N. Ninomiya, F. Kanai, S. Uchida, S. Moriya, *J. Appl. Phys.* **66**, (1989) 109.
4. Y. Nishi, F. Kanai, N. Ninomiya, S. Uchida, K. Oguri, S.Moriya, *J. Mater. Sci. Lett.* **8**, (1989) 1395.
5. N. Nucker, J. Fink, C. J. Fuggle, J. P. Durham, M. W. Temmerman, *Phys. Rev. B.* **38**, (1988) 5158.
6. Y. Tokura, B. J. Torrance, C. T. Huang, I. A. Nazzal, *Phys. Rev. B.* **38**, (1988) 7156.
7. C. Kendziora, D. Mandrus, L. Mihaly, L. Forro, *Phys. Rev. B.* **46**, (1992) 14297.
8. Y. Nishi, N. Ninomiya, K. Oguri, S. Tokunaga, *J. Appl. Phys.* **65**, (1989) 3972.
9. Y. Nishi, K. Oguri, H. Ohinata, K. Tanioka, Y. Kita, N. Ninomiya, *Phys. Rev. B.* **41**, (1990) 6520.
10. K. Lark-Horowitz, V. A. Johnson, *Phys. Rev.* **69** ,(1946) 258.
11. K. Lark-Horowitz, V. A. Johnson, *Phys. Rev.* **71** ,(1947) 374.
12. K. Lark-Horowitz, V. A. Johnson, *Phys. Rev.* **71** ,(1947) 909.
13. F. Devaux, A. Manthiram, J. B. Goodenough, *Phys. Rev.* **41** ,(1990) 8723.
14. I. Felner, U. Yaron, I. Nowik, E. R. Bauminger, *Phys. Rev.* **40**, (1989) 6739.
15. Y. Nishi, A, Igarashi, S. Moriya, N. Ninomiya, S. Tookunaga, *J. Mater. Sci*, **25**, (1990) 1.
16. M. Imada, *J. Phys. Soc. Jpn.*, **56**, (1987) 3793.
17. M. Ogata, H, Shiba, *J. Phys. Soc. Jpn.*, **57**, (1988) 3074.

IMPROVEMENT OF DUCTILITY BY Hf ADDITION FOR LIQUID-QUENCHED HIGH T_c Bi OXIDE

SATOSHI UCHIDA*, KAI SHIRAISHI**, MAMORU SHIMABARA**, SHIGEMI WATANABE**, TAKASHI OHWAKI**, KAZUO NOZAKI**, MAMORU KAWAKAMI, KAZUYA OGURI** and YOSHITAKE NISHI,

Department of Materials Science, Tokai University, 1117 Kitakaname, Hiratsuka, Kanagawa, 259-12, JAPAN

*Tokyo Metropolitan Industrial Technology Center, 3-13-10, Nishigaoka, Kita-ku, Tokyo 115 Japan.

**Student of Materials Science, Tokai University

The high T_c oxide is brittle. Using ultra micro-Vickers hardness tester effects of Hf substitutions for Bi, Pb, Sr, Ca and Cu atoms on ductility are investigated for the liquid-quench BiPbSrCaCuO glasse. The Hf substitution increases ductility. When the Hf atoms substitute for the Sr atoms, it strongly enhances ductility.

1. INTRODUCTION

High T_c has been reported above 100 K for BiPbSrCaCuO.[1-3] It is important to know effects of solutes on T_c. From an engineering point of view, it is important to know mechanical properties. Most high T_c superconductors are brittle. Influences of the Hf addition on ductility are investigated for the oxide at the limit composition that the Hf addition doesn't decrease the T_c value for BiPbSrCaCuO crystal sample.

From a science point of view, the Hf atom site in the crystal and Hf distribution coefficient in the high and low T_c phases are serious problems for the high T_c sample. The glassy sample doesn't have the problem except the low T_c below boiling point of helium. Thus, the glassy sample, obtained by the liquid- quenching is used.

The ionic radius (r) and electronegativity (Z_e) may be useful indicators to deduce the ductility and are summarized in Table 1. The ionic radius of Pb, Sr and Ca are larger than Bi and Cu in BiPbSrCaCuO sample. The difference of ionic radii likely induces the local residual strain, which is easy to generate the microcrack. It is well know that the Hf ionic radius is smaller than those of the elements. If Hf atom substitutes for Sr, Ca and Pb, it is possible to decrease the local residual strain. It may improve ductility.

Table 1 Ionic radii and electronegativity value

	Bi	Pb	Sr	Ca	Cu	A.V.	O	Hf
Valency	3+	2+	2+	2+	2+	—	2-	4+
Ionic radius	96	132	127	118	90	107	132	84
Electronegativity	1.9	1.8	1.0	1.0	1.9	1.5	3.5	1.3

The electronegativity difference between the oxygen and metal is one of the convenient indicators to evaluate the ionic bonding force. Based on the Z_e value in Table 1, the bonding forces of Sr and Ca against oxygen ion are larger than those of Cu, Bi and Pb. It likely induces the local residual stress. It is easy to generate the microcrack. When the Hf atom, which Z_e value is 1.3, is substituted for Sr and Ca, it may decrease the local residual stress. Thus, it prevents to growth of crack and may improve ductility. Therefore, effects of the Hf substitution on ductility are investigated for the $Bi_{1.6}Pb_{0.4}Sr_2Ca_2Cu_3O_x$ glassy sample before crystallization.

2. EXPERIMENTAL PROCEDURE

Samples with nominal composition were prepared by high purity powders of CuO (99.9 mass %), HfO_2 (99.9 mass %), Bi_2O_3 (99.99 mass %), PbO (99.9 mass %), $SrCO_3$ (99.99 mass %), and $CaCO_3$ (99.9 mass %). The powders were mixed and sintered in air at 1023 K for 6 h and then air-cooled. After crushing, the sintered powders were resintered in air 1073 K for 18 h and then air-cooled. After crushing, a pelletized tablets, 1 mm thick and 13 mm diameter, was sintered for 12 h at 1123 K in air and air-cooled. Liquid-quenched glassy foils were prepared from the resintered tablets using a twin-type piston-anvil apparatus.[4,5]

Here, the z values are determined by the critical solubility limits to maintain 100 K of the T_c.[6] Namely, we determined the sample composition of the critical substitution limits.

To evaluate the brittlness, we use the new type of hardness teaster. It is ultra micro-Vickers hardness tester (DUH-50 Shimadzu Kyoto JAPAN). The traditional hardness tester has been mainly used for the measurement of the resistance to plastic deformation.[7] On the other hand, the new tester is used the measurement of the resistance to deformation. It includes elastic deformation as well as plastic deformation.

Figure 1 shows change in load ,(L) plotted as a of ultra micro-Vickers indentation depth (h). When crack forms around the ultra micro-Vickers impression at a certain load, it is suggested that a critical load (L_f, stress) and a critical depth (h_f, strain) are defined.

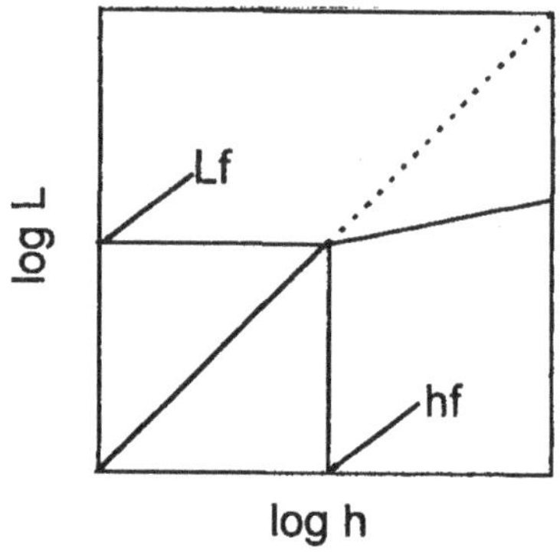

Fig. 1. Change in load (L) plotted as a function of ultra micro-Vickers indentation depth (h).

3. RESULTS AND DISCUSSIONS

Figure 2 shows changes in the critical depth (h_f) against hafnium concentration (z) for the Hf substituted BiPbSrCaCuO glassy sample. When the Hf atoms are substituted for Sr and Ca atoms, it is not easy to form the microcrack.

Figure 3 shows changes in the critical load (L_f) against hafnium concentration (z) for the BiPbSrCaCuO glassy sample. When the Hf atoms are substituted for Bi, Pb, Sr and Ca atoms, it is not easy to form the microcrack.

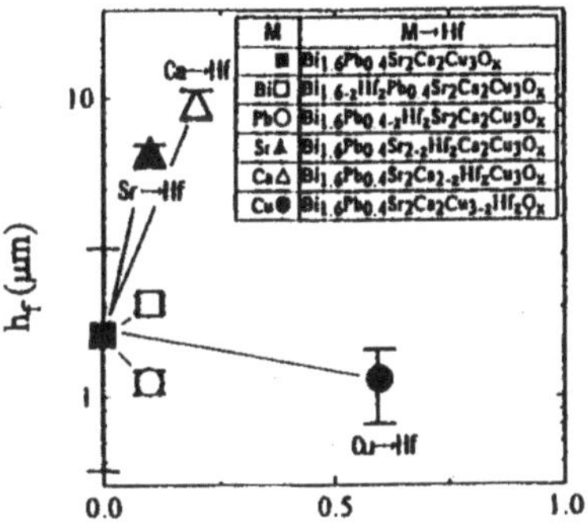

Fig. 2 Change in critical depth (h_f)against hafnium concentration (z) for BiPbSrCaCuO.

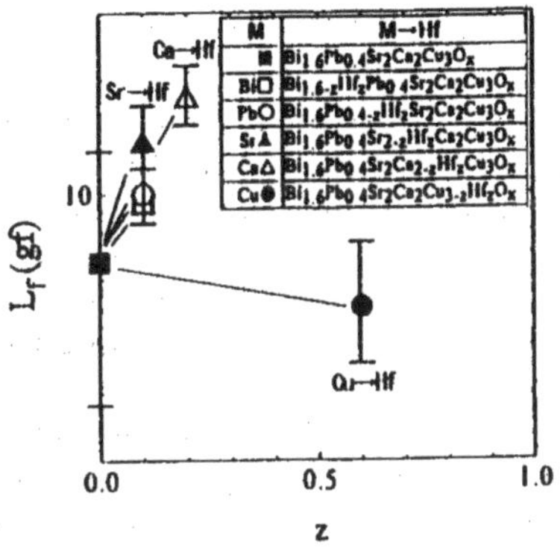

Fig. 3 Change in critical laod (L_f) against hafnium concentration (z) for BiPbSrCaCuO.

Furthermore, we suggest a deformation energy of fracture as a new parameter to evaluate the brittleness. The energy is defined as the work to move for the constant length on the constant force. The deformation energy (W) is assumed to be expressed by the following equation:

$$W = L\,V / S \text{ — (1)}$$

Here, L is the force to deform the hole. V/S is the mean deformed length, where V is the deformed volume of the triangular pyramidal hole and S is the triangle area of the basal plane of the triangular pyramidal hole of ultra micro-Vickers. Based on Eq. (1), the critical deformation energy (W_f) to fracture is expressed by the following equation:

$$W_f = L_f h_f / 3 \text{ — (2)}$$

Here h_f is the critical depth to traction of the triangular pyramidal hole. Using Eq. (2), the critical deformation energies (W_f) on fracture are estimated by the critical load (L_f) and critical depth (h_f).

Figure 4 shows the change in the critical deformation energy (W_f) against the hafnium concentration (z). When the Hf atoms substitute for Bi, Sr and Ca atoms, the large W_f values are obtained. Namely, it is not easy to form the crack.

The Hf ion is mainly substituted for metallic ions. If the substitution increases the difference between ions, it inhomogenizes the internal strain among ions in the sample. Thus, it is easy to break the sample. As shown in table 1, the Hf substitution decreases the difference. Thus, it decreases the internal strain and prevents to generate the crack, it increases the critical depth (h_f).

On the other hand, if the electronegativity difference increases between oxygen and metallic ions, it inhomogenizes the bonding force among ions. Thus, it is easy to break the sample. The Hf substitution inhomogenizes the bonding force among ions and it prevents to generate the crack. Thus, it increases the value of critical load (L_f).

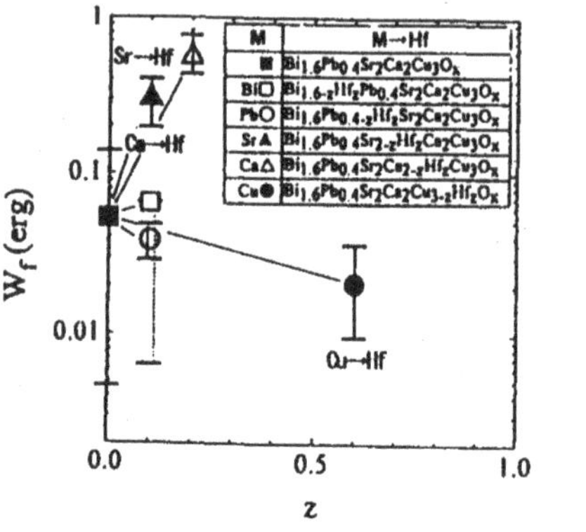

Fig. 4 Change in critical deformationenergy (W_f) against hafnium concentration (z) for BiPbSrCaCuO.

Considering with the deformation energy (W_f) to fracture, the Hf substitutions for Sr and Ca homogenize the ionic radius and electronegativity between oxygen and metallic ions. Therefore, they increase the W_f value.

The Hf substitutions for Cu increase the differences of ionic radius. Thus, it slightly decreases the W_f values.

Since the Hf substitutions for Pb and Bi are very little, they are not distinct difference in this work.

4. CONCLUSION

In summary, we evaluate ductility of the BiPbSrCaCuO glassy sample by the use of ultra micro-Vickers hardness tester. The Hf substitutions for Sr and Ca enhance ductility.

5. References

1. Y.Nishi, S.Moriya and T.Manabe, J. Appl. Phys., **65** (1989) 2389.
2. Y.Nishi, K.Oguri, H.Ohinata, K.Tanioka, Y.Kita and N.Ninomiya, Phys. Rev. B, **41** (1990) 6520.
3. Y.Nishi, Y.Kita and K.Tanioka, J. Mater. Sci. Lett., 7 (1990) 359.
4. N.Inoue, Y.Sekiguchi, T.Shima and Y.Nishi, Mater. Sci. Eng., **98** (1988) 425.
5. N.Inoue, T.Shima and Y.Nishi, ibid., **98** (1988) 429.
6. Y.Nishi, K.Shiraishi, S.Watanabe, T.Ohwaki, K.oguri and K.Nozaki, J. Adv. Sci., **5** (1993) 127.
7. Y.Nishi, T.Katagiri, T.Yamano, F.Kanai, N.Ninomiya, S.Uchida, K.Oguri, T.Morishita, T.Endo and M..Kawakami, Appl. Phys. Lett., 58(1991) 2084.

www.ingramcontent.com/pod-product-compliance
Lightning Source LLC
LaVergne TN
LVHW050548311026
851422LV00009B/32
* 9 7 8 9 8 1 0 2 2 2 4 8 2 *